Yasmeen Siddiqui • Kamil Kuca
Rajni Dhalaria
Editors

Sustainable Mycorrhizal Cultivation

Innovations for Hillside Farming Systems

Springer

Editors

Yasmeen Siddiqui
Department of Biological Sciences
College of Science
King Faisal University
Al Ahsa, Saudi Arabia

Kamil Kuca
Department of Chemistry
University of Hradec Kralove
Hradec Kralove, Czech Republic

Rajni Dhalaria
Department of Biosciences
Minerva P.G. College of Arts, Science and Commerce
Kangra, Himachal Pradesh, India

ISBN 978-981-95-1777-0 ISBN 978-981-95-1778-7 (eBook)
https://doi.org/10.1007/978-981-95-1778-7

This Springer imprint is published by the registered company Springer Nature Singapore Pte Ltd.
The registered company address is: 152 Beach Road, #21-01/04 Gateway East, Singapore 189721, Singapore

Foreword

Today, agriculture is at a pivotal juncture. As the world contends with the escalating impacts of climate change, environmental degradation, and increasing food insecurity, there is an urgent need to rethink how we cultivate our lands and feed our populations. This transformation is particularly crucial in hilly regions, where the convergence of rich biodiversity, fragile ecosystems, and traditional agricultural practices presents both significant challenges and unique opportunities. This timely book highlights the often overlooked yet transformative role of mycorrhizal fungi in promoting sustainable agriculture in vulnerable landscapes. With clarity and scientific rigor, the authors delved into the symbiotic relationships that mycorrhizae form with plants, enhancing soil fertility, crop yield, and ecological resilience. The strength of this work lies not only in its thorough research and analysis but also in its strong commitment to practical relevance. Farmers, researchers, extension workers, and policymakers will find valuable insights into how mycorrhizal technologies can be leveraged to improve soil health, conserve biodiversity, and mitigate the effects of climate change, which are now more urgent than ever. This book also offers a crucial economic perspective considering the feasibility and long-term sustainability of implementing mycorrhizal-based solutions in real-world farming systems. By advocating biologically sound practices, it lays the groundwork for more resilient and profitable farming systems that respect the integrity of hill environments and communities that rely on them. In championing the shift toward mycorrhiza-centered agricultural innovation, this publication makes a meaningful contribution to the global discourse on sustainability. It will serve as a cornerstone resource for anyone committed to building a more equitable, ecologically sound, and productive future for agriculture. It is with great pleasure that we commend this book to readers from across disciplines and sectors. It may inspire action, inform practice, and ignite a deeper appreciation for the symbiotic partnerships that help sustain life both above and below the soil.

Mohd. Sayeed Akhtar
Department of Botany
Gandhi Faiz-e-Aam College
Shahjahanpur, Uttar Pradesh, India

Preface

Agricultural sustainability has emerged as an urgent imperative in response to climate change, land degradation, and food insecurity. Hilly regions, characterized by abundant natural resources and diverse agroclimatic conditions, present optimal conditions for agriculture. However, these advantages are particularly evident in hill agriculture, where fragile ecosystems and challenging topographies test the resilience of farmers and the productivity of crops. Mycorrhizal symbiosis is increasingly recognized as a transformative factor in advancing sustainable agriculture amidst these challenges.

This book offers a comprehensive analysis of the influence of mycorrhizal fungi on crop productivity across diverse landscapes, emphasizing their complex and essential roles in plant health, soil fertility, and ecosystem functioning. The chapters commence with a foundational introduction to hill agriculture, delve into the intricate realm of mycorrhizal diversity, detail the methodologies and tools necessary for their detection and application, and highlight the significant potential of modifying agronomic practices to enhance soil health.

Each chapter of this book explores a critical dimension of mycorrhizal symbiosis, ranging from its role in enhancing the yield of vegetables, fruits, flowers, and medicinal crops to its influence on staple crops, including cereals, pulses, and oilseeds. The text also addresses the necessity of preserving mycorrhizal quality and regulating its production to achieve effective outcomes at the field level. Furthermore, the book evaluates the economic feasibility and environmental advantages of mycorrhizal technologies, highlighting their contributions to biodiversity promotion, soil conservation, and climate change mitigation.

This publication aspires to serve as a valuable resource for researchers, agronomists, policymakers, extension workers, and progressive farmers by bridging scientific research with practical applications. It offers guidance, evidence, and inspiration for developing more resilient and profitable hill farming systems, whether through leveraging symbiosis for increased yields, revitalizing soil health, or creating new market opportunities for bioinoculants. Ultimately, this book advocates for a transformation in farming practices in hilly regions. By embracing the potential of mycorrhiza and sustainable agriculture, a more robust and sustainable future can be constructed for these areas, benefiting both the environment and local communities. It is hoped that this book will foster greater awareness, broader adoption, and continuous innovation in the utilization of mycorrhizae, contributing to a more sustainable and equitable agricultural future.

Al Ahsa, Saudi Arabia — Yasmeen Siddiqui
Hradec Kralove, Czech Republic — Kamil Kuca
Kangra, Himachal Pradesh, India — Rajni Dhalaria

Acknowledgment

We would like to thank the Deanship of Scientific Research, Vice Presidency for Graduate Studies and Scientific Research, King Faisal University, Saudi Arabia, Grant No.: 252883 for supporting this project.

Contents

Editors and Contributors

About the Editors

Yasmeen Siddiqui is an Associate Professor at King Faisal University (KFU), Saudi Arabia. She is a leading figure in plant and postharvest pathology, with extensive expertise in minimizing pre and postharvest lossess of fruits and vegetables often through innovative applications like edible coatings, biostimulant, and biopesticides, managing Basal Stem Rot of oil palm and valorizing agro-industrial by-products. She has published over 100 scientific articles in peer-reviewed journals, with an H-index of 37 and more than 5000 citations. She has earned recognition for her interdisciplinary work that bridges basic research with practical agricultural applications.

Kamil Kuca is a distinguished scholar who has held professorial positions and served as an invited or adjunct faculty member at esteemed academic institutions worldwide. He is currently working at University of Hradec Kralove in Hradec Kralove, Czech Republic. Recognized for his scholarly impact, he boasts an impressive H-index of 70 and has accumulated over 20,000 citations in the Web of Science. In 2021, he was honored as a Highly Cited Researcher by Clarivate Analytics. Additionally, he holds five patents and serves as a scientific advisor to several technology companies, further demonstrating his leadership and expertise.

Rajni Dhalaria is an emerging botanist specializing in mycology, ethnobotany, and phytochemistry. She earned her PhD in Botany from Shoolini University of Biotechnology and Management Sciences, India, focusing on the mycorrhizal impact on phytochemicals and biological properties of medicinal plants. Currently serving as Assistant Professor at Minerva PG College, she has published 28 scientific papers, including research articles, reviews, and book chapters, with over 759 citations and an h-index of 10. She actively contributes to the scientific community as a peer reviewer for esteemed journals and has presented her work at various national and international conferences, showcasing her strong interdisciplinary expertise and innovative research capabilities.

Contributors

Elsayed Fathi Abd-Allah Plant Production Department, College of Food and Agricultural Sciences, King Saud University, Riyadh, Saudi Arabia

R. Aditya Chaudhary Sarwan Kumar Himachal Pradesh Krishi Vishvavidyalaya, Palampur, Himachal Pradesh, India

Bita Babakhani Department of Plant Pathology, North Dakota State University, Fargo, ND, USA

Subrata Bag ICAR-Central Research Institute for Dryland Agriculture, Hyderabad, Telangana, India

Swarnendra Banerjee Academy of Biology and Biotechnology, Southern Federal University, Rostov-on-Don, Russia

Anamika Barman ICAR-Indian Agricultural Research Institute, New Delhi, India

Indrani Baruah Department of Botany, Gauhati University, Guwahati, Assam, India

Jagadeesh Bathula Department of Forest Resource Management, Forest College and Research Institute, Mulugu, Siddipet, Telangana, India

Nitesh Boro Department of Botany, Gauhati University, Guwahati, Assam, India

Mahamanikya Brahma Department of Botany, Gauhati University, Guwahati, Assam, India

Chinnappan Deepa Joan of Arc Post Graduate and Research Centre in Biotechnology, School of Life Sciences, Arignar Anna College (Arts and Science), Krishnagiri, Tamil Nadu, India

Devendra Kumar Dadhich ICAR-National Dairy Research Institute, Karnal, Haryana, India

Parijat De Department of Agriculture, Brainware University, Kolkata, West Bengal, India

Khomdram Bijoya Devi Department of Botany, Gauhati University, Guwahati, Assam, India

Shipa Rani Dey Department of Agronomy, Faculty of Agricultural Sciences, GLA University, Chaumuhan, Uttar Pradesh, India

Rajni Dhalaria Department of Biosciences, Minerva P.G. College of Arts, Science and Commerce, Kangra, Himachal Pradesh, India

Sundarasamy Dhanapal Post Graduate and Research Centre in Biotechnology, School of Life Sciences, Arignar Anna College (Arts and Science), Krishnagiri, Tamil Nadu, India

Suman Dutta Ramkrishna Mission Vivekananda Educational and Research Institute, Kolkata, West Bengal, India

Muthukumar Elankumaran Department of Botany, Bharathiar University, Coimbatore, Tamil Nadu, India

M. A. Elinson Ufa University of Science and Technology, Ufa, Russia

Marcelo Carvalho Minhoto Teixeira Filho Department of Rural Engineering, Plant Health and Soils, São Paulo State University (UNESP), Ilha Solteira, Brazil

Saurabh Gangola Department of Microbiology, Graphic Era Deemed to be University, Dehradun, India

Kumar Gaurav School of Agriculture, Graphic Era Hill University, Dehradun, India

Subhadwip Ghorai Department of Agriculture, Brainware University, Kolkata, West Bengal, India

Shibani Gogoi Department of Botany, Gauhati University, Guwahati, Assam, India

Department of Biotechnology, Mizoram University, Aizawl, Mizoram, India

Nidhi Gupta University Institute of Biotechnology, Chandigarh University, Mohali, Punjab, India

Department of Biotechnology, Chandigarh Group of Colleges, Mohali, Punjab, India

Nur' Amira Hamid Faculty of Plantation and Agrotechnology, Universiti Teknologi MARA (UiTM), Malacca, Malaysia

Surmani Huidrom Microbial Resource Division, Institute of Bioresources and Sustainable Development (IBSD), Imphal, Manipur, India

Ajmila Islam Department of Biological and Physical Sciences, School of Arts and Natural Sciences, American University in Dubai, Dubai, UAE

Indu Kale Department of Forest Biology and Tree Improvement, College of Forestry, Navsari Agriculture University, Navsari, Gujarat, India

Subramanian Kaliyannan ICAR – National Institute of Biotic Stress Management, Raipur, Chhattisgarh, India

Katyayni Kanth College of Basic Sciences and Humanities (CBS&H), Dr Rajendra Prasad Central Agricultural University, Samastipur, Bihar, India

Nur Qursyna Boll Kassim Faculty of Plantation and Agrotechnology, Universiti Teknologi MARA (UiTM), Malacca, Malaysia

Soil Conservation and Management Research Interest Group (RIG), Universiti Teknologi MARA (UiTM), Shah Alam, Selangor, Malaysia

Ajay Katoch Department of Biosciences, MCM DAV College Kangra (Himachal Pradesh), Kangra, Himachal Pradesh, India

Nagarajan Kiruthiga Post Graduate and Research Centre in Biotechnology, School of Life Sciences, Arignar Anna College (Arts and Science), Krishnagiri, Tamil Nadu, India

Appani Laxman Kumar Sri Konda Laxman Telangana Horticultural University, Siddipet, Telangana, India

Harsh Kumar Centre of Advanced Technologies, Faculty of Science, University of Hradec Kralove, Hradec Králové, Czech Republic

Prasann Kumar Department of Agronomy, School of Agriculture, Lovely Professional University, Phagwara, Punjab, India

Vikas Kumar University Institute of Biotechnology, Chandigarh University, Mohali, Punjab, India

Diksha Kumari University Institute of Biotechnology, Chandigarh University, Mohali, Punjab, India

Payal Maiti Department of Post-Harvest Management, Bidhan Chandra Krishi Viswavidyalaya, Nadia, West Bengal, India

Fadzilah Adibah Abdul Majid Institute of Climate Adaptation and Marine Biotechnology (ICAMB), Universiti Malaysia Terengganu, Kuala Nerus, Terengganu, Malaysia

Luis Del Rio Mendoza Department of Plant Pathology, North Dakota State University, Fargo, ND, USA

Rajdeep Mohanta Department of Agriculture, Brainware University, Kolkata, West Bengal, India

Yogesh Dashrath Naik Department of Agricultural Biotechnology and Molecular Biology, College of Basic Sciences and Humanities, Dr Rajendra Prasad Central Agricultural University, Samastipur, Bihar, India

Nuraini Mohd Noor Faculty of Plantation and Agrotechnology, Universiti Teknologi MARA (UiTM), Malacca, Malaysia

Nur Maizatul Idayu Othman Faculty of Plantation and Agrotechnology, Universiti Teknologi MARA (UiTM), Malacca, Malaysia

Soil Conservation and Management Research Interest Group (RIG), Universiti Teknologi MARA (UiTM), Shah Alam, Selangor, Malaysia

Abhishekh Patidar Institute of Agricultural Sciences, Banaras Hindu University, Varanasi, Uttar Pradesh, India

Kandeeparoopan Prasannath Department of Agricultural Biology, Faculty of Agriculture, Eastern University, Chenkalady, Sri Lanka

Vanitha Prasannath Department of Biosystems Technology, Faculty of Technology, Eastern University, Chenkalady, Sri Lanka

Amit Raj Department of Agronomy, School of Agriculture, Lovely Professional University, Phagwara, Punjab, India

Pandian Rajendran ICAR – National Institute of Biotic Stress Management, Raipur, Chhattisgarh, India

Vishnu D. Rajput Academy of Biology and Biotechnology, Southern Federal University, Rostov-on-Don, Russia

Thibisha Rakulan Department of Agricultural Biology, Faculty of Agriculture, Eastern University, Chenkalady, Sri Lanka

Nurul Wahida Ramli Faculty of Plantation and Agrotechnology, Universiti Teknologi MARA (UiTM), Malacca, Malaysia

Hasnain Razzaq Department of Plant Pathology, University of Agriculture Faisalabad, Faisalabad, Punjab, Pakistan

Khizar Razzaq Department of Plant Pathology, North Dakota State University, Fargo, ND, USA

Purumandla Vennela Reddy Department of Agroforestry, Bundelkhand University, Jhansi, Uttar Pradesh, India

Sougata Roy International Potato Centre (CGIAR-CIP), Patna, Bihar, India

Soumik Dey Roy Department of Agriculture, Brainware University, Kolkata, West Bengal, India

Priyanka Saha ICAR-Research Complex for North Eastern Hilly Region, ICAR-KVK West Garo Hills, Barapani, Meghalaya, India

Chinmaya Sahoo Dr. Rajendra Prasad Central Agricultural University, Samastipur, Bihar, India

Smrutirekha Sahu ICAR – National Institute of Biotic Stress Management, Raipur, Chhattisgarh, India

Ankit Saini Chaudhary Sarwan Kumar Himachal Pradesh Krishi Vishvavidyalaya, Palampur, Himachal Pradesh, India

Dr Khem Singh Gill Akal College of Agriculture, Eternal University, Sirmaur, Himachal Pradesh, India

Rishika Saini Chaudhary Sarwan Kumar Himachal Pradesh Krishi Vishvavidyalaya, Palampur, Himachal Pradesh, India

Amit Samanta Department of Agriculture, Brainware University, Kolkata, West Bengal, India

Krishnagowdu Saravanan Post Graduate and Research Centre in Biotechnology, School of Life Sciences, Arignar Anna College (Arts and Science), Krishnagiri, Tamil Nadu, India

Piyali Sarkar Bidhan Chandra Krishi Viswavidyalaya, Mohanpur, West Bengal, India

Keerthika Shakthivel Department of Agricultural Biology, Faculty of Agriculture, Eastern University, Chenkalady, Sri Lanka

Nitin Sharma Department of Biotechnology, Chandigarh Group of Colleges, Mohali, Punjab, India

Yasmeen Siddiqui Department of Biological Sciences, College of Science, King Faisal University, Al Ahsa, Saudi Arabia

Ashish Kumar Singh Department of Vegetable Science, CHF, Acharya Narendra Deva University of Agriculture & Technology, Kumarganj, Ayodhya, UP, India

Joginder Singh Department of Botany, Nagaland University, Lumami, Nagaland, India

Khomdram Niren Singh Department of Botany, Gauhati University, Guwahati, Assam, India

Viabhav Kumar Upadhayay Department of Microbiology, College of Basic Sciences and Humanities, Dr. Rajendra Prasad Central Agricultural University, Pusa, Samastipur, Bihar, India

Rajeev Kumar Yadav Departement of Horticulture, PGCA, Dr. Rajendra Prasad Central Agricultural University, Pusa, Samastipur, Bihar, India

Nur Hafizah Zakaria Institute of Climate Adaptation and Marine Biotechnology (ICAMB), Universiti Malaysia Terengganu, Kuala Nerus, Terengganu, Malaysia

Overview of Hill Agriculture: Challenges, Opportunities, and Sustainability

Prasann Kumar

Abstract

Hill agriculture is crucial for sustaining livelihoods, conserving biodiversity, and maintaining the ecological balance in mountainous regions. However, it faces several challenges, including rugged terrain, climate variability, soil erosion, and limited access to markets and technologies. This chapter overviews hill agriculture and highlights its unique characteristics, significant constraints, and emerging opportunities for sustainable development. It discusses traditional and modern agricultural practices, innovative solutions for enhancing productivity, and policy interventions to promote environmental sustainability and economic resilience. Additionally, this chapter explores the role of agroecological approaches, climate-smart agriculture, and integrated farming systems in addressing the challenges faced by hill farmers. Moreover, it provides insights into achieving long-term agricultural sustainability while ensuring food security and rural development by identifying key strategies and conducting case studies from different hill regions.

Keywords

Hill agriculture · Biodiversity · Farming · Sustainability · Food · Rural · Emerging · Policy

P. Kumar (✉)
Department of Agronomy, School of Agriculture, Lovely Professional University, Phagwara, Punjab, India

Y. Siddiqui et al. (eds.), *Sustainable Mycorrhizal Cultivation*,
https://doi.org/10.1007/978-981-95-1778-7_1

Abbreviations

CSA	Climate-Smart Agriculture
GIS	Geographic Information System
IoT	Internet of Things
IPM	Integrated Pest Management
CA	Conservation Agriculture
VRT	Variable Rate Technology
GPS	Global Positioning System
IFS	Integrated Farming Systems
MFSI	Making Farming Successful and Integrated
FPO	Farmer Producer Organization
SHG	Self-Help Group
NGO	Non-governmental Organisation
FAO	Food and Agriculture Organization
IFAD	International Fund for Agricultural Development
R&D	Research and Development
PES	Payments for Ecosystem Services
GMOs	Genetically Modified Organisms
AI	Artificial Intelligence
IPCC	Intergovernmental Panel on Climate Change (implied reference)
CFM	Community Forest Management

1 Introduction

Hill agriculture refers to farming in mountainous or hilly areas. These areas present challenges and opportunities, owing to their unique terrain and climatic conditions. Soil properties vary widely depending on location, influencing agricultural potential. Fields are typically small because water for irrigation must often be transported over long distances, either through aqueducts or underground piping systems. Despite their limited field sizes, large bodies of standing water, such as marshes and lakes near town centres or rural villages, are sometimes used for livestock waste disposal. While the conditions are not consistent across all regions, hill agriculture is practised in many parts of the world, including the Himalayan region, the Andes, and European mountain ranges such as the Alps and the Cumberland Mountains (Cesco et al. 2023; Rai et al. 2024).

Given topographic constraints, hill agriculture often requires unique technologies such as terracing, contour farming, and agroforestry to prevent soil erosion, make the best possible land use, and maintain fertility (land). Hill agriculture plays a vital role in food production, supporting biodiversity, managing watersheds, and enhancing climate resilience. Practices such as using organic fertilisers contribute to these benefits by creating habitats for insects and other beneficial organisms.

The ecosystems of the mountain areas are primary sources of clean water and diversified agricultural practices that sustainably regulate flow rates and avoid landslides or other extreme natural phenomena. Furthermore, hill agriculture ensures the preservation of traditional wisdom and knowledge that have bedecked rural life for centuries (De Corato et al. 2024; Pant et al. 2024). Many important crops, especially staple crops such as rice, maize, and wheat, have wild relatives in the mountainous regions. Such areas are significant for crop genetic diversity and breeding within conservation programmes designed to cultivate different crop plants.

Because of its unique topographical and climatic conditions, hill farming differs from conventional plain-based farming systems, which agronomists can understand. Rugged terrain is a defining characteristic of hill agriculture, limiting the use of mechanised equipment and demanding intensive manual labour. The steep slopes make it impractical to operate large-scale machinery, making human labour and animal husbandry crucial components of farm operations. Terracing is the primary cultivation method in these regions; these horizontal steps carved into the slopes help prevent soil erosion and improve water retention, creating a distinctive agricultural landscape. Hill landscapes are divided into small plots with varying gradients—typically two, three, or four slope levels. These diverse elevations create microclimatic variations that allow the cultivation of a wide range of crops (Guindo et al. 2024; Qiu et al. 2024).

One hill can produce several types of crops. It grows lemon trees and linseed at higher altitudes, whereas bananas and tobacco grow below the valley. To date, farmers continue to plant fruit trees near their fields. Hill agriculture is closely integrated into forest ecosystems. Farmers often engage in agroforestry. Trees and shrubs grow alongside crops, improve soil fertility, and reduce environmental stress. They also bring more animals to the food. Farm households living near woods obtain some of their income from non-member forest products such as medicinal plants, mushrooms, and honey. In addition, hill agriculture is oriented towards small-scale production (Ahammad et al. 2023; Tran et al. 2023). Since most peasants own small plots of land, they often practise intercropping and multi-cropping to maximise productivity and preserve soil health. Organic farming is the status quo in the mountains; traditional farming methods rely on natural compost, farmyard manure, and minimal inputs for agricultural use. A rich ecosystem such as that in mountain landscapes requires adopting sustainable farming techniques to protect the environment while ensuring long-term agrarian productivity.

Hill agriculture is an economic and environmental cornerstone for millions living in mountainous regions worldwide. From the perspective of rural economics, hill agriculture is a lifeline that offers vast jobs for a significant part of the population, food security, and a home. Subsistence farming communities in mountainous countries depend, for their daily nourishment, on surplus produce from what they grow. Hill peasants create added opportunities for wealth generation by growing high-value crops such as fruits, spices, tea, and coffee, and collecting medicinal plants from mountainsides to develop new pharmaceuticals (Geddes et al. 2024; Subba et al. 2024). These speciality crops, often indigenous to particular regions, command high prices in home and international markets, enabling hill farmers to participate in agribusiness to produce export goods. In addition, hill agriculture creates social harmony and cherishes cultural heritage by preserving the traditional farming know-how handed down for generations. Indigenous peoples of hilly regions have developed highly sophisticated farming systems tailored to their environment, featuring crop rotation, organic fertilisation, and biological pest control techniques. These ancient ways of making food ensured that production was self-supported. They form an integral part of the cultural style and resistance in facing modern agricultural challenges to secure hill farmers' essential food supplies (Li et al. 2024; Shi et al. 2024). Hill agriculture, however, is threatened economically on two sides: one is by emigration from the countryside, and the other is by rural out-migration and an increase in farmer suicides. Because farming in mountainous areas is physically demanding, many young people leave the countryside to search for work in towns, resulting in labour shortages and farmland abandonment by many farmers. In addition, smallholder farmers typically face barriers to market entry, are cut off from being able to borrow, and must purchase their equipment, making it difficult for them to go head-to-head with big farms that take advantage of modern agricultural science in fertile lowland regions. For these problems, government policy, cooperative farming systems, and rural development programmes are key to realising an era in which hill farmers can make a living.

Hill agriculture is essential for ecological conservation and climate change mitigation, and the production methods developed in mountainous regions hinder soil erosion and conserve water. The terracing and contour ploughing of mountain slopes and agroforestry plantings stabilise the soil against erosion. When surface runoff flows over or floods into rocky terrain, it is impeded, and the drainage area increases. These practices prevent the land from losing fertility, so it can continue cultivating for many years. However, even more critically, hill agriculture helps to maintain biodiversity. Many hilly land areas have unique ecosystems with flora and fauna. Traditional farming systems in these areas encourage genetic diversity (Kittichotsatsawat et al. 2024; Vashishat et al. 2024). They preserve old crop species and livestock adapted to the local environment. Compared with monoculture-based industrial agriculture, diversified farming in mountain regions helps restore ecological balance. It also reduces reliance on synthetic fertilisers and pesticides. Hill agriculture also helps bind carbon and shore up the resilience of the climate. Agroforestry incorporates trees into growing crops and livestock, acting as a carbon sink by absorbing ambient carbon dioxide and increasing soil organic matter. Mountain forests associated with hill farming play a crucial role in regulating regional climates and, thus, reducing extreme weather occurrences. They maintain hydrological cycles, ensure adequate crop water supply, and replenish soil formation through litter that falls from above. Environmental sustainability of mountain agriculture is being challenged to the left and right (Tran et al. 2022; Pradhan et al. 2024). Human activities such as deforestation, land use changes, and climate change undermine the sustainability of mountain agriculture by degrading its soil. Climate change exacerbates these problems; temperature and precipitation patterns are altered, so weather cannot be predicted from one season to the next. This means that more pests infest hill farmers. In addition, there is less water available. Less resilient farms: if not destroyed by tidal waves first. To cope with these challenges, it is necessary to adopt comprehensive landscape management approaches that combine sustainable farming practices with reforestation and climate-resilient land-use planning. Policy measures aimed at soil conservation and afforestation, as well as strategies for adapting to climate change, can safeguard the living environment of mountain agricultural communities.

Hill agriculture is indispensable to food systems, rural economies, and environmental sustainability. It also has characteristics like small-family farmland husbandry and mixed farming, wind-slope privacy vineyards, and an ecological combination of animal husbandry and aquaculture. This makes hill agriculture different from lowland farming, but also offers new possibilities and constraints (Kumeh 2023; Komba et al. 2025). Despite rugged terrain and significant changes in climate and social development, hill agriculture is essential to the existence of mountain communities. By promoting sustainable agricultural methods, building market access, and enacting supportive policies, hill farming can guarantee food safety, drive wealth growth for residents, and help increase environmental resources in mountainous areas of vast complexity. As the world's attention to sustainability grows, a good example of hill agriculture is a key lesson in balancing agricultural output against environmental constraints.

2 Challenges in Hill Agriculture

Agriculture in the hills of nature suffers from the physical and social economy of the environment, which affects sustainable production. First and foremost, the hill agriculture of regions with hilly, rugged land and rocky terrain makes an excellent demand for capital. Compared with expansive plains where cultivation is relatively easy, hilly land takes many forms: steep slopes or straight slope-by-steep levels, uneven topography (rolling hills and deep valleys), and fragmented landholdings, which are always a precondition. Abroad, the topography is characterised by the hardship of irrigation and a lack of demand for large-scale ditches. The gradients are too steep to use machines, and the fields are too small to install beds; therefore, people use traditional land preparation methods. Mountains tend to be particularly susceptible to soil erosion, and variations in agricultural productivity are high because of these conditions. Amongst the first uses, flat or gently sloping land is often converted into residential areas (Granco et al. 2019; Bardsley et al. 2022). This means little flat land for cultivation, leading to waste on the hilly sides looking west in zones prone to soil erosion. Agricultural suitability is highly limited in many mountainous areas, and only a few small areas are suitable for intensive farming. Thus, farmers must use marginal lands more prone to degradation in these areas. When large patches of forests are cleared for temporary agricultural use and then abandoned as soil fertility declines, a common form of shifting cultivation in some areas is associated with the problem of people's evil faith-turned-habitual subsistence guarantee.

Soil erosion is one of the most serious environmental threats to agriculture, resulting in land degradation and water scarcity. In many mountainous regions, steep slopes indicate heavy seasonal rainfalls. This causes accelerated soil erosion;

topsoil rich in organic matter and plant nutrients is washed away to the detriment of plant growth. Once vegetation cover is hard to find, even as the raw material turns brown and withers, the soil becomes even more prone to erosion by wind and water. In regions where deforestation is caused by the expansion of agricultural land, soil erosion rates are significantly higher (Cui et al. 2024; Schillings et al. 2024). This leads to decreased soil fertility, lower crop yields, and more mud in the downstream water bodies. Sediments eroded from mountains can accumulate in streams and lakes, reducing their storage capacity for water and increasing the risk of flooding and damage to ecosystems. Land degradation in hill agriculture is partially due to unsustainable farming practices, such as excessive land ploughing and excessive grazing of livestock on pastures, which causes overgrazing. If a farmer chooses to work on only one kind of product over many years, the soil does not provide enough nutrients for plants in that future year, so yields are reduced (Ramakrishnan et al. 2024; Sahoo et al. 2025). The organic matter then flows down mountainsides into rivers, intensifying erosion and reducing productivity. Farmers must use fertilisers to maintain yields. However, fertilisers are not easy to obtain in highland regions because of the problems in transporting them and their high price. Therefore, to maintain soil health, it is essential to prescribe cultivation techniques, such as terracing and cultivation on the slopes of terraces or mulching. Water scarcity is another significant problem in hilly agriculture. Many mountainous areas grow only during the rain during the growing season. Although some regions have spring floods from glaciers or natural springs, water resources are distributed unevenly and are subject to such seasonal fluctuations that there is frequently insufficient water available when crops need it the most. With climate change and varying rain patterns, water scarcity has further intensified. Therefore, management practices must become more water-efficient for rippled irrigation and watershed conservation (Laudari et al. 2024; Rasul and Gurung 2024).

Climate change and unpredictable weather conditions pose serious challenges to mountain farming. Crops wither when it is too hot; if there is unseasonal frost, cows eventually wander into danger. An increase in altitude resulted in rising temperatures. Mountain regions have experienced twice the temperature increase rate compared to lowland areas. At the same time, temperatures are rising, and climate change is causing shifts in crop growing seasons and changing patterns of pests and diseases. There has also been an increase in extreme events such as unseasonable frosts, droughts, and extreme flooding. Where commonly seen, there should be fine days and disastrous weather, as old folk think of turning everything over in the fields. Weather forecasters can no longer predict the future; farmers base their planting and harvesting cycles on unreliable reports. Unpredictable rainfall patterns have given rise to excessive moisture, which causes fungal infections, whereas prolonged dry spells and occasional frost inhibit crop growth (Kharel et al. 2023; Phung and Dao 2024; Supe et al. 2024). The increasing frequency and ferocity of landslides or flash floods add to the damage caused by climate change, and agriculture is threatened yet again. Concurrently, climate change has altered the range of insect pests and diseases, so specific bugs that once lived only in the tropics can move up into different altitudinal bands. Higher temperatures also make it more difficult for animals to breed and raise offspring. For example, milk production drops steeply as heat stress sets in. On many highland farms, livestock provides manure, added income, and pulling power. The effects of climate change on livestock have the potential to be catastrophic. Due to their vulnerability, small farmers and others who rely on it in China, like themselves, are deeply concerned about what lies ahead. To reduce risk, climate-resilient crop varieties are planted, alternative farming systems are developed, and agroecological practices are integrated. However, the adoption of such measures is often hampered by socioeconomic constraints such as a lack of access to good seeds, agricultural extension services, or credit facilities.

Weakness in farming technology and soil degradation worsens the problems hill farmers face. For hill agriculture in China, the yield per hectare is generally lower than that in the plains because money and resources are scarce for developing more prosperous farmland. The natural limitations of mountain agriculture, such as growing seasons of lesser length, poor soil quality, and meagre amounts of mechanisation, lead to reductions in productivity. Owing to long-term cultivation without good soil management, farmland loses fertility. Then, farmers of counties located in the Huang-Huai-Hai Plain have it without irrigation, and the situation can lead to cultivation only once every two years (Kamboj et al. 2024; Majidian et al. 2024; Yang et al. 2025). These traditional seed varieties passed down from generation to generation in various areas do not always suit contemporary climatic conditions and, thus, have lower average crop yields than higher-yielding inter-hybrid strains. Relatively poor access to these means of replenishing soil fertility, organic compost, manure, and other additions to the soil serves to maintain soil fertility to some extent. Nevertheless, this is not always sufficient to offset the impact of high-intensity land use. Problems with declining crop yields have significant social and economic implications. With low farm yields, farm households lack food and income, making life difficult. Smallholders dominate the vast majority of people engaged in hill agriculture; individual living units are fragile and vulnerable to changes in production as they lack any price or other buffer source—interviews with farmers, government officials, and ministers of marketing (Deng et al. 2024; Pakeerathan 2025).

For instance, many rural areas where the hills are located remote from lowland areas do not have access to modern seeds, machines, or advanced farming techniques. Agrarian inputs are expensive to carry up to mountainous areas. If, on the other hand, services such as credit and insurance for hill farmers are not available through conventional commercial channels because they lack collateral or relationships with formal banks, this becomes a significant bottleneck for innovation. Mineral fertiliser, hybrid seeds, and pesticides are currently used by hill farmers at a low level of intensification. Market access is continually constrained by constricting factors such as high transport costs in and out of hill areas, bad roads, or no network at all from village to city. Perennial problems on these fronts make it difficult for farmers, let alone smallholder ones such as those we are considering here—to sell their produce. In many places outside the cities, all buying and selling involves middlemen who control prices. Another problem is that perishable high-value products such as fruit and vegetables or dairy items simply cannot be kept for any length of time and suffer significant losses from the moment they leave the field. If some progress in building the infrastructure of the hills is made, and if farmer cooperatives and industries processing raw materials into finished products are built in hillside regions, it may help bridge, even if only temporarily, the gap between small farmers and large markets (Samoraj et al. 2024; An et al. 2025).

Labour migration and demographic changes have further affected the viability of Hill Agriculture. Farming in mountainous areas is a challenging task. Young people have few financial opportunities; therefore, many migrants go to towns with better work prospects. This has led to a gradual reduction in the number of farm workers, with those left increasingly older family members. Youth outmigration leads to labour shortages, abandoned land, and reduced agricultural investments. In some cases, remittances from migrating workers provide farmhouses with a financial lifeline but do not compensate for the decline in the rural labour force. The feminisation of agriculture, where women, in the face of male outmigration, take on a more significant responsibility to keep their families fed and clothed, has become a notable trend in many rural areas (Piras et al. 2024; Nkinyam et al. 2025). There may have been some success here in empowering women while still causing them an extra workload that does not necessarily lead to access to resources, training opportunities, or improvement in the levels of financial support available to a family. Dealing with the challenges of labour migration calls for efforts to integrate rural livelihood diversification skill development programmes and policies that can make agricultural goods profitable for future generations at Johnston Shy-Hill Farm. Hill agriculture has many problems, from natural features such as hilly ground and environmental degradation to socio-economic conditions and demographic shifts. Overcoming these challenges will require a combined approach that combines methods for sustainable land management and advocacy by hill farmers with strengthened market access and policies supportive of those living on the hills. By utilising hardy farming practices, incorporating new technologies, and increasing financial and institutional support, the sustainability and productivity of hill agriculture can be strengthened for generations (Table 1).

3 Traditional Agricultural Practices in Hill Regions

For thousands of years, however, indigenous farming techniques and agroecology have provided lifeblood to hilly and mountainous regions. With roots deeply embedded in local traditions, these methods have arisen from the unique environmental conditions and delicate resource base of hill agriculture. In contrast to conventional industrial farming, indigenous techniques place great emphasis on ecological stability, biodiversity conservation, and sustainable use of resources. Farmers in hilly regions have bargained their way to obtain an intimate understanding of microclimates, different types of soil, and local water availability. In doing so, they generated practices that allowed for higher production while protecting ecological integrity. Many indigenous farming systems are based on polyculture, agroforestry, and natural soil fertility management. This makes it possible to reduce dependence

Table 1 Different challenges in hill agriculture

Category	Specific challenges	Causes	Impacts	Current mitigation efforts	Gaps in current strategies	Proposed solutions
Environmental	Soil erosion and land degradation	Steep slopes, deforestation, overgrazing, poor land management	Reduced soil fertility, lower crop yields, loss of arable land	Terracing, contour farming, afforestation programmes	Limited adoption of soil conservation techniques	Large-scale agroforestry, community-based erosion control programmes
	Water scarcity and erratic rainfall	Climate change, deforestation, poor water management	Crop failure, water stress for irrigation, drinking	Rainwater harvesting, watershed management	Inadequate infrastructure for water storage and distribution	Integrated water resource management, advanced micro-irrigation
	Biodiversity loss and ecosystem degradation	Habitat destruction, monocropping, pesticide overuse	Reduced pollination, lower genetic diversity in crops	Agroforestry, organic farming, protected area programmes	Weak enforcement of conservation policies	Incentives for biodiversity-friendly farming, community conservation programmes
Economic	High transportation and market access costs	Poor road infrastructure, remote locations	Increased post-harvest losses, lower income for farmers	Rural road development, farmer cooperatives	Limited investment in infrastructure, high dependency on middlemen	Expansion of rural connectivity, digital market access for farmers
	Low farm productivity and profitability	Small landholdings, lack of modern inputs, low mechanisation	Increased poverty, migration of youth from farming	Subsidies for fertilisers and seeds, financial aid programmes	High input costs, lack of access to affordable credit	Promotion of high-value crops, agri-business entrepreneurship
Technological	Limited mechanisation and outdated farming methods	Hilly terrain, high cost of machinery, lack of training	Low efficiency, increased labour demand, high drudgery	Introduction of small-scale machinery, training programmes	Machines not suited for hill terrain, limited access to technology	Development of hill-friendly agricultural machinery, R&D investment

(continued)

Table 1 (continued)

Category	Specific challenges	Causes	Impacts	Current mitigation efforts	Gaps in current strategies	Proposed solutions
	Poor digital penetration in agriculture	Lack of internet access, low awareness of digital tools	Limited access to market information, poor financial inclusion	Farmer mobile apps, extension services	Low adoption rates, high initial investment	Expansion of digital literacy programmes, mobile-based extension services
Social	Migration of youth to urban areas	Lack of job opportunities, low profitability in farming	Ageing farming population, loss of traditional knowledge	Skill development programmes, rural employment schemes	Limited long-term employment opportunities	Promotion of agri-tourism, agripreneurship programmes for youth
	Gender disparities in land ownership and decision-making	Patriarchal land laws, cultural norms	Limited economic empowerment for women, reduced farm productivity	Women empowerment programmes, legal reforms	Poor implementation of gender-inclusive policies	Equal land ownership rights, training for women farmers
Policy related	Weak enforcement of land-use policies	Fragmented policies, lack of monitoring mechanisms	Unregulated expansion of settlements, deforestation	Land tenure reforms, government subsidies	Corruption, lack of transparency	Strengthening governance, community-based land management policies
	Insufficient support for climate-resilient farming	Short-term agricultural policies, lack of funding	Reduced adaptation capacity, higher vulnerability to climate change	Climate-smart agriculture promotion, adaptation projects	Limited financial resources, poor farmer engagement	Long-term investment in climate adaptation programmes
Institutional	Lack of research and extension services	Low investment in R&D, inadequate staffing in the agriculture sector	Slow adoption of improved technologies, poor farm productivity	Agricultural extension programmes, farmer field schools	Limited outreach, lack of region-specific research	Establishment of local research centres, farmer-led innovation platforms

	Weak farmer organisations and cooperatives	Lack of leadership, poor financial management	Farmers' inability to negotiate fair prices, low collective bargaining power	Formation of cooperatives, financial support	Poor management, lack of trust in cooperatives	Capacity building for cooperatives, incentivising farmer participation
Infrastructural	Poor road connectivity and market access	Remote terrain, lack of government investment	Post-harvest losses, high transportation costs	Road construction projects, mobile market facilities	Limited coverage, high maintenance costs	Development of alternative transportation (cable cars, aerial ropeways)
	Limited irrigation facilities	High construction cost, lack of water storage systems	Seasonal water shortages, reduced cropping intensity	Canal building, groundwater recharge programmes	Poor maintenance, water wastage	Adoption of advanced irrigation technologies, community water management

Source: Kamboj et al. (2024), Ramakrishnan et al. (2024), Samoraj et al. (2024), Supe et al. (2024) and Yang et al. (2025)

on external inputs (Bardsley et al. 2022; Tran et al. 2022; Cui et al. 2024; Laudari et al. 2024; Shi et al. 2024). Imported materials, such as synthetic fertilisers and an array of pesticides, in the short term provide labour-saving methods or pest prevention but are counterproductive over the long run. Both systems ensure food security for local communities and contribute to climate resilience by retaining plant species diversity, improving soil health, and conserving water resources. Moreover, indigenous knowledge plays a significant role in preserving traditional varieties of seeds. These are often better able to weather changes regarding climate and pests than modern hybrid varieties. The importance of agroecology in hill farming cannot be overstated; agroecology integrates ecological principles with agricultural practices to promote long-term sustainability. Agroecological approaches emphasise the importance of biodiversity, soil health, and ecosystem services, making them particularly suitable for rugged landscapes in hilly regions and delicate ecosystems. By mimicking natural processes, using traditional ecological knowledge, and expanding to more distant territories, agroecology maintains yield levels while reducing the number of harmful substances involved in agriculture.

Compound planting is one of hill agriculture's most widespread indigenous farming techniques. This means planting several kinds of crops together in the same place to obtain the maximum benefit from water and land resources. It also has the advantage of reducing the risk associated with climate change and pest outbreaks. This enhances soil fertility, stores water, and prevents erosion by maintaining a continuous plant canopy (Cesco et al. 2023; Tran et al. 2023; Pant et al. 2024; Vashishat et al. 2024). Farmers sometimes choose crops that complement each other regarding growth habits, root structures, and nutritional needs to use the available resources best. For example, leguminous crops, such as beans and lentils, are often intercropped with cereals, such as maize or millet, to fix atmospheric nitrogen and naturally enrich soil fertility. Likewise, deep- and shallow-rooted combinations of crops can be harvested together in one field to reduce plant water availability and land erosion. A tried-and-tested method in hilly areas is terracing, which increases arable areas and prevents soil erosion on steep slopes. Terracing is the practice of cutting many steps into the surface of mountain slopes to create flat or gently sloping land for cultivation. This slows the speed at which water runs off the land while allowing more moisture to sink into the earth, which reduces nutrient loss. Traditional terraced fields are often supplemented by stone walls or vegetation on their slopes to reinforce them and prevent landslides. In addition to preventing erosion, terraces help maintain water resources more effectively. They store rainwater and obviate the need for irrigation. In combination with terraced systems, contour farming is another means many mountain community members use to avoid soil degradation: crops are grown along natural contours. Soil productivity has been preserved using these ancient methods, and the land will continue to provide fertility (Guindo et al. 2024; Pradhan et al. 2024; Subba et al. 2024; Komba et al. 2025).

Organic farming and low-input sustainable practices have recently been recognised as viable approaches to this problem. Due to transportation costs and economic constraints, organic farming has evolved naturally as a possible medium, given the relative scarcity of chemical inputs in many hilly areas. Organic farming uses natural fertilisers such as compost, manure, and green cover crops to maintain soil fertility and simultaneously prevent environmental damage. Standard practices include composting organic waste, adding livestock manure, and using biofertilisers from beneficial microorganisms. The soil structure is optimised, microbial activity is enhanced, and plant health is promoted. Insect pests and diseases can be managed using biological control methods. For example, natural predators are introduced, and neem-based pesticides are used. Crop rotation breaks the pest cycles. Organic farming encourages traditional seed varieties, often better adapted to local climatic conditions and more resistant to pests than commercial hybrids. Because of such emphasis on low-input sustainable practices, organic farming complements rather than replaces hill agriculture in many ways. Reducing reliance on synthetic chemicals and organic husbandry contributes to the conservation of biological diversity and has helped arrest the decline of pollinators. This also implies less soil and water contamination (Deng et al. 2024; Kamboj et al. 2024; Samoraj et al. 2024; An et al. 2025; Pakeerathan 2025).

An essential aspect of sustainable hill farming is integrating agroforestry systems, in which trees, shrubs, and crops are grown together to form a diverse and resilient agricultural landscape. These systems deliver a wide range of commercial benefits, including improved infertile fields, absorption of carbon dioxide, and protection against extreme weather events. In many hilly areas, farmers incorporate fruit-bearing trees, nitrogen-fixing plants, and medicinal herbs for cultivation next to their main crops to increase economic returns while maintaining ecological balance. Trees stabilise slopes, prevent erosion, and shelter the ground crop strip beneath them in the shade. New microclimates form and are suitable for multiple crop types (Tran et al. 2023; Kittichotsatsawat et al. 2024; Pant et al. 2024; Subba et al. 2024). Agroforestry systems help to sequester carbon by removing, capturing, and storing atmospheric carbon dioxide through trees or other vegetation growth. Therefore, combating global climate change is indispensable. Traditional agroforestry practices, such as planting tea and coffee beneath tall shade trees or integrating tree fodder with livestock systems, have been typical in many hill regions worldwide for centuries. Such systems offer an ecologically sustainable and productive means of increasing farm production. Water conservation is another crucial part of sustainable hill farming, particularly in areas with unreliable rainfall and water shortages during specific periods of the year. Over the years, traditional water harvesting methods, such as constructing small reservoirs, check dams, and roof-water collection systems, have kept water available for agriculture and household living costs. Contour bunding, where embankments are built along natural contours to break the water flow and allow groundwater recharge, works on the same principles. This is similar to mulch, cover crops, no-till farming practices, and other methods that preserve soil moisture. In areas where artificial watering is necessary, low-cost and efficient irrigation techniques, such as drip and gravity-fed systems, are becoming increasingly popular for saving water in agriculture. These sustainable water management practices are indispensable tools for maintaining hill farming productivity in the era of climate change, particularly in light of the shifts in global precipitation patterns attributable to this phenomenon (Cesco et al. 2023; Tran et al. 2023; Pradhan et al. 2024; Qiu et al. 2024; Shi et al. 2024; Komba et al. 2025).

Livestock integration is another vital aspect of traditional hill-farming systems. Yes, animals are present to help recycle nutrients and improve soil fertility. Manure from these means household food for farmers. Mixed farming of crops and animals is widespread in mountainous areas, such as cattle, goats, and poultry, providing livestock and animal traction during normal cultivation operations. In addition, dairy and beef production backed further market income. Livestock manure is an organic fertiliser high in essential nutrients that enrich the soil while acting as the mother of the plant roots. Herbal practices include rotational grazing and protection against the sowing of fodder crops, maintenance of soil fertility, and cessation of land degradation caused by overgrazing. Traditional herding techniques that work within the constraints of mountainous terrain often involve striking a balance between production output and environmental sustainability (Cui et al. 2024; Deng et al. 2024; Rasul and Gurung, 2024; Yang et al. 2025). However, despite the multiple advantages of indigenous techniques for farming, mixed farming terracing, and organic agriculture, they are subject to widespread adoption problems from county restrictions owing to financial constraints and the hallmarks of the influence of industrial agriculture. Organic farmers face numerous challenges as they try to gain market share, ranging from applying for certification on their produce to seeking financial backing for sustainable farming initiatives. However, government policies promoting monoculture and rewarding chemical inputs have encouraged neglecting traditional and ecological farming techniques. Furthermore, young adults are increasingly opting out of agriculture because they blame their lack of success on having left what 'people' now consider a viable way forward in favour of labour migration, new economic opportunities, and old-fashioned notions that it is still less profitable to farm with conventional methods than to do anything else sine qua non in a society where agriculture was once used but is now a subsidiary occupation. Although these problems pose formidable challenges, tackling them requires increasing investment in research and extension services and promoting policies supporting sustainable hill farming methods. By strengthening farmer cooperatives, providing financial incentives for organic certification, and encouraging community initiatives in other ecological agriculture models, we hope to continue these ancient farming traditions (Table 2).

4 Modern Innovations and Technological Interventions

In particular, climate-smart agriculture and precision farming are two key strategies that improve agricultural systems' resilience and productivity in hill and mountain lands where the environmental conditions are pretty unrelenting. Although innovation in farming aimed at realising higher productivity and products in harmony with nature is still relatively new, this concept embraces practices and technologies that increase agricultural productivity to adapt to changing climatic conditions and mitigate some of its costs. These include improved land and water management, soil

Table 2 Traditional agricultural practices in hilly regions

Category	Traditional practice	Description	Benefits	Challenges	Modern adaptations	Sustainability potential
Soil and water management	Terrace farming	The land is shaped into step-like structures along the slopes to reduce erosion and retain moisture	It prevents soil erosion, conserves water, and improves land productivity	Labour-intensive and requires regular maintenance	Stone and concrete reinforcements and mechanised levelling were used	It is highly sustainable and supports long-term agricultural use
	Contour ploughing	Ploughing along natural elevation contours to minimise runoff and improve water absorption	Reduces soil erosion and increases water infiltration	Difficult to implement on steep terrains	Integration with precision farming techniques and GIS mapping	Effective for erosion control, especially with modern enhancements
	Zabo system (Nagaland)	Integrates water harvesting, livestock farming, and crop cultivation	Ensures water availability and promotes soil fertility	Requires a large land area and complex management	Improved water storage systems and scientific livestock integration	It supports multi-functional land use and is highly sustainable
Crop cultivation	Shifting cultivation (jhum farming)	Farmers clear and burn forest patches, cultivate for a few years, and then move to a new plot	It allows land to regenerate naturally and maintains soil nutrients	It causes deforestation, carbon emissions, and declining productivity	Agroforestry-based shifting cultivation and controlled burn techniques	It can be made sustainable with proper follow-up management

(continued)

Table 2 (continued)

Category	Traditional practice	Description	Benefits	Challenges	Modern adaptations	Sustainability potential
	Mixed cropping	Growing multiple crops in the same field optimises land use and nutrient sharing	It enhances biodiversity and improves soil fertility	Requires detailed planning and knowledge of crop compatibility	Incorporation of modern polyculture systems and precision agriculture	Increases resilience to pests and climate variability
Agroforestry	Alley cropping	Planting trees or shrubs between crop rows improves soil fertility and provides shade	Reduces soil erosion, increases biodiversity, and enhances carbon sequestration	Slow initial tree growth requires careful species selection	Use of nitrogen-fixing trees and improved pruning techniques	High sustainability due to soil conservation and carbon sequestration
	Silvopasture	Combining forestry with livestock grazing for integrated land use	Improves land productivity and provides multiple income streams	The risk of overgrazing requires land management skills	Rotational grazing systems use native fodder species	Climate-resilient and maintains ecological balance
Livestock management	Transhumance (seasonal livestock migration)	Herds are moved between high-altitude summer pastures and low-altitude winter pastures	Utilises seasonal pasture availability and prevents overgrazing	It requires extensive land access and risks livestock loss	GPS tracking for herds, structured pasture rotation	Maintains ecological balance and prevents overuse of single grazing areas
	Ethnoveterinary medicine	Use of local herbs and natural remedies to treat livestock diseases	Low-cost, locally available, and minimal side effects	Lacks scientific validation and is inconsistently effective	Scientific validation and integration into veterinary care	Reduces dependency on chemical medicines and promotes biodiversity

Pest control	Neem-based pest control	Neem extracts are used to repel pests and prevent crop damage	Eco-friendly, non-toxic, and enhances plant health	Limited effectiveness against large infestations	Formulation of commercial neem biopesticides, integrated pest management	Sustainable alternative to synthetic pesticides
	Companion planting	Plant certain crops together to repel pests (e.g., marigolds with vegetables) naturally	Reduces pest populations and enhances biodiversity	Requires knowledge of plant compatibility	AI-driven planting guides and the use of bioengineered pest-resistant plants	Supports sustainable pest management and improves soil and crop health
Seed preservation	Traditional seed banks	Farmers save and exchange heirloom seeds to maintain crop diversity	Preserves genetic diversity and ensures seed availability	Vulnerable to extreme weather and seed deterioration	Community-managed seed banks with cryopreservation techniques	Vital for long-term food security and climate resilience
	Seed pelleting with ash and clay	Seeds are coated with natural materials to improve germination and protection	Enhances seed viability and prevents fungal infections	Time-consuming, inconsistent coating thickness	Biopolymer coatings and mechanised seed pelleting are used	Increases germination rates and protects against environmental stressors
Harvesting and storage	Traditional granaries (mud and bamboo storage bins)	Crops are stored in raised, ventilated structures to prevent spoilage	Protects against rodents and moisture and is a low-cost storage solution	Limited capacity, vulnerable to extreme weather	Use of improved storage materials, controlled-atmosphere storage	Supports food security and reduces post-harvest losses
	Sun drying and smoke drying	Grains, vegetables, and meat are dried using solar heat or smoke to extend shelf life	Preserves food without electricity and prevent spoilage	Weather-dependent, time-consuming	Solar dryers, temperature-controlled dehydration techniques	Reduces food waste and supports year-round availability

Source: Bardsley et al. (2022), Kamboj et al. (2024), Laudari et al. (2024), Samoraj et al. (2024), Supe et al. (2024) and Yang et al. (2025)

practices sensitive to sustainability principles, and modern climate-resilient crop varieties (Rasul and Gurung 2024; Shahzad et al. 2024; Supe et al. 2024; Pakeerathan 2025). Advanced technologies such as remote sensing, geographic information systems (GIS), and artificial intelligence (AI) aim to increase resource use efficiency and crop yield while reducing inputs. Using real-time data on soil moisture content, temperature, and nutrients, farmers can schedule irrigation, apply fertilisers, or determine which pests need to be addressed. This minimises environmental pollution while maximising productivity. In hilly areas, precision farming techniques, such as variable rate technology (VRT) and satellite field monitoring, help overcome some difficulties created by steep terrain, microclimatic variability, and soil differences. These technological interventions increase resource use efficiency and save inputs by optimising their applications (Laudari et al. 2024; Phung and Dao 2024; An et al. 2025; Sahoo et al. 2025). Consequently, greenhouse gas emissions were minimised. Internet of Things (IoT) devices, such as soil moisture sensors and automated irrigation systems, are incorporated into climate-smart arable farming to enhance efficiency. Instead of sitting back and weathering these erratic weather patterns, prolonged droughts, or unpredictable rainfall without relief, hill farmers who practise climate-smart agriculture (CSA) and precision farming get ahead of whatever may come.

Agroforestry and conservation agriculture (CA) are indispensable for sustainable hill farms, offering ecological and economic benefits that promote healthy soil, more effectively retain water in the land, and conserve the biodiversity of native species. Agroforestry, the practice of combining trees with fields in a mutually beneficial way to provide land use management, is especially suitable for hilly lands where both soil erosion and land degradation occur. Trees grown using these systems have numerous functions; stabilising slopes, preventing surface runoff, storing carbon, and providing timber, fruit, and medicinal plant income are a few. Traditional agroforestry systems exist in many hilly areas, such as homestead gardens and silvopastures. They show how agroforestry can play a role in making environmentally sustainable lifestyles more common for future generations. Modern agroforestry innovations, such as alley and multilayer cropping, maximise land use efficiency while nurturing soil fertility (Kharel et al. 2023; Piras et al. 2024; Pakeerathan 2025; Sahoo et al. 2025; Yang et al. 2025). Along with agroforestry, conservation agriculture (CA) is essential for maintaining the productivity of hill farm and preventing environmental degradation. CA is based on three core principles: no-till farming (minimising soil disturbance), permanent soil cover such as mulching and cover crops, and diversified crop rotations. These features combine to improve the soil structure, increase the organic matter content, and promote water infiltration. As a result, CA practices are particularly beneficial in hilly terrain areas where soil erosion is probable and surface water runoff must be controlled. For example, no-till farming reduces soil compaction and enables soil microbial communities to persist throughout the year. This indicates improved soil fertility and a higher crop yield. Mulching the soil with organic residues protects it from erosion and enhances the moisture content, thus diminishing the need for frequent irrigation. Another aspect of conservation agriculture is cover cropping, which fixes nitrogen from the air, suppresses weeds, and maintains soil health. Implementing agroforestry and conservation agriculture increases the agricultural sustainability of hill farming, mitigating climate change by increasing carbon sequestration and reducing greenhouse gas emissions.

In the last 10 years, hill crops have been completely changed by using biofertilisers, the latest techniques to ensure high-quality soil fertility, and machines that work in place of human labour. Organic liquid fertilisers contain beneficial microorganisms, such as nitrogen-fixing bacteria, phosphate-solubilising fungi, and plant growth-promoting rhizobacteria. They help improve the fertility of our soil through an eco-friendly approach by replenishing nutrients lost during crop growth from nearby mineral substances, but seldom en route via such organic means that they might ever hope for survival in future times (much like ourselves). However, unlike chemical fertilisers, biofertilisers can restore soil health without causing pollution or the loss of essential nutrients. Where soil degradation and nutrient loss are serious concerns in hill agriculture, biofertilisers can restore soil microbial balance, improve nutrient supply to plants, and enhance their resilience against specific abiotic stresses such as drought or fluctuations in temperature (Deng et al. 2024; Laudari et al. 2024; Piras et al. 2024; Shahzad et al. 2024; Yang et al. 2025). Furthermore, enhanced crop varieties, including drought-resistant, pest-resistant, and high-yield cultivars, have significantly increased agricultural production capabilities in hilly areas. More advanced breeding techniques, such as marker-assisted selection and genetic engineering, can improve the quality of crops to help them withstand challenging environmental conditions while maintaining their key nutritional value. For example, upland rice and maize (corn) have improved water-saving varieties with greater water-use efficiency in pastoral farming systems, which have performed well. Traditional crops, such as millets and buckwheat, raised by nature on the Qinghai-Tibet Plateau at over 3000 m above sea level, have been improved by selective breeding programmes to higher-yielding, disease-resistant versions. These improved varieties of traditional crops can be combined with sound farming practices to achieve higher agricultural productivity in a way that is also ecologically sustainable.

Mechanisation has been challenging in mountain agriculture because of the terrain and scattered land parcels. The development of small-scale terrain-specific equipment is a revolution. In hilly regions, farming is traditionally labour-intensive. Using labour in this high-demand environment means relying on human power or oxen, not machine tools. Minitillers, power weeders, and precision seeders have entirely transformed modern agriculture, reducing workforce dependency, improving the precision of planting operations, and reducing costs. Mechanisation promotes the perfection of every detail involved: land preparation, planting, and harvesting operations can all be self-sustained and spread continuously. If inputs are distributed evenly in precise amounts, labour will return efficiently to its previous leisure pace. Drone technology controls pests, monitors crops remotely, and provides precision spraying to serve agriculture in harsh terrain better. Multispectral sensor-equipped drones help maintain crop health in real time, encouraging interventions that use as few pesticides as possible while yielding results from compound means. On the other hand, remote-controlled, satellite-navigation machinery has made precision farming possible in areas that are not only difficult to access but also located on steep sites. This addresses a significant challenge for mountain agriculture (Kharel et al. 2023; Cui et al. 2024; Deng et al. 2024; An et al. 2025; Nkinyam et al. 2025; Yang et al. 2025). Researchers have attempted to develop stable crop machinery on slopes in the autonomous regions of Tibet, Yunnan, and other regions. Significant advances have been made, but there is still room for improvement in certain areas, such as the technology associated with machine seeding. Research institutions and governments are encouraging the production and distribution of hill-specific machines to help smallholder farmers avoid the traps inherent in traditional manual farming.

For various reasons, it is challenging to realise the widespread use of such modalities in hilly farming systems. Pollution has been stopped by disease control and has reached a convincing level, but it is unclear where all the poisons go. Compared with conventional cultivation methods, the higher costs associated with modern farming inputs and technology make them unpopular for small farmers. Credit problems prevent smallholder farmers from adopting new strategies and sustaining their livelihoods. Farm machinery's heavy expense and the shortage of equipment suited to mountainous areas are equally fatal to the mechanisation promise. Such innovations must be implemented with practical capacity-building schemes and support systems if their integration into farming systems is to occur soon. Only through research and development, extension services, infrastructure, and farmer training can the most advanced climate-smart and environment-friendly farming practices be extended to remote farming communities. To impart knowledge and encourage good practices, farmer training courses may use demonstration crab pots and participatory research programmes. In addition, by using digital platforms, one can not only exchange information but also render calls for service and respond in real time, for example, on soil health, crop varieties, or the best crops to plant (Kamboj et al. 2024; Phung and Dao 2024; Ramakrishnan et al. 2024; An et al. 2025; Nkinyam et al. 2025).

Incorporating climate-smart farming, precision farming, agroforestry, conservation agriculture, and modern technology has enormous potential for turning mountain agriculture into a more resilient or sustainable pattern. These measures address the key problems of soil pollution, fertility reduction, and water scarcity, which can increase output, alleviate poverty, and help environmental conservation efforts. Biofertilisers, improved varieties of crops, and mechanisation are other ways in which hill farmers' capacity to adapt to climate variations can still be further increased (Deng et al. 2024; Laudari et al. 2024; Ramakrishnan et al. 2024; Ibe and Kollur, 2025; Yang et al. 2025). All these remarkable advances call for a new policy of constant support: efforts to promote technological progress in basic and applied research materials, humane or people-oriented loan financing, and innovations for smallholder farmers. Only by taking these sustainable agriculture practices as a norm can the farming community in hilly areas narrow the gap to long-term food security, economic stability, and ecological security, so that all things are sustainable for generations yet unborn (Table 3).

Table 3 Modern innovations and technological interventions in agriculture

Category	Innovation/ technology	Description	Benefits	Challenges	Current adoption trends	Future potential
Precision agriculture	GPS-guided tractors	GPS-enabled tractors allow precise navigation, reducing input waste and optimising field operations	Increased efficiency, reduced fuel and labour costs, minimal environmental impact	High initial cost, need for technical expertise, and limited accessibility in developing regions	Widely adopted in developed countries, growing interest in emerging economies	High scalability with advancements in AI and automation
Soil and water management	Drip irrigation systems	Delivers water directly to plant roots, reducing wastage and improving water-use efficiency	Significant water savings, improved crop yields, and reduced soil erosion	High installation costs, clogging issues, and maintenance requirements	Extensively used in water-scarce regions; adoption incentives are available in many countries	Crucial for sustainable agriculture in arid and semi-arid areas
Crop protection	AI-powered pest detection	AI-driven image recognition and remote sensing detect pest infestations early	Reduced pesticide use, early intervention, and lower crop losses	Requires infrastructure (drones, cameras), initial investment, and AI training	Growing adoption of precision farming, particularly in high-value crops	It is expected to become an industry standard with IoT and big data integration
Biotechnology	Genetically modified crops (GMOs)	Crops are engineered for higher yields, pest resistance, and drought tolerance	Increased productivity, reduced pesticide dependency, and improved resilience	Public resistance, regulatory restrictions, and potential ecological concerns	Widely used in countries like the USA, Brazil, and Argentina; restricted in others	Potential for tailored solutions to climate change and food security challenges
Climate resilience	Drought-resistant crop varieties	Crops are genetically or traditionally bred to withstand water scarcity	Ensures food security, reduces irrigation dependency, and stabilises yields	Long breeding cycles, regulatory approvals, and potential loss of biodiversity	Increasing global demand in response to climate change	The key solution for ensuring agricultural resilience in dry regions

(continued)

Table 3 (continued)

Category	Innovation/ technology	Description	Benefits	Challenges	Current adoption trends	Future potential
Post-harvest management	Cold chain logistics	Temperature-controlled storage and transport to reduce post-harvest losses	Extends shelf life, minimises spoilage, and ensures food safety	High energy consumption, infrastructure costs, and logistical challenges	Advanced adoption in developed nations and emerging markets is investing heavily	Essential for reducing food waste and improving global food distribution
Smart farming	IoT-enabled sensors	Sensors collect real-time soil, weather, and crop health data for decision-making	Precision monitoring, efficient resource use, improved yields	High setup costs, data management challenges, and connectivity issues in rural areas	Increasing adoption with innovative agriculture initiatives and 5G expansion	Future integration with AI for fully automated farming systems

Source: Cui et al. (2024), Laudari et al. (2024), Ramakrishnan et al. (2024), Nkinyam et al. (2025), Pakeerathan (2025) and Yang et al. (2025)

5 Opportunities for Sustainable Hill Agriculture

Strengthening local value chains and agribusiness in hilly agriculture is essential for creating business opportunities, improving farmers' living standards, and, ultimately, keeping these extra income streams coming into future generations. Due to the particular conditions of farming in hilly areas, the local value chains that ensure that people's meat and vegetables reach the market must add real value to farm products by reducing post-harvest wastage rates and connecting them with markets at a distance. The key to developing these value chains lies in organising local processing, storage, and transport systems so that distant farmers are linked to urban markets. Hill farmers can win better prices by adding value through primary processing, such as drying, diversity of grades, and packaging. They also began to lessen their reliance on mezzanine traders (e.g., buyers who transport foliage plants from the mountains to the lowlands require upfront payment and guarantee a price for what they collect). Encouraging agro-based industries, such as small-scale food processing operations and traditional craft industries, can create jobs while retaining economic benefits within communities. In addition, digital platforms and e-commerce solutions can open direct marketing routes for highland products, enabling hill farmers to reach more consumers (Cui et al. 2024; Deng et al. 2024; Kamboj et al. 2024; Supe et al. 2024; Nkinyam et al. 2025; Yang et al. 2025). Government interventions, public-private partnerships (PPPs), and financial institutions must combine to provide the financial support, infrastructure, and training necessary to build efficient and sustainable agricultural value chains. Integrating digital technologies, such as innovative packaging and cold storage facilities, will help local agri-food enterprises become more efficient. For hillies, this means less market volatility and a higher realisation price.

In hilly areas, where small-scale farming generally limits productivity, the roles of cooperatives and farmers' associations are indispensable. Individually, farmers are often powerless during negotiations. They need a cooperative organisation that empowers them to group to obtain better terms of trade, liaise with customers, and reduce transaction costs. In the supply of inputs and credit, postharvest handling, and marketing, cooperatives are indispensable in allowing small farmers to participate in more significant markets. Farmer producer organisations (FPOs) and self-help groups (SHGs) also help with knowledge sharing, technical training, and skill acquisition. They provide farmers with a set of better agricultural and business management practices. These organisations can buy inputs (e.g., seeds, fertilisers) and implement the necessary equipment at a lower cost due to economies of scale. This means that farmers' costs decrease directly with the amount they produce. Contract farming arrangements for hill agricultural products, such as fruit and vegetables, are helped by these organisations when interacting with agribusiness enterprises, food processing industries, retailers, and exporters. Cooperatives can set up direct marketing arrangements to ensure fair-trade ethics and create added value at the farm level. The cooperatively owned and operated by women and the agro-enterprise led by youth could considerably change the employment pattern, be more favourable to inclusive growth, and reduce the population out of rural areas. Finally, policy support, training projects to strengthen cooperatives' capacity to manage overall sustainability, and financing incentives from public block grant sources will go a long way to improving hill agriculture (Kharel et al. 2023; Laudari et al. 2024; Majidian et al. 2024; Komba et al. 2025; Sahoo et al. 2025).

Opportunities are significant for high-value crops, horticulture, and medicinal plants, and provide considerable promise for hill farmers to diversify their income sources and strengthen agricultural profitability. In addition, high-value crops are now suitable for planting in these highlands, thanks to their unique agro-climatic conditions, including spices such as cardamom, tea, coffee, and fruit trees rich in aroma and medicinal herbs such as gentian root that Cathy dislikes. Instead of traditional staple crops, a move into high-value horticulture can increase farmers' returns per unit of land, and land use can become sustainable. Saffron, ginger, and turmeric have high market demand and export potential. In other words, they are profitable for small-scale producers. In addition, developing more meat processing will result in high returns, which is where the fruit industry should move. Cherries, peaches, apricots, and kiwis are medicinal and aromatic plants, including ashwagandha, aloe vera, rosemary, and lavender, which enjoy great demand in the pharmaceutical and cosmetic circles, thus offering an additional outlet for earning credit. The organic cultivation of these high-value crops makes them more commercially acceptable. It also brings higher prices, especially in international markets, where people are becoming more health-conscious and are looking for organically grown products. Improving plant material quality and ensuring disease-resistant, high-yield varieties for farmers by setting up high-value crop nurseries, tissue culture labs, and research stations can improve these aspects. In addition, governments and research institutions should mediate between hill farmers and pharmaceutical, herbal, and meat processing industries. In this way, they ensure value addition and market expansion for niche crops. Suppose that rural regions incorporate high-value crops, horticulture, and medicinal plants. In this case, it is significant that these provide ways to diversify the sources of farmers' income and improve agricultural income (Granco et al. 2019; Geddes et al. 2024; Pradhan et al. 2024; Rasul and Gurung 2024; Shi et al. 2024; Sahoo et al. 2025).

Promoting ecological tourism and eco-agriculture in hilly regions can offer new ways to increase rural incomes and foster sustainable development. Eco-tourism or agri-tourism, a combination of agriculture and tourism, is another way for farmers to earn income from their land. They can provide tourists with farm stays, guide experiences, and encourage experiential farming activities. Hills naturally offer beautiful scenery in addition to eco-tourism activities, such as trekking, bird watching, and experiences in organic farming. Therefore, visitors also visit these extra attractions. Agritourism programmes result in jobs in the hospitality industry. They also create markets for local businesses and people that did not exist before, reducing the erstwhile labour drain in rural areas. By encouraging farm-based tourism, we can save traditional farming skills, folk knowledge, and indigenous cultural heritage, allowing the farmer to reap the fruit of his labour (Ahammad et al. 2023; Cesco et al. 2023; Kumeh 2023; Geddes et al. 2024; Li et al. 2024; Pradhan et al. 2024; Qiu et al. 2024). This community spirit in developing rural areas also comes from farm-based tourism. Additionally, agribusiness investment attracts rural women and young people to engage in new ways of life. This can include learning about organic agricultural farming or herbal beauty products that are produced from homemade macadamia nut oil, teaching at dairy farms, and learning about beekeeping. In addition to providing training on how to create one's business, obtaining small business incubators and providing a skilled labour force through finance helps speed up the transformation of the rural economy. Moreover, the integration of clean energy practices in agriculture, the construction of environmentally friendly facilities for farming, and waste recycling ideas can further expand the circle of agricultural business. All these factors can ensure both environmental sustainability and growing profits. To achieve this, government agencies, local governments, non-governmental organisations (NGOs), and private entrepreneurs must work together to set out policy frameworks, facilities, or incentives for development, and promotional methods for green development businesses, making agri-tourism sustainable in the long run (Kumeh 2023; Tran et al. 2023; Geddes et al. 2024; Kittichotsatsawat et al. 2024; Li et al. 2024; Pant et al. 2024).

How can the best agricultural efforts be supported? Multi-stakeholder approaches, such as those suggested in the illustration, are required for a successful implementation. Policymakers, researchers, private sector companies, and local communities should play a role in making new practices a reality. Funds for rural infrastructure, such as roads, electricity, information, and logistics links, are vital to help develop agribusiness and improve efficient value chains. Measures of agrarian reform familiar to hill farmers also need to be applied: direct production subsidies for high-yielding kinds (fish, crawfish, rice, mulberries, etc.), tax breaks and loan guarantees on agribusiness start-ups, and export promotion campaigns conducted jointly by international, central, and regional authorities. Three climate risk-reducing strategies were highlighted: water collection systems, precision irrigation, and agroforest integration. These should be aligned with local agribusiness models for environmental reasons (Granco et al. 2019; Tran et al. 2022; Geddes et al. 2024; Guindo et al. 2024; Shi et al. 2024; Vashishat et al. 2024). Helping farmers to obtain online shopping platforms, advisory staff able to communicate through mobiles, and online training can effectively reduce this knowledge gap and help with decision-making. Mobile payment companies are now reaching their most profound rural areas. The collection of village 'enterprises' in agribusiness runs to hundreds, while in the grain industry, this is just starting. Finally, in forming hill chain models, there is an infrastructure of cooperatives and farmers' organisations working together to gather produce from the farm, buy equipment, finance development, and set rules to govern trading practices. High-value crops, horticulture, and supplementary planting of medicinal plants are all viable options that mean earning more money for the hill tribes so they live in harmony with nature. Gentlemen farmers and ecotourists present villas with safe residences, bridges, and picnic facilities. They help give villages something which provides them with an income and means that spring-planting courses are taken as seriously now as ever. Of course, this is only possible when help is provided from all quarters: the environment, village-level development planning, and communities themselves, with support from outside parties. By exploiting these opportunities, hill agriculture can ascend to a more resilient and dynamic system with brighter prospects. This will ensure food security, incredible wealth, and cultural preservation in hill regions for generations (Table 4).

Table 4 Opportunities for sustainable hill agriculture

Category	Opportunity	Description	Environmental benefits	Economic benefits	Challenges in implementation	Scalability and future potential
Agroecology	Terrace farming	Involves creating stepped levels on hilly terrain to reduce erosion, manage water flow, and optimise land use	Reduces soil erosion, conserves water, and improves soil fertility	Enhances productivity and supports diverse crops	High initial labour, maintenance, and soil fertility management	Scalable with community involvement and potential for diverse crop cultivation
Water resource management	Rainwater harvesting	Collects and stores rainwater for agricultural use, reducing dependence on external water sources	Reduces surface runoff and improves groundwater recharge	Lowers water costs and ensures supply during dry periods	Requires infrastructure and periodic maintenance	High scalability, especially in areas with variable rainfall
Climate resilient crops	Introduction of drought-resistant crops	Cultivation of crops like millets and sorghum that require less water and are resilient to changing climatic conditions	Conserves water and reduces the need for chemical inputs	Stable yields, market demand for resilient crops	Limited awareness and seed availability require research support	The high potential of climate change adaptation can be replicated in various hill regions
Organic farming	Vermicomposting	Utilising earthworms to decompose organic waste creates rich, organic crop fertiliser	Improves soil structure and increases microbial activity	Reduces fertiliser costs and creates a new market for organic produce	Requires training, initial setup, and ongoing management	High scalability and growing demand for organic products
Agri-tourism	Farm stays and eco-tourism	Integrates tourism with agriculture, offering experiences like farm stays, guided tours, and local cuisine	Encourages conservation and promotes local flora and fauna	An additional income stream supports local culture	Requires marketing, infrastructure, and regulatory compliance	High scalability with growing interest in sustainable tourism

Value-added products	Processing of local produce	Converts raw agricultural products into value-added items like jams, pickles, and herbal teas	Reduces food waste and promotes biodiversity by using diverse crops	Higher profit margins open new markets and support the local economy	Requires processing knowledge, equipment, and market access	Scalable with proper training and support, with export potential
Policy and institutional support	Government subsidies and incentives	Financial assistance and support programmes for sustainable practices, organic certification, and infrastructure development	Encourages adoption of sustainable practices and supports conservation	Reduces financial burden and improves access to resources	Bureaucratic hurdles, awareness, and timely access to funds	High potential with effective policy implementation is key for widespread adoption

6 Policy Interventions and Institutional Support

Government initiatives and rural development programmes have shaped the agricultural landscape of the hills. These measures focus on crucial undertakings such as limited arable land, inadequate infrastructure, and climate fragility. Numerous national and regional governments have implemented policies and programmes to increase agricultural output, improve rural livelihoods, and protect the environment. Often, the emphasis is on providing financial aid to infrastructure development projects, farmers' capacity-building initiatives, and financial support for farmers' introduction of new technologies. Governments have also established subsidies and incentives to benefit hill farmers, such as climate-resilient agricultural practices, improved irrigation systems, and the introduction of mechanical tools (Laudari et al. 2024; Samoraj et al. 2024; Supe et al. 2024; Nkinyam et al. 2025; Pakeerathan 2025; Yang et al. 2025). Governments can exert influence on the market by encouraging self-help groups for farmers, producer groups (FPOs) of farmers, or farmer cooperatives to strengthen linkages to markets and enhance access among peasants to inputs such as livestock products, feed, fertilisers, and credit. Programmes designed to improve soil health management, foster organic farming activities, and promote integrated watershed development have been launched for hill areas where soil is poor and land degradation is rife. Digital platforms and mobile-based agricultural advisory services provide real-time weather forecasts, market prices, and recommendations for particular farming practices. Government-supported infrastructure investment, such as cold storage and transport facilities or improved supply chains, is indispensable if hill agriculture is to be strengthened. Post-harvest losses can be reduced, market access improved, and farmers can be assured a fair price. Moreover, skill development and vocational training programmes have taken young people up in large numbers for agriculture, entrepreneurship, and allied businesses such as agroforestry, beekeeping, and non-chemical farming agricultural practices (Granco et al. 2019; Tran et al. 2022; Cui et al. 2024; Li et al. 2024; Rasul and Gurung 2024). Credit access is another area where the government can help microfinance institutions, rural banks, and cooperative societies to trust small farmers. Small marginal peasants need money for sustainable agricultural practices to develop and flourish.

Non-governmental organisations, research institutions, and international collaboration are vital to furthering sustainable hill agriculture. Many NGOs actively work with rural villages by teaching new agricultural techniques, promoting sustainable livelihoods, and increasing their resilience against climate change. Over the years, many NGOs have built the capacity for agriculture, agroecology, and water development for hill tribes. Once water sources are protected, they train local people to understand the science of soil conservation. They employ both men and women to perform these tasks. Many even have projects to empower female farmers, hoping that future female workers might have the same opportunities as men. Research institutions are responsible for developing improved high-yield crops, disease-resistant seeds to suit the local environment, and climatically responsive agricultural programmes (Granco et al. 2019; Tran et al. 2022; Cui et al. 2024; Kittichotsatsawat et al. 2024; Laudari et al. 2024; Subba et al. 2024; Sahoo et al.

2025). They also carry out experimental work, advise on best practices, and advise farmers to appreciate how they can maximise productivity while considering the ecological balance. Local governments and foreign NGOs frequently call upon universities and agricultural research institutions to offer technical expertise, make policy recommendations, and train farmers. Moreover, international cooperation with organisations such as the Food and Agriculture Organization (FAO), the International Fund for Agricultural Development (IFAD), and the World Bank has provided channels for knowledge exchange, financial support, and the implementation of sizeable rural development projects. These collaborations offer technical training, development of measures to adapt to climate change, and market access for hill area smallholder farmers. They also help to maintain a wide range of biodiversity, promote agroforestry, and implement watershed management across borders. Moreover, investment organisations and global philanthropic foundations use 'grants' and 'sponsorship programmes' in big cities to support sustainable rural development and well-being projects so that the world's food supply can better satisfy human needs.

Crafting policies for sustainable land and water management is essential if hill agriculture is to survive long-term without causing environmental degradation. Our policies must stress soil erosion control, water resource conservation and development, and wildlife habitat protection for fragile hill ecosystems. Government agricultural departments and environmental protection authorities are introducing afforestation programmes, contour farming, and agroforestry models. They aim to protect the structure and fertility of the soil. Sustainable land-use planning policies advocate mixed cropping, terracing, and minimum tillage practices to prevent soil degradation and increase production efficiency (Kumeh 2023; Tran et al. 2023; De Corato et al. 2024; Guindo et al. 2024; Pradhan et al. 2024; Shi et al. 2024). Watershed management programmes are required to adjust water levels, regulate water flow into groundwater reserves, and make the most sensible use of limited water resources. Check dams, rainwater harvesting construction designs, and micro-irrigation systems have been promoted to address the need to solve water shortages and enhance water use efficiency. The government has incentivised farmers to adopt sustainable practices such as organic certification, agroforestry integration for raising crops, and regenerative agriculture techniques. This hopes to encourage long-term commitment as a steward of the environment. Moreover, community water management initiatives, including participatory irrigation management and decentralised water governance, stress the need for equal and efficient distribution among hill farming communities (Bardsley et al. 2022; Tran et al. 2022; Ahammad et al. 2023; Pant et al. 2024; Shi et al. 2024; Vashishat et al. 2024; Komba et al. 2025).

For instance, climate adaptation policies have been incorporated into sustainable land and water management to address the increasing daily pressures of climate change. Policies that foster the development and transfer of drought-resistant and flood-tolerant crop varieties mean that farmers are better prepared to cope with unpredictable weather patterns. By growing trees and farming in the same place, agroforestry encourages carbon storage, ameliorates the soil structure, and provides farmers with an extra source of income. Conservation agriculture, which pursues minimum soil disturbance, retention of crop residues, and crop rotation, has been advocated as a sustainable way to maintain soil health and increase the water-holding capacity. Additionally, integrated pest management (IPM) policies promote fewer pesticides by biologically controlling pests and promoting organic alternatives to protect the environment and human health. Policies for sustainable livestock management, such as controlled grazing systems, improved fodder production, alleviated land degradation, and providing households with nutritional security in rural regions are necessary. Strong governance mechanisms, financial support, and community participation are required to execute these programmes successfully (Ahammad et al. 2023; Kumeh 2023; Kittichotsatsawat et al. 2024; Li et al. 2024; Pant et al. 2024). Governments must collaborate closely with local institutions, farmer associations, and research organisations, taking care not just to promulgate policies but also to tailor them according to hill farmers' particular needs and constraints. Decentralised governance systems that empower local communities to manage land and water resources sustainably tend to be more effective than top-down models. The environmental and production traditions of sustainable land and water management initiatives will be considerably enhanced by rewarding and adopting climate-smart agricultural practices, providing financial aid for farm diversification, and incorporating traditional ecological knowledge into policy frameworks. Furthermore, encouraging green finance mechanisms such as payments for ecosystem services (PES), carbon credits, and climate-resilient agricultural insurance can positively affect sustainable hill farming (Granco et al. 2019; Tran et al. 2022; Cesco et al. 2023; Kumeh 2023; Geddes et al. 2024; Li et al. 2024; Pradhan et al. 2024; Qiu et al. 2024).

The success of government initiatives, NGO interventions, and sustainable land management policies largely depends on the active participation of farmers. This is true for any action taken in rural economies, such as those upstream or directly off China's main rivers. Community engagement in agricultural production and life sciences should be enhanced. Finally, continued innovation in the agricultural sector contributes to long-term economic sustainability. However, the logic of such an obvious connection between ecology and monetary policy is such that the public cannot penetrate deep enough. Future strategies should integrate digital agricultural solutions and expand market access to high-value crops. Therefore, it is necessary to strengthen climate resilience. Approaches based on kind and adaptable ecosystems can do no less than reverse environmental extermination, gradually transforming all natural conditions into artificial forms. (Granco et al. 2019; Ahammad et al. 2023; Kumeh 2023; Pant et al. 2024; Pradhan et al. 2024; Subba et al. 2024; Vashishat et al. 2024). Collaborative efforts between policymakers, researchers, private sector stakeholders, and grassroots organisations can ensure that a holistic solution addresses hill agriculture's economic, social, and environmental problems. By leveraging technology, policy support, and community-driven initiatives, the sustainability of hill agriculture can be significantly enhanced, leading to improved food security, economic prosperity, and ecological balance in these fragile locations.

7 Case Studies and Best Practices from Different Hill Regions

Integrated Farming Systems (IFS) are a sustainable, multi-functional management approach that is good for farming but also beneficial to rural communities. The highest priority is to combine various resources and reduce environmental pollution while maintaining economic benefits to form a balanced farming production system. Making Farming Successful and Integrated (MFSI) is the guiding principle of IFS. Thus, it should combine and mutually reinforce the strengths of components such as food crop cultivation, animal husbandry, aquaculture fisheries, forestry, and food processing. However, successful IFS models have been established in all countries, confirming that they can sustain resources over the long term. The Rice-Fish Farming System in Southeast Asia is an outstanding case: rice fields are also used as fishing ponds. This increases the farm output and reduces the need for chemical pesticides and fertilisers. Fish help control pests and provide farmers with an extra source of income and protein. Similarly, the Duck-Rice System in China and Japan integrates ducks into rice fields, where they eat insects and weeds and fertilise the shoots through droppings. Another successful IFS model is the dairy-based integrated system, particularly in India. Smallholders combined dairy farming with crop production and biogas generation. Minimal waste is processed into biogas for household use, and the slurry is used as an organic fertiliser, reducing the need for synthetic inputs and improving soil fertility. The Silvopastoral System is another successful IFS method that uses trees to provide shade and grazing land. With high-mortar soil fertility improved from livestock manure treatment so quickly obtained, the model has long been practised extensively in Latin America because it offers the benefits of biodiversity enhancement. Moreover, aquaponics and hydroponics-based IFS are now starting to catch on in urban and peri-urban areas, where fish and plants grow together in a symbiotic relationship, with minimum water use and land resources required. IFS involves technical dissemination, policy amendments, and strategic development. Governments and research institutions build these systems by creating an enabling context. The proven practice of the IFS guarantees food security and long-term sustainable development against global warming.

Efforts to conserve communities are critical because they help protect biodiversity and restore the land's health and resources. They also enable community members to use modern conservation methods that work with ancient wisdom. Unlike top-down policies, community-led conservation programmes can integrate local knowledge into modern environmental protection measures to benefit the people. Chen Yunguo, former director of the Remote Sensing Division at the Chinese Academy of Forestry Science in Beijing, argues that under the Community Forest Management (CFM) mode of forestry resource management in India and Nepal, people form organisations among themselves where necessary for deciding conservational rules, collecting dues for timber extraction on a cooperative basis, which guarantees sustainable use of forests. This CFM model has increased forest coverage, reduced some forms of soil erosion, and created livelihoods for forest-dependent groups. Community-driven water resource management suits regions suffering from soil degradation or water shortages. In India and Ethiopia, local people have taken the lead in everything, from building check dams to contour trenches and trees to enclose degraded watersheds. Suitable management has helped groundwater recovery rates improve significantly, soil erosion has been curbed drastically, and crop yields have increased appreciably. Marine conservation benefits from community-driven efforts. Consequently, another way to restore land is through the agroecological approach, which uses agroforestry, permaculture, and natural farming techniques to restore land. The Great Green Wall Initiative in Africa suggests that many local people vigorously participate in tree planting and hillside reforestation work to fight desertification and climate change. The success of community-driven conservation efforts depends on governmental governance mechanisms, training programmes that build skills, and money. Non-governmental organisations (NGOs), research institutions, and governmental agencies supply the technical knowledge that community conservation work needs, and policies support this approach. The daunting challenges climate change presents to global food security demand the adoption of climate-resilient agricultural techniques. These techniques make farming systems flexible under extreme weather conditions, reduce greenhouse gas emissions, and enhance sustainable land management practices. One of the most effective climate-resilient techniques is conservation agriculture (CA). This approach involves little or no soil disturbance, crop rotation, or permanent soil cover. Owing to these measures, soil fertility will be preserved, soil water-retaining capability will be improved, and carbon emissions from farming activities will decrease. Countries such as Brazil, Argentina, and Zambia have been using CA with great success. Crop yields have increased, and farmers are now more resilient to drought. Another crucial approach is the system of rice intensification or SRI. This method makes rice growing more water-efficient and productive while lowering methane emissions. SRI techniques such as alternate wetting and drying have been adopted widely in South Asia and Africa, resulting in higher yields and lower costs.

Drought-tolerant crops and improved seed varieties are necessary to address climate risks. Developments in biotechnology and plant breeding have resulted in the cultivation of genetically modified (GM) and conventionally bred crops that can withstand droughts, pests, and diseases. For example, drought-resistant varieties of maize introduced to sub-Saharan Africa have helped farmers deal with unreliable rainfall and decreasing soil fertility, resulting in poorer yields. Agroforestry practices contribute significantly to boosting an area's climatic resilience. Trees provide shade, promote a good soil structure, absorb carbon, and improve water circulation. Introducing agroforestry into Southeast Asia and Africa has increased farm yields while reducing the risks caused by climate change. Smart Irrigation Technologies, such as drip irrigation and precision farming, minimise water waste and increase crop yields. These technologies enhance water-use efficiency and achieve high yields, especially in dry or semi-arid regions. Solar-powered irrigation systems generate income, help save energy, and make farming more sustainable.

Management strategies for climate-smart livestock, such as rotational grazing, improved diet formulations, and methane recovery technologies, can contribute to reducing greenhouse gas emissions in the livestock sector. Techniques for livestock management developed in countries such as New Zealand and Denmark have reduced methane emissions while maintaining production levels. However, they are not endemic to the majority of the world population. In some areas, establishing climate-resilient agricultural techniques must be supported by strong institutions, education programmes for farmers, and incentives for policy development. The three parties of government, international organisations, and the private sector must pool financial resources, provide research funds, and open markets to realise the shift towards climate-resilient agriculture. Successful models of integrated farming systems, community-based conservation efforts, and climate-resilient agricultural techniques all help ensure sustainable agriculture and ecological conservation. IFS integrates diverse agrarian businesses, resulting in synergistic resource savings, economic resilience, and food security. Organisations initiated by local communities to conserve the environment give the native inhabitants themselves a stake in natural resources and ensure that the ecological balance will be maintained in the long term. With climate-resilient agricultural techniques, the harmful impacts of climate change can be mitigated, and the sustainability of food production systems can be increased. To realise global food security and environmental sustainability, these models must be extended through policy support, research breakthroughs, and joint local, national, and international efforts.

8 Conclusion and Future Perspectives

When discussing farming problems, we must emphasise hill agriculture because it must deal with unique issues. Hill agriculture not only grapples with changing and unpredictable weather patterns but is also hampered by erosion of valuable topsoil, has little land to cultivate anyway, and is the hardest of all farming systems, suffering decreases in output in the face of climate change. Finally, traditional techniques and modern technologies have also brought success stories. Integrating these two approaches can significantly enhance the resilience and productivity of hilly agriculture. Agroecological practices, diverse cultivation systems, and holistic watershed management have brought rich returns from successful agricultural interventions in hill regions. Therefore, these should be considered crucial focus areas for policymakers and development efforts in highland areas. For example, one critical policy suggestion is the promotion of climate-smart farming practices such as conservation tillage, intercropping, and organic agriculture. These methods increase soil fertility and retain water, thus reducing the adverse environmental effects on the Earth's systems where they are applied. While members of hilly communities still want to preserve their local customs and resources, policies must be introduced to support these efforts. Another area that requires attention is maintaining biodiversity and ecosystem services, which are crucial for agriculture in hills today. Thus, policies need to meet the demands of hilly communities to save their own local cultures and resources and, simultaneously, ensure enough biodiversity is left for sustainable agriculture in those areas. They state that increased investment in rural infrastructure can help boost productivity, such as better road networks, which make it easier for goods to get from one farm to another until they reach their destination elsewhere. Government agencies and non-governmental organisations must collaborate to organise training sessions and workshops for farmers, translating local knowledge and experiences into a form that ordinary people can understand. Government and non-governmental organisations must cooperate to provide capacity-building programmes and technical training for farmers so that valuable knowledge can be passed on and best practices adopted. Financial incentives should also be provided. Features like subsidies and insurance schemes might help small-scale farmers survive climate change or inter-annual weather variability. Farming on hills will face significant changes because new consumer preferences, technological advancements, and climate change adaptation measures will shape this field's development in the future. One development in hill agriculture is the growing dependence on precision agriculture: systems that use satellite communication and other technologies to streamline production and clean out waste from any inputs or processes with virtually no loss of output.

In future hill farming systems, agroforestry will be essential, combining trees, crops, and livestock to ensure long-term productivity and ecosystem stability. It is predicted that organic farming, permaculture, and similar model farms where soil fertility is the primary concern will soon become significantly more popular as the worldwide demand for clean food products increases. Sustainable water management, including rainwater harvesting and micro-irrigation, is essential to address water scarcity in mountain farming.

Efforts to improve soil health through bio-fertilisers, vermicomposting, and crop rotation are also increasing. By incorporating time-honoured "numeracy" techniques into the process of developing new crop varieties that are locally adapted, robust against climate change, and tailored to diverse ecologies, there is a need to strengthen the collective bargaining power and market access of smallholder farmers through policies and programmes favouring cooperation among them and producer organisations.

Rapidly expanding digital platforms and mobile-based advisory services allow farmers to access real-time weather forecasts, pest management strategies, and up-to-date market information on their mobile phones. International cooperation for exchanging technology and sustainable development efforts will strengthen hill farming systems. Research and innovation are key to the resilience of mountain farming, helping farmers cope with threats stemming from environmental changes, such as extreme weather events (heat, cold) or pests. Developing biotechnologically stress-tolerant varieties that yield well is vital for food security and the stability of the economic life of hill communities. Innovations in agroecological practices combined with traditional controls such as integrated pest management (IPM), zero-budget farming, and polyculture may allow hills to recapture their resilience without chemical inputs. Artificial intelligence (AI) and machine learning in agriculture will enable more precise prediction of pest outbreaks, soil health diagnosis, and yield prediction. This will place the farmer in a position to act proactively.

Universities and research institutions must collaborate with local farming communities to conduct participatory investigations. Otherwise, new findings will not be helpful or relevant to the local circumstances. Investment in climate research, including long-term studies on how temperature differences and irregular weather events affect the natural environment, will contribute to developing targeted adaptation plans. The contribution of public-private partnerships (PPPs) to promoting innovation is significant. Businesses specialising in agri-tech can collaborate with government agencies and research organisations to develop low-cost solutions suitable for implementation in mountainous areas such as those described by Hill Farmers: From Poverty to Peace. Sustainable finance models, including micro-credit systems and impact investment funds, encourage the application of resilient agricultural technologies. Capacity-building programmes focusing on technology transfer and educating farmers are essential for translating research results into usable products. The government must fund research and innovation centres in mountainous areas, facilitating trial verification and popularising new farming methods. Finally, hill agriculture sustainability requires a more comprehensive approach that includes policy promotion, innovative research, and community participation. By promoting resilience through climate-smart practices, sustainable water use, and farmer-oriented inventions, hill agriculture may overcome environmental or socio-economic adversity.

References

Ahammad R, Hossain MK, Sobhan I, Hasan R, Biswas SR, Mukul SA (2023) Social-ecological and institutional factors affecting forest and landscape restoration in the Chittagong Hill Tracts of Bangladesh. Land Use Policy 125:106478. https://doi.org/10.1016/j.landusepol.2022.106478

An L, Turner BL, Liu J, Grimm V, Zhang Q, Wang Z, Huang R (2025) Complex adaptive systems science in the era of global sustainability crisis. Geogr Sustain 6:100250. https://doi.org/10.1016/j.geosus.2024.09.011

Bardsley DK, Cedamon E, Paudel NS, Nuberg I (2022) Education and sustainable forest management in the mid-hills of Nepal. J Environ Manag 319:115698. https://doi.org/10.1016/j.jenvman.2022.115698

Cesco S, Sambo P, Borin M, Basso B, Orzes G, Mazzetto F (2023) Smart agriculture and digital twins: applications and challenges in a vision of sustainability. Eur J Agron 146:126809. https://doi.org/10.1016/j.eja.2023.126809

Cui H, Ke S, Lu X (2024) An empirical study on the impact of green transition of farmland use on agricultural economic growth: a case of Hubei Province of China. Habitat Int 148:103090. https://doi.org/10.1016/j.habitatint.2024.103090

De Corato U, Viola E, Keswani C, Minkina T (2024) Impact of the sustainable agricultural practices for governing soil health from the perspective of a rising agri-based circular bioeconomy. Appl Soil Ecol 194:105199. https://doi.org/10.1016/j.apsoil.2023.105199

Deng J, Liu K, Feng X, Ye J, Harrison MT, de Voil P, Hussain T, Huang L, Tian X, Zhou M, Zhang Y (2024) Exploring strategies for agricultural sustainability in super hybrid rice using the food-carbon-nitrogen-water-energy-profit nexus framework. J Integr Agric. https://doi.org/10.1016/j.jia.2024.07.025

Geddes E, Morgan-Davies C, McLaren A, Skuce PJ, Duncan JM, Sargison N, Kenyon F (2024) Investigating the perceived versus actual gastrointestinal nematode challenge on extensive sheep farms. Vet Parasitol 327:110148. https://doi.org/10.1016/j.vetpar.2024.110148

Granco G, Heier Stamm JL, Bergtold JS, Daniels MD, Sanderson MR, Sheshukov AY, Mather ME, Caldas MM, Ramsey SM, Lehrter RJ II, Haukos DA, Gao J, Chatterjee S, Nifong JC, Aistrup

JA (2019) Evaluating environmental change and behavioral decision-making for sustainability policy using an agent-based model: a case study for the Smoky Hill River Watershed. Kansas Sci Total Environ 695:133769. https://doi.org/10.1016/j.scitotenv.2019.133769

Guindo M, Traore B, Birhanu BZ (2024) Biomass and nutrient flow dynamics and sustainability practices to de-risk environmental challenges in the sub-saharan Africa farming system. Farming Syst 2:100109. https://doi.org/10.1016/j.farsys.2024.100109

Ibe KK, Kollur SP (2025) Livelihood and household clean cooking challenges in rural India: a community-based participatory approach for solutions. Energy Res Soc Sci 119:103846. https://doi.org/10.1016/j.erss.2024.103846

Kamboj A, Sadh PK, Chawla P, Saharan BS, Seth CS, Sridhar K, Duhan JS, Sharma M (2024) Sustainable management of rice by-products: environmental challenges, industrial applications, and circular bio-economy innovations. Biocatal Agric Biotechnol 62:103430. https://doi.org/10.1016/j.bcab.2024.103430

Kharel M, Raut N, Dahal BM (2023) An assessment of good agriculture practices for safe and sustainable vegetable production in mid-hills of Nepal. J Agric Food Res 11:100518. https://doi.org/10.1016/j.jafr.2023.100518

Kittichotsatsawat Y, Rauch E, Tippayawong KY (2024) Designing sustainability measurement of a Thai coffee supply chain using axiomatic design and business model canvas. Results Eng 24:103443. https://doi.org/10.1016/j.rineng.2024.103443

Komba BP, Chattha MWA, Fatima A, Thomson PC, González LA, Hyder MF (2025) Assessing the environmental, economic, and social impacts of industrial mining on agricultural communities in Sierra Leone: a sustainable approach using PLS-SEM. Clean Responsible Consum 16:100250. https://doi.org/10.1016/j.clrc.2025.100250

Kumeh EM (2023) Contestations, counteractions and equitable conservation—a case study of Ghana's Krokosua Hills Forest Reserve. Forest Policy Econ 157:103090. https://doi.org/10.1016/j.forpol.2023.103090

Laudari HK, Sapkota LM, Maraseni T, Subedi P, Pariyar S, Kaini TR, Lopchan SB, Weston C, Volkova L (2024) Community forestry in a changing context: a perspective from Nepal's mid-hill. Land Use Policy 138:107018. https://doi.org/10.1016/j.landusepol.2023.107018

Li N, Zhao H, Luo Z, Wang T, Yang J, Li L, Que S (2024) Soil erosion prediction in multiple scenarios based on climate change and land use regulation policies in context of sustainable agriculture. Catena 247:108525. https://doi.org/10.1016/j.catena.2024.108525

Majidian P, Ghorbani HR, Farajpour M (2024) Achieving agricultural sustainability through soybean production in Iran: potential and challenges. Heliyon 10:e26389. https://doi.org/10.1016/j.heliyon.2024.e26389

Nkinyam CM, Ujah CO, Asadu CO, Kallon DVV (2025) Exploring geothermal energy as a sustainable source of energy: a systemic review. Unconv Resour 6:100149. https://doi.org/10.1016/j.uncres.2025.100149

Pakeerathan K (2025) Chapter 1 – A global overview and the fundamentals of sustainable agriculture. In: Singh S, Sood V, Srivastav AL, Ampatzidis YBT-H in PA (eds). Academic, pp 3–13

Pant J, Teoh SJ, Gomes S, Pereira A, Pereira M, de Jesus LS, Du CADF, Bhujel RC (2024) Sustainable intensification of genetically improved farmed tilapia (GIFT) in Timor-Leste's farming systems: challenges and opportunities. Agric Syst 216:103874. https://doi.org/10.1016/j.agsy.2024.103874

Phung QA, Dao N (2024) Farmers' perceptions of sustainable agriculture in the Red River Delta. Vietnam Heliyon 10:e28576. https://doi.org/10.1016/j.heliyon.2024.e28576

Piras F, Allegri M, Fiore B, Agnoletti M, Santoro A (2024) Cultural landscape transformations in the "Prosecco hills of Conegliano and Valdobbiadene" UNESCO World Heritage Site in the context of terrain morphology. J Rural Stud 112:103452. https://doi.org/10.1016/j.jrurstud.2024.103452

Pradhan P, Subedi DR, Dahal K, Hu Y, Gurung P, Pokharel S, Kafle S, Khatri B, Basyal S, Gurung M, Joshi A (2024) Urban agriculture matters for sustainable development. Cell Reports Sustain 1:100217. https://doi.org/10.1016/j.crsus.2024.100217

Qiu H, Feng M, Zhang X, Song Z, Luo M, Wang J, Ye F (2024) Exploring the poverty-reduction benefits of agricultural machinery socialization services in China: implications for the sustainable development of farmers. Heliyon 10:e32636. https://doi.org/10.1016/j.heliyon.2024.e32636

Rai S, Gurung A, Bhakta Sharma H, Prakash Ranjan V, Ravi Sankar Cheela V (2024) Sustainable solid waste management challenges in hill cities of developing countries: insights from eastern Himalayan smart cities of Sikkim, India. Waste Manag Bull 2:1–18. https://doi.org/10.1016/j.wmb.2024.02.009

Ramakrishnan R, Rajendrakumar S, Kothurkar NK (2024) Regional sustainability of the Kattunayakan tribe in Kerala, India through the enhancement of agricultural, livestock, and livelihood options. Agric Syst 217:103929. https://doi.org/10.1016/j.agsy.2024.103929

Rasul G, Gurung P (2024) Unlocking the potentials of sustainable livelihoods in Chattogram Hill Tracts of Bangladesh. Nature-Based Solut 5:100108. https://doi.org/10.1016/j.nbsj.2023.100108

Sahoo S, Singha C, Govind A, Moghimi A (2025) Review of climate-resilient agriculture for ensuring food security: sustainability opportunities and challenges of India. Environ Sustain Indic 25:100544. https://doi.org/10.1016/j.indic.2024.100544

Samoraj M, Çalış D, Trzaska K, Mironiuk M, Chojnacka K (2024) Advancements in algal biorefineries for sustainable agriculture: biofuels, high-value products, and environmental solutions. Biocatal Agric Biotechnol 58:103224. https://doi.org/10.1016/j.bcab.2024.103224

Schillings J, Holohan C, Lively F, Arnott G, Russell T (2024) The potential of virtual fencing technology to facilitate sustainable livestock grazing management. Animal 18:101231. https://doi.org/10.1016/j.animal.2024.101231

Shahzad HMA, Almomani F, Shahzad A, Mahmoud KA, Rasool K (2024) Challenges and opportunities in biogas conversion to microbial protein: a pathway for sustainable resource recovery from organic waste. Process Saf Environ Prot 185:644–659. https://doi.org/10.1016/j.psep.2024.03.055

Shi Z, Zhang L, Xiong X, Hao Y, Zhang P, Pang M, Yang Y (2024) Environmental burden of excessive food intake: exploring new opportunities to enhance food and environmental sustainability in China. Resour Conserv Recycl 209:107820. https://doi.org/10.1016/j.resconrec.2024.107820

Subba B, Mondal S, Mandal UK, Tamang L (2024) Small-scale tea farming sustainability: a case study of Darjeeling hill region. India Farming Syst 2:100080. https://doi.org/10.1016/j.farsys.2024.100080

Supe H, Abhishek A, Avtar R (2024) Assessment of the solar energy–agriculture–water nexus in the expanding solar energy industry of India: an initiative for sustainable resource management. Heliyon 10:e23125. https://doi.org/10.1016/j.heliyon.2023.e23125

Tran DX, Pearson D, Palmer A, Gray D, Lowry J, Dominati EJ (2022) A comprehensive spatially-explicit analysis of agricultural landscape multifunctionality using a New Zealand hill country farm case study. Agric Syst 203:103494. https://doi.org/10.1016/j.agsy.2022.103494

Tran DX, Pearson D, Palmer A, Dominati EJ, Gray D, Lowry J (2023) Integrating ecosystem services with geodesign to create multifunctional agricultural landscapes: a case study of a New Zealand hill country farm. Ecol Indic 146:109762. https://doi.org/10.1016/j.ecolind.2022.109762

Vashishat RK, Sharma S, Chandel RS, Dev I, Chauhan GS, Divyanshu GA, Gupta M, Laishram C (2024) Sustainable agriculture made easy: CETARA-NF's self-certification. J Agric Food Res 18:101404. https://doi.org/10.1016/j.jafr.2024.101404

Yang H, Tan T, Ren G, Liu Y, Liu Z, Xia S, Wu Z, Zhang Y (2025) The dual nature of water hyacinth (Pontederia crassipes): environmental threats and sustainable solutions. Water Biol Secur 100359. https://doi.org/10.1016/j.watbs.2025.100359

Agronomic Opportunities to Alter Cropping Systems and Soil Health

R. Aditya, Rishika Saini, Chinmaya Sahoo, and Ankit Saini

Abstract

The world has seen an incredible population boom since 1950, more than tripling. This growth has pushed agriculture to evolve rapidly to feed our expanding populations. This rapid agricultural development has brought its own set of challenges, such as less variety in our crops, loss of nutrients in our soil and declining overall soil health. To tackle these issues while keeping farms productive, it has become necessary to rethink how to grow crops and manage the soil. This chapter explores different ways of improving farming practices, like changing crop rotation patterns, bringing in new types of crops and incorporating rotational grazing, which will help to build better soil health and reduce reliance on chemical fertilizers. Growing multiple crops together (intercropping), using diverse cropping systems and including cover crops can lead to better yields and healthier soils by creating farming systems that work more like natural ecosystems. This chapter also discusses conservation agriculture, which focuses on disturbing the soil as little as possible while keeping it covered. It also focuses on agroforestry and the practice of incorporating trees into farming systems through methods such as windbreaks and alley cropping, which can improve soil fertility,

R. Aditya · R. Saini
Chaudhary Sarwan Kumar Himachal Pradesh Krishi Vishvavidyalaya,
Palampur, Himachal Pradesh, India

C. Sahoo
Dr. Rajendra Prasad Central Agricultural University, Samastipur, Bihar, India

A. Saini (✉)
Chaudhary Sarwan Kumar Himachal Pradesh Krishi Vishvavidyalaya,
Palampur, Himachal Pradesh, India

Dr Khem Singh Gill Akal College of Agriculture, Eternal University,
Sirmaur, Himachal Pradesh, India

Y. Siddiqui et al. (eds.), *Sustainable Mycorrhizal Cultivation*,
https://doi.org/10.1007/978-981-95-1778-7_2

provide shade and support wildlife. Sustainable practices such as organic farming, smart irrigation systems and integrated pest management (IPM) are explored in this work as key strategies for long-term agricultural sustainability.

Keywords

Sustainable agriculture · Soil health restoration · Crop diversification · Conservation practice

Abbreviations

IPM	Integrated Pest Management
INM	Integrated Nutrient Management
CSA	Climate-Smart Agriculture
CA	Conservation Agriculture
SOC	Soil Organic Carbon
SOM	Soil Organic Matter
FFS	Farmer Field Schools
Bt	*Bacillus thuringiensis*

1 Introduction

1.1 Background and Importance

Management of farming systems plays a crucial role in maintaining healthy soil and sustainable agriculture. Over the past century, agricultural practices have undergone significant changes, driven by technological advancements and the pressing need to feed a growing global population (Sahoo et al. 2023). These changes have not always benefited the soil and environment in the long term. Traditional farming methods, such as growing single crops annually (monocropping), excessive ploughing and heavy use of chemical fertilizers, have taken their toll on agricultural land. These practices, while initially boosting productivity, have led to serious consequences such as poor soil quality, reduced fertility and declining crop yields. More concerning is how they have stripped essential nutrients from the soil, disrupted natural ecological processes and contributed to serious environmental problems such as soil erosion and loss of biodiversity. The impact extends beyond just farm fields—it affects water quality, wildlife habitats and even local climate patterns. Research has shown that scientists can do better and scientists worldwide are discovering promising alternatives. Diverse and sustainable farming methods, including crop rotation, intercropping, agroforestry and cover cropping, offer solutions that work with nature rather than against it. For instance, Tilman et al. (2002) found that increasing biodiversity in farming systems improved essential ecological services, including nutrient cycling and natural pest control. This leads to farming systems that are both more resilient and more productive. Their research demonstrated that diverse agricultural systems can maintain high yields while requiring fewer external inputs, making them economically viable in the long term.

When we create more biodiverse farming systems, we help nature to maintain its own balance in remarkable ways. Beneficial organisms such as pollinators, predatory insects and soil microbes thrive and naturally help control pests and diseases, thereby reducing the need for chemical pesticides. These systems are particularly effective for the efficient use of nutrients, especially in soils that are not naturally rich in nutrients. The intricate web of plant and animal interactions in diverse farming systems creates natural resilience against pests, diseases and environmental stress. However, the role of legumes in sustainable agriculture cannot be overlooked. Drinkwater et al. (1998) discovered that rotating crops with legumes (soybean, clover, or alfalfa) is especially beneficial for soil health. Legumes have the unique ability to convert nitrogen from the air into a form that plants can use through a symbiotic relationship with soil bacteria. This natural process reduces the need for synthetic fertilizers, which not only improves soil fertility but also helps prevent water pollution and reduces greenhouse gas emissions that often result from chemical fertilizer use. Economic benefits are also significant, as farmers can reduce their fertilizer costs while maintaining or even improving yields.

Conservation agriculture (CA) practices have shown tremendous promise for maintaining soil health, as highlighted by Lal (2004). When farmers minimize ploughing and keep their soil covered with plants, it helps preserve soil structure, increases organic matter content and helps the soil hold more water. These practices are particularly important for preventing soil erosion, which is a major problem in conventional farming. In particular, no-till farming has revolutionized soil management. By disturbing the soil less, farmers can maintain the natural soil structure, preserve beneficial organisms and reduce fuel costs associated with traditional tillage. The benefits of sustainable farming practices extend far beyond those of the individual farms. Conservation agriculture plays a crucial role in fighting climate change by capturing carbon from the atmosphere and storing it in soil. This process of carbon sequestration not only improves soil health but also helps reduce greenhouse gas levels in the atmosphere. Research suggests that improved agricultural practices could sequester significant amounts of carbon annually, making agriculture a part of the solution to climate change rather than part of the problem.

Modern farmers are increasingly recognizing that sustainable practices are both environmental and economic. While the transition to more sustainable methods may require initial investments in knowledge and equipment, long-term benefits often outweigh costs. Reduced input costs, improved soil health, better water retention and more stable yields have contributed to more profitable and resilient farming operations. Additionally, the growing consumer demand for sustainably produced food creates new market opportunities for farmers who adopt these practices.

The challenge lies in scaling up these sustainable practices while ensuring that they remain economically viable for farmers. This requires support from agricultural policies, research institutions and market mechanisms that recognize and reward sustainable farming practices. Education and knowledge sharing among farmers also play a crucial role in spreading these beneficial practices.

This chapter explores and highlights four key areas. (1) Understanding how cropping systems affect soil health: we will examine how different farming approaches, such as crop rotation, intercropping and agroforestry, impact important soil characteristics, including organic matter content, nutrient availability and microbial activity. This will help us to understand how sustainable farming practices influence soil health and productivity. (2) Exploring practices that improve soil health: specific farming techniques such as conservation tillage, cover cropping, intercropping and using organic amendments are highlighted. (3) Learning from real-world success stories: we will examine actual cases where farmers have modified their cropping systems and improved their soil health. These real-world examples from different farming environments will show how these techniques work in practice. (4) Assessing environmental and economic effects: we will evaluate the long-term benefits for the environment, such as carbon storage and reduced chemical use, while also considering the economic aspects of sustainable farming systems. We will look at both small farms and large agricultural operations to understand how these practices can work at different scales.

By addressing these objectives, we provided a thorough overview of how farmers can modify their cropping systems to improve soil health, make agriculture more sustainable and contribute to broader environmental goals.

2 Understanding Cropping Systems

2.1 Definition and Components

When we discuss a cropping system, we refer to the specific choices farmers make about which crops to grow, when to grow them and how to arrange them in their fields over time. This includes decisions about managing water, nutrients and pest control. Studying these systems usually requires teams of scientists from different fields to work together to solve problems and find ways to make farming both more productive and sustainable.

2.2 Components of the Crop Production Systems

2.2.1 Crop Rotation

When different crops are grown sequentially on the same piece of land, it is called crop rotation. This is a smart way to maintain soil health, reduce pest and disease problems and manage weeds. For example, farmers may plant soybeans, which add nitrogen to the soil (Sahoo et al. 2024; Manhas et al. 2024), followed by wheat, which uses nitrogen. Liu et al. (2018) found that this practice can improve soil quality while reducing the need for chemical fertilizers.

2.2.2 Monocropping and Polycropping

Monocropping—growing the same crop repeatedly—can lead to problems such as increased pest issues, deteriorating soil and loss of biodiversity. Polycropping, where multiple crops grow together, offers better sustainability by making more efficient use of resources, preventing soil erosion and helping the nutrient cycle better through the system.

2.2.3 Mixed Cropping

This involves growing several different crops simultaneously, which helps protect against crop failure. If extreme weather occurs, some crops might survive, while others do not.

2.2.4 Agroforestry

This approach combines trees and shrubs with crop farming. Trees provide shade, help prevent soil erosion and create homes for wildlife. Sanchez's research team (2019) found that farmers in Kenya who used agroforestry improved their soil, obtained better crop yields and earned extra money from selling timber and fruit.

2.2.5 Relay Cropping

Here, farmers plant a new crop before harvesting the previous one. This is an efficient use of land and helps ensure continuous production, especially in places with short growing seasons.

2.2.6 Tillage and No-Till Systems

How farmers prepare their soil can make a significant difference. Traditional tillage involves disturbing the soil through ploughing, whereas no-till farming minimizes the soil disturbance. Hobbs et al. (2008) found that no-till farming helps the soil retain moisture and improves its quality, especially in dry areas.

These different approaches to farming show that there is no one-size-fits-all solution. Instead, farmers need to choose the methods that work best for their specific situations while keeping in mind the long-term health of their soil and environment.

2.3 Current Trends and Practices in Cropping Systems

The agricultural landscape is undergoing a remarkable transformation as farmers worldwide adapt to meet the increasing food demands, address climate challenges and embrace sustainable farming methods. Today's agricultural practices represent an innovative blend of traditional farming wisdom and modern technological advancements, thereby creating more efficient and environmentally conscious cropping systems.

2.3.1 Sustainable Intensification

Sustainable intensification has emerged as a cornerstone of modern agriculture, emphasizing the need to optimize crop production from existing farmland while minimizing environmental impact. This comprehensive approach incorporates various strategies ranging from advanced precision farming methods to integrated pest management solutions. The results have been particularly encouraging in regions such as sub-Saharan Africa, where smallholder farms have achieved substantial yield improvements while simultaneously reducing their reliance on synthetic inputs. By focusing on resource efficiency and sustainable practices, these farms demonstrate the potential to balance productivity with environmental stewardship.

2.3.2 Climate-Smart Agriculture

Climate-smart agriculture (CSA) stands at the forefront of agricultural innovation, offering a robust framework for addressing climate-related challenges while maintaining productive farming systems. This approach emphasizes three key objectives: enhancing farm productivity, building climate resilience and reducing greenhouse gas emissions. Farmers implementing CSA strategies, such as diversifying crop selection and adopting conservation tillage practices, have witnessed remarkable improvements in both yield stability and environmental performance, particularly in the East African agricultural communities.

2.3.3 Organic and Regenerative Agriculture

The growing adoption of organic and regenerative agriculture reflects a profound shift toward nature-aligned farming practices. These methods prioritize soil health enhancement, biodiversity conservation and reduced dependence on synthetic agricultural inputs. While organic farming systems might initially produce slightly lower yields than conventional methods, research indicates that they demonstrate superior resilience during extreme weather events and offer enhanced long-term sustainability (Parida et al. 2023; Kumari et al. 2024). Regenerative agriculture further takes this philosophy by actively working to restore degraded soil ecosystems and promote carbon sequestration, contributing to both agricultural productivity and environmental health.

2.3.4 Precision Agriculture

Precision agriculture represents the technological vanguard of modern farming practices. By leveraging sophisticated data collection tools and analytical systems, farmers can make highly informed decisions about crop management, soil health and resource allocation. This technology-driven approach has shown impressive results in improving resource efficiency, with studies documenting significant enhancements in water and nutrient utilization while reducing operational costs and improving harvest yields.

2.3.5 Conservation Agriculture

Conservation agriculture (CA) continues to gain prominence with its emphasis on minimal soil disturbance, continuous soil coverage and strategic crop rotation patterns. These practices have demonstrated remarkable effectiveness in improving water retention, controlling soil erosion and building the organic matter content (Sahoo et al. 2023). Research has shown significant increases in soil carbon levels across various farming systems, with particularly impressive results in semiarid regions. The transformation of Brazil's Cerrado region serves as a compelling example of CA's potential to revolutionize agricultural productivity.

2.4 Challenges in Cropping Systems

Despite these innovative approaches, the agricultural sector faces numerous hurdles that impede the widespread adoption and effectiveness of sustainable farming practices. These challenges encompass the biological, environmental, economic and social dimensions that affect both immediate crop productivity and long-term sustainability goals (Table 1).

2.4.1 Soil Degradation and Erosion

Soil degradation remains a critical concern in modern agriculture. The cumulative effects of intensive farming practices, excessive tillage and heavy reliance on chemical fertilizers have resulted in the widespread deterioration of soil quality. Global research indicates that soil erosion alone causes substantial annual crop losses, while the continuous depletion of soil nutrients often creates a destructive cycle of increased dependence on chemical fertilizers.

2.4.2 Water Scarcity and Management

Water scarcity has emerged as a pressing challenge, particularly affecting regions that depend on irrigation systems or rainfall patterns. This situation is especially concerning in areas that practise water-intensive farming methods, such as rice and wheat cultivation systems. In many agricultural regions, particularly in India, intensive irrigation practices have led to alarming decreases in groundwater levels in recent decades, raising serious concerns about the sustainability of current agricultural practices.

Table 1 Challenges in sustainable cropping systems

Challenge	Description	Impact on cropping systems
Soil degradation and erosion	Loss of soil fertility and structure due to continuous cropping and excessive tillage	Reduces crop productivity, increases dependency on chemical inputs and threatens soil health
Water scarcity and management	Limited water availability, especially for heavily irrigated crops like rice and wheat	Increases risk of crop failure, promotes unsustainable water usage and impacts food security
Climate change	Unpredictable weather patterns affect plant and harvesting	Alters crop yields, increases pest pressure and affects seasonal farming practices
Pest and disease management	Increased pest resistance and disease pressure in monocropping systems	Heightens dependency on chemical pesticides, risks environmental harm and reduces biodiversity

2.4.3 Changing Climate and Erratic Weather Conditions

Climate change continues to disrupt traditional farming patterns by introducing greater weather unpredictability, which affects crucial aspects such as planting schedules, water availability and pest dynamics. Scientific projections suggest that staple crop yields could face significant declines by the mid-century due to climate-related impacts, with rain-dependent farming regions particularly vulnerable to extreme weather events.

2.4.4 Pest and Disease Management

The management of pests and diseases has become increasingly complex, especially in monoculture systems where limited crop diversity creates optimal conditions for pest proliferation. This often results in excessive pesticide use, which leads to environmental degradation and the development of resistant pest populations. Research indicates that pest-related crop losses continue to pose a significant challenge, accounting for a substantial portion of the potential agricultural production.

2.4.5 Socio-economic Constraints

Socio-economic barriers present significant challenges, particularly for small-scale farmers in developing regions. Limited access to financial resources, technical knowledge and modern infrastructure often prevents the adoption of more sustainable farming practices. Market price fluctuations and demand uncertainty further complicate the transition into improved cropping systems. Pingali (2012) noted that smallholder farmers often lack the capital and technical support to adopt sustainable cropping systems.

2.4.6 Nutrient Management and Fertilizer Dependency

The challenges of nutrient management and fertilizer dependency persist as critical issues affecting agricultural sustainability. Current farming practices frequently result in poor nutrient utilization efficiency, with significant amounts of applied fertilizers being lost to the environment, contributing to both economic waste and environmental degradation. This inefficiency not only impacts farm profitability, but also poses serious environmental concerns, including water pollution and greenhouse gas emissions. Zhang et al. (2015) found that global nitrogen use efficiency is less than 50%, meaning that more than half of the applied nitrogen is lost to the environment.

3 Soil Health: Concepts and Indicators

A healthy soil ecosystem serves as the foundation for sustainable agriculture, functioning as a dynamic living system that nurtures plants, animals and human life. This concept extends beyond the traditional measures of soil fertility, encompassing a holistic view that integrates physical properties, chemical composition and biological activity. Understanding soil health requires recognizing its vital role in

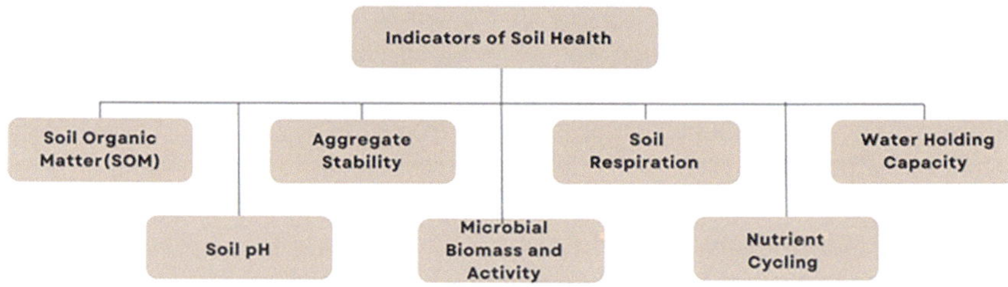

Fig. 1 Indicators of soil health

supporting plant growth, managing water resources, purifying contaminants and facilitating essential nutrient cycles.

3.1 Indicators of Soil Health

To effectively assess soil health, scientists and land managers rely on specific measurable properties that reveal the functionality of the soil as a living ecosystem. These indicators span physical, chemical and biological dimensions, providing crucial insights into the overall condition of the soil and its capacity to sustain plant life, maintain biodiversity and deliver essential ecosystem services (Fig. 1).

3.1.1 Organic Matter in Soil

Soil organic matter (SOM) is a fundamental indicator of soil health and comprises decomposed plant and animal materials that form the backbone of soil fertility. This vital component enhances the soil structure, improves water retention capabilities and stimulates microbial activity. When present in optimal amounts, SOM facilitates efficient nutrient cycling, minimizes soil erosion risks and plays a crucial role in carbon storage. Research has consistently demonstrated that maintaining adequate SOM levels is essential to prevent soil degradation and ensure long-term agricultural sustainability.

3.1.2 Soil pH

The pH level is a critical chemical indicator that influences multiple aspects of soil health, including nutrient availability, microbial activity and overall plant development. Most agricultural crops thrive in soils with pH values of 6.0 and 7.5. When soil becomes too acidic (below 5.5), plants may suffer from aluminium toxicity, whereas highly alkaline conditions (above 8.0) can restrict access to essential nutrients. Understanding and managing soil pH is crucial for optimizing fertilizer effectiveness and maximizing crop yields.

3.1.3 Aggregate Stability

The formation and stability of soil aggregates play a vital role in maintaining a healthy soil structure. Strong aggregate stability helps prevent erosion, reduces soil compaction risk and prevents surface crusting. These well-structured soils promote better root development, enhance water infiltration and ensure proper air exchange within the soil profile. Conservation practices, such as minimal tillage, cover crop implementation and organic matter addition, have proven effective in improving aggregate stability.

3.1.4 Microbial Biomass and Activity

The living component of soil organic matter, comprising diverse microorganisms, represents a key indicator of soil health. These microscopic organisms drive nutrient-cycling processes, facilitate organic matter decomposition and help to suppress plant diseases. The abundance and activity levels of soil microbes provide valuable insights into soil fertility and resilience to environmental stress. Studies have consistently shown that soils with robust microbial communities demonstrate superior fertility and resistance to disturbances.

3.1.5 Soil Respiration

Measuring soil respiration provides valuable insights into microbial activity and the overall biological health. This process, in which soil organisms break down organic matter and release carbon dioxide, serves as a key indicator of soil biological activity. Higher respiration rates typically correlate with increased organic matter content and microbial presence, whereas lower rates might indicate soil compaction or poor health conditions. Environmental factors such as temperature, moisture levels and organic matter availability significantly influence soil respiration rates.

3.1.6 Nutrient Cycling (N, P, K)

The efficient movement of essential nutrients, such as nitrogen, phosphorus and potassium, through the soil ecosystem characterizes healthy soils. Proper nutrient cycling ensures that plants receive necessary nutrients while minimizing losses to the environment. Inefficient cycling can lead to nutrient depletion, environmental pollution and reduced agricultural productivity. Implementing sustainable practices, such as organic amendment application and cover cropping, has shown significant improvements in nutrient cycling efficiency, reducing dependence on synthetic fertilizers.

3.1.7 Water-Holding Capacity

A soil's ability to retain and provide water to plants represents another crucial health indicator. Well-structured soils rich in organic matter exhibit superior water retention capabilities, enhanced drought resistance and reduced irrigation requirements. Conversely, soils with a poor water-holding capacity often experience rapid moisture loss, leading to plant stress and reduced productivity. Research has indicated that even modest increases in organic matter content can significantly improve soil water retention properties, highlighting the interconnected nature of various soil health indicators. Hudson (1994) found that a 1% increase in organic matter could improve soil water retention by 25%.

3.2 Factors Affecting Soil Health

The health of soil ecosystems depends on the complex interplay between factors that shape their physical, chemical and biological characteristics. These elements can either enhance or diminish soil health based on their management approaches. A thorough understanding of these factors is essential to implement sustainable agricultural methods and ensure long-term soil productivity.

3.2.1 Soil Management Practices

The methods used to manage the soil significantly influence its overall health. Key practices, including tillage techniques, crop rotation strategies and organic matter management, play vital roles in the maintenance and degradation of soil quality. Research has consistently shown that conservation tillage and no-till farming methods help preserve soil health by maintaining organic matter levels, improving soil structure and minimizing erosion risks. Reicosky et al. (1995) demonstrated that no-till and conservation tillage practices improve soil health by preserving organic matter, enhancing soil structure and reducing erosion.

3.2.2 Inputs of Organic Matter

The presence and management of soil organic matter are fundamental determinants of soil health. Incorporating organic materials through various methods, such as composting programmes, cover crop implementation and crop residue management, enhances soil structure, boosts nutrient availability and promotes microbial activity. Studies have demonstrated that increased organic matter incorporation leads to improved soil aggregate stability, enhanced microbial populations and more efficient nutrient cycling. Six et al. (2002) showed that increasing organic matter input enhanced soil aggregate stability, microbial biomass and nutrient cycling.

3.2.3 Soil Texture and Structure

The inherent physical characteristics of soil, particularly its texture and structural composition, play crucial roles in determining the overall soil health. Although sandy soils typically struggle with water and nutrient retention, clay-rich soils often face challenges with compaction and poor aeration. Research has indicated that soils with well-developed aggregate structures demonstrate superior water retention capabilities, enhanced nutrient availability and better support for root development. Zou et al. (2010) highlighted that well-aggregated structures exhibit better water retention, higher nutrient availability and improved root growth.

3.2.4 Chemical Inputs and Fertilization

Agricultural chemical applications, including fertilizers and pesticides, significantly affect soil health. Although fertilizers play an essential role in providing the necessary nutrients, their excessive or unbalanced application can result in nutrient leaching, increased soil acidity and the depletion of vital trace elements. Scientific studies have shown that the overuse of nitrogen- and phosphorus-based fertilizers often leads to nutrient imbalances and an increased risk of water contamination.

3.2.5 Water Management

Proper water management is critical for maintaining soil health. Ineffective irrigation practices can lead to several problems, including waterlogged conditions, salt accumulation and excessive nutrient leaching. Research has demonstrated that improper drainage combined with excessive irrigation often results in soil salinization, ultimately reducing fertility and crop productivity. Rhoades et al. (1999) examined the effects of salinization and found that excessive irrigation without proper drainage led to salt accumulation in soils, reducing soil fertility and crop productivity.

3.2.6 Climate and Environmental Factors

Environmental conditions, including temperature variations, precipitation patterns and wind exposure, significantly influence soil health through their effects on erosion rates, organic matter decomposition and microbial activities. Current research indicates that climate change accelerates soil degradation by increasing erosion rates and reducing organic matter content. Lal (2010) reported that climate change exacerbates soil degradation by increasing the rate of erosion and reducing soil organic matter content.

3.2.7 Biological Factors (Soil Biota)

The diverse communities of organisms within soil ecosystems play an essential role in maintaining soil health. These organisms facilitate nutrient cycling, support organic matter decomposition and contribute to soil structure development. Modern agricultural research emphasizes the critical importance of soil biodiversity in promoting sustainable farming practices.

4 Agronomic Practices for Altering Cropping Systems

Modern agricultural systems employ various agronomic practices, including crop rotation, cover cropping, intercropping and conservation tillage, to enhance soil health, increase agricultural productivity and ensure sustainable farming operations (Fig. 2). These methods aim to reduce the environmental impact, preserve natural resources and build resilient agricultural systems capable of withstanding challenges such as climate change and pest invasions.

4.1 Crop Rotation

Crop rotation is a time-tested agricultural practice that involves growing different crops in sequence on the same land area across multiple seasons. This approach effectively disrupts the pest and disease cycles, enhances soil nutrient profiles and promotes overall soil health.

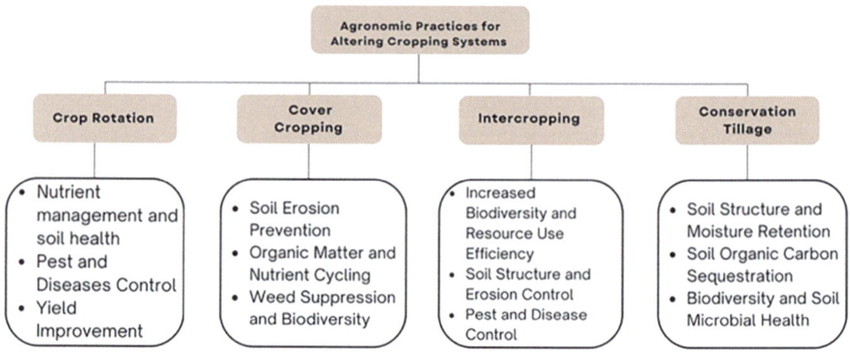

Fig. 2 Agronomic practices for altering cropping systems

4.1.1 Nutrient Management and Soil Health

Implementing crop rotation enhances soil fertility by altering crops with varying nutrient requirements. Legume crops, such as peas and beans, contribute to the soil nitrogen content through their symbiotic relationships with nitrogen-fixing bacteria. Drinkwater et al. (1998) found that incorporating legumes into rotation systems significantly reduces the need for synthetic fertilizers while improving soil fertility and decreasing greenhouse gas emissions. Additionally, alternation between crops with different root structures improves soil composition by developing diverse root zones, enhancing organic matter content and water infiltration.

4.1.2 Pest and Disease Control

Crop rotation serves as an effective strategy for managing pest and disease pressures that affect specific crops. According to Larkin et al. (2010), rotating susceptible crops with non-host species can substantially reduce the incidence of soil-borne diseases, offering a natural alternative to chemical pest control methods and promoting sustainable farming practices.

4.1.3 Yield Improvement

Systematic crop rotation has demonstrated significant benefits for crop yield. Smith et al. (2008) reported that alternating crops like corn and soybeans can lead to substantial yield increases compared to continuous monoculture systems, attributed to improved soil fertility and reduced pest pressure.

4.2 Cover Cropping

Cover cropping involves cultivating specific plants primarily to benefit soil health, rather than harvesting. These crops, which may include various legumes, grasses and brassicas, protect and enhance soil during the off-season or between the main crop cycles (Ancy and Sahoo 2024).

4.2.1 Soil Erosion Prevention

Cover crops provide essential soil protection by establishing root systems that hold soil. A study by Dabney et al. (2001) demonstrated that implementing cover crops such as rye and clover can dramatically reduce soil erosion in areas prone to water runoff, helping to maintain the soil structure and preserve valuable nutrients.

4.2.2 Organic Matter and Nutrient Cycling

These crops significantly contribute to the soil organic matter content, enhancing overall fertility and nutrient retention capacity. As cover crops decompose, nutrients are released into the soil system. Leguminous cover crops, in particular, can provide substantial amounts of natural nitrogen, reducing dependency on synthetic fertilizers and promoting sustainable nutrient management. According to Sustainable Agriculture Research & Education (SARE 2017), planting leguminous cover crops, such as hairy vetch or crimson clover, can provide up to 110 pounds of nitrogen per acre. This naturally fixed nitrogen reduces the need for synthetic fertilizers and promotes a more sustainable nutrient cycle.

4.2.3 Weed Suppression and Biodiversity

Many cover crops effectively suppress weed growth by competing for essential resources, such as light, water and nutrients. Teasdale et al. (2007) found that certain cover crops can significantly reduce weed presence compared to uncovered systems. The increased biodiversity introduced through cover cropping supports beneficial soil organisms, enhancing overall soil health and nutrient cycling processes.

4.3 Intercropping

Intercropping is an innovative agricultural practice in which farmers grow multiple crop species simultaneously in the same field. This strategic approach maximizes land use while enhancing nutrient efficiency and promoting overall soil wellness.

4.3.1 Increased Biodiversity and Resource-Use Efficiency

The implementation of intercropping significantly enhances field biodiversity, fostering a more resilient and stable ecosystem. A prime example is the partnership between legumes and cereal crops, where leguminous plants naturally fix nitrogen from the atmosphere while cereal crops utilize excess nitrogen in the soil. Brooker et al. (2015) found that combining barley with pea crops can boost nitrogen availability and increase overall yield productivity by approximately 15% compared to traditional single-crop farming methods.

4.3.2 Soil Structure and Erosion Control

The diverse root networks created by multiple crop species in intercropping systems significantly improve soil quality. Different root structures work together to reduce soil compaction and enhance the airflow throughout the soil profile. This is particularly beneficial in semi-arid regions, where combining corn cultivation with leguminous crops has shown remarkable improvements in organic content and moisture retention capabilities. These benefits are crucial for preventing soil degradation and maintaining agricultural productivity, especially in areas with limited water availability.

4.3.3 Pest and Disease Control

Intercropping serves as a natural defence mechanism against agricultural pests. This method effectively disrupts the habitat and feeding patterns of pests by creating diverse plant communities. Boudreau (2013) found that integrating herbs with vegetable crops could decrease pest problems by as much as 60%. This reduction occurs because the variety of plants creates confusion among pest species, making it more difficult for them to locate their preferred host crops. Such natural pest control significantly reduces the need for chemical pesticides, making intercropping an essential component of environmentally friendly pest management strategies.

4.4 Conservation Tillage

Conservation tillage encompasses various agricultural practices designed to minimize soil disruption. These techniques, including no-till and reduced-till farming approaches, help preserve soil integrity, improve water retention and enhance carbon storage.

4.4.1 Soil Structure and Moisture Retention

By maintaining crop residues on field surfaces, conservation tillage helps to protect the soil structure and prevent erosion while preserving valuable moisture. Six et al. (2002) found that no-till farming practices can improve soil particle binding, reduce erosion rates by half and enhance water penetration, particularly in dry climate regions. This approach offers additional benefits by reducing the energy required for field operations, resulting in lower fuel consumption and greenhouse gas emissions.

4.4.2 Soil Organic Carbon Sequestration

Conservation tillage plays a vital role in carbon storage by promoting the accumulation of organic matter over time. According to Lal (2004), these systems can capture approximately 1.2 tons of carbon per hectare annually, significantly contributing to climate change mitigation efforts. Increased organic carbon levels enhance soil fertility and structure, leading to improved long-term agricultural productivity.

4.4.3 Biodiversity and Soil Microbial Health

Reduced tillage practices promote thriving soil ecosystems by supporting diverse microbial communities. Tiemann et al. (2015) demonstrated conservation tillage can boost microbial diversity by 40%, enhancing nutrient cycling and overall soil health. These robust microbial populations naturally suppress plant diseases, reduce the dependence on chemical fungicides and improve crop resilience.

Incorporating agronomic practices such as crop rotation, cover cropping, intercropping and conservation tillage into cropping systems offers numerous benefits, including improved soil health, enhanced biodiversity and increased agricultural sustainability. These practices contribute to the long-term productivity of farming systems by improving nutrient cycling, reducing the need for chemical inputs and promoting an ecological balance. Scientific evidence strongly supports the adoption of these methods, as they provide both agronomic and environmental benefits while mitigating the negative impacts of conventional farming systems. By embracing sustainable practices, farmers can ensure the resilience and productivity of their cropping systems for future generations.

5 Enhancing Soil Health Through Agronomy

5.1 Organic Matter Management

The Management of organic materials is the cornerstone of sustainable soil health practices. This approach significantly influences the soil structure, fertility levels, microbial populations and nutrient availability. Organic matter, including plant debris, composted materials, animal manure and various organic inputs, provides essential carbon and nutrients for soil organisms. Effective organic matter management enhances water retention, prevents soil erosion and provides optimal conditions for plant development.

5.1.1 Role of Organic Matter in Soil Structure

Organic matter serves as a crucial binding agent in the soil, creating stable aggregates that improve soil porosity and reduce compaction risks. These well-structured soils facilitate better water movement and root growth, thereby establishing ideal conditions for plant development. Six et al. (2000) demonstrated that higher organic matter levels significantly improved soil stability, leading to better erosion resistance and enhanced moisture retention capabilities.

5.1.2 Nutrient Cycling and Organic Matter

Organic matter functions as a natural nutrient bank in soil systems. Its presence enhances nutrient retention and prevents valuable elements from leaching, while ensuring steady availability for plant uptake. Drinkwater et al. (1998) demonstrated that implementing organic matter management strategies, such as cover cropping and organic amendments, significantly improved soil nitrogen retention and reduced reliance on synthetic fertilizers.

5.1.3 Enhancing Microbial Activity Through Organic Matter

Soil microorganisms depend heavily on organic matter for energy production and sustenance. The breakdown of organic materials by microscopic organisms creates stable soil compounds that contribute to long-term fertility. Healthy microbial populations improve nutrient availability, naturally suppress diseases and enhance overall soil quality.

5.1.4 Carbon Sequestration and Climate Mitigation

Proper organic matter management significantly contributes to carbon sequestration. Agricultural lands can help combat climate change by increasing SOC levels through reduced greenhouse gas emissions. Lal (2004) highlighted the role of organic matter in carbon sequestration, showing that soils managed with organic inputs and conservation practices effectively store more carbon, helping reduce agriculture's environmental impact.

5.1.5 Organic Matter and Water Retention

The presence of organic matter significantly improves soil water retention capabilities, particularly in dry farming regions and areas that experience water scarcity. Enhanced moisture retention reduces irrigation requirements while improving plant resilience during dry periods. Hudson (1994) showed that even a modest 1% increase in organic matter content can boost soil water retention capacity by 25%.

5.1.6 Organic Matter and Erosion Control

Soils enriched with organic matter demonstrated superior resistance to erosion. Improved soil structure and reduced surface crusting help maintain valuable topsoil and prevent organic carbon loss. Implementing practices such as cover cropping, mulching and organic amendment applications effectively controls erosion while preserving soil organic matter, particularly in challenging tropical and semi-arid environments. Lal (1998) found that organic matter management plays a key role in controlling soil erosion, particularly in tropical and semi-arid regions.

5.2 Soil Amendments

To improve soil quality, farmers and gardeners have a variety of soil amendments at their disposal. These valuable additions can transform poor soil into productive soil by enhancing its physical structure, chemical composition and biological activity. Think of soil amendments as nature's toolkit for creating a perfect growing environment—they help the soil retain water better, make nutrients more accessible to plants and create a thriving ecosystem for beneficial microorganisms.

5.2.1 Organic Amendments

Nature provides numerous organic materials that can breathe new life into tired soil. These amendments come from plant and animal sources and are particularly effective at building up soil organic matter, which is essential for healthy plant growth.

5.2.1.1 Compost

Compost is one of the most versatile organic amendments. Picture it as a superfood for your soil; it is packed with nutrients and treated with beneficial microorganisms.

When compost is mixed with soil, it works like a sponge, helping the soil to hold onto water while slowly releasing nutrients to the growing plants. It is like creating a well-stocked pantry that plants can access whenever they require nourishment (Lal 2004).

5.2.1.2 Manure

Animal manure brings its own benefits to the soil. It is rich in nutrients and acts as a probiotic for the soil, introducing beneficial microorganisms that keep the soil ecosystem balanced and healthy.

5.2.1.3 Biochar

Biochar represents an innovative approach to soil amendment. This charcoal-like material was produced by heating organic matter in an oxygen-limited environment. It is similar to creating tiny apartments for beneficial microorganisms while also helping the soil retain water and nutrients. As an added bonus, biochar helps lock carbon in the soil, contributing to climate change mitigation (Lehmann et al. 2006).

5.2.1.4 Green Manure Crops

Green manure crops offer a dynamic method to improve soil health. These plants are grown to be incorporated back into the soil, acting as natural fertilizers and soil conditioners. Popular choices include legumes, which have a remarkable ability to fix nitrogen from the air, essentially creating their own fertilizer (Vanek and Matejka 2014).

5.2.2 Inorganic Amendments

Sometimes, soil requires specific mineral-based amendments to address particular issues. These inorganic amendments are particularly effective at adjusting the soil pH and providing essential minerals.

5.2.2.1 Lime

Lime is a go-to amendment for acidic soils. It works like an antacid for the soil, raising the pH to make nutrients more available to plants. This is particularly important in regions with naturally acidic soils or in areas affected by acid rain (McLaughlin and Parker 1994).

5.2.2.2 Gypsum

Gypsum plays a crucial role in improving clay soils. It helps break up compacted soil particles, creates better drainage and allows roots to penetrate more easily. It is a natural soil conditioner that helps to create a perfect structure for plant growth (Carter and Gregorich 2007).

5.2.2.3 Sulfur

Sulfur serves as the opposite of lime, helping to lower pH in alkaline soils. This is particularly important for plants that prefer slightly acidic conditions, such as blueberries and azaleas. The key to successful soil amendment lies in proper soil testing and the careful selection of amendments based on specific needs. It is not about applying everything available; it is about choosing the right amendments to address your soil's particular challenges.

5.3 Integrated Pest Management

Modern agriculture has evolved from the 'spray and pray' approach to embrace Integrated Pest Management (IPM), a scientifically grounded strategy that combines multiple pest control techniques. This paradigm shift emerged from mounting evidence of pesticide resistance, environmental damage and economic inefficiencies in prophylactic pesticide applications (Kogan 1998).

5.3.1 Key Principles of IPM

5.3.1.1 Monitoring and Identification

Regular pest monitoring forms IPM's foundation. Pedigo et al. (1986) demonstrated that systematic monitoring reduces unnecessary pesticide applications by 40–60% while maintaining crop protection. Modern programmes integrate pheromone traps, visual scouting and degree-day models to accurately time interventions.

5.3.1.2 Economic Thresholds

Rather than treating every pest sighting, IPM uses science-based thresholds. Stern et al. (1959) established that treatments should occur only when pest densities reach economically damaging levels. For example, soybean aphid thresholds of 250 aphids per plant reduce insecticide use by 60% compared to calendar-based spraying (Hutchins et al. 1988).

5.3.1.3 Cultural Control

Crop rotation, resistant varieties and field sanitation prevent pest establishment. Gray et al. (2009) showed corn-soybean rotation reduces corn rootworm populations by 95%, eliminating soil insecticide needs.

5.3.1.4 Biological Control

Natural enemies provide sustainable pest suppression. The California citrus industry's vedalia beetle programme continues providing complete cottony cushion scale control after 135 years, with annual benefits exceeding $100 million (DeBach and Rosen 1991). Modern programmes use mass-released beneficial insects and microbial agents like *Bacillus thuringiensis*.

5.3.1.5 Physical Control

Barriers, traps and mating disruption offer chemical-free solutions. Pheromone-based mating disruption achieves 85–95% codling moth control in apple orchards, eliminating conventional insecticide needs (Witzgall et al. 2010).

5.3.1.6 Selective Chemical Control

When necessary, IPM emphasizes selective materials and precision timing. Selective insecticides like spinosad provide 80–90% pest control while impacting less than 10% of beneficial insects (Nauen and Denholm 2005).

5.3.2 Research-Documented Benefits

5.3.2.1 Environmental Impact

Norton et al. (2000) documented 50–70% pesticide use reductions across IPM farms while maintaining equivalent pest control. Water quality studies show 60–80% reductions in pesticide detections in agricultural watersheds following IPM adoption.

5.3.2.2 Economic Returns

IPM systems generate 15–25% higher net returns than conventional approaches through reduced input costs and improved yields. Biological control programmes alone provide benefit-cost ratios of 20–30:1 (Norton and Mullen 1994).

5.3.2.3 Food Safety

IPM-produced crops contain 40–60% lower pesticide residues while maintaining equivalent quality and nutritional value.

5.3.3 Implementation Challenges

IPM requires significant knowledge investment and intensive monitoring. However, comprehensive training programmes generate 4–6:1 benefit-cost ratios through improved decision-making. Climate change is altering pest dynamics, requiring adaptive management strategies and updated thresholds (Deutsch et al. 2018).

5.4 Fertilization Strategies

Effective fertilization requires a thoughtful approach that goes beyond simply applying nutrients to the soil. It is about understanding and working with natural systems to optimize plant growth while minimizing environmental impact.

5.4.1 Key Considerations for Fertilization Include

5.4.1.1 Soil Testing

Regular soil testing provides the foundation for informed decision-making. It is like getting a health checkup for your soil, revealing which nutrients are abundant and which need supplementation.

5.4.1.2 Crop Nutrient Requirements

Understanding crop nutrient requirements is crucial. Different plants have varying nutritional needs and these requirements change throughout their growth cycle. For instance, leafy greens might need more nitrogen, while fruiting plants require additional phosphorus and potassium.

5.4.1.3 Fertilizer Type and Application

Choosing the right fertilizer type and application method can significantly impact effectiveness. Whether using organic or synthetic fertilizers, timing and application rates must be carefully considered to maximize benefits while minimizing waste.

5.4.1.4 Nutrient Cycling

Natural nutrient cycling plays a vital role in sustainable fertilization. This involves understanding how nutrients move through the soil system and working to enhance these natural processes rather than relying solely on external inputs.

5.4.1.5 Environmental Impacts

Environmental impacts: Minimizing nutrient leaching and environmental pollution through proper fertilizer application and management.

5.4.2 Common Fertilization Strategies

5.4.2.1 Broadcast Application

Broadcast application spreads fertilizer evenly across the field, providing uniform nutrient distribution. This method works well for large areas but may not be the most efficient for all situations.

5.4.2.2 Localized Application

Localized application places nutrients directly where plants need them, reducing waste and improving efficiency. This precision approach is particularly valuable for high-value crops or when conserving resources is crucial.

5.4.2.3 Fertigation

Fertigation combines irrigation and fertilization, delivering nutrients through the irrigation system. This method offers precise control over nutrient delivery and can be particularly effective in water-limited environments.

5.4.2.4 Foliar Fertilization

Foliar fertilization involves applying nutrients directly to plant leaves, providing a quick response when plants need immediate nutrient support. This method works well for addressing specific deficiencies but typically complements rather than replaces soil-based fertilization.

5.4.2.5 Organic Fertilization

Organic fertilization emphasizes the use of natural materials like compost and manure. While these materials may release nutrients more slowly than synthetic fertilizers, they offer additional benefits like improved soil structure and enhanced microbial activity.

5.4.3 Challenges and Considerations

5.4.3.1 Nutrient Leaching

Excessive fertilizer application can lead to nutrient leaching into groundwater and surface water.

5.4.3.2 Soil Acidification

The use of nitrogen fertilizers can contribute to soil acidification.

5.4.3.3 Pest and Disease Outbreaks

Imbalances in nutrient availability can create conditions that favour pest and disease infestations.

6 Challenges and Opportunities

The path to sustainable agriculture is complex and multifaceted. While the ultimate rewards are substantial, farmers encounter numerous obstacles as they work to transform their operations. Let us delve into these challenges and examine how cutting-edge innovations are reshaping the agricultural landscape.

6.1 Economic and Social Barriers

The adoption of sustainable cropping systems, such as crop rotation, cover cropping, intercropping and conservation tillage, is hindered by several economic and social barriers. These barriers range from initial costs and knowledge gaps to resistance to change within farming communities.

6.1.1 Economic Costs

Converting to sustainable farming methods represents a significant financial commitment, much like undertaking a major home improvement project. While the long-term benefits are clear, the upfront expenses can be overwhelming for many farmers. They must invest in specialized equipment, purchase different varieties of seeds and often completely restructure their farming operations.

Consider the implementation of cover crops as an example. Despite their proven benefits for soil enhancement, the initial investment is substantial. Farmers typically spend between $25 and $50 per acre on seeds alone. When managing vast tracts of land spanning hundreds or thousands of acres, these expenses become considerably challenging to manage.

The situation is particularly dire for farmers operating small holdings, especially in developing nations. These agriculturists frequently struggle to secure conventional financing options, as financial institutions tend to view agricultural transitions as high-risk ventures. This creates an unfortunate paradox—farmers require financial backing to implement sustainable practices, yet banks are reluctant to provide funding without demonstrated success stories.

6.1.2 Social Resistance

Implementing change in agriculture presents unique challenges, particularly because farming practices are often deeply rooted in generational knowledge. For many farmers, their methods represent a legacy passed down through families, making the adoption of new techniques feel like a departure from valued ancestral wisdom.

This hesitation extends beyond mere sentiment—it is grounded in practical considerations. While farmers may recognize the theoretical advantages of sustainable methods, they harbour legitimate concerns about potential impacts on crop yields and their ability to maintain profitable operations. The prospect of experimenting with novel approaches can seem particularly daunting when responsible for sustaining both family and business.

Encouragingly, successful transitions are becoming more common through innovative programmes like Farmer Field Schools (FFS), particularly in nations such as Kenya and Uganda (Doss, 2006). These initiatives showcase sustainable practices in real-world settings, enabling farmers to witness tangible benefits and learn directly from peers who have successfully navigated the transition.

6.1.3 Market Challenges

Successfully implementing sustainable practices does not guarantee market success. Despite growing consumer interest in sustainably produced foods, market structures do not always support this agricultural transformation. The price premiums commanded by sustainably grown products frequently fail to compensate for increased production costs, creating economic disincentives for farmers contemplating the switch.

6.2 Technological Innovations

Despite these obstacles, technological advancements are creating unprecedented opportunities in sustainable agriculture. These innovations are making sustainable farming increasingly viable, efficient and economically rewarding.

6.2.1 Precision Agriculture

Contemporary farming has evolved into a sophisticated, data-driven endeavour. Picture having access to detailed health metrics for every segment of your farmland—this effectively describes precision agriculture. Through an integrated system of sensors, drone technology and satellite imaging, farmers can now monitor crucial factors like soil moisture content, nutrient levels and crop health with unprecedented accuracy.

This technological revolution enables farmers to make highly informed decisions about irrigation, fertilization and pest management. Rather than applying uniform treatments across entire fields, farmers can now address specific areas based on their unique needs, significantly reducing waste while maximizing efficiency. According to Tilman et al. (2002), these advanced practices could potentially increase crop yields by up to 70% while simultaneously minimizing environmental impact.

6.2.2 Biofertilizers and Biopesticides

Natural agricultural solutions are experiencing a technological renaissance through advanced biofertilizers and biopesticides. These innovative products leverage beneficial microorganisms and organic materials to enhance soil fertility and manage pest problems without relying on synthetic chemicals.

Biofertilizers function similarly to digestive probiotics, but for soil—they introduce beneficial microorganisms that enhance nutrient availability and overall soil health. Concurrently, biopesticides derived from natural sources offer effective pest control while preserving beneficial insects and maintaining soil ecosystem balance. These microorganisms help fix nitrogen, solubilize phosphates and decompose organic matter, enriching the soil and promoting healthy crop growth (Drinkwater et al. 1998).

6.2.3 Climate-Smart Technologies

As climate change increasingly impacts agriculture, innovative technologies are emerging to help farmers adapt. This includes the development of resilient crop varieties capable of withstanding extreme weather conditions and the implementation of water-conserving irrigation systems. These advancements are particularly vital for sustainable farming operations, which aim to maintain productive yields while reducing environmental impact.

6.3 Future Research Directions

While technological innovations provide a path forward, more research is needed to fully understand how to optimize cropping systems for sustainability and soil health. Future research should focus on the following areas:

6.3.1 Improved Crop Breeding for Diverse Systems

Sustainable agriculture's future success partially hinges on developing crop varieties specifically adapted for diversified farming systems. Unlike conventional breeding programmes focused on maximizing yields in monoculture environments, contemporary research must consider plant performance in mixed cropping systems and varied environmental conditions. Current research is already exploring the development of cover crops and legume species that can better support soil health and improve productivity in organic systems (Lal 2004).

6.3.2 Long-Term Studies on Soil Health

While we understand many immediate advantages of sustainable practices, their extended impact on soil health requires further investigation. Scientists are now conducting extensive, multi-year studies to comprehend how different farming methods influence soil structure, carbon sequestration and nutrient cycling over extended periods.

6.3.3 Integration of Technologies with Traditional Knowledge

One of the most promising research areas involves harmonizing modern technology with traditional farming wisdom. Many traditional agricultural practices have sustained communities for generations and combining this time-tested knowledge with modern innovations could create more resilient and effective farming systems.

The shift toward sustainable agriculture faces significant challenges, particularly in economic and social spheres. The convergence of technological innovation and ongoing research provides powerful tools for overcoming these obstacles. As our understanding of sustainable farming practices deepens and new solutions emerge, the path forward becomes increasingly clear. Agriculture's future lies in successfully balancing traditional wisdom with modern innovation, creating systems that benefit both farmers and the environment.

Achieving success in sustainable agriculture demands a comprehensive approach that addresses immediate practical challenges while pursuing long-term sustainability goals. By maintaining investment in research, supporting farmers through transitional periods and developing market mechanisms that properly reward sustainable practices, we can establish an agricultural system that meets current demands while preserving resources for future generations.

7 Conclusion

The future of global food security hinges on our capacity to transition from extractive farming methods to regenerative systems that work with nature rather than against it. Practices such as diversified crop rotations, cover cropping, intercropping, conservation tillage, agroforestry, and rotational grazing collectively rebuild soil structure, enrich organic matter, and foster a thriving soil microbiome. Numerous studies demonstrate that these approaches not only stabilize yields and sequester carbon but also curb the need for synthetic fertilizers and pesticides, thereby mitigating water contamination, greenhouse-gas emissions, and biodiversity loss. Realizing these benefits at scale, however, demands more than on-farm innovation; it requires a coordinated, systemic effort. Interdisciplinary research can refine region-specific best practices, while targeted policies—such as incentive payments, crop-insurance reforms, and carbon markets—can help farmers overcome financial barriers. Equally vital are robust extension services and farmer-to-farmer knowledge networks that close information gaps and respect local traditions. By investing in education, incentive-based support programmes, and locally adapted technologies, governments, industry, and civil society can accelerate the adoption of sustainable practices. As agriculture confronts intensifying pressures from climate change, land degradation, and growing populations, the strategies outlined here offer a scientifically grounded roadmap for resilient food systems and ecological stewardship. Embracing them will not only safeguard our soils and ecosystems but also ensure that farming remains productive, profitable, and sustainable for generations to come.

References

Ancy GM, Sahoo C (2024) Cover crops and green manure crops: sustainable crop. ND Global Publication, pp 150–166

Boudreau MA (2013) Diseases in intercropping systems. Annu Rev Phytopathol 51:499–519

Brooker RW, Bennett AE, Cong WF, Daniell TJ, George TS, Hallett PD, White PJ (2015) Improving intercropping: a synthesis of research in agronomy, plant physiology and ecology. New Phytol 206(1):107–117

Carter MR, Gregorich ME (2007) The use of gypsum improves soil structure and reduces compaction. Can J Soil Sci 87(3):321–332

DeBach P, Rosen D (1991) Biological control by natural enemies. Cambridge University Press

Dabney SM, Delgado JA, Reeves DW (2001) Using winter cover crops to improve soil and water quality. Communications in Soil Science and Plant Analysis, 32(7–8):1221–1250

Deutsch CA, Tewksbury JJ, Tigchelaar M, Battisti DS, Merrill SC, Huey RB, Naylor RL (2018) Increase in crop losses to insect pests in a warming climate. Science 361(6405):916–919

Doss CR (2006) Analyzing technology adoption using microstudies: limitations, challenges and opportunities for improvement. Agric Econ 34(3):207–219

Drinkwater LE, Wagoner P, Sarrantonio M (1998) Legume-based cropping systems have reduced carbon and nitrogen losses. Nature 396(6708):262–265

Gray ME, Sappington TW, Miller NJ, Moeser J, Bohn MO (2009) Adaptation and invasiveness of western corn rootworm. Annu Rev Entomol 54:303–321

Hobbs PR, Sayre K, Gupta R (2008) The role of conservation agriculture in sustainable agriculture. Philos Trans R Soc B 363(1491):543–555

Hudson BD (1994) Soil organic matter and available water capacity. J Soil Water Conserv 49(2):189–194

Hutchins SH, Higley LG, Pedigo LP (1988) Injury equivalency as a basis for developing multiple-species economic injury levels. J Econ Entomol 81(1):1–8

Kogan M (1998) Integrated pest management: Historical perspectives and contemporary developments. Annual Review of Entomology, 43(1):243–270

Kumari G, Saini A, Lakshmi, Sahoo C (2024) Organic and natural farming: concept and principles: climate change and agriculture: its impact and mitigation potential. https://www.researchgate.net/publication/380879833_Organic_and_Natural_Farming_Concept_and_Principles

Lal R (1998) Soil erosion impact on agronomic productivity and environment quality. Crit Rev Plant Sci 17(4):319–464

Lal R (2004) Soil carbon sequestration impacts on global climate change and food security. Science 304(5677):1623–1627

Lal R (2010) Managing soils and ecosystems for mitigating anthropogenic carbon emissions and advancing global food security. Bioscience 60(9):708–721

Larkin RP, Honeycutt CW, Griffin TS (2010) Effect of crop rotation and irrigation on potato disease. Plant Dis 94:1491–1502

Lehmann J, Joseph S, Steiner C, Neumueller M, Amelung W, Zaehle S (2006) Biochar-induced changes in soil organic matter: a review. Mitig Adapt Strateg Glob Chang 11(1):375–397

Liu X, Zhang W, He X (2018) Impact of crop rotation on soil fertility and crop yield in northern China. Soil Sci Plant Nutr 64(2):201–209

Manhas S, Singh J, Manuja S, Saini A, Kumawat R, Dahiya P, Mehta S, Sahoo C, Johnson R, Puthur JT, Fayezizadeh MR (2024) Assessing the impact of tillage practices and nutrient levels on the growth and productivity of Ethopian mustard (*Brassica carinata* L.) – soybean (*Glycine max* (L.) Merr.) cropping system. BMC Plant Biol 24(1):1059

McLaughlin MJ, Parker DR (1994) Lime application rates for correcting soil acidity. Agron J 86(5):1183–1191

Nauen R, Denholm I (2005) Resistance of insect pests to neonicotinoid insecticides. Arch Insect Biochem Physiol 58(4):200–215

Norton GW, Mullen JD (1994) Economic evaluation of integrated pest management programmes. Department of Primary Industries

Norton GW, Rajotte EG, Gapud V (2000) Participatory research in integrated pest management. Agric Human Values 16(4):431–439

Parida PK, Sahoo C, Sowjanya J, Deb A (2023) Innovations in organic farming system: recent trends in agriculture (8). https://doi.org/10.22271/int.book.302

Pedigo LP, Hutchins SH, Higley LG (1986) Economic injury levels in theory and practice. Annu Rev Entomol 31(1):341–368

Pingali PL (2012) Green revolution: impacts, limits and the path ahead. Proc Natl Acad Sci 109(31):12302–12308

Reicosky DC, Kemper WD, Langdale GW, Douglas CL, Rasmussen PE (1995) Soil organic matter changes resulting from tillage and biomass production. J Soil Water Conserv 50(3):253–261

Rhoades JD, Kandiah A, Mashali AM (1999) The use of saline waters for crop production. Food and agriculture Organization of the United Nations (FAO), Rome

Sahoo C, Ameena M, Shalini PP, Susha VS, Ayisha R, Ankit Shilpa Sharma S, Verma A (2024) Impact of land configuration methods on growth parameters of soybean in summer rice fallows. Plant Sci Today 12(1):1 7

Sahoo C, Martin AG, Reddy CDK, Saini A (2023) Role of conservation agriculture in sustainable agriculture. Kripa Drishti Publications 1, pp 161–169

Six J, Conant RT, Paul EA, Paustian K (2002) Stabilization mechanisms of soil organic matter: implications for C-saturation of soils. Plant Soil 241(2):155–176

Six J, Elliott ET, Paustian K (2000) Soil structure and soil organic matter: II. A normalized stability index and the effect of mineralogy. Soil Sci Soc Am J 64(3):1042–1049

Smith RG, Gross KL, Robertson GP (2008) Effects of crop diversity on agroecosystem function. Ecosystems 11:355–366

Stern VM, Smith RF, Van Den Bosch R, Hagen KS (1959) The integrated control concept. Hilgardia 29(2):81–101

Sustainable Agriculture Research & Education (SARE) (2017) Cover crops and soil health

Teasdale JR, Brandsaeter LO, Calegari AD, Neto FS, Upadhyaya MK, Blackshaw RE (2007) Cover crops and weed management. In: Non-chemical weed management: principles, concepts and technology, pp 49–64

Tiemann LK, Grandy AS, Atkinson EE, Marin-Spiotta E, McDaniel MD (2015) Crop rotational diversity enhances belowground communities and functions in an agroecosystem. Ecol Lett 18(8):761–771

Tilman D, Cassman KG, Matson PA, Naylor R, Polasky S (2002) Agricultural sustainability and intensive production practices. Nature 418(6898):671–677

Vanek TA, Matejka R (2014) Green manure crops for soil health and nutrient management. Agron J 106(1):1–10

Witzgall P, Kirsch P, Cork A (2010) Sex pheromones and their impact on pest management. J Chem Ecol 36(1):80–100

Zhang X, Davidson EA, Mauzerall DL, Searchinger TD, Dumas P, Shen Y (2015) Managing nitrogen for sustainable development. Nature 528(7580):51–59

Zou C, Penfold C, Sands R, Misra RK, Hudson I, Newell G (2010) Effects of soil aeration on root growth and nutrient uptake of young *Eucalyptus grandis* and *E. globulus* in dense soil. Plant Soil 329(1):123–136

Unravelling the Hidden World Beneath: Exploring Mycorrhizal Diversity and Methods of Identification and Isolation

Ajmila Islam

Abstract

This chapter provides a review of mycorrhizal fungal diversity, which is beneficial for ecosystems owing to mutualistic symbiosis with the roots of various types of plants. This chapter also covers the different methods of isolation and identification of mycorrhizal fungi. These range from conventional methods to cutting–edge techniques, including biochemical, molecular, and functional genomics. The importance of different types of mycorrhizal fungi in ecological systems is well known; however, there are still ongoing efforts to effectively identify and isolate diverse species that play crucial roles in these symbiotic relationships. Current methods, although varied, often lack the precision and applicability needed to fully understand the complexity and diversity of mycorrhizal fungi. This chapter will also highlight the mycorrhizal diversity that will assist researchers in better understanding the symbiotic role of mycorrhizal fungi and their beneficial effects on the ecosystem. By the end of the chapter, readers will gain sufficient knowledge on both the diversity of mycorrhizal fungi and the various techniques needed to advance ecological research and use sustainable practices.

Keywords

Classification · Mycorrhizal diversity · Isolation methods · Molecular identification · Functional genomics

A. Islam (✉)
Department of Biological and Physical Sciences, School of Arts and Natural Sciences, American University in Dubai, Dubai, UAE
e-mail: aislam@aud.edu

Y. Siddiqui et al. (eds.), *Sustainable Mycorrhizal Cultivation*,
https://doi.org/10.1007/978-981-95-1778-7_3

Abbreviations

AMF	Arbuscular mycorrhizal fungi
ECMF	Ectomycorrhizal fungi
ERMF	Ericoid mycorrhizal fungi
ORMF	Orchid mycorrhizal fungi
SEM	Scanning electron microscopy
SDS	Sodium dodecyl sulfate
EDTA	Ethylenediaminetetraacetic acid
WST	Wet-sieving technique
UWST	Ultrasound-assisted wet-sieving
UCT	Ultrasound centrifugation
G6PDH	Glucose-6-phosphate dehydrogenase
NADPH	Nicotinamide adenine dinucleotide phosphate
SOD	Superoxide dismutase
NBT	Nitroblue tetrazolium
HPLC	High-performance liquid chromatography
GC	Gas chromatography
VOC	Volatile organic compounds
TLC	Thin layer chromatography
MS	Mass spectrometry
PCR	Polymerase chain reaction
ITS	Internal transcribed spacer
SSU	Small subunit
LSU	Large subunit
NGS	Next-generation sequencing
TEF	Translation elongation factor
PT	Phosphate transporter
AMT	Ammonium transporter

1 Introduction

Mycorrhizal symbiotic interactions between fungi and plant roots benefit over 85% of the terrestrial plants. There are different types of mycorrhizae, each possessing distinct characteristics and roles in the ecosystem, in which both the fungus and the plant exchange nutrients required for their growth and survival through this mutualistic partnership. For exploitative mycorrhizae, this relationship is symbiotic but not mutualistic; therefore, only plants benefit and obtain nutrients from the fungi. In addition to the important nutrient exchange required for growth and development, mycorrhizae also increase soil fertility, improve plant resilience to environmental stress, and help sequester carbon. Owing to their beneficial roles, mycorrhizal associations in agricultural systems offer greater potential to increase crop productivity, reduce chemical fertilizer usage, and enhance soil health.

The diversity of mycorrhizae and their unique ecological benefits highlight the importance of studying their roles in ecosystem balance, especially in conserving endangered plant species and their natural fungal interactions. Since the discovery of mycorrhiza, various identification methods have been developed to understand their underlying mechanisms of symbiotic interactions. This chapter focuses on the diversity of mycorrhizal fungi, different identification techniques, and their ecological significance.

2 Classification of Mycorrhizae

Depending on the orientation of fungal hyphae with plant roots, mycorrhizae are classified into two main types: ectomycorrhiza and endomycorrhiza. In ectomycorrhiza, hyphae expand outside the plant roots, whereas in endomycorrhiza, the fungal hyphae grow inside the plant roots. Their mutualistic associations benefit most terrestrial plants, as well as desert plants, whereas ectomycorrhiza primarily benefit woody plants, accounting for only 3% of mycorrhizal associations. The most common type of endomycorrhiza is arbuscular mycorrhizae, also known as vesicular-arbuscular mycorrhizae (Mahmoudi et al. 2019). Based on the morphological characteristics, mycorrhizae are further classified into four groups, including arbuscular mycorrhizas, ectomycorrhizas, ericoid mycorrhiza, and orchid mycorrhizas, which are discussed below in detail.

2.1 Arbuscular Mycorrhizae (AM)

Arbuscular mycorrhizae (AM) are found in over 71% of terrestrial plant species, including the most common crops such as wheat, corn, and rice. They are characterized by the presence of a specialized, highly branched structure known as arbuscules that normally form in the cortical cells of the plant's root. Depending on the orientation of hyphal morphology within the roots, two additional subtypes, linear and coiling, have been identified, each containing its variants. Significant divergence exists between arbuscular mycorrhizal fungi (AMF) and mycoheterotrophic plants depending on the nutritional needs of the arbuscular fungi (Mahmoudi et al. 2019). During symbiosis, the arbuscules penetrate the cortical cells of the plant root and facilitate the exchange of nutrients. Arbuscular fungi also exhibits other structures associated with plant roots, including vesicles, hyphal coils, and nonseptate hyphae. These structures perform various functions that benefit both the plant and the extensive fungal hyphal networks in the surrounding soil (Smith and Smith 2011). These functions include enhanced uptake of essential nutrients, such as phosphorus and nitrogen, from the soil, increased resistance to pathogens, and protection against toxic heavy metals in plants. In exchange, fungi receive carbohydrates produced by plants during photosynthesis (Bouffaud et al. 2016).

AMF are the most abundant fungi, adapted to a wide variety of environmental conditions, and are associated with over 200,000 plant species. However, only approximately 240 species have been identified. Ribosomal RNA (rRNA) analysis classified AMF into two main phyla, Glomeromycota and Zygomycotina. Recent gene sequencing and barcoding studies have further revealed the taxonomic ranking of the phylum Glomeromycota, which includes 17 orders, 19 families, 44 genera, and 342 species, and the phylum Zygomycotina, which includes 26 species (Brundrett and Tedersoo 2018).

2.2 Ectomycorrhizae (ECM)

Ectomycorrhizae (ECM) are commonly associated with woody plants and often link seedlings to mature plants. Ectomycorrhizal fungi (ECMF) do not penetrate the cortical cells of plant roots; instead, they create a dense sheath or mantle around the root tip, forming a thick multicellular outer layer that serves as a protective barrier. Inside the root, they form an extended hyphal network known as the Hartig net, which helps in nutrient exchange between the fungus and host plant. The Hartig net penetrates the root's epidermal and cortical cells but does not enter the plant cells. Both the Hartig net and mantle help fungi penetrate the soil, thereby increasing the surface area available for nutrient and water absorption (Chot and Reddy 2022).

Through symbiotic interactions, ECMF efficiently deliver organic forms of nitrogen and phosphorus from the soil to plants while obtaining carbohydrates produced by the plant's photosynthetic process. Their extended hyphal network expands the surface area for the absorption of sufficient water and keeps host plants hydrated, especially in arid and nutrient-deficient soils. The fungal mantle provides a physical barrier against soil pathogens, thereby reducing the risk of root infections. Some ECMF produce antimicrobial compounds against harmful microbes that offer additional benefits to plants (Ren et al. 2022).

Based on the morphological orientation of the mantle and the Hartig net, ECMF are further divided into several host-specific subtypes. These include Ectendo, which is characterized by a thin mantle with occasional intracellular colonization in Pinus and Larix hosts; Arbutoid and Pyroloid both exhibit a variable mantle, with Arbutoid having thin intracellular colonization in Arbuteae and Pyroloid showing the same in Pyroleae hosts; Monotropoid, which has a thick mantle and substantial intracellular colonization in Monotropaceae hosts; Pisonioid involves transferring cells in the Hartig net in Pisonia hosts; Gnetoid shows finger-like projections above the root epidermis in Gnetum hosts; and superficial Gnetoids have a thin mantle with either a poorly developed Hartig net or none at all.

ECMF comprise more than 200 genera belonging to two main phyla, approximately 83% in Basidiomycota and 13% in Ascomycota. Basidiomycota includes over 250 genera and 6000 species, whereas Ascomycota includes only a few (Tedersoo et al. 2010). Recent molecular techniques have estimated over 6000 known ectomycorrhizal species. The most diverse genera are Cortinarius and Russula, each containing several 100 ectomycorrhizal species (Tedersoo and Smith 2013). Other well-known genera include Amanita, Boletus, Lactarius, Suillus, Tuber, and Peziza (Hawksworth and Lücking 2017).

2.3 Ericoid Mycorrhizae (ERM)

Ericoid mycorrhizae (ERM) are specialized, and small groups of fungi are associated with a limited variety of plants. These fungi penetrate the fine roots of plants, consisting of a vascular bundle with an outer layer of cortical cells, and form characteristics of densely interwoven hyphal coils. As these fungi are associated with a specific family of hosts, they have been found to play a very important role in the evolution of ericoid mycorrhizal fungi (ERMF). For example, host specialization in ERMF belonging to the phyla Ascomycota and Basidiomycota has led to the evolution of other types of mycorrhizae, such as arbutoids and monotropoids. Arbutoid mycorrhizas have a thick fungal mantle and a Hartig net like Ectomycorrhizas, consisting of unique dense hyphal complexes located within root epidermal cells. Monotropoid mycorrhizas form symbiotic relationships with nonphotosynthetic members of the Monotropaceae and Basidiomycetes, helping in the exchange of carbohydrates from the photosynthetic plant to their myco-heterotrophic host plants. These mycorrhizas possess a unique 'peg' of fungal hyphae that extend within the epidermis of the root, in addition to the fungal mantle and Hartig (Cairney and Meharg 2003).

Unlike other mycorrhizae, the exact number of ERMF species has not been well established. Traditional culturing methods have identified dozens of species from various genera, including Oidiodendron, Rhizoscyphus, Meliniomyces, and Cryptosporiopsis. Recent molecular technologies have identified an additional 100 species that have been found to be nonculturable (Perotto et al. 2002).

2.4 Orchid Mycorrhizae (ORM)

Orchid mycorrhizae (ORM) are endomycorrhizal fungi with symbiotic associations with the roots and seeds of plants in the Orchidaceae family. During symbiosis, fungi develop specialized structures known as pelotons, which are coiled or densely branched formations within the cortical cells of orchid plants (Rasmussen 2002). Since most orchids are myco-heterotrophic and lack sufficient nutrients for growth, they sequester carbon for germination through their symbiotic association with ORM.

Orchidaceae comprises approximately 28,000 species and is one of the largest plant families (Chase et al. 2015). These saprotrophic fungi belong mainly to the genera Thelephora, Tomentella, and Russula. A few green orchids associated with Rhizoctonia fungi usually belong to the Basidiomycetes family, which includes the genera Ceratobasidium (Rhizoctonia), Sebacina, Tulasnella, and Russula (Yagame et al. 2012).

3 Habitat Diversity of Mycorrhizae

The habitat diversity of mycorrhizae is a result of their ecological adaptability and their symbiotic relations with numerous plant hosts. Across a wide range of habitats, mycorrhizal fungi are associated with approximately 92% of plant families, which include approximately 80% of all plant species. Mycorrhizal fungi are found in nearly every terrestrial ecosystem, as well as in some aquatic environments. The presence of mycorrhizae in different habitats impacts not only plant biodiversity but also the functionality of ecosystems. In each habitat, mycorrhizal fungi perform vital roles through specific mechanisms that establish symbiotic relationships with plants.

AMF is predominantly found in grasslands and savannas, which helps the plants acquire phosphorus and water from nutrient-deficient soils. They are commonly utilized to enhance crop productivity in both agricultural areas and wetlands. Their ability to survive in arid environments also helps plants in tolerating various abiotic stresses, including drought, and promotes root development in areas with limited vegetation by improving soil quality and reducing environmental challenges. In urban green areas, these fungi play an important role in supporting plant health by enhancing nutrient absorption, drought resistance, and soil quality. Although AMF have diverse functions, in temperate and boreal ecosystems, ECMF dominate through large mycelial networks that influence both nutrient cycling and plant diversity, as well as in some tropical and Mediterranean ecosystems (Wu et al. 2024).

ERMF are primarily associated with plants grown in acidic, nutrient-poor habitats like heathlands, bogs, and coniferous forests. They aid soil stabilization by enhancing organic matter and nutrient cycling, thereby improving host plant tolerance against abiotic stresses like metal toxicity and drought (Singavarapu et al. 2022). Found in diverse habitats including tropical rainforests, temperate grasslands, and semi-arid, orchid mycorrhizal fungi (ORMF) supports orchid seedlings' development by supplying essential carbon and nutrients. This symbiosis helps orchids to survive in nutrient-deficient soils and shaded environments by substituting photosynthetic carbon uptake (Willing et al. 2024). Table 1 summarizes the diversity of the mycorrhizae and their beneficial roles in symbiotic associations.

4 Methods of Isolation

4.1 Field Sampling, Soil and Root Collection for Identification

The isolation and identification of mycorrhizae require proper sampling, including aseptic collection of soil and root samples from plants. Proper storage conditions and timely processing are essential for preserving the integrity of the fungi (Trejo-Aguilar and Banuelos 2020). Depending on the purpose of the study, field sampling can be random, selective, or systematic to ensure representative coverage of the site. The most common approaches include laying a grid or setting up transects across the study area for direct sample collection.

Table 1 Diversity of mycorrhizae and their benefits

Ecosystem	Plant species	Fungi species	Symbiotic benefit
Arbuscular mycorrhizae			
Tropical and temperate forests	Maples, hardwoods	*Glomus* and *Rhizophagus*	Essential for phosphorus absorption especially in nutrient-depleted forest soils
Grasslands and savannas	Oak, Festuca, Bouteloua (e.g., buffalo grass) Forbs, Shrubs.	*Glomus*, *Gigaspora*, *Acaulospora*, Scutellospora	Efficient absorption of phosphorus, nitrogen, and water from the soil
Aquatic and wetland	Aquatic macrophytes, riparian trees, shrubs (willows and alder)	*Glomus*, *Acaulospora*, *Gigaspora*	Enhance uptake of phosphorus, nitrogen, and trace minerals
Extreme harsh environment	Cacti, desert shrubs halophytes, alpine grasses, pine trees, olive trees	*Glomus*, *Funneliformis* *Claroideoglomus* *Septoglomus* *Glomus*	Enhance uptake of essential micronutrients (e.g., zinc, copper) in harsh soil, conserve water, and increase resilience to oxidative stress
Ectomycorrhizae			
Boreal and temperate forests	Pine, oak, birch, spruce	*Russula, Boletus, Amanita*	Facilitate the uptake of nitrogen and phosphorus in nutrient-poor soils
Grasslands and savannas	Acacia, eucalyptus, quercus (oak)	*Pisolithus*, *Amanita*, *Cortinarius*	Enhance phosphorus and nitrogen absorption and protection against drought
Aquatic and wetland	Riparian trees, shrubs	*Cenococcum*, *Pisolithus*, *Amanita*	Increase the absorption of water and nutrients
Extreme harsh environment	Oak, Juniper birch, spruce	*Cenococcum, Pisolithus, Paxillus, Russula*	Improve plant resilience to nutrient-deficient soils, cold temperatures, drought, and high salinity
Ericoid mycorrhizae			
Heathlands, coniferous and boreal forests	Blueberries, cranberries	*Oidiodendron, Pezizella*	Build resilience to acidic soils by the acquisition of organic nitrogen and phosphorus
Grasslands and savannas	Heathers, blueberries	*Hymenoscyphus,* ascomycete fungi	Decompose organic matter in soils and release inaccessible nutrients
Aquatic and wetland	Shrubs, bogs	*Hymenoscyphus, Oidiodendron*	Decompose organic matter in nutrient and water-deficient environment
Extreme harsh environment	Ericaceous plants, heathers, cranberries, crowberry, bearberry	*Oidiodendron* *Meliniomyces, Cryptosporiopsis*	Promote plant growth in low nutrients, high acidity, and harsh climate

(continued)

Table 1 (continued)

Ecosystem	Plant species	Fungi species	Symbiotic benefit
Orchid mycorrhizae			
Tropical and temperate forest	Lady slipper rattlesnake plantain, coralroot	*Rhizoctonia, Ceratobasidium, Sebacina, Tulasnella*	Facilitate nutrient absorption in nutrient-deficient soil and acidic soils
Grassland and savannas	Spiranthes, Disa species, Eulophia species, Cynorkis	*Ceratobasidium, Sebacina, Tulasnella*	Facilitate nutrient and water absorption in harsh environments, support seed germination
Aquatic and wetland	Marsh Helleborine, water spider orchid, slender ladies' tresses	*Tulasnella, Ceratobasidium, Sebacinacium*	Facilitate nutrient absorption, in nutrient-deficient, acidic environments where nitrogen and phosphorus are limited
Extreme harsh environment	Marsh orchids, butterfly orchids, slipper orchids Ladies' tresses	*Tulasnella Rhizoctonia Tulasnella*	Facilitate nutrient absorption in nutrient-deficient soil and cold regions, particularly in high-altitude regions

For accurate sampling, it is recommended to use soil cores from the plant root zone. Soil samples are collected at different depths using sterile tools, such as soil corers, spades, and shovels, to avoid contamination between samples. Soil samples were collected by digging at a depth of 10–20 cm, maintaining a distance of at least 10 m from each sampled plant. To account for the variability in mycorrhizal distribution, it is necessary to collect multiple replicate samples, typically 100–200 grams from each site. Because fine roots are highly likely to contain mycorrhizal structures, it is important to carefully dig the fine roots from host plants while ensuring minimal disturbance to the roots. After collection, proper handling and processing of the samples are essential for accurate identification and isolation of fungal diversity. (Mahmoudi et al. 2019).

4.2 Sample Handling and Preservation

To prevent contamination, it is recommended to carefully handle samples using sterile gloves, tools, and containers. It is also important to change gloves and sterilize tools between sample collection. The roots were carefully placed in sterile plastic bags or containers to avoid damage to delicate mycorrhizal structures. Each sample should be labeled with a date, collection site, depth, plant species, and any other relevant details to avoid confusion during analysis.

For short-term storage, particularly when processing requires soon after collection, the samples are stored in sealed plastic bags at 4 °C in a refrigerator or cooler to prevent DNA degradation and preserve fungal functionality. Root samples must also be kept at 4 °C in a moist environment either by using moist paper towels or by placing them in plastic bags with a small amount of water to prevent drying. For long-term storage, root samples are either preserved in 50% ethanol or frozen depending on the type of analysis. For molecular analysis, root samples are kept frozen in liquid nitrogen to preserve DNA integrity. Root samples for spore isolation must be air-dried or refrigerated, as freezing may reduce spore viability (Stürmer et al. 2021).

4.3 Spore Isolation Techniques

Spore isolation has traditionally been carried out by extracting fungal spores from the root samples of soil or natural plants using the standard wet-sieving technique (WST) followed by germination under controlled conditions such as temperature, light, nutrients, and humidity either on water agar with host roots or in modified Murashige and Skoog medium (Ijdo et al. 2011). The wet-sieving technique (WST), combined with sucrose solution centrifugation, is a commonly used traditional method for isolating mycorrhizal spores. Spores between 50 and 750 μm (up to 1 mm) are extracted using a 45 μm sieve. The sieved sample is centrifuged at 4000 rpm for 5 min and refiltered through a 45 μm sieve. To further purify, the retained sample is mixed with a 55% sucrose solution and centrifuged at 1500 rpm for 7 min. The supernatant is filtered through a 45 μm sieve, thoroughly rinsed with tap water to remove sucrose residues and placed into a sterile tube with 20 mL of water.

Researchers have recently developed other effective techniques, such as ultrasound-assisted wet sieving (UWST) and ultrasound centrifugation (UCT), to increase spore yield, reduce extraction times, and decrease water consumption (Boyno et al. 2023). The ultrasound wet-sieving technique (UWST) is occasionally used in mycorrhizal spore isolation to improve the separation of spores from soil and root debris. The samples are exposed to high-frequency ultrasonic waves (28 kHz for 30 s) to break apart soil aggregates and release spores trapped within soil particles. This process follows the same general procedure as the traditional WST, in addition to incorporating ultrasound vibrations in the sieve mesh to reduce sieve clogging and repeated filtering through a 45 μm mesh to further increase spore yield and purity. The ultrasound centrifuge technique (UCT) combines ultrasonic treatment with high-speed centrifugation to enhance the efficiency of mycorrhizal spore isolation. Ultrasonic waves (28 kHz) disaggregate soil particles, releasing trapped spores, while centrifugation at 3000 rpm for 3 min separates spores from debris. Before transferring to a new collection tube, the supernatant was passed through a 1 mm strainer. This method significantly reduces contamination, improves spore yield, and enhances the purity of samples, especially in soils with high organic content or density (Boyno et al. 2023).

4.4 Root Staining and Processing

The root sample used for microscopic analysis requires the removal of root debris to eliminate cellular content and pigments that might obscure fungal structures.

Roots are typically cleared by boiling for 30–60 min in 10% potassium hydroxide (KOH) solution. The roots are stained with either trypan blue, Sheaffer blue, or Parker blue solutions. After staining, the roots were thoroughly rinsed and mounted on microscope slides for observation of fungal structures, including arbuscules, vesicles, and hyphae under light microscopy. A recent study (Wilkes 2023) compared the efficacy of Parker blue and Parker washable blue inks as substitutes for the commonly used Sheaffer blue ink in staining arbuscular fungal structures within plant roots. The study showed that Parker inks were equally effective as traditional dyes in visualizing fungal vesicles and arbuscules under the microscope.

If scanning electron microscopy (SEM) is used, then root samples require specific preparation, that is, drying and coating the root samples with a thin conductive material, such as gold or carbon, and mounting on specialized holders to maintain stability and prevent degradation by the high-vacuum environment of the microscope. For molecular analysis, proper sample preservation, processing, and extraction are needed for maintaining DNA integrity. To prevent degradation, root and soil samples are often stored at −80 °C. Additionally, grinding root tissues under liquid nitrogen into a fine powder and incubating at 65 °C for 30 min with lysis buffer (1% sodium dodecyl sulfate (SDS)) or a commercial lysis buffer helps to release DNA for extraction. The DNA was subsequently purified using a silica-based column method and eluted with Tris-ethylenediaminetetraacetic acid (EDTA) buffer or nuclease-free water (Yang et al. 2024).

5 Methods of Culturing

Mycorrhizal fungi depend on symbiotic associations with host plant growth and development; therefore, culturing them is quite challenging. Various in vivo and in vitro methods have been used to culture fungi. The methods are outlined as follows: Table 2 highlights the benefits and limitations of these methods.

5.1 Pot Culture

Pot culture is widely used in vivo techniques for cultivating and examining mycorrhizal fungi. In this method, host plants are grown in pots containing sterile soil with adequate nutrients. Plants are inoculated with fungal spores or mycelia and allowed to grow for several weeks or months in a controlled environment until the fungi are completely colonized. Following colonization, fungal hyphae and spores are extracted from both the soil and plants' roots (Gadkar et al. 2006). A recent study focused on improving this method to enhance both the quality and quantity of fungal inoculum (Praveen et al. 2023). Using substrate-based pot culture that includes all arbuscular fungal structures, the study observed that different fungal species display unique microbial preferences and proliferation patterns within the cultures, emphasizing the importance of including diverse mycorrhizal species in inoculants to optimize plant productivity.

Table 2 The benefits and limitations of various culturing methods

Culturing methods	Benefits	Limitations
Pot culture	Closely mimics natural conditions enabling mycorrhizal fungi to grow and produce spores effectively	Time-consuming and labor-intensive and requires constant care of the host plant
Pure culture	Creates a controlled environment, reduces the risk of contamination, and is useful for studying fungal morphology and genetics	May not fully mimic natural conditions, resulting in slower fungal growth
Dual culture	Very closely mimics natural conditions, promoting the growth and colonization of fungi through optimized, nutrient-rich environments	Requires thriving both plant and fungus simultaneously, which presents technical challenges, particularly in small-scale production and the maintenance of sterile techniques and lab equipment
Spore extraction and germination	Enhances propagation of fungi directly from spores	Challenging due to difficulties in maintaining sterility and varying germination of spores of different types of fungi
Hydroponic system	Allows precise control over environmental conditions and nutrient availability of mycorrhizae	Colonization of AMF in hydroponic systems poses a challenge compared to soil-based systems
Open field culturing	Completely mimics the natural environment and facilitates the large-scale propagation of mycorrhizal fungi	Difficult to control and presents a greater risk of contamination by other soil microorganisms

5.2 Trap Culture

The trap culture is an in vivo method for mycorrhizal fungal isolation and proliferation from soil or root samples. This method involves growing a host plant in soil containing mycorrhizal spores, allowing the fungi to colonize the plant roots. Over time, when the fungi fully proliferate, spores are harvested from the soil. All trap cultures are stored in plastic bags for at least 30 days before spores are extracted, seedling inoculation, and maintained as monospecific cultures for inoculum preservation (Ksanbok and Highland 2019). A study by Tenzin et al. (2022) suggests that in trap cultures, the use of dry-season soil and co-cultivation of host plants e.g., maize (*Zea mays*) and sorghum (*Sorghum bicolor*) produce AMF spore at higher number compared to monocultures of either plant; hence, this method is useful for significant production of this fungal inoculant.

5.3 Pure (Axenic) Culture

Pure (axenic) culture is an in vitro method for studying the growth, behavior, and interactions of a specific fungal species without any influence from other microorganisms. In this method, plant root extracts are grown in petri dishes or flasks containing nutrient-rich media (Abdellatif et al. 2019). In case of AMF, hyphae or spores are usually submerged in the nutrient-rich media under sterile conditions for colonization into the roots. In contrast, ECMF are grown on nutrient agar plates, such as Modified Melin-Norkrans (MMN), without a host plant as they can independently colonize and establish a hyphal network (Laatikainen and Heinonen-Tanski 2002).

It has been observed that under asymbiotic conditions, nutrients like palmitoleic acid increase AMF growth and induce the generation of infection-competent secondary spores (Srinivasan et al. 2014). A recent study found that the addition of myristic acid and organic nitrogen stimulated the production of large number of secondary spores and jasmonic acid resulted in higher spore mass and larger spore size (Tanaka et al. 2022).

5.4 Dual Culture

Dual culture is an in vitro co-culture method of studying symbiotic interactions under controlled conditions. In this method, both the mycorrhizal fungus (hyphae or spores) and the host plant (roots) grow either in sterile soil or growth media, typically agar medium or liquid broth under controlled conditions. The desired fungi were extracted after a complete symbiosis was established. Studies have shown that in the absence of host plants, co-culture of AMF *Rhizophagus irregularis* and mycorrhiza-helper bacterium *Paenibacillus validus* successfully enables completion of their life cycle (Hildebrandt et al. 2006). This method has also been used to isolate soluble metabolites released from the hyphae of AMF (Gadkar et al. 2006). A recent study (Teste et al. 2020) highlights that dual-mycorrhizal plants associate with both AMF and ECMF, enhancing nutrient uptake and adaptability. This study highlights valuable insights into plant-fungal interactions and offers promising applications in restoring and stabilizing ecosystems.

5.5 Hydroponic Culture

Hydroponic culture methods have recently been developed to support mycorrhizal symbiosis, particularly designed for higher production of commercial inoculum. This method involves growing plants on a solid substrate for several weeks to facilitate fungal precolonization, followed by transfer to a hydroponic system. Das et al. (2020) used a highly flexible, cost-effective method that introduced direct inoculation of fungal spores into plants grown in a hydroponic system. This method allowed easy visual monitoring of growth of plant roots into the nutrient-rich growth media, easily harvested without damaging them (Das et al. 2020).

In hydroponics, owing to the frequent movement of water and suspension of roots in solution, it is recommended to apply multiple doses of mycorrhizal inoculation during the early stages of plant development. Mycorrhizal inoculums increase crop growth and productivity in hydroponics, but their effectiveness is influenced by factors such as phosphorus levels in the nutrient solution, pH levels, and the type of hydroponic system used. Therefore, to obtain complete root colonization and maximize the benefits of mycorrhizae, it is recommended to inoculate spores with each water change. Additionally, optimizing environmental conditions and refining hydroponic system designs can further support effective integration of mycorrhizae and enhance the plant–fungus symbiotic relationship (Othman et al. 2024).

5.6 Open Field Culture

Open field culture is a method of increasing crop productivity and quality in open-field agriculture. This method supports natural mycorrhizal symbiosis by growing host plants in the soil, either in large outdoor containers or in natural fields. This open field culture improves adaptation of mycorrhizal plants to the native microflora of the soil, contributing to soil fertility, nutritional benefits, and overall productivity of symbiosis (Raklami et al. 2019). Following the establishment of complete natural symbiosis, spores or hyphae are harvested for the desired application. Despite being common, this technique can be challenging due to potential disruption of the symbiosis process by factors such as environmental contamination, slower growth, and the specificity of the host interaction. In addition, the effectiveness of mycorrhizal application also depends on soil fertility, inoculation timing, and environmental conditions such as temperature, pH, and nutrient content. A recent study by Prettl et al. (2024) suggests that mycorrhizal inoculation might be beneficial in nutrient-deficient soil with limited fertilization; however, more field trials across varying fertilization levels and soil types are needed to optimize application methods and to fully understand complex interactions among climatic, soil, and agricultural factors.

6 Morphological Identification of Mycorrhizal Diversity

6.1 Light Microscopy

Light microscopy is an effective method for morphological identification of both AMF and ECMF, as it can clearly distinguish colonization patterns, hyphal networks, and root interactions at the cellular level. Under light microscopy, arbuscules of AMF are visualized as small, tree-like projections, and hyphae appear as thin, thread-like structures, whereas thick-sheath mantle of ECMF is identified as a dark layer surrounding the root, and Hartig net as a dense, web-like structure (Brundrett and Tedersoo 2018).

In a study conducted by Chakraborty et al. (2019), the colonization patterns of AMF and dark septate endophytic (DSE) fungi in rice (*Oryza sativa*) roots were examined under field conditions. Using a light microscope, mycorrhizal colonization in plant roots, as well as other key features including arbuscules, vesicles, and hyphal coils in arbuscular mycorrhizae, and melanized septate hyphae and microsclerotia in dark septate endophytic (DSE) fungi were visualized over time. The study revealed that colonization of both AMF and DSE varied across different types of plant life forms as herbaceous plants showed higher colonization rates than woody plants, suggesting a significant influence of plant life form on mycorrhizal associations. Although light microscopes are cheaper and easier to use than confocal or electron microscopes, their limited resolution often makes it challenging to identify delicate fungal structures. Moreover, the necessary preparations for light microscopic analysis, including clearing and staining of root samples for better visibility, can be time-consuming and labor-intensive.

6.2 Scanning Electron Microscopy (SEM)

Scanning electron microscopy (SEM) produces high-resolution images that show detailed fungal structures at the micro and nanometer levels, enabling visualization of fungal diversity. High-resolution SEM imaging facilitates identification from various perspectives, including detailed fungal-root examination that reveals fungal hyphae, spores, and root colonization crucial for differentiating mycorrhizal species, capturing reproductive structures such as sporangia, septa, and vesicles, supporting mycorrhizal species identification, and visualization of biofilms that showcase the ecological interactions of mycorrhizal fungi.

SEM was utilized in a study by Palenzuela et al. (2011) to examine the spore morphology of *Ambispora granatensis*, a newly identified AMF species associated with *Asparagus officinalis* in Andalucía, Spain. This SEM analysis revealed distinctive features, such as the presence of two spore morphs (acaulosporoid and glomoid), each exhibiting unique wall structures and surface ornamentations. These detailed morphological observations were crucial for the accurate taxonomic classification of the species. The use of SEM provides valuable insights into the ultrastructural external features of AMF, aiding in the differentiation of closely related species. Its high-resolution power reveals detailed surface structures of spores and hyphae, making it a valuable tool for the morphological characterization of AMF; however, it is not typically used for routine identification due to its cost and complexity. Its limitations in analyzing internal structures require the use of more advanced techniques, such as molecular methods, to achieve a comprehensive assessment of mycorrhizal diversity (Yang et al. 2024).

Overall, the high-resolution capabilities of SEM provide surface topography and three-dimensional images that reveal the unique spatial arrangement of fungal structures. SEM also investigates the complex networks formed by mycorrhizal fungi, which improve the soil structure and fertility for ecological benefits.

7 Enzymatic Biochemical Techniques: Functional Profiling of Metabolites

Enzymatic assays are the simplest and most convenient methods for understanding various biochemical processes associated with mycorrhizal symbiosis, including metabolism, nutrient cycling, and stress responses. Table 3 summarizes the catalytic and functional activities of the key enzymes involved in this symbiotic relationship.

Table 3 Key enzymes and their role in mycorrhizal symbiosis

Enzymes	Catalytic activity	Functional activity
Invertase	Hydrolysis of sucrose to glucose	Higher activity indicates enhanced metabolic activity
Sucrase	Hydrolysis of sucrose into glucose and fructose	Higher activity indicates active carbohydrate metabolism that promotes plant growth and symbiosis
GDPDH	Convert glucose-6-phosphate into 6-phosphogluconate while reducing $NADP^+$ to NADPH	Higher activity indicates enhanced biosynthesis, redox balance, and defense against oxidative stress
Laccase	Oxidation of substrate 2,2′-azinobis by transferring electrons from the substrate to molecular oxygen (O_2)	Higher activity indicates a higher rate of decomposition of complex organic molecules
Lignin peroxidase	Oxidation of veratryl alcohol into veratraldehyde while reducing H_2O_2	igher activity indicates a higher rate of lignin degradation enhancing its availability for symbiosis
Cellulase	Hydrolysis of cellulose into simpler sugars	Higher activity indicates effective decomposition of cellulose
Nitrate reductase	Reduction of sodium nitrite ($NaNO_2$) into nitrite	Higher activity indicates enhanced uptake and metabolism of nitrogen
Glutamine synthetase	Formation of glutamylhydroxamate from hydroxylamine	Higher activity indicates enhanced nitrogen metabolism through the assimilation of ammonium from nitrogen fixation or nitrate reduction
Glucanase	Decomposition of either laminarin or lichenan into reducing sugar	Higher activity indicates active fungal growth enhanced defense against pathogens
Chitinases	Breaks down colloidal chitin to N-acetylglucosamine (GlcNAc)	Higher activity indicates enhanced defense against pathogens
Superoxide dismutase	Formation of superoxide radical through oxidative reaction	Higher activity indicates a greater mycorrhizal association and reduced oxidative stress in plants
Catalase	Decomposition of H_2O_2 into water and oxygen	Higher activity indicates a greater mycorrhizal association and reduced oxidative stress in plants
Peroxidase	Oxidation of guaiacol into tetraguaiacol	Higher activity indicates greater defense against oxidative stress
Polyphenol oxidase	The oxidation of catechol (or pyrogallol) to quinone	Higher activity indicates greater defense against pathogens
Phosphatase	Hydrolysis p-nitrophenyl phosphate into p-nitrophenol	Higher activity indicates enhanced breakdown of organic phosphorus, improving its availability for plant growth

7.1 Enzymes Regulating Carbohydrate Metabolism

The measurement of key carbohydrate-metabolizing enzyme activities including sucrase, invertase, and glucose-6-phosphate dehydrogenase (G6PDH) revealed the distribution and significance of sugar and its derivatives in mycorrhizal symbiosis. The spectrophotometric method effectively measures the amount of glucose produced by sucrase and invertase, and the nicotinamide adenine dinucleotide phosphate (NADPH) produced by G6PDH, which directly reflects enzyme activity (Wang and Wu 2023).

Moreover, the measurement of enzyme activities such as laccase (Arregui et al. 2019), lignin peroxidase (Chen et al. 2018), and cellulase (Yang et al. 2020) demonstrates their roles in supporting the nutritional needs of symbiosis by decomposing complex organic matter. The enzyme activity was quantified spectrophotometrically by measuring the absorbance resulting from their interactions with the respective substrate molecules.

7.2 Enzymes Regulating Nitrogen Cycle

Assessing the activity of enzymes associated with the nitrogen cycle is important, particularly nitrate reductase for nitrogen assimilation and glutamine synthetase for nitrogen uptake. Colorimetric detection of nitrite reduction by nitrate reductase and the transfer of γ-glutamate by glutamine synthetase enables quantification, as these processes are directly related to the regulatory mechanism of these enzymes (Wang et al. 2024). The enzyme glutamine synthetase is observed to be a key regulator for nitrogen uptake and assimilation for AMF symbiosis, particularly in cotton (Zhang et al. 2025). This enzyme plays a crucial role, enhancing the plant's resistance to pathogens by modulating inorganic nitrogen assimilation.

It has been identified that AMF produces two important enzymes, β-1,4-N-acetylglucosaminidase and leucine aminopeptidase for nitrogen cycling, thus playing crucial roles in decomposing soil organic matter and facilitating nutrient availability to plants (Bortolot et al. 2024). The study by Xiaomei Gou et al. (2024) also demonstrates that AMF significantly contributes to nitrogen retention and assimilation by increasing nitrogen-cycling enzyme activity and upregulating genes involved in nitrogen fixation and nitrification, especially in maize.

7.3 Enzymes Regulating Hydrolysis

Assessing hydrolytic enzyme activities, such as glucanase and chitinase, explains the defense mechanisms that plants use against pathogens offered by mycorrhizal fungal colonization. Spectrophotometric measurements of the enzymatic breakdown of glucans into 4-methylumbelliferyl β-D-glucan by glucanase and chitin into 4-nitrophenyl N-acetyl-β-D-glucosaminide by chitinase (Patil and Solanki 2016) directly demonstrated the efficiency of these enzymes (Pozo et al. 1999).

A recent study (Mei et al. 2024) observed that for efficient cellulose degradation, AMF produces a variety of hydrolytic enzymes, including β-1,4-glucosidase, α-1,4-glucosidase, β-1,4-xylosidase, and cellobiohydrolase, and the activity of these hydrolytic enzymes is greatly influenced by nitrogen levels and the temperature of the soil. The study also highlights that the production of hydrolytic enzymes differs among different AMF species; for example, Glomus versiforme has higher activities of protease, polyphenol oxidase, and urease in certain soils, compared to *Glomus diaphanum*, which influences their root colonization capability and their role in altering soil enzyme activity. The production of hydrolytic enzymes by certain mycorrhiza-helper bacteria contributes to increased fungal colonization and soil nutrient status by facilitating spore germination and hyphal extension (Bortolot et al. 2024).

7.4 Enzymes Regulating Defense Mechanism

Spectrophotometric analysis of certain antioxidant enzymes, particularly superoxide dismutase (SOD), catalase, peroxidase, and polyphenol oxidase, is essential for understanding the defense mechanisms of mycorrhizal fungi under environmental stress. These enzymes significantly reduce oxidative stress by scavenging reactive oxygen species (ROS), thereby protecting plant cells from damage. The activities of these enzymes are measured through chemical reactions such as the reduction of nitroblue tetrazolium (NBT) by SOD, the decomposition of hydrogen peroxide (H_2O_2) by catalase, the degradation of guaiacol by peroxidase, and the oxidation of phenolic compounds by polyphenol oxidase (Delaeter et al. 2024).

7.5 Enzymes Regulating Phosphorus Cycle

Determining the activity of both alkaline and acidic phosphatases reveals the mechanism of phosphorus uptake and mobilization in both plants and soil. The activity of these enzymes can be measured spectrophotometrically by the hydrolysis of p-nitrophenyl phosphate to p-nitrophenol (Lu et al. 2023). A study by Chen et al. (2024) observed that phosphatases secreted by mycorrhizal fungi strengthen plant defense by several mechanisms, i.e., by improving phosphorus uptake, activating antioxidant systems, regulating hormone signaling pathways, disrupting the quorum-sensing mechanisms of pathogens, and activating plant immune responses. Although the enzyme phosphatase is indirectly involved, it is shown to be very crucial for strengthening plant resistance to pathogens.

In summary, to understand the practical impact of mycorrhizal symbiosis on plant metabolic pathways, enzymatic assays can be combined with metabolomics methods such as liquid or gas chromatography mass spectrometry (LC/GC-MS) which will be discussed in the following sections.

8 Chromatographic Methods for Metabolite Profiling

Chromatographic methods are designed based on the chemical and physical properties of metabolites to study the nature of mycorrhizal symbiosis. Several chromatographic methods have been employed to analyze specific metabolites required for mycorrhizal interactions. High-performance liquid chromatography (HPLC) is often used to identify primary nonvolatile and polar metabolites such as sugars, amino acids, and secondary metabolites involved in mycorrhizal symbiotic interactions (Matsubara and Ishii 2015). Gas chromatography (GC) is used for identifying volatile organic compounds (VOCs) such as strigolactones, a plant endogenous hormone and signaling molecule involved in symbiotic interactions (Kaur et al. 2022). Thin-layer chromatography (TLC) is a cost-effective method for extracting lipids and hydrophobic compounds such as neutral lipids, phospholipids, fatty acids, and sterols (e.g., ergosterol) that are essential for understanding arbuscular mycorrhizal symbiotic interactions, nutrient exchange, signaling, and energy storage (HaiRu and Xiangy 2012). Ion exchange chromatography, an effective method for identifying charged molecules such as nitrates, amino acids, and phosphates; (Liu et al. 2000) and size exclusion chromatography for isolating large molecules, mainly phosphate transporter, exopolysaccharides, glycoproteins as shown in studies on mycorrhizal sulfur transfer and plant metabolic response (Sieh et al. 2013).

Researchers have employed various chromatographic techniques to investigate the roles of different metabolites involved in mycorrhizal symbiosis. When combined with mass spectrometry (MS), these methods allow for more comprehensive analysis of the metabolites involved.

9 Mass Spectrometry (MS) for Metabolite Profiling

Mass spectrometry (MS) has been a preferred method for understanding the detailed biochemical processes of mycorrhizal symbiosis, such as the mechanism of primary and secondary metabolite formation, nutrient exchange, and molecular signaling. MS is often combined with liquid chromatographic (LC) and gas chromatographic (GC) methods for successful detection and quantification of different types of primary and secondary metabolites, proteins, and lipids involved in the complex symbiotic process. MS combined with LC quantifies water-soluble metabolites such as sugars and organic acids, while MS combined with GC detects volatile compounds like terpenes and fatty acids (Bravo et al. 2016). A notable example is the study by Ranner et al. (2022), which employed untargeted metabolomics using high-resolution liquid chromatography–mass spectrometry (LC-MS) to investigate the metabolic changes in *Lotus japonicus* roots during AMF colonization. The researchers discovered that AMF symbiosis led to the accumulation of novel methylated polyphenols, some of which exhibited antifungal properties. This finding suggests a role for these compounds in modulating the symbiotic interaction, potentially balancing mutualistic benefits with defense mechanisms.

Watts-Williams and Cavagnaro (2020) highlight in their review that under phosphate-deficient conditions, plant roots release signaling molecules strigolactone that promote symbiotic interactions by stimulating fungal AMF spore germination and hyphal branching. Moreover, strigolactone also interacts with other plant hormones such as gibberellins and auxins to strengthen the colonization process.

The combined spectrophotometric methods have shown significant value in the detection of specific secondary metabolites, including terpenoids, phenolic compounds, and strigolactone, which have a significant role in symbiosis (Ranner et al. 2022; Luginbuehl et al. 2017). Such studies underscore the utility of MS-based metabolomics in uncovering previously uncharacterized metabolites that play crucial roles in plant-fungal interactions. By capturing the dynamic metabolic shifts associated with symbiosis, MS facilitates a deeper understanding of the functional mechanisms underpinning these complex relationships.

10 Molecular Methods of Identification

10.1 Polymerase Chain Reaction (PCR), Quantitative PCR (qPCR) and Nested PCR

Molecular techniques such as polymerase chain reaction (PCR), quantitative real-time PCR (qPCR) and recent nested PCR have incorporated added benefits in investigating the genetic mechanisms of mycorrhizal symbiosis. PCR-based high-throughput sequencing of internal transcribed spacer (ITS) region and large subunit (LSU) rRNA genes has revealed strong host preferences among AMF taxa, especially in tropical and alpine ecosystems. A recent study developed a global database of AMF occurrences using sequencing data derived from AMF taxon barcoding regions located in the small subunit (SSU) rRNA gene, the internal transcribed spacer 2 (ITS2) region, and the LSU rRNA gene. This study not only documented the distribution of AMF species accessible for diverse users but also enabled access to detailed information on the ecology of these fungal taxa (Větrovský et al. 2023).

PCR is a commonly used method to amplify specific genes like sym or Myc from fungal tissues or spores that are involved in signaling pathways, nutrient transport (such as phosphate transporters), or spore germination. The most widely used genetic markers include 18S ribosomal RNA (rRNA) and ITS regions. Quantitative PCR (qPCR) through amplicon sequencing identifies the abundance of specific genes such as phosphate transporters or signaling pathway regulators and evaluates community composition (Schoch et al. 2012, Bodenhausen et al. 2021). In qPCR, amplification is monitored in real-time using specific primers and SYBR Green dyes or fluorescent probes that bind to the target gene of interest. qPCR remains a preferred choice for quantifying and evaluating root colonization by the fungal community as this method reduces the number of repetitive analyses required to assess mycorrhizal symbiosis(Corona Ramírez et al. 2023).

Recently, Manga et al. (2025) applied nested PCR to amplify the SSU region, enhancing the specificity of DNA amplification of AMF. This method involved an initial amplification using common fungal primers, followed by a second-round amplification with AMF-specific primers. This innovative approach enabled the effective detection of low-abundance AMF taxa in complex soil and root samples.

10.2 DNA Sequencing and Barcoding

Conventional Sanger DNA sequencing is still considered a reliable method for identifying specific gene sequences of mycorrhizal fungi. This method focuses on the noncoding ITS region of fungal rRNA, crucial for mycorrhizal colonization, nutrient transport, and signal transduction. In closely related species of AMF, the ITS region is either absent or highly variable (Davison et al. 2015); therefore, the ITS region serves as a key marker for distinguishing AMF from ECMF.

The length of the ITS region varies between 450 and 700 base pairs (bp) and comprises two subregions known as ITS1 and ITS2. These subregions are separated by a highly conserved, 150 bp long 5.8S rRNA gene. This ITS is particularly located between the 18S rRNA, SSU (18S rRNA) LSU (28 rRNA). Particularly, ITS1 is located upstream of 18S rRNA and downstream of 5.8S rRNA, while ITS2 is located upstream of 5.8S rRNA and downstream of 28S rRNA (Schoch et al. 2012). Both ITS1 and ITS2 subregions are valuable for species identification, while conserved 5.8S rRNA is useful for comparative analysis through sequence alignment.

The ITS region is recognized as a useful marker for barcoding and studying mycorrhizal communities in the environment (Krüger et al. 2012; Stockinger et al. 2010). However, its use for AMF remains challenging because of its high variability and the inability to differentiate closely related species. Thus, to ensure accurate identification, a complete 1500 bp fragment that includes SSU, ITS region, and a partial LSU containing the two variable subregions, referred to as D1 and D2 (Stockinger et al. 2010).

A recent study by Delavaux et al. (2024) introduces significant advancements in the molecular identification of AMF by sequencing environmental DNA (eDNA). The study successfully expanded and refined the LSU reference database, enabling more accurate phylogenetic analysis of AMF sequences obtained from eDNA. This method is particularly valuable for identifying newly described or poorly characterized AMF genera.

Using a newly designed primer specific to fine root endophytes (FRE), Seeliger et al. (2024) demonstrate that primer choice has a significant impact on the accurate detection and analysis of AMF communities. Notably, the use of this primer revealed that nitrogen addition reduced mucoromycotina (M-AMF) diversity and modified community structure, effects that were not detected with standard AMF primers.

10.3 Next-Generation Sequencing (NGS)

Next-Generation Sequencing (NGS) is a cutting-edge technology that targets different genomic regions depending on the type of mycorrhizal fungi studied. ITS is recognized as a universal gene for NGS because of its variability, the availability of comprehensive reference databases, and its capability to generate consistent data output (Schoch et al. 2012). Nonetheless, ITS may not be effective for all types of mycorrhizal fungi as its substantial variability often complicates successful amplification with universal primers designed for this region.

Wu et al. (2024) proposed a mechanism of AMF in soil organic matter (SOM) dynamics and carbon cycling. The study shows that AMF utilizes interconnecting pathways to generate SOM diversity by producing a variety of organic compounds including exudates and biomass. AMF also alters the physical properties of soil, triggering SOM transport, transformation, and storage; finally, it stabilizes SOM through mineral weathering and organo–mineral interactions.

A study by Kaur and Reddy (2023) investigated the diversity of AMF in seleniferous soils and their role in plant growth and development. Using Illumina MiSeq next-generation system which sequences DNA fragments and produces millions of short reads (typically 50–600 base pairs long) for further analysis and assembly, they identified 17 AMF species on maize roots in seleniferous soils and 18 in non-seleniferous soils. This research provided insights into the diversity of AMF and their ability to tolerate selenium.

10.4 Amplicon-Based Sequencing

Amplicon-based sequencing focuses on amplifying specific regions, such as ITS, 18S rRNA, or 28S rRNA. It utilizes specially designed oligonucleotide probes that are created to capture and amplify the regions intended for analysis. After amplification, high-throughput sequencing methods, such as Illumina, Ion Torrent, or PacBio, are employed to determine the sequences of the amplified genes, referred to as amplicons. The amplicon is subsequently analyzed using bioinformatics and is compared with reference databases like UNITE for mapping and identifying higher-order taxonomy, including species-level classification. Using a recent amplicon-based sequencing technique, researchers examined arbuscular mycorrhizal fungal communities on maize roots under various cover cropping and tillage systems to support the practical applications of these methods (Higo et al. 2020). The use of Illumina amplicon sequences from the small-subunit nuclear ribosomal RNA gene greatly illuminated the composition of both intra- and extraradical arbuscular mycorrhizal communities (Djotan et al. 2022). These studies support the effectiveness of amplicon sequencing in delivering detailed identification and characterization of fungal variants.

10.5 Molecular Cloning

For molecular cloning, highly conserved or species-specific regions such as 18S ribosomal RNA, internal transcribed spacer (ITS) regions, or genes involved in symbiosis are required. These regions are ideal for the characterization and functional analysis of mycorrhizal fungi. In molecular cloning, genes are amplified and then inserted into a vector for expression. The most widely used vector is the recombinant plasmid of *Escherichia coli*. Following cloning, heterologous expression systems are employed to investigate the comprehensive roles of these genes in symbiosis. This typically involves expressing mycorrhizal fungal genes in model organisms like yeast (*Saccharomyces cerevisiae*) or plants (*Arabidopsis thaliana*), based on the gene of interest (Gao et al. 2023).

11 Functional Genomics

Functional genomics has recently been used for analyzing gene functions and their interactions within mycorrhizae. The advancement of functional genomics offered revolutionary discoveries, but they also have some limitations. Table 4 summarizes the strengths and weaknesses of these techniques.

11.1 Metabolomics

Metabolomics is a potential tool to discover complex biochemical mechanisms of symbiosis under environmental stress conditions such as cold and drought. Metabolomics has revealed that certain primary and secondary metabolites, including lipids, flavonoids, and organic acids, enhance nutrient absorption and increase tolerance against cold and drought. Mycorrhizal fungi alter lipid composition and activate antioxidant enzymes during extreme cold, thereby helping plants to survive. It has been shown that AMF improves the cold resistance of Gramineae grasses by stimulating metabolism, activating antioxidant enzymes, and increasing the concentration of antioxidant-related metabolites such as flavonoids and lipids (Zhang et al. 2023). AMF also helps to alter the root lipid compositions of drought-tolerant plants. Higher concentrations of saturated fatty acids contribute to the stabilization of cellular structures during dehydration, allowing plants to endure severe drought conditions. Additionally, the modified synthesis of flavonoids and the expression of specific phytohormone family genes are identified as key regulators of symbiotic interaction (Gao et al. 2023).

A study by Ge et al. (2025) demonstrated that inoculation with AMF (*Rhizophagus irregularis*) increased flavonoid and free amino acid content in tea plant roots (*Camellia sinensis*) by upregulating the genes involved in these biosynthetic pathways, suggesting that AMF can improve tea quality by modulating secondary metabolite pathways. Sarathambal et al. (2024) conducted a study to evaluate the synergistic effects of AMF and vermicompost on the nutrient uptake,

Table 4 Strengths and weaknesses of different methods

Methods	Strength	Weakness
Metabolomics	Comprehensive analysis of biochemical pathways Detailed insight into stress tolerance mechanism	Data interpretation is challenging Requires advanced computational tools
Transcriptomics	Comprehensive analysis of gene expression. Discovery of novel genes and their expression Analysis of gene expression at different stages of fungal colonization Greater sensitivity and higher resolution	Expensive and data-intensive, especially for large-scale study Information is limited to gene expression and hence requires combining with other methods for a better understanding Challenging due to high variability in gene expression across different environmental conditions
Microarray	Allow high-throughput analysis of gene expression Cost-effective for large-scale gene expression analysis	Less sensitive compared to RNA sequencing Increased chance of overlooking less abundant genes during analysis Rely on pre-designed probes, hence challenging for novel gene identification
Gene editing	Rapid functional analysis Precise gene alteration CRISPR can create plants with desired traits	Risk of unintentional, random gene editing Limited to mycorrhizal fungal gene editing May not be feasible for crops
Proteomics	Wide range of protein identifications both in symbiotic and non-symbiotic associations Enable identifying posttranslational modifications such as phosphorylated, glycosylated, and ubiquitinated protein Allow high-throughput analysis of large-scale proteins Highly sensitive toward dynamic range abundant regulatory protein expression analysis	Sample preparation is technically challenging Detection of low-abundance transcription factors and signaling proteins is challenging Difficult to accurately identify rare species due to the lack of availability of protein databases The method is labor-intensive, expensive, and time-consuming

photosynthetic efficiency, secondary metabolite production, and growth in ginger plants (*Zingiber officinale*) and observed that both the quantity and quality of ginger crops significantly improved when biostimulants were applied together.

11.2 Transcriptomics (RNA Sequence)

Transcriptomics is a very important method of detection of mRNA sequence focusing on special protein-coding mRNA or portion of sequenced mRNA that have distinct functions in mycorrhizae. The mRNA sequence has been utilized to study the physiology of arbuscular mycorrhizal fungi, which are not culturable and difficult to study genetically (Abdelsattar et al. 2024) To analyze the large amount of raw data generated by the mRNA, transcriptomics often combines with bioinformatic tools that focus on library trimming, read mapping, normalization, and differential expression calculations. Transcriptomics has revealed the mechanisms by which AMF enhance salt tolerance in rice plants. It indicates that this symbiotic association involves the upregulation of several genes responsible for the synthesis of inositol-3-phosphate and starch, modifications of the cell wall, and lipid ion homeostasis necessary for growth in salt-stressed environments. Transcriptome analysis also suggests that AMF promote the natural regeneration of a forest tree (*Toona ciliata var. pubescens)* seedling through the upregulation of certain genes involved in macromolecule localization, intracellular protein transport, and signal transduction (Jiang et al. 2024).

Combining single-cell RNA sequencing and spatial transcriptomics, Somoza et al. (2024) have delivered novel insights into the cellular processes of symbiotic development, identifying distinct stages of arbuscule formation and disintegration within plant root cells. While single-cell RNA sequencing provides detailed high-resolution data, it lacks spatial context. The authors underline that rather than a single method alone using a combined method provides a more accurate and comprehensive understanding of AMF symbiosis.

11.3 Microarray Analysis

Microarray technology has been used for large-scale analysis of gene expression involved in symbiotic interactions between mycorrhizal fungi and plant roots, providing information about the early evolution of symbiotic traits (Miyauchi et al. 2020). This large-scale analysis has revealed the detailed role of genes at the transcriptional level during stress responses; for example, expansion-like proteins have been found to play a significant role in both the presymbiosis and symbiosis stages of AMF. In the presymbiosis stage, these proteins modify the cell wall components to facilitate the penetration of fungal hyphae into the roots of the host plant, whereas in the symbiosis stage, they support the development and maintenance of hyphal network within root cells and facilitate host colonization (Abdelsattar et al. 2024). This large-scale analysis also improves our understanding of the ecological roles of these fungi, which is essential for future applications in sustainable agriculture practices.

11.4 Gene Editing (CRISPR/Cas9)

CRISPR/Cas9 gene editing has recently been utilized to investigate the complex mechanisms of symbiosis between mycorrhizal fungi and host plants. This approach has revealed that specific genes involved in plant growth, nutrient uptake, and defense against biotic stresses are altered to promote phosphorus absorption and increase stress tolerance in plants. Furthermore, CRISPR/Cas9 gene editing has been used in crops like soybean and grapevine to create varieties with enhanced resilience to environmental stress. Although this approach is promising for plant transformation, its routine application for analysis remains challenging (Tsakirpaloglou et al. 2023).

11.5 Proteomics

Proteomics is a powerful method of understanding the significance of proteins in plant growth, development, and photosynthesis under both normal conditions and water deficit stress during symbiosis. These studies have revealed that proteins involved in various metabolic pathways are vital for the survival of plants under water deficit stress. These proteins are involved in various processes including energy metabolism (e.g., ATP synthase), glucose metabolism (e.g., malate dehydrogenase, triosephosphate isomerase, sucrose-phosphatase), oxidative phosphorylation (e.g., mitochondrial-processing peptidase), and sulfur metabolism (e.g., thiosulfate/3-mercaptopyruvate sulfurtransferase) (Olalde-Portugal et al. 2020). These proteins were identified and quantified through liquid chromatography-tandem mass spectrometry (LC-MS/MS) using both label-free and isotope-labeling methods (Yu et al. 2023). Following identification, proteins were annotated by using specific databases such as UniProt and Gene Ontology (GO). Proteomic analysis has demonstrated that specific proteins are upregulated in response to certain needs, such as plant phosphate transporters (e.g., PT4) being upregulated during phosphate uptake and ammonium transporters (e.g., AMT3) being upregulated during nitrogen uptake from fungal hyphae for transfer to the plant. Moreover, proteomic analysis has confirmed several key symbiotic benefits, including the upregulation of signaling proteins, the role of strigolactones in colonization, the function of effector proteins in immune responses, and the synthesis of plant growth-promoting proteins under oxidative stress (Domingo et al. 2023).

12 Conclusion

This chapter explores the diversity and classification of mycorrhizal associations, along with both traditional and advanced methods for their isolation and identification, to reveal their ecological benefits. Based on the morphological characteristics, mycorrhizae are classified into four groups including arbuscular mycorrhizas (AM), ectomycorrhizas (ECM), ericoid mycorrhiza (ERM), and orchid mycorrhizas (ORM) each of which exhibits distinct ecological interactions across various natural ecosystems. Isolation of mycorrhizal fungi is beneficial for exploring the detailed processes involved in fungal–plant root symbiosis. Different types of isolation methods have been used to understand their roles in nutrient exchange, plant development, and soil ecology. Despite their individual advantages and limitations, these methods are essential for restoring natural ecosystems and advancing sustainable agricultural practices. Traditional mycorrhizal identification relies on morphological analysis of spores and root structures through various microscopic techniques. Culturing methods differ depending on the type of mycorrhizal fungi and their ecological preferences. AMF are typically cultivated using trap plants in pot cultures due to their obligate symbiotic nature, whereas ECMF can be grown on synthetic media under laboratory conditions. Advanced cultivation methods, such as co-culture and hydroponic systems, now provide more efficient, sterile, and scalable methods of growing mycorrhizal fungi successfully. In contrast, cutting-edge techniques like DNA barcoding, PCR, metagenomics, and proteomics enable faster, more accurate identification of mycorrhizal species to expand our knowledge on their ecological interactions and functional dynamics.

References

Abdelsattar M, Soliman MS, Mohamed RA, Radwan KH, El-Mahdy MM, Mousa KH, Khalil SRM, Osman E, Alameldin HF, Hussein A, Hassanein SE, Abdallah NA, Alsamman AM, Osama O (2024) Transcriptomic insights into mycorrhizal interactions with tomato root: a comparative study of short- and long-term post-inoculation responses. Front Genet 15:1434761

Abdellatif L, Lokuruge P, Hamel C (2019) Axenic growth of the arbuscular mycorrhizal fungus Rhizophagus irregularis and growth stimulation by coculture with plant growth-promoting rhizobacteria. Mycorrhiza 29(6):591–598

Arregui L, Ayala M, Gómez-Gil X, Gutiérrez-Soto G, Hernández-Luna CE, Herrera de los Santos M, Levin L, Rojo-Domínguez A, Romero-Martínez D, Saparrat MCN, Trujillo-Roldán MA, Valdez-Cruz NA (2019) Laccases: structure, function, and potential application in water bioremediation. Microb Cell Factories 18:1–33

Bodenhausen N, Deslandes-Hérold G, Waelchli J, Held A, van der Heijden MGA, Schlaeppi K (2021) Relative qPCR to quantify colonization of plant roots by arbuscular mycorrhizal fungi. Mycorrhiza 31(2):137–148

Bouffaud ML, Bernaud E, Colombet A, Van Tuinen D, Wipf D, Redecker D (2016) Regional-scale analysis of arbuscular mycorrhizal fungi: the case of Burgundy vineyards. Oeno One 50(1):1–8

Bortolot M, Buffoni B, Mazzarino S, Hoff G, Martino E, Fiorilli V, Salvioli Di Fossalunga A (2024) The importance of mycorrhizal fungi and their associated bacteria in promoting crops' performance: an applicative perspective. Horticulturae 10(12):1326–1352

Boyno G, Demir S, Rezaee Danesh Y, Durak ED, Çevik R, Farda B, Djebaili R, Pellegrini M (2023) A new technique for the extraction of arbuscular mycorrhizae fungal spores from rhizosphere. J Fungi 9(8):845–852

Bravo A, York T, Pumplin N, Mueller LA, Harrison MJ (2016) Genes conserved for arbuscular mycorrhizal symbiosis identified through phylogenomics. Nat Plants 2(2):1–6

Brundrett MC, Tedersoo L (2018) Evolutionary history of mycorrhizal symbioses and global host plant diversity. New Phytol 220(4):1108–1115

Cairney JWG, Meharg AA (2003) Ericoid mycorrhiza: a partnership that exploits harsh edaphic conditions. Eur J Soil Sci 54(4):735–740

Chakraborty K, Roy Das A, Saha AK, Das P (2019) Mycorrhiza in plants of different lifeforms. J Bot Soc Bengal 73:35–41

Chase MW, Cameron KM, Freudenstein JV, Pridgeon AM, Salazar G, van den Berg C, Schuiteman A (2015) An updated classification of Orchidaceae. Bot J Linn Soc 177(2):151–174

Chen A, Gu M, Wang S, Chen J, Xu G (2018) Transport properties and regulatory roles of nitrogen in arbuscular mycorrhizal symbiosis. Semin Cell Dev Biol 74:80–88

Chen L, Zhang X, Li Q, Yang X, Huang Y, Zhang B, Ye L, Li X (2024) Phosphatases decoding the role of mycorrhizal fungi in plant disease resistance. Int J Mol Sci 25(17):9491–9503

Chot E, Reddy MS (2022) Role of ectomycorrhizal symbiosis behind the host plants ameliorated tolerance against heavy metal stress. Front Microbiol 13:855473

Corona Ramírez A, Symanczik S, Gallusser T, Bodenhausen N (2023) Quantification of arbuscular mycorrhizal fungi root colonization in wheat, tomato, and leek using absolute qPCR. Mycorrhiza 33(5–6):387–397

Das D, Torabi S, Chapman P, Gutjahr C (2020) A flexible, low-cost hydroponic co-cultivation system for studying arbuscular mycorrhiza symbiosis. Front Plant Sci 11:63

Davison J, Moora M, Öpik M, Adholeya A, Ainsaar L, Bâ A, Burla S, Diedhiou AG, Hiiesalu I, Jairus T, Johnson NC, Kane A, Koorem K, Kochar M, Ndiaye C, Pärtel M, Reier Ü, Saks Ü, Singh R, Zobel M (2015) Global assessment of arbuscular mycorrhizal fungus diversity reveals very low endemism. Science 349(6251):970–973

Delaeter M, Magnin-Robert M, Randoux B, Lounès-Hadj Sahraoui A (2024) Arbuscular mycorrhizal fungi as biostimulant and biocontrol agents: a review. Microorganisms 12(7):1281–1309

Delavaux CS, Ramosv RJ, Stürmer SL, Bever JD (2024) An updated LSU database and pipeline for environmental DNA identification of arbuscular mycorrhizal fungi. Mycorrhiza 34(4):369–373

Djotan AKG, Matsushita N, Fukuda K (2022) Amplicon sequencing reveals the arbuscular mycorrhizal fungal community composition in *Cryptomeria japonica* at one local site. J For Res 27(5):399–407

Domingo G, Vannini C, Bracale M, Bonfante P (2023) Proteomics as a tool to decipher plant responses in arbuscular mycorrhizal interactions: a meta-analysis. Proteomics 23(6):2200108

Gadkar V, Driver JD, Rillig MC (2006) A novel in vitro cultivation system to produce and isolate soluble factors released from hyphae of arbuscular mycorrhizal fungi. Biotechnol Lett 28(14):1071–1076

Gao Y, Huang S, Wang Y, Lin H, Pan Z, Zhang S, Zhang J, Wang W, Cheng S, Chen Y (2023) Analysis of the molecular and biochemical mechanisms involved in the symbiotic relationship between arbuscular mycorrhiza fungi and Manihot esculenta Crantz. Front Plant Sci 14:1130924–1130940

Ge S, Zhang Z, Hu Q, Wang Q, Gong X, Huang F, Zhang L, Han W, Luo F, Li X (2025) Metabolomics analysis reveals crucial effects of arbuscular mycorrhizal fungi on the metabolism of quality compounds in shoots and roots of Camellia sinensis L. Plant Physiol Biochem 219:109426–109436

Gou X, Hu Y, Ni H, Wang X, Qiu L, Chang X, Shao M, Wei G, Wei X (2024) Arbuscular mycorrhizal fungi alleviate erosional soil nitrogen loss by regulating nitrogen cycling genes and enzymes in experimental agro-ecosystems. Sci Total Environ 906(1):167425–167435

HaiRu J, Xiangy J (2012) Chromatographic analysis of nitrogen utilization and transport in arbuscular mycorrhizal fungal symbiosis. In: Chromatography - The most versatile method of chemical analysis, vol 24, pp 231–251

Hawksworth DL, Lücking R (2017) Fungal diversity revisited: 2.2 to 3.8 million species. Microbiol Spectr 5(4):1110–1128

Higo M, Tatewaki Y, Iida K, Yokota K, Isobe K (2020) Amplicon sequencing analysis of arbuscular mycorrhizal fungal communities colonizing maize roots in different cover cropping and tillage systems. Sci Rep 10(1):6039–6042

Hildebrandt U, Ouziad F, Marner FJ, Bothe H (2006) The bacterium Paenibacillus validus stimulates growth of the arbuscular mycorrhizal fungus Glomus intraradices up to the formation of fertile spores. FEMS Microbiol Lett 254:258–267

Ijdo M, Cranenbrouck S, Declerck S (2011) Methods for large-scale production of AM fungi: past, present, and future. Mycorrhiza 21(1):1–16

Jiang XR, Pan JF, Zhao M, Guo XY, Wang Q, Zhang L, Liu W (2024) Transcriptome analysis reveals the impact of arbuscular mycorrhizal symbiosis on Toona ciliata var. pubescens seedlings. Forests 15(4):673–687

Kaur T, Reddy MS (2023) Diversity of arbuscular mycorrhizal fungi in seleniferous soils and their role in plant growth promotion. Biotech 13(11):369–384

Kaur S, Campbell BJ, Suseela V (2022) Root metabolome of plant–arbuscular mycorrhizal symbiosis mirrors the mutualistic or parasitic mycorrhizal phenotype. New Phytol 234(2):672–687

Krüger M, Krüger C, Walker C, Stockinger H, Schüßler A (2012) Phylogenetic reference data for systematics and phylotaxonomy of arbuscular mycorrhizal fungi from phylum to species level. New Phytol 193(4):970–984

Ksanbok M, Highland K (2019) Diversity of arbuscular mycorrhizal fungi in trap cultures prepared from abandoned coalmine overburden spoils. J Pure Appl Microbiol 13(1):629–636

Laatikainen T, Heinonen-Tanski H (2002) Mycorrhizal growth in pure cultures in the presence of pesticides. Microbiol Res 157(2):127–137

Liu A, Hamel C, Hamilton RI, Ma B, Smith DL (2000) Acquisition of phosphorus and nitrogen in field-grown wheat cultivars inoculated with mycorrhizal fungi. Agron J 92(5): 983-990

Lu Y, Yan Y, Qin J, Ou L, Yang X, Liu F, Xu Y (2023) Arbuscular mycorrhizal fungi enhance phosphate uptake and alter bacterial communities in maize rhizosphere soil. Front Plant Sci 14:1206870–1206883

Luginbuehl LH, Menard GN, Kurup S, Van Erp H, Radhakrishnan GV, Breakspear A, Oldroyd GED, Eastmond PJ (2017) Fatty acids in arbuscular mycorrhizal fungi are synthesized by the host plant. Science 356(6343):1175–1178

Mahmoudi N, Cruz C, Mahdhi M, Mars M, Caeiro MF (2019) Arbuscular mycorrhizal fungi in soil, roots and rhizosphere of *Medicago truncatula*: diversity and heterogeneity under semi-arid conditions. PeerJ 7:6401–6419

Manga AGB, Sene G, Diatta AA, Diop TA, Barroso G, Tuinen DV (2025) Molecular diversity of arbuscular mycorrhizal fungi associated with the rhizosphere of Vachellia seyal Del. From selected saline soils in Senegal. Microbiol Res 16(1):19–37

Matsubara T, Ishii T (2015) Development of preparative chromatography for proteomic approach of mycorrhizal symbiosis. Adv Biol Chem 5(1):16–23

Mei L, Li A, Xu L, Lin Y, Yang X, Liu H, Zhao H (2024) The effects of warming and nitrogen application on the stoichiometric characteristics of arbuscular mycorrhizal fungi in forest ecosystems. Forests 15(12):2121–2136

Miyauchi S, Kiss E, Kuo A, Drula E, Kohler A, Sánchez-García M, Morin E, Andreopoulos B, Barry KW, Bonito G, Buée M, Carver A, Chen C, Cichocki N, Clum A, Culley D, Crous PW, Fauchery L, Girlanda M, Martin FM (2020) Large-scale genome sequencing of mycorrhizal fungi provides insights into the early evolution of symbiotic traits. Nat Commun 11(1):5125–5142

Olalde-Portugal V, Cabrera-Ponce JL, Gastelum-Arellanez A, Guerrero-Rangel A, Winkler R, Valdés-Rodríguez S (2020) Proteomic analysis and interactions network in leaves of mycorrhizal and nonmycorrhizal sorghum plants under water deficit. Peer J 8:8991–9011

Othman YA, Alananbeh KM, Tahat MM (2024) Can AMF enhance crop productivity and quality in hydroponics? A Meta-Analysis Sustainability 16(9):3662–3683

Palenzuela J, Barea J-M, Ferrol N, Oehl F (2011) Ambispora granatensis, a new arbuscular mycorrhizal fungus associated with Asparagus officinalis in Andalucía (Spain). Mycologia 103(2): 333–340

Patil P, Solanki R (2016) β-1,3-Glucanase and chitinase enzyme activity assay: A potential approach for plant defense against fungi. J Plant Biochem Biotech 25(3):301–304

Perotto S, Girlanda M, Martino E (2002) Ericoid mycorrhizal fungi: some new perspectives on old acquaintances. Plant Soil 244:41–53

Pozo MJ, Azcón-Aguilar C, Dumas-Gaudot E, Barea JM (1999) β-1,3-Glucanase activities in tomato roots inoculated with arbuscular mycorrhizal fungi and/or Phytophthora parasitica and their possible involvement in bioprotection. Plant Sci 141(2):149–157

Praveen R, Ashwin R, Bagyaraj D (2023) Quantifying microbial communities in pot cultures of arbuscular mycorrhizal fungi: a comprehensive analysis. J Soil Biol Ecol 43(2):63–69

Prettl N, Biró B, Nugroho PA, Kotroczó Z, Kabalan S, Kovács F, Papdi E, Juhos K (2024) Limited effect of mycorrhizal inoculation depending on soil type and fertilization level in a Central European field trial. Plant Growth Regul 104(4):1669–1681

Raklami A, Bechtaoui N, Tahiri A, Anli M, Meddich A, Oufdou K (2019) Use of rhizobacteria and mycorrhizae consortium in the open field as a strategy for improving crop nutrition, productivity and soil fertility. Front Microbiol 10:1106–1117

Ranner J, Stabl G, Piller A, Paries M, Sharma S, Zeng T, Spaccasassi A, Stark TD, Gutjahr C, Dawid C (2022) Molecular assessment of metabolome alterations in Lotus japonicus roots induced by arbuscular mycorrhiza. Land Degrad Dev 34(10): 1241–1254

Rasmussen HN (2002) Recent developments in the study of orchid mycorrhiza. Plant Soil 244:149–163

Ren Y, Guo M, Ding G, Wang Y (2022) Ectomycorrhizal fungi associated with Pinus sylvestris var. Mongolica were altered by soil environments with aging plantation in a Semi-arid Desert. Front Environ Sci 10:858452–858464

Sarathambal C, Srinivasan V, Jeevalatha A, Sivaranjani R, Alagupalamuthirsolai M, Peeran MF, Mukesh Sankar S, George P, Dilkush F (2024) Unravelling the synergistic effects of arbuscular mycorrhizal fungi and vermicompost on improving plant growth, nutrient absorption, and secondary metabolite production in ginger (*Zingiber officinale* Rosc.). Front Sustain Food Syst 25(8):1412610–1412625

Schoch CL, Seifert KA, Huhndorf S, Robert V, Spouge JL, Levesque CA, Chen W, Bolchacova E, Voigt K, Crous PW, Miller AN, Wingfield MJ, Aime MC, An K-D, Bai F-Y, Barreto RW, Begerow D, Bergeron M-J, Blackwell M, Schindel D (2012) Nuclear ribosomal internal transcribed spacer (ITS) region as a universal DNA barcode marker for fungi. Proc Natl Acad Sci 109(16):6241–6246

Seeliger M, Hilton S, Muscatt G, Walker C, Bass D, Albornoz F, Standish RJ, Gray ND, Mercy L, Rempelos L, Schneider C, Ryan MH, Bilsborrow PE, Bending GD (2024) New fungal primers reveal the diversity of Mucoromycotinian arbuscular mycorrhizal fungi and their response to nitrogen application. Environ Microbiome 19(1):71–86

Sieh D, Watanabe M, Devers EA, Brueckner F, Hoefgen R, Krajinski F (2013) The arbuscular mycorrhizal symbiosis influences sulfur starvation responses of *Medicago truncatula*. New Phytol 197(2):606–616

Singavarapu B, Beugnon R, Bruelheide H, Cesarz S, Du J, Eisenhauer N, Guo LD, Nawaz A, Wang Y, Xue K, Wubet T (2022) Tree mycorrhizal type identity and diversity shape forest root-associated mycobiomes. Environ Microbiol 24(9):4236–4255

Smith SE, Smith FA (2011) Roles of arbuscular mycorrhizas in plant nutrition and growth: new paradigms from cellular to ecosystem scales. Annu Rev Plant Biol 62(1):227–250

Somoza SC, Bonfante P, Giovannetti M (2024) Breaking barriers: improving time and space resolution of arbuscular mycorrhizal symbiosis with single-cell sequencing approaches. Biol Direct 19(1):67–79

Srinivasan M, Kumar K, Kumutha K, Marimuthu P (2014) Establishing monoxenic culture of arbuscular mycorrhizal fungus *Glomus intraradices* through root organ culture. J Appl Nat Sci 6:290–293

Stockinger H, Krüger M, Schüßler A (2010) DNA barcoding of arbuscular mycorrhizal fungi. New Phytol 187(2):461–474

Stürmer SL, Bever JD, Schultz PA, Bentivenga SP (2021) Celebrating INVAM: 35 years of the largest living culture collection of arbuscular mycorrhizal fungi. Mycorrhiza 1:117–126

Tanaka S, Hashimoto K, Kobayashi Y (2022) Asymbiotic mass production of the arbuscular mycorrhizal fungus *Rhizophagus clarus*. Commun Biol 5:43–51

Tedersoo L, Smith ME (2013) Lineages of Ectomycorrhizal Fungi and Their Evolution. New Phytol 200(4): 1106–1127

Tedersoo L, May TW, Smith ME (2010) Ectomycorrhizal lifestyle in fungi: global diversity, distribution, and evolution of phylogenetic lineages. Mycorrhiza 20(4): 217–263

Tenzin UW, Noirungsee N, Runsaeng P, Noppradit P, Klinnawee L (2022) Dry-season soil and co-cultivated host plants enhanced propagation of arbuscular mycorrhizal fungal spores from sand dune vegetation in trap culture. J Fungi 8(10):1061–1079

Teste FP, Jones MD, Dickie IA (2020) Dual-mycorrhizal plants: their ecology and relevance. New Phytol 225(5):1835–1851

Trejo-Aguilar D, Banuelos J (2020) Isolation and culture of AMF from field samples. Methods Mol Biol 2146:1–18

Tsakirpaloglou N, Septiningsih EM, Thomson MJ (2023) Guidelines for performing CRISPR/Cas9 genome editing for gene validation and trait improvement in crops. Plants 12(20):3564–3582

Větrovský T, Kolaříková Z, Lepinay C, Awokunle Holla S, Davison J, Fleyberková A, Gromyko A, Jelínková B, Kolařík M, Krüger M, Lejsková R (2023) Global AM fungi: a global database of arbuscular mycorrhizal fungal occurrences from high-throughput sequencing metabarcoding studies. New Phytol 240(5):2151–2163

Wang Y-J, Wu Q-S (2023) Influence of sugar metabolism on the dialogue between arbuscular mycorrhizal fungi and plants. Horticult Adv 1(1):2–14

Wang Z, Lian J, Liang J, Wei H, Chen H, Hu W, Tang M (2024) Arbuscular mycorrhizal symbiosis modulates nitrogen uptake and assimilation to enhance drought tolerance of *Populus cathayana*. Plant Physiol Biochem 210:108648–108658

Watts-Williams SJ, Cavagnaro TR (2020) Hormonal regulation of arbuscular mycorrhizal symbioses: recent discoveries and future directions. New Phytol 225:2192–2197

Wilkes TI (2023) Alternative blue inks for staining arbuscular mycorrhizal fungal structures in plant roots. Appl Plant Sci 11(8):e11510

Willing CE, Wan J, Yeam JJ, Cessna AM, Peay KG (2024) Arbuscular mycorrhizal fungi equalize differences in plant fitness and facilitate plant species coexistence through niche differentiation. Nat Ecol Evol 8(11):2058–2071

Wu S, Fu W, Rillig MC, Chen B, Zhu YG (2024) Soil organic matter dynamics mediated by arbuscular mycorrhizal fungi: An updated conceptual framework. New Phytol, 242(4): 1417–1425

Yagame T, Orihara T, Selosse M, Yamato M, Iwase K (2012) Mixotrophy of *Platanthera minor*, an orchid associated with ectomycorrhiza-forming *Ceratobasidiaceae* fungi. New Phytol 193(1):178–187

Yang H, Zhao X, Li L, Zhang J (2020) Detecting the colonization of ericoid mycorrhizal fungi in *Vaccinium uliginosum* using in situ polymerase chain reaction and green fluorescent protein. Plant Methods 16:1–8

Yang M, Song Y, Ma H, Li Z, Ding J, Yin T, Niu K, Sun S, Qi J, Lu G, Fazal A (2024) Unveiling the hidden world: how AMF and its regulated core fungi modify the composition and metabolism of soybean rhizosphere microbiome. Environ Microbiome 9(1):78–96

Yu H, Ji C, Zheng Z, Yu M, Liu Y, Xiao S, Pan Z (2023) Comparative proteomic analysis identifies proteins associated with arbuscular mycorrhizal symbiosis in *Poncirus trifoliata*. Front Plant Sci 14:1294086–1294102

Zhang H, Qi H, Lu G, Zhou X, Wang J, Li J, Zheng K, Fan Y, Zhou H, Wang J, Wu C (2023) Non-targeted metabolomics analysis reveals the mechanism of arbuscular mycorrhizal symbiosis regulating the cold-resistance of *Elymus nutans*. Front Microbiol 14:1134585–1113498

Zhang Y, Liu L, Wang Y (2025) Arbuscular mycorrhizal fungal communities and promoting the growth of alfalfa under salt stress. Front Plant Sci 15: 1438771

Role of Mycorrhiza and Its Application to Crop Production: A Path Ahead for Sustainable Agriculture

Sougata Roy, Subrata Bag, Priyanka Saha, Piyali Sarkar, Devendra Kumar Dadhich, Abhishekh Patidar, Anamika Barman, and Suman Dutta

Abstract

The twentieth century saw significant advancements in agricultural production driven by pesticides, fertilizers, and genetic improvements. However, the excessive use of chemical fertilizers has reduced the natural presence and benefits of arbuscular mycorrhizal fungi (AMF) in agricultural fields. AMF form a symbiotic relationship with approximately 80% of terrestrial plants, significantly enhancing nutrient acquisition, particularly phosphorus, and providing protection against pathogens. This relationship is facilitated by the growth of mycelium within the plant roots (intraradical) and extending out into the soil (extraradical).

S. Roy (✉)
International Potato Centre (CGIAR-CIP), Patna, Bihar, India

S. Bag
ICAR-Central Research Institute for Dryland Agriculture, Hyderabad, Telangana, India

P. Saha
ICAR-Research Complex for North Eastern Hilly Region, ICAR-KVK West Garo Hills, Barapani, Meghalaya, India

P. Sarkar
Bidhan Chandra Krishi Viswavidyalaya, Mohanpur, West Bengal, India

D. K. Dadhich
ICAR-National Dairy Research Institute, Karnal, Haryana, India

A. Patidar
Institute of Agricultural Sciences, Banaras Hindu University, Varanasi, Uttar Pradesh, India

A. Barman
ICAR-Indian Agricultural Research Institute, New Delhi, India

S. Dutta
Ramkrishna Mission Vivekananda Educational and Research Institute, Kolkata, West Bengal, India

Y. Siddiqui et al. (eds.), *Sustainable Mycorrhizal Cultivation*,
https://doi.org/10.1007/978-981-95-1778-7_4

Studies show that AMF have ability to increase crop growth by up to 30% and phosphorus uptake by 20–70% compared to non-mycorrhizal plants. There are seven types of mycorrhizae, but AMF and ectomycorrhizae (ECM) are the most abundant and widely studied. AMF are especially efficient in nutrient-poor environments where nutrients are tightly bound to soil particles. They play a critical role in disease suppression, heavy metal stress alleviation, and increased grain production, thereby enhancing sustainable agricultural practices. For example, the co-inoculation of AMF with beneficial saprophytic fungi has been discovered to improve plant growth and root colonization and enhance plant resilience against diseases. This synergy helps to transform disturbed ecosystems into productive land, emphasizing the importance of reintroducing mycorrhizal systems for soil health restoration and bioremediation of contaminated soils. Recent research has highlighted the molecular mechanisms underlying nutrient exchange and symbiotic signalling between AMF and host plants. The fungus releases 'Myc factors', whereas plant roots release strigolactones, triggering mutualistic symbiosis. Understanding these processes offers a pathway for leveraging these fungi for crop production and crop improvement, supporting the shift toward sustainable agriculture under diverse environmental conditions.

Keywords

Mycorrhiza · Nutrient uptake · Soil health · Stresses · Sustainable agriculture

1 Introduction

The global population is expected to reach 9.7 billion by 2050, which will significantly increase the demand for food and water. To satisfy this demand, food production must increase by 70% by 2050 (Poveda 2021). Intensive farming practices and excessive use of mineral fertilizers, agrochemicals, and water have led to land degradation, environmental pollution, and the depletion of natural resources (Roy et al. 2023). Sustainable agriculture aims to produce food with minimal impact on the environment and reduce the residual effects of chemicals in the food chain (Suman et al. 2022). Soil health is essential for plants to produce food, fibre and fuel to meet the needs of the growing global population. Over the past 50 years, cultivated land has expanded by over 500%, fertilizer use has increased by 700%, and pesticide use has risen significantly (Banerjee et al. 2018). However, a decline in soil organic carbon (SOC) and the production of low biomass in crop fields pose major challenges to sustainable agriculture globally. The conversion of natural ecosystems to agricultural systems has resulted in the loss of 90–135 Pg of soil carbon due to intensive tillage, residue removal or burning, and soil degradation from erosion and other processes (Lal 2018; Paustian et al. 2016). The soil carbon pool and its dynamics play crucial roles in the global carbon cycle and atmospheric chemistry by regulating greenhouse gas emissions (Ladha et al. 2016). Soil microbes provide vital ecosystem services, such as cycling carbon and other nutrients, including nitrogen, phosphorus and support plant growth. Organic and natural soil fertility management methods primarily rely on the application of organic fertilizers and beneficial microbial communities to improve soil health (Silveira et al. 2021). The effectiveness of nutrient release from organic fertilizers is influenced by the type of fertilizer used, soil physicochemical properties, and existing microbial communities (Ye et al. 2020). However, widespread adoption is limited by the high cost and low availability of organic fertilizers, which prevents farmers from meeting the full nutrient demands of their crops (Bergstrand 2022). Additionally, in some cases, organic management can result in negative environmental impacts such as nitrate leaching into groundwater, soil acidification, and denitrification, contributing to soil fertility loss and degradation.

To address these challenges, continuous monitoring of soil microflora is crucial to understand their roles in nutrient cycling and soil fertility (Reyes-Sanchez et al. 2022). The scientific management of these microbial communities, particularly through the use of biofertilizers, can offer a sustainable alternative to chemical fertilizers (Mrunalini et al. 2022). Biofertilizers enriched with crop-friendly microorganisms are essential for the evolution of organic agriculture (Yang et al. 2021). Among the microbial communities in soil, arbuscular mycorrhizal fungi (AMF) play significant roles in promoting soil fertility and enhancing crop productivity (Giovannini et al. 2020). AMF improve nutrient and water uptake in plants, especially in nutrient-poor soils (Kobae 2019), and positively influence other soil microbiota (Bhale et al. 2018).

Organic farms tend to support higher diversity and abundance of AMF than chemically managed farms, which often lack key AMF species (Jiang et al. 2021). These fungi contribute to the improvement of soil structure, increase plant tolerance to heavy metals, and enhance disease resistance through microbial-assisted immune pathways (Riaz et al. 2021). As such, AMF are integral to the sustainability of organic farming systems and long-term health of agroecosystems (Kuila and Ghosh 2022).

2 Types of Mycorrhizae and Their Characteristics

Based on their structure and nature of root colonization, mycorrhizae are broadly classified into two types.

2.1 Arbuscular Mycorrhizal (AM)

The ubiquitous soil fungi known as AMF can form symbioses with over 80% of terrestrial plants. They can help host plants acquire nutrients, such as the macronutrients nitrogen (N) and phosphorus (P), in exchange for carbohydrates and lipids from the plants (Bravo et al. 2017; Jiang et al. 2017; Luginbuehl et al. 2017). AMF create intracellular structures in the root cortex of host plants, called arbuscules that help the plant and fungus exchange nutrients (Smith and Read 2010). According to numerous reports (Tisdall 1991; Rillig 2004), AMF produce organic binding agents that aid the formation, stabilization, and aggregate growth of soil organic matter (SOM). Through enhanced nutrient acquisition, including phosphorus, water, and mineral uptake, AM promotes plant development. AM increases a plant's ability to withstand and bounce back from abiotic stresses, such as salt, drought, and heavy metal toxicity. The arbuscular mycorrhizal interface, which permits the flow of nutrients, signalling molecules, and protective chemical substances between fungal and plant partners, is responsible for these advantages. AM infestation also affects osmotic adjustment, hormone control, and antioxidant defence systems of plants. Under abiotic stress, these reactions enhance the biomass output, photosynthetic efficiency, and plant performance. AM helps in maintaining resilient ecosystems because of its beneficial effects on carbon sequestration, nitrogen cycling, and soil structure. AMFs have different effects on different species and environments in terms of plant growth and ecological stability (Wahab et al. 2023).

2.2 Ectomycorrhiza

Ectomycorrhizae are symbiotic relationships that occur between soil fungi and plant roots, primarily those of woody species such as birch, oak, and conifers. Ectomycorrhizal fungi surround plant roots in a sheath or mantle without entering the root cells, in contrast to other mycorrhizal associations. Rather, they create a 'Hartig net' between the cortical root cells and epidermal cells, which allows the fungus and host plant to exchange nutrients. The health and productivity of temperate forests and boreal regions depend heavily on ectomycorrhizal relationships, which are particularly common in these areas (Smith and Read 2010). They may obtain nutrients that plant roots alone cannot obtain, especially in nutrient-poor soils, by increasing the root surface area via their hyphal networks (Kleinert et al. 2018). Better soil structure and water retention are encouraged by the hyphae of ectomycorrhizal fungi, which bind soil particles and improve soil aggregation and stability. According to Leifheit et al. (2014), this structural enhancement is essential to reduce erosion, increase soil porosity, and promote water penetration. Additionally, these fungi cause host plants to produce specific defence-related substances, which increase the resistance of plants to root infections. Ectomycorrhizal fungi improve the stability of soil organic matter through their vast hyphal networks, which greatly aid in soil carbon sequestration. They aid in the cycling and storage of carbon by contributing organic materials to the soil through their interactions with root exudates and litter breakdown (Clemmensen et al. 2013). Because they improve soil health, increase plant growth, improve nutrient and water acquisition, and provide resilience against environmental stressors, ectomycorrhizal fungi are vital components of forest ecosystems.

3 Symbiotic Relationships and Mechanisms

Mycorrhizal symbiosis, the most important biological interactions (Genre et al. 2020), is formed by mycorrhizal fungi in the soil and the roots of over 90% of plant species, from liverworts to angiosperms (Smith and Read 2010). Although they can spend part of their life cycle in the rhizosphere, mycorrhizal fungi are always associated with the roots of plants, including forest trees, wild grasses, and many crops, and colonize environments, such as alpine and boreal zones, tropical forests, grasslands, and croplands. These fungi improve the nutrient status of host plants and influence mineral nutrition, water absorption, growth and disease resistance. In contrast, the host plants are necessary for fungal growth and reproduction (Khan et al. 2022; Smith and Read 2010).

Most terrestrial plants engage in mutually beneficial symbiosis with AMF (Garg and Chandel 2011). AM symbiosis is the most common mutualistic relationship between fungi and plants (Qiang-Sheng et al. 2016). The operation of low-input settings depends on AMF, which are found in association with all plant phyla, including those that are relevant to agriculture. This relationship is initiated through chemical signalling, where both partners communicate via diffusible molecules such as lipo-chitooligosaccharides from fungi and strigolactones from plants (Gao et al. 2023; Choi et al. 2018). These signals, known as 'Myc factors', trigger a series of morphological and physiological changes, including increased root branching and enhanced nutrient uptake capabilities of the host plant (Garg and Chandel 2011; Gao et al. 2023). Furthermore, the development of arbuscular mycorrhizal symbiosis is regulated by various phytohormones including strigolactones, gibberellic acids and auxins (Liao et al. 2018). Strigolactones are plant-derived signalling compounds that facilitate fungal recognition and colonization. They not only attract AMF but also play a major role in regulating plant growth and development under nutrient-deficient conditions. Auxin, on the other hand, is required for both the early steps of fungal growth and the differentiation of arbuscules, the fungal structures within the plant root cells (Liao et al. 2018).

The primary function of mycorrhizal associations is bidirectional exchange of nutrients. Mycorrhizal fungi promote the absorption and utilization of water and mineral elements (Qiu et al. 2022) and significantly reduce soil N and P losses (Parniske 2008) and affects the mineralization of N and P (Hodge et al. 2001; Jing et al. 2022). These fungi are particularly effective in mobilizing phosphorus from soil particles because of their extensive hyphal networks that extend beyond the depletion zones around plant roots (Wang et al. 2022). In return for phosphorus, plants supply fungi with carbohydrates derived from photosynthesis (Salmeron-Santiago et al. 2021). This exchange is vital for plant growth, especially in nutrient-poor soils where phosphorus availability is limited (Rui et al. 2022). In addition to phosphorus, mycorrhizal fungi facilitate nitrogen uptake (Bücking and Kafle 2015). They can access forms of organic nitrogen that are typically unavailable to plants directly (Hodge and Storer 2015). This capability is especially critical in ecosystems where nitrogen is a limiting factor for plant growth. Studies have shown that plants associated with AMF exhibit improved growth rates and resilience against environmental stresses because of their enhanced nutrient acquisition capabilities (Wahab et al. 2023; Bahadur et al. 2019; Thangavel et al. 2022). Furthermore, mycorrhizae contribute to other nutrient dynamics by enhancing the availability of micronutrients such as zinc and copper (Ruytinx et al. 2020; Lambert et al. 1979; Moreno Jiménez et al. 2024). The physical presence of fungal hyphae not only expands the root surface area but also improves soil structure, promoting better water retention and aeration factors that collectively enhance nutrient availability for plants.

The symbiotic relationships between plants and mycorrhizal fungi represent a complex interplay between biochemical signalling, structural adaptations, and nutrient exchange processes. Understanding these interactions is essential for improving agricultural practices and ecosystem management strategies to enhance plant health and productivity. Future research should continue to explore these mechanisms at the molecular level to provide further insight into optimizing these beneficial relationships in various ecological contexts.

4 Benefits of Mycorrhizae in Crop Production

Mycorrhizal fungi enhance the ability of plants to absorb nutrients and water from the soil, boosting their resilience to various environmental stressors. Additionally, these fungi contribute significantly to soil aggregation and promote microbial activity. Depending on the plant species, cultivation practices, and environmental conditions, mycorrhizae offer diverse advantages to plants and their ecosystems.

- Promote the growth of stronger, healthier plants.
- Plant establishment percentage and survival rate of seedlings enhanced.
- Increase yield and quality of produce.
- Enhance flowering and fruiting.
- Improve drought and salinity tolerance.
- Optimize fertilizers use, especially phosphorus.
- Maintain soil quality and nutrient cycle.
- Disease tolerance.

Several studies have reported that fungal symbiosis improves resilience to a range of stressors, such as temperature, metals, herbivory, drought, salt, and diseases (Ahanger et al. 2013; Abdel-Salam et al. 2018). Approximately 90% of plant species, including ferns, bryophytes, and flowering plants, can establish interdependent interactions with AMF (Zhu et al. 2010; Ahanger et al. 2013). The prominent benefits of mycorrhizal fungi in terms of nutrient uptake, disease suppression and abiotic stress tolerance are discussed.

4.1 Enhancing Nutrient Uptake and Soil Health

The advantages of colonizing roots with AMF include enhanced uptake of copper and zinc (Li et al. 1991; Azaizeh et al. 1995), as well as nitrogen, phosphorus, potassium, and manganese (Smith and Read 2010; Seleiman et al. 2013). Up to 20% of the organic carbon required for the development and operation of AMF is supplied by host plants (Smith and Read 2010; Smith and Smith 2011). AMF can improve the efficiency with which plants absorb and use phosphorus (Koide et al. 2000), which can lead to an increase in the growth and yield of plant species (Smith et al. 2011).

Plants absorb phosphorus from the rhizosphere through direct and indirect pathways. Phosphorus is directly absorbed by plant roots via the transporters in the rhizosphere (root zones) in the direct pathway. In the indirect or mycorrhizal pathway, fungal phosphorus transporters (behind the root apex) can take phosphorus and transfer it into hyphae, which can subsequently translocate it into arbuscules. Phosphorus is then readily translocated from the interfacial apoplast to plant cortical cells via plant phosphorus transporters (Seleiman and Hardan 2021). Mycorrhizae have the ability to absorb nitrogen from various organic sources, convert it into mineral nitrogen, and then deliver it to the host plant through its roots (Hodge et al. 2001). Nitrogen in the form of amino acids (Hawkins et al. 2000) and NH_4^+ and NO_3^- (Johansen et al. 1996; Smith and Read 2010) can be taken up by hyphae from their environment and transferred into the roots of their host plants.

4.2 Disease Suppression and Plant Protection

One of the upcoming advancements in plant pathology is biological control technology, which holds great promise. This has gained considerable attention in recent years. Mycorrhiza is one of the potential agents of pest and disease control in plant. AMF-mediated biocontrol across a range of crops, diseases, and modes of action has recently been studied. Mycorrhizal symbiosis creates a microbial community that is governed by multitrophic interactions. AMF provide effective biocontrol by competing with rhizospheric pathogens for root space and host carbohydrates (Singh et al. 2020).

As bioprotective agents, AMF suppress fungal pathogens such as *Alternaria* (Song et al. 2015), *Fusarium* (Pozo et al. 2010), *Colletotrichum* (Saldajeno and Hyakumachi 2011), and *Phytophthora* (Gallou et al. 2011). For example, AMF prevent *Sclerotium rolfsii* in groundnut, and preinoculation with *Funneliformis mosseae* increased tomato resistance to *Alternaria*-caused early blight (Song et al. 2015). AMF increase plant resistance to nematodes, especially *Meloidogyne* species, with effectiveness varying depending on the AMF and host species (Da Silva Campos 2020). In *Coffea arabica*, mycorrhizal inoculation with *Meloidogyne* spp. reduced root-knot nematode infection by 52.5%, 38.5%, and 38.3%, respectively, confirming protective effects of AMF (Vallejos-Torres et al. 2021). AMF have shown significant protective effects against various plant pathogens including bacteria, phytoplasma and actinomycetes. Research evidence highlights several mechanisms and outcomes related to role of AMF in plant pathogen defence. AMF protect plants against pathogens, such as *Xanthomonas* and *Pseudomonas syringae*. Studies have shown that *Rhizophagus irregularis* reduces necrotic lesions in alfalfa and tomato, whereas *F. mosseae* suppresses pathogen density in soybean and chrysanthemum, thereby enhancing plant resistance through mechanisms supported by omics data (Fiorilli et al. 2018). mycorrhizal inoculation mitigates bacterial wilt in tomatoes caused by *P. solanacearum*, a globally prevalent soil-borne pathogen. Additionally, Kamble et al. (2014) reported that phosphorus application and inoculation with *Glomus fasciculatum* or *F. mosseae* significantly reduced *P. syringae pv.* incidence in mulberry. AMF, plants, harmful bacteria, and other microbes interact in a variety of intricate ways and are affected by different materials and experimental setups. Accordingly, there is inconsistent evidence of AMF-induced disease resistance in plants among studies (Vos et al. 2013).

4.3 Stress Tolerance: Heavy Metal and Drought Resistance

Abiotic stresses, such as drought, salinity, heavy metals, and heat, significantly reduce crop productivity. Mycorrhizal fungi, especially AMF, improve plant tolerance to such conditions by enhancing nutrient and water uptake, improving water-use efficiency, and increasing drought resistance. For instance, the local bacteria-associated strain of drought-tolerant *Glomus intraradices* decreased water requirements for *Retama sphaerocarpa* by 42% (Marulanda et al. 2006).

Heavy metals are naturally found in the Earth's crust, but they are primarily present as trace elements due to paedogenic processes (Wuana and Okieimen 2011). Mechanisms of heavy metals exposure and drought tolerance.

4.3.1 Heavy Metal Tolerance

- *Metal Chelation and Immobilization*: Heavy metals (such as cadmium, lead and zinc) can be immobilized by mycorrhizal fungi within their hyphal structures, which lowers the uptake of metals in plant tissues. They generate substances such as glomalin, which can bind heavy metals in the soil and keep them from getting to the plant.
- *Antioxidant Production*: AMF increase the production of antioxidants in plants, which helps mitigate the oxidative damage caused by metal stress.
- *Enhanced Nutrient Uptake*: AMF counteract the imbalance of heavy metals by improving the uptake of essential nutrients such as phosphorus.

4.3.2 Drought Tolerance

- *Improved Water Uptake*: To access water in soil micropores that roots alone cannot reach, mycorrhizal fungi expanded hyphal networks and expanded the surface area available for water absorption.

- *Osmolyte Production*: Under drought stress conditions, mycorrhizal plants accumulate osmolytes such as proline, which helps maintain cell turgor and reduces dehydration.
- *Hormonal Modulation*: AMF have impact on plant hormones, such as abscisic acid (ABA) levels, which regulate stomata and reduce water loss during drought.

In one study, AMF species richness and root colonization capacity decreased as soil Cu concentration increased, according to greenhouse experiments on grapevine plantlets of the 1103 Paulsen rootstock (*Vitis berlandieri* × *Vitis rupestris*) exposed to progressive Cu doses (from 50 to 150 mg/kg) and a native AMF species mix (with *Glomus ambisporum* as the dominant species) (Betancur-Agudelo et al. 2023). Several studies have shown an improvement in the chlorophyll content in AMF-inoculated plants under different heavy metal stress conditions which signifies the ability of AMF to reduce the uptake of heavy metals (Crişan et al. 2024).

5 Mycorrhiza in Sustainable Agriculture

Mycorrhiza play several crucial roles in agriculture, particularly in organic farming. According to a review by George and Ray (2023), mycorrhiza have several advantages in organic agriculture which includes:

5.1 Soil Fertility and Health

Mycorrhizae increase the absorption of nutrients from the soil by plants, particularly phosphorus, nitrogen, and other micronutrients. This contributes to the soil fertility and overall health. In addition, they contribute to the structure of the soil by secreting glomalin-related soil proteins (GRSP), which are capable of assisting in the aggregation and stability of the soil. Additionally, AMF support microbial diversity and foster a healthy soil environment that inhibits pathogens, thus assisting in the preservation of balanced soil ecosystems. AMF boost long-term soil production and resilience via these methods, making them crucial to sustainable agriculture (George and Ray 2023)..

5.2 Crop Productivity

Mycorrhizae are responsible for increasing agricultural output because they encourage the development and yield of plants. They make it easier for plants to thrive in soils that are deficient in nutrients and increase the amount of water they take up, which is particularly beneficial during times of drought. In nutrient-deficient soils, AMF increase the surface area of crop roots via symbiotic partnerships, thereby improving plant nutrient uptake. This connection may result in increased biomass and better crop yields, particularly in organic and sustainable agricultural systems.

5.3 Disease Resistance

Mycorrhizae contribute to the development of a plant's resistance to diseases and pests by stimulating the plant's innate immune system and producing protective secondary metabolites. This is in addition to the fact that they physically hinder root pathogens. AMF significantly enhance plant disease resistance through a multifaceted set of strategies that modulate the immune system of the host plant. Despite extensive colonization of root tissues, AMF are well tolerated by the host because they actively suppress or modulate the plant's typical defense responses. This is primarily due to the release of fungal signalling molecules, such as Myc factors, which inhibit immune activation and facilitate symbiosis (Chen et al. 2018). More importantly, AMF induce systemic resistance in plants, particularly against soil-borne pathogens, by priming the plant defense system. This priming results in a faster and stronger activation of defense mechanisms upon subsequent pathogen attack. AMF stimulate both systemic acquired resistance (SAR), typically linked with salicylic acid signalling, and induced systemic resistance (ISR), associated with the jasmonic acid and ethylene pathways (Pozo and Azcón-Aguilar 2007; Jung et al. 2012). For example, Chen et al. (2018) demonstrated that in *Oryza sativa* (rice), AMF colonization upregulated defense-related genes, such as OsNPR1, OsAP2, OsEREBP, and OsJAmyb, as well as key signalling components, such as OsMPK6, OsCBP, and OsCaM, contributing to enhanced resistance against *Magnaporthe oryzae* (the rice blast fungus). This highlights the role of AMF in priming host immunity, thereby enhancing resistance without the energetic cost of constitutive defense expression.

5.4 Heavy Metal Tolerance

Mycorrhiza contribute to the capacity of plants to endure heavy metal stress by attaching heavy metals to hyphal walls or spores. This reduces the amount of metal available to plants and prevents them from becoming poisonous. According to Abdelhameed and Metwally (2019), AMF play a significant protective role in mitigating the detrimental effects of heavy metal stress, particularly cadmium (Cd) and mercury (Hg), in *Trigonella* species. This study highlights that AMF colonization enhances the ability of plants to tolerate and adapt to these toxic conditions through a multifaceted defense mechanism. A key strategy employed by AMF is the upregulation of antioxidant enzymes such as superoxide dismutase (SOD), catalase (CAT), and peroxidases (POD). These enzymes are part of the plant's intrinsic oxidative stress defense system, which neutralizes reactive oxygen species (ROS) generated under heavy metal exposure. Cd and Hg toxicity typically leads to excessive ROS production, which damages lipids, proteins, and nucleic acids, ultimately impairing plant growth and physiological function. By enhancing antioxidant enzyme activities, AMF help maintain the cellular redox balance, protect cellular structures, and mitigate oxidative damage.

5.5 Water Relations

Mycorrhiza helps to keep soil moisture levels higher and promotes root water absorption, both of which are vital for the survival of plants in situations that are water-deficient. Along with other organisms in an organic environment, AMF form a distinct biotic community that helps to explain the high concentration of microbial residues such as GRSP in soil. This helps to improve the general water regime of the soils by improving soil aggregate stability to a considerable extent (Sun et al. 2016).

5.6 Biodiversity and Ecosystem Health

Mycorrhiza improve the variety of soil microbes, which is beneficial for the stability of ecosystems and the overall health of the soil. This is due to the fact that Mycorrhiza increases biodiversity. Additionally, they contribute to the growth of beneficial bacteria found in the soil. For instance, Wang et al. (2023) investigated the impact of AMF on the rhizosphere bacterial community of tomato plants infected with the soil-borne pathogen, *Ralstonia solanacearum*. These results demonstrated that AMF inoculation significantly increased the diversity and stability of the bacterial community in the rhizosphere. Specifically, there was enrichment of beneficial bacterial genera, such as *Flavobacterium*, *Flavisolibacter*, *Cytophaga*, and *Bryobacter*. These changes were associated with reduced abundance of the pathogen and enhanced resistance of tomato plants to bacterial wilt.

6 Role of Mycorrhiza in Improving Soil Structure and Carbon Sequestration

Mycorrhizae directly contribute to the maintenance and structural integrity of natural soils. One of the components that enter the soil structure is glomalin, which is produced by the extraradical hyphae of AMF (Muneer et al. 2020). Globulin-related soil proteins (GRSPs) are a class of glycoproteins produced in soil by the hyphae and spore walls of AMF. Hydrophobic and thermostable glomalin is also produced when hyphae disintegrate and AMF spores are discharged into the soil. GRSP cements and shields soil organic matter and particles as surface substances that repel water and promote soil aggregation. According to Yang et al. (2017), organic carbon and soil organic matter generally as general cementing elements, in addition to helping with soil aggregation. AMF regulates plant root exudation, which enhances soil aggregate stability. Agnihotri et al. (2022) claimed that GRSP released into the soil by AMF may increase the amount of time that carbon stays in soils on average. This can help soils build water-stable aggregates that can improve the stability of organic matter, stop soil deterioration, and provide other benefits. According to experimental research, AMF dramatically boost the ability of soil macroaggregates to stabilize rhizo-carbon (Jeewani et al. 2021). Particularly in organic farms, AMF are inextricably related to C buildup and cycling in cultivated fields. Further experimental studies are required to completely comprehend the role of AMF and their participation in optimal C maintenance in agricultural areas.

6.1 Contribution to Ecosystem Services

Mycorrhizal fungi play a crucial role in various ecosystems, including grasslands, forests, deserts, and agroecosystems. AMF positively affect plant nutrition and fitness by helping plants uptake phosphorus from the soil and translocate water and other mineral nutrients. They also boost the effectiveness of phytoremediation and aid in the establishment of plants in challenging environments. It can directly influence the diversity and composition of the aboveground plant community, potentially promoting the seedling establishment of secondary plant species. They also play a critical role in soil aggregation, with glycoproteins secreted by AMF potentially contributing to carbon sequestration in soil ecosystems (Table 1). This potential capability has led to the recognition of the importance of AMF in climate-change mitigation processes (Berruti et al. 2014).

7 Application of Mycorrhiza in Modern Agricultural Practices

The application of AMF in modern agriculture offers a holistic approach for improving crop performance, soil health, and environmental sustainability. Their multifaceted benefits make them a valuable tool for the development of resilient and sustainable farming systems.

Table 1 Ecosystem services provided by mycorrhiza (Gianinazzi et al. 2010)

Mycorrhiza function	Ecosystem services provided
Changes in the morphology of roots and the development of an intricate, ramifying mycelial network in soil	Boost soil stability (binding effect) and plant/soil adhesion (improving soil structure)
Increasing plant absorption of minerals, nutrients, and water	Encourages plant growth while using less fertilizer
Protection from abiotic stresses	Enhanced plant resilience to salt, heavy metal contamination, drought, and depletion of vital nutrients
Protecting against root pathogens	Enhanced plant resilience to biotic stressors while lowering the amount of phytochemical input
Secretion of 'glomalin' into the soil	Increased soil stability and water retention
Modification of plant metabolism and physiology	Plant development bioregulation and improved plant quality for human health

7.1 Mycorrhizae-Based Biofertilizers

The use of microbial inoculants, often known as 'biofertilizers' (Alori et al. 2017), in sustainable food production enhances the ability of plants to efficiently utilize mineral elements such as nitrogen and phosphorus. Plants require phosphorus, and its shortage may impede plant growth. AM connections enable many plants to acquire adequate phosphorus from rhizosphere soils. The arbuscular mycorrhizal symbiotic connection is particularly vital for plants under phosphorus-deficient conditions because it promotes plant development and increases phosphorus concentrations within plants. The production and utilization of microbial fertilizers are experiencing a global rise because of the adverse effects of excessive or improper chemical fertilizer use and heightened awareness of the relationship between rhizosphere microorganisms and plants (Babalola et al. 2007; Babalola 2010; Hou and Oluranti 2013; Alori et al. 2017).

The primary categories of mycorrhizae utilized in biofertilizers are ectomycorrhiza, which are generally present in arboreal species, including pine, eucalyptus, peach, and oak, whereas AMF are present in grasses, shrubs, and certain tree species. AMF serve as crucial connections between plants and soil minerals. Consequently, there is increasing interest in natural fertilizers. AMF are obligate symbionts within the phylum Glomeromycota (Schüßler et al. 2001), which establish mutualistic relationships with approximately 80% of terrestrial plant species, including various crops. They supply the host plant with mineral nutrients and water in return for photosynthetic products (Smith and Read 2010). The mycelium of AMF, that develops from the root system, can obtain nutrients from soil regions that are unreachable by roots (Smith et al. 2000). Furthermore, fungal hyphae possess greater slenderness than roots, enabling them to infiltrate tiny pores (Allen 2011). At the interface between the plant and fungus, minerals and carbohydrates are successfully exchanged within the roots. The root cortex is mostly inhabited by AM fungal hyphae, which forms branching structures within the cells called arbuscules, which serve as the functional sites for nutrition exchange (Balestrini et al. 2018). Consequently, AMF can mitigate the constraints of plant growth resulting from insufficient nutrient availability (Nouri et al. 2014). Recent studies have indicated that a non-mycorrhizal state should be considered atypical for most species in natural environments (Smith and Smith 2012), despite the significant diversity among arbuscular mycorrhizal fungal communities in the soil, influenced by plant species diversity, soil type, season, or a combination of these factors (Smith and Smith 2012). Furthermore, a role in controlling arbuscular morphogenesis and sustaining symbiosis has been established with the enhancement of plant Pi acquisition (Javot et al. 2007, 2011; Yang et al. 2012; Xie et al. 2013). Recent findings indicate that the expression of the AM-induced Pi transporters MtPT4 and LjPT4 in *Medicago truncatula and Lotus japonicus* occurs at the root tips of non-colonized plants, suggesting a role for PT4 genes as elements of the Pi-sensing mechanism in the root tips (Volpe et al. 2016).

In addition to supplying nutrients to plants, mycorrhiza also enhance soil health and offer various other advantages, including:

(a) The filaments of mycorrhizal fungi enhance drought resistance in the associated plants by augmenting the capacity of the soil to retain water.
(b) Fungi specifically prevent the passive uptake of toxic substances that diminish the partner plant's sensitivity to heavy metals, such as lead and cadmium.
(c) Mycorrhizal fungi decompose and assimilate nutrients from primary rock surfaces in high-latitude, high-altitude, and rocky environments.
(d) Mycorrhizal fungi can protect plants both directly and by enhancing their vigour against pests including worms and diseases.
(e) These fungi may protect companion plants from elevated salt concentrations in saline soils.
(f) The exterior walls of the filaments secrete adhesive chemicals that aggregate fine soil particles, thereby enhancing soil structure and reducing erosion susceptibility.

7.2 Co-inoculation with Beneficial Microorganisms

Mycorrhizal inoculation enhances growth, photosynthesis and nutrient absorption in several host plants. The definitive impacts of AM symbiosis differ depending on the specific plant and fungus species engaged in the relationship. The method employed to exhibit a similar effect on cereal crops, primarily concerning nitrogen nutrition, involves the inoculation of plant growth-promoting rhizobacteria (PGPR) and AMF, either separately or in conjunction (Raklami et al. 2019; Bechtaoui et al. 2019; Ben-Laouane et al. 2020). These PGPRs can indirectly improve plant growth and health by producing metabolites that inhibit phytopathogens (like antibiotics, hydrogen cyanide, siderophores, and antifungal agents), causing systemic resistance, or through spatial and temporal competition. They can also directly improve plant growth and health by fixing atmospheric nitrogen, solubilizing phosphate and potassium, producing auxin, and facilitating nutrient transfer (Vacheron et al. 2013; Ahemad and Kibret 2014).

7.3 Mycorrhizal Management in Crop Rotation and Intercropping

The practice of growing two or more crops at the same time and location is known as intercropping (Vandermeer et al. 1998). An increase in living roots, a decrease in spatiotemporal barren soil and an increase in plant diversity are the outcomes of this method, which can take many different forms. Companion crops can be sown and harvested simultaneously or sequentially. This format is occasionally referred to as relay cropping. Relay cropping is characterized by the overlap of crops, distinguishing it from crop rotation, in which one crop is cultivated from seeding to harvest before another crop is sown. Crop rotation promotes temporal variation in agricultural systems, whereas intercropping enhances crop diversity. Intercropping can entail simultaneous cultivation of three or more crops. AMF significantly contribute to intercropping systems by safeguarding plant health and enhancing plant development and output. It specifically entails managing the transfer, distribution, and acquisition of resources (e.g., nitrogen and phosphorus), affecting crop relationships; and mitigating biotic (plant diseases) and abiotic (metals, drought) stresses. AMF are essential for enhancing the use of resources in agricultural intercropping systems, particularly in soils with limited resources. Intercropping systems demonstrate a growth advantage for intercrop species compared to monoculture systems (Hauggaard-Nielsen and Jensen 2005; Chapagain and Riseman 2014), whereas the inoculation of AMF in intercropping systems, such as rice/mung bean, further enhances the biomass of the intercrop (mung bean) (Xiao et al. 2010). AMF can link many organisms, including distinct plant species, by establishing underground mycelial networks through hyphal branches (Ren et al. 2015; Giovannetti et al. 2004). Mycorrhizal networks can affect the nutrient distribution among intercropped species (Hu et al. 2019) and provide feedback that enhances plant health.

Intercropping cereals and legumes with AMF inoculation usually results in overyielding in low-input agriculture. Mycorrhizal networks enhance overyielding in intercropping by facilitating nitrogen absorption in mycorrhiza-dependent plant species. AM fungal hyphal networks facilitate nutrient transfer among intercrops and enhance phosphorus uptake by plants. However, anticipated benefits during the intercropping of eggplant (*Solanum melongena* L.) with maize are lost in the absence of AMF (Vandermeer 1989). In a maize-grass pea intercropping system, AMF inoculation enhanced the symbiotic relationship between maize and grass peas under low phosphorus and low moisture conditions. However, with increasing resource availability, maize benefited while grass pea played a neutral or facilitative role (Zhu et al. 2023). In an intercropping system of wheat and faba bean, inoculation with AMF *F. mosseae* improved resource complementarity for intercrops under nitrogen and phosphorus-deficient conditions (Qiao et al. 2015).

8 Molecular Insights and Genetic Advances in Mycorrhizal Research

8.1 Arbuscular Mycorrhizal Fungi

As of November 2023, the JGI MycoCosm database includes 2701 fungal genomes, with over 220 mycorrhizal fungi, and GenBank has annotated 4770 genomes, reflecting significant progress in fungal genomics (Grigoriev et al. 2014; Martin et al. 2011). The genomes of 181 Mucoromycota, 696 Basidiomycota, and 1693 Ascomycota are among the major fungal clades represented by these sequenced species (Martin and van Der Heijden 2024). The evolution of mycorrhizal symbionts has been better understood because of comparisons with saprotrophic and pathogenic fungi made possible by the genomes of the AMF *R. irregularis* and ECM fungi *Laccaria bicolor* and *Tuber melanosporum* (Kobayashi et al. 2023). The ecological functions, evolutionary links, and pragmatic considerations of mycorrhizal fungi are driving forces behind genome sequencing. Genomic similarities and divergence across millions of years have been demonstrated by studies that have examined evolutionary trends, genome features, and genes involved in mycorrhizal symbiosis, especially in families such as Boletales (Martin and van Der Heijden 2024; Roy et al. 2024). Tracking lifestyle changes, connecting genomes to ecological roles, and discovering genes essential for mycorrhizal connections are among the main goals. Researchers have been fascinated by AMF because of their 500-million-year-old beginnings. Although some have raised Glomeromycotina to a distinct phylum, AMF are widely regarded as belonging to Mucoromycota, sister to Dikarya (Ascomycota and Basidiomycota), despite some ambiguities (Spatafora et al. 2016). AMF reproduce by producing large, multinucleated spores and coenocytic hyphae with thousands of nuclei. Unusually large genomes, ranging from 150 Mb in *Rhizophagus* species to 784 Mb in *Gigaspora* species, have been revealed by genome assemblies from AMF orders, including Glomerales and Diversisporales. These genomes have abundant repetitive sequences (Tisserant et al. 2013; Venice et al. 2020). Compared with recently evolved AMF, early diverging AMF, such as *Paraglomus occultum*, appear to be less advantageous for plant growth because of their smaller genomes. Owing to their similar evolutionary histories, *Geosiphon pyriformis* and obligatory AMF biotrophy share characteristics (Malar et al. 2021). According to genomic studies, AMF rely on host plants for nutrition because they have lost the genes necessary for thiamine synthesis, plant cell wall breakdown, and fatty acid production (Martin and van Der Heijden 2024). In AMF heterokaryons, Hi-C and HiFi sequencing have identified two different sets of homologous chromosomes with varying gene expression and epigenetic regulation that are dependent on host plants and environmental factors (Sperschneider et al. 2023).

Due to the discovery of genes and features associated with sexual reproduction, genomic investigations have cast doubt on the notion that AMF are ancient asexual creatures (Martin and van Der Heijden 2024). Although their genomes have not yet been sequenced, Mucoromycotina fungi also have symbiotic relationships with liverworts and other plants that resemble AMFs (Albornoz et al. 2022). These fungi may supplement the Glomeromycotinian AMF in ecosystems because they are more prevalent in agricultural fields than in native biomes (Mansfield et al. 2023). In the formation of AMF symbiosis, coordinated gene expression and signalling pathways—like the conserved 'shared symbiotic signalling pathway' (CSSP)—are essential (Delaux et al. 2015). According to phylogenomic research, CSSP helped form the earliest AM connections and predated terrestrial plants. Symbiosis-responsive genes, such as those related to phosphorus metabolism and effector-like proteins, have been discovered in AMF using RNA sequencing (Lanfranco and Bonfante 2023). Numerous AMF effector proteins have been found to control the expression of host genes. Further research is necessary to fully understand the role of endobacteria, virome, and related microbes in AMF symbiosis, as little is known about how genetic variation in plants and fungi affects transcriptional responses (Lanfranco and Bonfante 2023).

8.2 Ectomycorrhizal Fungi

Throughout their evolutionary history, fungi have undergone several transitions from saprotrophism to ectomycorrhizal (ECM) symbiosis (Strullu-Derrien et al. 2018; Tedersoo et al. 2010). Saprotrophic fungi colonizing plant roots and progressively developing mutualistic connections are probably the origin of this change (Barman et al. 2024; Martin and van Der Heijden 2024). This shift may have been influenced by elements including the availability of soluble carbohydrates, particularly in nutrient-poor settings, where ECM fungi can profit from their host plants (Martin et al. 2016). The capacity of ECM fungi to form partnerships with different tree species is a reflection of their niche specialization and adaptive evolution. To expand into new ecological niches, ECM fungal lineages that separated during the Late Cretaceous may have co-evolved with angiosperms, establishing host recognition systems and inhibiting plant defences. The ECM lifestyle has undergone extensive evolution, with no reversal to the saprotrophic state, underscoring its adaptive value (Wolfe et al. 2012; Lebreton et al. 2021). Because of their symbiotic interactions, ECM fungi have a reduced requirement for plant cell wall-degrading enzymes (PCWDEs), as evidenced by the lowering of PCWDE gene families (Floudas et al. 2020; Kohler et al. 2015). Nevertheless, many ECM fungi still contain some PCWDEs, such as laccases and endoglucanases, which enable them to scavenge nutrients from soil debris (Mondal et al. 2025). Controlling these genes is essential for preventing the activation of plant defense mechanisms (Martin and van Der Heijden 2024). Overall, selective factors that encourage symbiosis and ecological adaptation have moulded the evolutionary success of ECM fungi.

8.3 Orchid Mycorrhizal Fungi

Orchids, which comprise approximately 30,000 species, are an important source of information for studying the adaptation and evolution of plants (Martin and van Der Heijden 2024). Orchids, which first appeared approximately 112 million years ago, are highly dependent on mycorrhizal fungi, particularly for germination and seedling establishment (Givnish et al. 2015; Zhang et al. 2017). These fungi are the only sources of protocorm growth during the initial stage of germination. Without this mutually beneficial interaction, orchids would not be able to survive. We now have a better understanding of orchid-mycorrhizal interactions based on recent genomic studies. Gene loss and higher substitution rates are associated with mycoheterotrophy or a fungus's need for nutrition, according to genome comparisons of orchid species. For example, the fully mycoheterotrophic orchid *Platanthera guangdongensis* exhibits expanded trehalase genes, indicating reliance on fungi for trehalose during early development. Key orchid mycorrhizal fungi, such as *Sebacina vermifera*, *Tulasnella calospora* and *Ceratobasidium* species, are the focus of genomic efforts; these fungi have a large set of plant cell wall-degrading enzymes (PCWDEs) that support their dual saprotrophic/symbiotic lifestyles and contribute to the evolution of mycorrhizal symbiosis, dating back to the Early Devonian (393–419 Ma)

(Martin and van Der Heijden 2024). Ongoing genomic research will deepen our knowledge of orchid-fungal symbiosis, aiding conservation, horticulture and restoration efforts (Martin and van Der Heijden 2024).

9 Challenges and Future Perspectives

The application of AMF in agriculture presents promising benefits; however, its large-scale adoption faces significant challenges. The mass production of AMF inoculants that are viable and consistent across diverse farming conditions is complex. Environmental variables, from soil type to climate, impact AMF efficacy, making it difficult to standardize products that work reliably in different regions. Quality control is another hurdle, as inconsistent standards lead to variable product quality and outcomes, which can discourage farmer adoption due to unpredictable results. Ensuring that AMF inoculants survive storage and transport while retaining their functionality in the field requires refined production and quality-assessment practices. The integration of AMF with modern agricultural technologies offers a pathway for broader adoption. Precision agriculture can optimize AMF application by identifying the field zones where it would be most effective, adjusting doses to match soil needs, and integrating AMF into current nutrient management systems. Emerging tools, such as GIS and Internet of Things (IoT) sensors could enhance the real-time monitoring of AMF performance, allowing farmers to adapt practices as needed. Additionally, drones and robotics could make AMF application more efficient on large farms, reducing labour costs and increasing accuracy. Formulating AMF products that are compatible with other inputs, such as chemical fertilizers, could enable farmers to gradually reduce their chemical reliance while maintaining productivity. Research is essential to address the current knowledge gaps regarding the application of AMF. Future studies should focus on classifying AMF species based on host specificity and environmental resilience, enabling tailored inoculants for specific crops and climates. Advanced molecular tools, such as next-generation sequencing, can reveal how AMF interact with other soil microbes, thereby guiding the development of multi-microbial inoculants. As these gaps are addressed, AMF have the potential to revolutionize sustainable agriculture, especially in tropical regions with high fertilizer costs and fragile soils, fostering resilient farming systems and reducing the environmental impacts associated with synthetic fertilizers.

10 Conclusion

In conclusion, mycorrhizal fungi, particularly AMF, play a vital role in enhancing crop production through their symbiotic relationships with plants. These fungi significantly improve nutrient and water uptake, leading to enhanced resilience to different stresses such as drought and salinity. For instance, mycorrhizal inoculation has been linked to a 42% reduction in water requirements for drought-tolerant plants, such as *Retama sphaerocarpa*. Presence of mycorrhizae boosts the absorption of essential nutrients, particularly nitrogen and phosphorus. Research indicates that AMF can improve phosphorus uptake efficiency by 50% or more, especially in phosphorus-deficient soils. Furthermore, mycorrhizae contribute to improved soil aggregation, with GRSP increasing soil stability and organic matter retention, crucial for long-term soil health. Mycorrhizal associations also enhance plant defenses against pathogens, with studies showing a 52.5% reduction in root-knot nematode infection in *Coffea arabica* when inoculated with AMF. Additionally, plants colonized by AMF exhibit a greater tolerance to heavy metals, with research highlighting a significant improvement in chlorophyll content and reduced uptake of harmful metals under stress conditions.

Given the rising challenges of nutrient depletion, soil degradation, and climate change, integrating mycorrhizal fungi into agricultural practices is a potential strategy for improving crop productivity and sustainability. Continued research is essential to elucidate the molecular mechanisms underlying mycorrhizal interactions, paving the way for optimized agricultural practices that harness the benefits of this critical ecological partnership.

References

Abdelhameed RE, Metwally RA (2019) Alleviation of cadmium stress by arbuscular mycorrhizal symbiosis. Int J Phytoremediation 21(7):663–671

Abdel-Salam E, Alatar A, El-Sheikh MA (2018) Inoculation with arbuscular mycorrhizal fungi alleviates harmful effects of drought stress on damask rose. Saudi J Biol Sci 25(8):1772–1780

Agnihotri R, Sharma MP, Prakash A, Ramesh A, Bhattacharjya S, Patra AK, Manna MC, Kurganova I, Kuzyakov Y (2022) Glycoproteins of arbuscular mycorrhiza for soil carbon sequestration: review of mechanisms and controls. Sci Total Environ 806:150571

Ahanger MA, Tyagi SR, Wani MR, Ahmad P (2013) Drought tolerance: role of organic osmolytes, growth regulators, and mineral nutrients. In: Physiological mechanisms and adaptation strategies in plants under changing environment: Volume 1. Springer, New York, pp 25–55

Ahemad M, Kibret M (2014) Mechanisms and applications of plant growth promoting rhizobacteria: current perspective. J King Saud Univ-Sci 26(1):1–20

Albornoz FE, Ryan MH, Bending GD, Hilton S, Dickie IA, Gleeson DB, Standish RJ (2022) Agricultural land-use favours Mucoromycotinian, but not Glomeromycotinian, arbuscular mycorrhizal fungi across ten biomes. New Phytol 233:1369–1382

Allen MF (2011) Linking water and nutrients through the vadose zone: a fungal interface between the soil and plant systems. J Arid Land 3(3):155–163

Alori ET, Dare MO, Babalola OO (2017) Microbial inoculants for soil quality and plant health. Sustain Agric Rev 22:281–307

Azaizeh HA, Marschner H, Römheld V, Wittenmayer L (1995) Effects of a vesicular-arbuscular mycorrhizal fungus and other soil microorganisms on growth, mineral nutrient acquisition and root exudation of soil-grown maize plants. Mycorrhiza 5:321–327

Babalola OO (2010) Beneficial bacteria of agricultural importance. Biotechnol Lett 32:1559–1570

Balestrini R, Lumini E (2018) Focus on mycorrhizal symbioses. Applied soil ecology 123:299–304

Babalola OO, Sanni AI, Odhiambo GD, Torto B (2007) Plant growth-promoting rhizobacteria do not pose any deleterious effect on cowpeas and detectable amounts of ethylene are produced. World J Microbiol Biotechnol 23:747–752

Bahadur A, Batool A, Nasir F, Jiang S, Mingsen Q, Zhang Q, Pan J, Liu Y, Feng H (2019) Mechanistic insights into arbuscular mycorrhizal fungi-mediated drought stress tolerance in plants. Int J Mol Sci 20(17):4199

Banerjee S, Schlaeppi K, van der Heijden MG (2018) Keystone taxa as drivers of microbiome structure and functioning. Nat Rev Microbiol 16(9):567–576

Barman A, Dutta S, Bera A, Saha P, Roy J, Roy Choudhury M, Bera M, Das S (2024) Synergizing sustainability: a critical review on harnessing agroforestry for biomass, carbon sequestration, and water-food-energy nexus. Energy Ecol Environ 9:1–35

Bechtaoui N, Raklami A, Tahiri AI, Benidire L, El Alaoui A, Meddich A, Göttfert M, Oufdou K (2019) Characterization of plant growth promoting rhizobacteria and their benefits on growth and phosphate nutrition of faba bean and wheat. Biol Open 8(7):bio043968

Ben-Laouane R, Baslam M, Ait-El-Mokhtar M, Anli M, Boutasknit A, Ait-Rahou Y, Toubali S, Mitsui T, Oufdou K, Wahbi S, Meddich A (2020) Potential of native arbuscular mycorrhizal fungi, rhizobia, and/or green compost as alfalfa (*Medicago sativa*) enhancers under salinity. Microorganisms 8(11):1695

Bergstrand KJ (2022) Organic fertilizers in greenhouse production systems–a review. Sci Hortic 295:110855

Berruti A, Borriello R, Orgiazzi A, Barbera AC, Lumini E, Bianciotto V (2014) Arbuscular mycorrhizal fungi and their value for ecosystem management. In: Biodiversity: the dynamic balance of the planet. InTech, Rijeta, pp 159–191

Betancur-Agudelo M, Meyer E, Lovato PE (2023) Increased copper concentrations in soil affect indigenous arbuscular mycorrhizal fungi and physiology of grapevine plantlets. Rhizosphere 27:100711

Bhale UN, Bansode SA, Singh S (2018) Multifactorial role of arbuscular mycorrhizae in agroecosystem. In: Fungi and their role in sustainable development: current perspective, pp 205–220

Bravo A, Brands M, Wewer V, Dörmann P, Harrison MJ (2017) Arbuscular mycorrhiza-specific enzymes FatM and RAM 2 fine-tune lipid biosynthesis to promote development of arbuscular mycorrhiza. New Phytol 214(4):1631–1645

Bücking H, Kafle A (2015) Role of arbuscular mycorrhizal fungi in the nitrogen uptake of plants: current knowledge and research gaps. Agronomy 5(4):587–612

Chapagain T, Riseman A (2014) Barley–pea intercropping: effects on land productivity, carbon and nitrogen transformations. Field Crop Res 166:18–25

Chen M, Arato M, Borghi L, Nouri E, Reinhardt D (2018) Beneficial services of arbuscular mycorrhizal fungi–from ecology to application. Front Plant Sci 9:1270

Choi J, Summers W, Paszkowski U (2018) Mechanisms underlying establishment of arbuscular mycorrhizal symbioses. Annu Rev Phytopathol 56(1):135–160

Clemmensen KE, Bahr A, Ovaskainen O, Dahlberg A, Ekblad A, Wallander H, Stenlid J, Finlay RD, Wardle DA, Lindahl B (2013) Roots and associated fungi drive long-term carbon sequestration in boreal forest. Science 339(6127):1615–1618

Crişan I, Balestrini R, Pagliarani C (2024) The current view on heavy metal remediation: the relevance of the plant interaction with arbuscular mycorrhizal fungi. Plant Stress 12:100439

Da Silva Campos MA (2020) Bioprotection by arbuscular mycorrhizal fungi in plants infected with Meloidogyne nematodes: a sustainable alternative. Crop Protect 135:105203

Delaux PM, Radhakrishnan GV, Jayaraman D, Cheema J, Malbreil M, Volkening JD, Sekimoto H, Nishiyama T, Melkonian M, Pokorny L, Rothfels CJ (2015) Algal ancestor of land plants was preadapted for symbiosis. Proc Natl Acad Sci USA 112:13390–13395

Fiorilli V, Vannini C, Ortolani F, Garcia-Seco D, Chiapello M, Novero M, Domingo G, Terzi V, Morcia C, Bagnaresi P, Moulin L (2018) Omics approaches revealed how arbuscular mycorrhizal symbiosis enhances yield and resistance to leaf pathogen in wheat. Sci Rep 8(1):9625

Floudas D, Bentzer J, Ahrén D, Johansson T, Persson P, Tunlid A (2020) Uncovering the hidden diversity of litter-decomposition mechanisms in mushroom-forming fungi. ISME J 14:2046–2059

Gallou A, Mosquera HPL, Cranenbrouck S, Suárez JP, Declerck S (2011) Mycorrhiza induced resistance in potato plantlets challenged by Phytophthora infestans. Physiol Mol Plant Pathol 76(1):20–26

Gao Y, Huang S, Wang Y, Lin H, Pan Z, Zhang S, Zhang J, Wang W, Cheng S, Chen Y (2023) Analysis of the molecular and biochemical mechanisms involved in the symbiotic relationship between Arbuscular mycorrhiza fungi and *Manihot esculenta* Crantz. Front Plant Sci 14:1130924

Garg N, Chandel S (2011) Arbuscular mycorrhizal networks: process and functions. Sustain Agric 2:907–930

Genre A, Lanfranco L, Perotto S, Bonfante P (2020) Unique and common traits in mycorrhizal symbioses. Nat Rev Microbiol 18(11):649–660

George NP, Ray JG (2023) The inevitability of arbuscular mycorrhiza for sustainability in organic agriculture—a critical review. Front Sustain Food Syst 7:1124688

Gianinazzi S, Gollotte A, Binet MN, van Tuinen D, Redecker D, Wipf D (2010) Agroecology: the key role of arbuscular mycorrhizas in ecosystem services. Mycorrhiza 20(8):519–530

Giovannetti M, Sbrana C, Avio L, Strani P (2004) Patterns of below-ground plant interconnections established by means of arbuscular mycorrhizal networks. New Phytol 164:175–181

Giovannini L, Palla M, Agnolucci M, Avio L, Sbrana C, Turrini A, Govannetti M (2020) Arbuscular mycorrhizal fungi and associated microbiota as plant biostimulants: research strategies for the selection of the best performing inocula. Agronomy 10(1):106

Givnish TJ, Spalink D, Ames M, Lyon SP, Hunter SJ, Zuluaga A, Iles WJ, Clements MA, Arroyo MT, Leebens-Mack J, Endara L (2015) Orchid phylogenomics and multiple drivers of their extraordinary diversification. Proc Roy Soc B Biol Sci 282:20151553

Grigoriev IV, Nikitin R, Haridas S, Kuo A, Ohm R, Otillar R, Riley R, Salamov A, Zhao X, Korzeniewski F, Smirnova T (2014) MycoCosm portal: gearing up for 1000 fungal genomes. Nucleic Acids Res 42:D699–D704

Hauggaard-Nielsen H, Jensen ES (2005) Facilitative root interactions in intercrops. In: Root physiology: From gene to function, pp 237–250

Hodge A, Storer K (2015) Arbuscular mycorrhiza and nitrogen: implications for individual plants through to ecosystems. Plant Soil 386:1–19

Hodge A, Campbell CD, Fitter AH (2001) An arbuscular mycorrhizal fungus accelerates decomposition and acquires nitrogen directly from organic material. Nature 413(6853):297–299

Hou MP, Oluranti BO (2013) Evaluation of plant growth promoting potential of four rhizobacterial species for indigenous system. J Cent South Univ 20:164–171

Hu J, Li M, Liu H, Zhao Q, Lin X (2019) Intercropping with sweet corn (*Zea mays* L. var. rugosa Bonaf.) expands P acquisition channels of chili pepper (*Capsicum annuum* L.) via arbuscular mycorrhizal hyphal networks. J Soils Sediments 19:1632–1639

Hawkins HJ, Johansen A, George E (2000) Uptake and transport of organic and inorganic nitrogen by arbuscular mycorrhizal fungi. Plant and Soil 226(2):275–285

Javot H, Penmetsa RV, Terzaghi N, Cook DR, Harrison MJ (2007) A *Medicago truncatula* phosphate transporter indispensable for the arbuscular mycorrhizal symbiosis. Proc Natl Acad Sci 104(5):1720–1725

Javot H, Penmetsa RV, Breuillin F, Bhattarai KK, Noar RD, Gomez SK, Zhang Q, Cook DR, Harrison MJ (2011) *Medicago truncatula* mtpt4 mutants reveal a role for nitrogen in the regulation of arbuscule degeneration in arbuscular mycorrhizal symbiosis. Plant J 68(6):954–965

Jeewani PH, Luo Y, Yu G, Fu Y, He X, Van Zwieten L, Liang C, Kumar A, He Y, Kuzyakov Y, Qin H (2021) Arbuscular mycorrhizal fungi and goethite promote carbon sequestration via hyphal-aggregate mineral interactions. Soil Biol Biochem 162:108417

Jiang Y, Wang W, Xie Q, Liu NA, Liu L, Wang D, Zhang X, Yang C, Chen X, Tang D, Wang E (2017) Plants transfer lipids to sustain colonization by mutualistic mycorrhizal and parasitic fungi. Science 356(6343):1172–1175

Johansen A, Finlay RD, Olsson PA (1996) Nitrogen metabolism of external hyphae of the arbuscular mycorrhizal fungus Glornus intraradices. New Phytologist 133(4):705–712

Jiang S, An X, Shao Y, Kang Y, Chen T, Mei X, Dong C, Xu Y, Shen Q (2021) Responses of arbuscular mycorrhizal fungi occurrence to organic fertilizer: a meta-analysis of field studies. Plant Soil 469:89–105

Jing S, Li Y, Zhu L, Su J, Yang T, Liu B, Zhang, M (2022) Transcriptomics and metabolomics reveal effect of arbuscular mycorrhizal fungi on growth and development of apple plants. Frontiers in Plant Science 13:1052464

Jung SC, Martinez-Medina A, Lopez-Raez JA, Pozo MJ (2012) Mycorrhiza-induced resistance and priming of plant defenses. J Chem Biol 38:651–664

Khan Y, Shah S, Tian H (2022) The roles of arbuscular mycorrhizal fungi in influencing plant nutrients, photosynthesis, and metabolites of cereal crops—a review. Agronomy 12(9):2191

Kleinert A, Benedito VA, Morcillo RJ, Dames J, Cornejo-Rivas P, Zuniga-Feest A, Delgado M, Muñoz G (2018) Morphological and symbiotic root modifications for mineral acquisition from nutrient-poor soils. In: Root Biology, pp 85–142

Kobae Y (2019) Dynamic phosphate uptake in arbuscular mycorrhizal roots under field conditions. Front Environ Sci 6:159

Kobayashi Y, Shibata TF, Hirakawa H, Nishiyama T, Yamada A, Hasebe M, Shigenobu S, Kawaguchi M (2023) The genome of *Lyophyllum shimeji* provides insight into the initial evolution of ectomycorrhizal fungal genomes. DNA Res 30:dsac053

Kohler A, Kuo A, Nagy LG, Morin E, Barry KW, Buscot F, Canbäck B, Choi C, Cichocki N, Clum A, Colpaert J (2015) Convergent losses of decay mechanisms and rapid turnover of symbiosis genes in mycorrhizal mutualists. Nat Genet 47:410–415

Koide RT, Goff MD, Dickie IA (2000) Component growth efficiencies of mycorrhizal and nonmycorrhizal plants. New Phytol 148(1):163–168

Kamble VR, Agre DG (2014) New report on AMF colonization in root parasite Striga gesnerioides and its host Lepidagathis hamiltoniana from high altitude region of Maharashtra. International Multidisciplinary Research Journal 3(2):27–31

Kuila D, Ghosh S (2022) Aspects, problems and utilization of Arbuscular Mycorrhizal (AM) application as bio-fertilizer in sustainable agriculture. Curr Res Microb Sci 3:100107

Ladha JK, Rao AN, Raman AK, Padre AT, Dobermann A, Gathala M, Kumar V, Saharawat Y, Sharma S, Piepho HP, Alam MM (2016) Agronomic improvements can make future cereal systems in South Asia far more productive and result in a lower environmental footprint. Glob Chang Biol 22(3):1054–1074

Lal R (2018) Digging deeper: a holistic perspective of factors affecting soil organic carbon sequestration in agroecosystems. Glob Chang Biol 24(8):3285–3301

Lambert DH, Baker DE, Cole H Jr (1979) The role of mycorrhizae in the interactions of phosphorus with zinc, copper, and other elements. Soil Sci Soc Am J 43(5):976–980

Lanfranco L, Bonfante P (2023) Lessons from arbuscular mycorrhizal fungal genomes. Curr Opin Microbiol 75:102357

Lebreton A, Zeng Q, Miyauchi S, Kohler A, Dai Y-C, Martin FM (2021) Evolution of the mode of nutrition in symbiotic and saprotrophic fungi in forest ecosystems. Ann Rev Ecol Evol Syst 52:385–404

Leifheit EF, Veresoglou SD, Lehmann A, Morris EK, Rillig MC (2014) Multiple factors influence the role of arbuscular mycorrhizal fungi in soil aggregation—a meta-analysis. Plant Soil 374:523–537

Li XL, Marschner H, George E (1991) Acquisition of phosphorus and copper by VA-mycorrhizal hyphae and root-to-shoot transport in white clover. Plant Soil 136:49–57

Liao D, Wang S, Cui M, Liu J, Chen A, Xu G (2018) Phytohormones regulate the development of arbuscular mycorrhizal symbiosis. Int J Mol Sci 19(10):3146

Luginbuehl LH, Menard GN, Kurup S, Van Erp H, Radhakrishnan GV, Breakspear A, Oldroyd GE, Eastmond PJ (2017) Fatty acids in arbuscular mycorrhizal fungi are synthesized by the host plant. Science 356(6343):1175–1178

Mansfield TM, Albornoz FE, Ryan MH, Bending GD, Standish RJ (2023) Niche differentiation of Mucoromycotinian and Glomeromycotinian arbuscular mycorrhizal fungi along a 2-million-year soil chronosequence. Mycorrhiza 33:1–14

Martin FM, van Der Heijden MG (2024) The mycorrhizal symbiosis: research frontiers in genomics, ecology, and agricultural application. New Phytol 242(4):1486–1506

Martin F, Cullen D, Hibbett D, Pisabarro A, Spatafora JW, Baker SE, Grigoriev IV (2011) Sequencing the fungal tree of life. New Phytol 190:818–821

Martin F, Kohler A, Murat C, Veneault-Fourrey C, Hibbett DS (2016) Unearthing the roots of ectomycorrhizal symbioses. Nat Rev Microbiol 14:760–773

Malar CM, Roux C, Corradi N (2021) Regulation of mating genes during arbuscular mycorrhizal isolate coexistence—where is the evidence?. The ISME Journal 15(8):2173–2179

Marulanda A, Barea JM, Azcón R (2006) An indigenous drought-tolerant strain of *Glomus intraradices* associated with a native bacterium improves water transport and root development in *Retama sphaerocarpa*. Microb Ecol 52:670–678

Mondal BP, Raj GB, Dixit M, Sadhu T, Pandit V, Prasanna L, Chattopadhyay A, Dutta S, Kumar S (2025) Role of soils for satisfying global demands for food, water, and bioenergy. In: Environmental nexus for resource management. CRC Press, pp 302–316

Moreno Jiménez E, Ferrol N, Corradi N, Peñalosa JM, Rillig MC (2024) The potential of arbuscular mycorrhizal fungi to enhance metallic micronutrient uptake and mitigate food contamination in agriculture: prospects and challenges. New Phytol 242(4):1441–1447

Mrunalini K, Behera B, Jayaraman S, Abhilash PC, Dubey PK, Swamy GN, Prasad JV, Rao KV, Krishnan P, Pratibha G, Srinivasa Rao C (2022) Nature-based solutions in soil restoration for improving agricultural productivity. Land Degrad Dev 33(8):1269–1289

Muneer MA, Wang P, Lin C, Ji B (2020) Potential role of common mycorrhizal networks in improving plant growth and soil physicochemical properties under varying nitrogen levels in a grassland ecosystem. Glob Ecol Conserv 24:e01352

Nouri E, Breuillin-Sessoms F, Feller U, Reinhardt D (2014) Phosphorus and nitrogen regulate arbuscular mycorrhizal symbiosis in Petunia hybrida. PLoS One 9(3):e90841

Parniske M (2008) Arbuscular mycorrhiza: the mother of plant root endosymbioses. Nat Rev Microbiol 6(10):763–775

Paustian K, Lehmann J, Ogle S, Reay D, Robertson GP, Smith P (2016) Climate-smart soils. Nature 532(7597):49–57

Poveda J (2021) Insect frass in the development of sustainable agriculture. A review. Agron Sustain Dev 41(1):5

Pozo MJ, Azcón-Aguilar C (2007) Unraveling mycorrhiza-induced resistance. Curr Opin Plant Biol 10(4):393–398

Pozo MJ, Jung SC, López-Ráez JA, Azcón-Aguilar C (2010) Impact of arbuscular mycorrhizal symbiosis on plant response to biotic stress: the role of plant defence mechanisms. In: Arbuscular mycorrhizas: physiology and function, pp 193–207

Qiang-Sheng W, Ming-Qin C, Ying-Ning Z, Chu W, Xin-Hua H (2016) Mycorrhizal colonization represents functional equilibrium on root morphology and carbon distribution of trifoliate orange grown in a split-root system. Sci Hortic 199:95–102

Qiao X, Bei S, Li C, Dong Y, Li H, Christie P, Zhang F, Zhang J (2015) Enhancement of faba bean competitive ability by arbuscular mycorrhizal fungi is highly correlated with dynamic nutrient acquisition by competing wheat. Sci Rep 5(1):8122

Qiu Q, Bender SF, Mgelwa AS, Hu Y (2022) Arbuscular mycorrhizal fungi mitigate soil nitrogen and phosphorus losses: a meta-analysis. Sci Total Environ 807:150857

Raklami A, Bechtaoui N, Tahiri AI, Anli M, Meddich A, Oufdou K (2019) Use of rhizobacteria and mycorrhizae consortium in the open field as a strategy for improving crop nutrition, productivity and soil fertility. Front Microbiol 10:1106

Ren L, Zhang N, Wu P, Huo H, Xu G, Wu G (2015) Arbuscular mycorrhizal colonization alleviates Fusarium wilt in watermelon and modulates the composition of root exudates. Plant Growth Regul 77:77–85

Reyes Sanchez LB, Horn R, Costantini EA (2022) Sustainable soil management as a key to preserve soil biodiversity and stop its degradation. International Union of Soil Sciences

Riaz M, Kamran M, Fang Y, Wang Q, Cao H, Yang G, Deng L, Wang Y, Zhou Y, Anastopoulos I, Wang X (2021) Arbuscular mycorrhizal fungi-induced mitigation of heavy metal phytotoxicity in metal contaminated soils: a critical review. J Hazard Mater 402:123919

Rillig MC (2004) Arbuscular mycorrhizae, glomalin, and soil aggregation. Can J Soil Sci 84(4):355–363

Roy S, Barman A, Saha P (2023) Diminishing rice productivity in North Bengal's flood-prone region: a critical analysis on impact of sediment deposition on Rice. Sci Rep 4(10):41–46

Roy A, Dutta S, Das S, Choudhury MR (2024) Next-generation sequencing in the development of climate-resilient and stress-responsive crops–a review. Open Biotechnol J 18(1)
Rui W, Mao Z, Li Z (2022) The roles of phosphorus and nitrogen nutrient transporters in the arbuscular mycorrhizal symbiosis. Int J Mol Sci 23(19):11027
Ruytinx J, Kafle A, Usman M, Coninx L, Zimmermann SD, Garcia K (2020) Micronutrient transport in mycorrhizal symbiosis; zinc steals the show. Fungal Biol Rev 34(1):1–9
Saldajeno M, Hyakumachi M (2011) The plant growth-promoting fungus Fusarium equiseti and the arbuscular mycorrhizal fungus *Glomus mosseae* stimulate plant growth and reduce severity of anthracnose and dAMFping-off diseases in cucumber (*Cucumis sativus*) seedlings. Ann Appl Biol 159(1):28–40. https://doi.org/10.1111/j.1744-7348.2011.00471.x
Salmeron-Santiago IA, Martínez-Trujillo M, Valdez-Alarcón JJ, Pedraza-Santos ME, Santoyo G, Pozo MJ, Chávez-Bárcenas AT (2021) An updated review on the modulation of carbon partitioning and allocation in arbuscular mycorrhizal plants. Microorganisms 10(1):75
Schüßler A, Schwarzott D, Walker C (2001) A new fungal phylum, the Glomeromycota: phylogeny and evolution. Mycol Res 105(12):1413–1421
Seleiman MF, Hardan AN (2021) Importance of mycorrhizae in crop productivity. In: Mitigating environmental stresses for agricultural sustainability in Egypt, pp 471–484
Seleiman MF, Santanen A, Kleemola J, Stoddard FL, Mäkelä PS (2013) Improved sustainability of feedstock production with sludge and interacting mycorrhiza. Chemosphere 91(9):1236–1242
Silveira R, de Mello TD, Sartori MR, Alves GS, de Assis Fonseca FC, Vizzotto CS, Krüger RH, da Cunha Bustamante MM (2021) Seasonal and long-term effects of nutrient additions and liming on the nifH gene in cerrado soils under native vegetation. iScience 24(4):Article 102349
Singh M, Mishra M, Srivastava DK, Singh PK (2020) Arbuscular mycorrhiza-associated rhizobacteria and biocontrol of soilborne phytopathogens. In: Biostimulants in plant science. IntechOpen
Smith SE, Read DJ (2010) Mycorrhizal symbiosis. Academic
Smith SE, Smith FA (2011) Roles of arbuscular mycorrhizas in plant nutrition and growth: new paradigms from cellular to ecosystem scales. Annu Rev Plant Biol 62(1):227–250
Smith SE, Smith FA (2012) Fresh perspectives on the roles of arbuscular mycorrhizal fungi in plant nutrition and growth. Mycologia 104(1):1–13
Smith FA, Jakobsen I, Smith SE (2000) Spatial differences in acquisition of soil phosphate between two arbuscular mycorrhizal fungi in symbiosis with *Medicago truncatula*. New Phytol 147(2):357–366
Smith SE, Jakobsen I, Grønlund M, Smith FA (2011) Roles of arbuscular mycorrhizas in plant phosphorus nutrition: interactions between pathways of phosphorus uptake in arbuscular mycorrhizal roots have important implications for understanding and manipulating plant phosphorus acquisition. Plant Physiol 156(3):1050–1057
Song Y, Chen D, Lu K, Sun Z, Zeng R (2015) Enhanced tomato disease resistance primed by arbuscular mycorrhizal fungus. Front Plant Sci 6:786
Spatafora JW, Chang Y, Benny GL, Lazarus K, Smith ME, Berbee ML, Bonito G, Corradi N, Grigoriev I, Gryganskyi A, James TY (2016) A phylum-level phylogenetic classification of zygomycete fungi based on genome-scale data. Mycologia 108:1028–1046
Sperschneider J, Yildirir G, Rizzi YS, Malar CM, Mayrand Nicol A, Sorwar E, Villeneuve-Laroche M, Chen EC, Iwasaki W, Brauer EK, Bosnich W (2023) Arbuscular mycorrhizal fungi heterokaryons have two nuclear populations with distinct roles in host–plant interactions. Nat Microbiol 8:2142–2153
Strullu-Derrien C, Selosse M-A, Kenrick P, Martin FM (2018) The origin and evolution of mycorrhizal symbioses: from palaeomycology to phylogenomics. New Phytol 220:1012–1030
Suman J, Rakshit A, Ogireddy SD, Singh S, Gupta C, Chandrakala J (2022) Microbiome as a key player in sustainable agriculture and human health. Front Soil Sci 2:821589
Sun H, Koal P, Liu D, Gerl G, Schroll R, Gattinger A, Joergensen RG, Munch JC (2016) Soil microbial community and microbial residues respond positively to minimum tillage under organic farming in southern Germany. Appl Soil Ecol 108:16–24

Tedersoo L, May TW, Smith ME (2010) Ectomycorrhizal lifestyle in fungi: global diversity, distribution, and evolution of phylogenetic lineages. Mycorrhiza 20:217–263

Thangavel P, Anjum NA, Muthukumar T, Sridevi G, Vasudhevan P, Maruthupandian A (2022) Arbuscular mycorrhizae: natural modulators of plant–nutrient relation and growth in stressful environments. Arch Microbiol 204(5):264

Tisdall JM (1991) Fungal hyphae and structural stability of soil. Soil Res 29(6):729–743

Tisserant E, Malbreil M, Kuo A, Kohler A, Symeonidi A, Balestrini R, Charron P, Duensing N, Frei dit Frey N, Gianinazzi-Pearson V, Gilbert LB (2013) Genome of an arbuscular mycorrhizal fungus provides insight into the oldest plant symbiosis. Proc Natl Acad Sci USA 110:20117–20122

Vacheron J, Desbrosses G, Bouffaud ML, Touraine B, Moënne-Loccoz Y, Muller D, Legendre L, Wisniewski-Dyé F, Prigent-Combaret C (2013) Plant growth-promoting rhizobacteria and root system functioning. Front Plant Sci 4:356

Vallejos-Torres G, Espinoza E, Marín-Díaz J, Solis R, Arévalo LA (2021) The role of arbuscular mycorrhizal fungi against root-knot nematode infections in coffee plants. J Soil Sci Plant Nutr 21(1):364–373

Vandermeer J (1989) The ecology of intercropping. Cambridge University Press, New York

Vandermeer J, van Noordwijk M, Anderson J, Ong C, Perfecto I (1998) Global change and multispecies agroecosystems: concepts and issues. Agric Ecosyst Environ 67(1):1–22

Venice F, Ghignone S, Salvioli di Fossalunga A, Amselem J, Novero M, Xianan X, Sędzielewska Toro K, Morin E, Lipzen A, Grigoriev IV, Henrissat B (2020) At the nexus of three kingdoms: the genome of the mycorrhizal fungus *Gigaspora margarita* provides insights into plant, endobacterial and fungal interactions. Environ Microbiol 22:122–141

Volpe V, Giovannetti M, Sun XG, Fiorilli V, Bonfante P (2016) The phosphate transporters LjPT4 and MtPT4 mediate early root responses to phosphate status in non-mycorrhizal roots. Plant Cell Environ 39(3):660–671

Vos C, Schouteden N, van Tuinen D, Chatagnier O, Elsen A, De Waele D, Panis B, Gianinazzi-Pearson V (2013) Mycorrhiza-induced resistance against the root–knot nematode Meloidogyne incognita involves priming of defense gene responses in tomato. Soil Biol Biochem 60:45–54

Wahab A, Muhammad M, Munir A, Abdi G, Zaman W, Ayaz A, Khizar C, Reddy SP (2023) Role of arbuscular mycorrhizal fungi in regulating growth, enhancing productivity, and potentially influencing ecosystems under abiotic and biotic stresses. Plants 12(17):3102

Wang F, Zhang L, Zhou J, Rengel Z, George TS, Feng G (2022) Exploring the secrets of hyphosphere of arbuscular mycorrhizal fungi: processes and ecological functions. Plant Soil 481(1):1–22

Wang Q, Song J, Zhang J, Qin X, Kang Y, Huang S, Zhou S, Chen T (2023) Effects of microbial agent application on the bacterial community in ginger rhizosphere soil under different planting years. Front Microbiol 14:1203796

Wolfe BE, Tulloss RR, Pringle A (2012) The irreversible loss of a decomposition pathway marks the single origin of an ectomycorrhizal symbiosis. PLoS One 7:e39597

Wuana RA, Okieimen FE (2011) Heavy metals in contaminated soils: a review of sources, chemistry, risks and best available strategies for remediation. Int Sch Res Notices 2011(1):402647

Xiao T, Yang Q, Ran W, Xu G, Shen Q (2010) Effect of inoculation with arbuscular mycorrhizal fungus on nitrogen and phosphorus utilization in upland rice-mungbean intercropping system. Agric Sci China 9(4):528–535

Xie X, Huang W, Liu F, Tang N, Liu Y, Lin H, Zhao B (2013) Functional analysis of the novel mycorrhiza-specific phosphate transporter AsPT1 and PHT1 family from *Astragalus sinicus* during the arbuscular mycorrhizal symbiosis. New Phytol 198(3):836–852

Yang SY, Grønlund M, Jakobsen I, Grotemeyer MS, Rentsch D, Miyao A, Hirochika H, Kumar CS, Sundaresan V, Salamin N, Catausan S (2012) Nonredundant regulation of rice arbuscular mycorrhizal symbiosis by two members of the PHOSPHATE TRANSPORTER1 gene family. Plant Cell 24(10):4236–4251

Yang Y, He C, Huang L, Ban Y, Tang M (2017) The effects of arbuscular mycorrhizal fungi on glomalin-related soil protein distribution, aggregate stability and their relationships with soil properties at different soil depths in lead-zinc contaminated area. PLoS One 12(8):e0182264

Yang T, Lupwayi N, Marc SA, Siddique KH, Bainard LD (2021) Anthropogenic drivers of soil microbial communities and impacts on soil biological functions in agroecosystems. Glob Ecol Conserv 27:e01521

Ye L, Zhao X, Bao E, Li J, Zou Z, Cao K (2020) Bio-organic fertilizer with reduced rates of chemical fertilization improves soil fertility and enhances tomato yield and quality. Sci Rep 10(1):177

Zhang GQ, Liu KW, Li Z, Lohaus R, Hsiao YY, Niu SC, Wang JY, Lin YC, Xu Q, Chen LJ, Yoshida K (2017) The Apostasia genome and the evolution of orchids. Nature 549:379–383

Zhu XC, Song FB, Xu HW (2010) Arbuscular mycorrhizae improve low temperature stress in maize via alterations in host water status and photosynthesis. Plant Soil 331:129–137

Zhu SG, Duan HX, Tao HY, Zhu L, Zhou R, Yang YM, Zhang XL, Wang WY, Zhu H, Zhang W, Wang RQ (2023) Arbuscular mycorrhiza changes plant facilitation patterns and increases resource use efficiency in intercropped annual plants. Appl Soil Ecol 191:105030

Mycorrhizal Influence on Vegetable Crops: Enhancing Productivity and Sustainability

Viabhav Kumar Upadhayay, Kumar Gaurav, Rajeev Kumar Yadav, Ashish Kumar Singh, Katyayni Kanth, Saurabh Gangola, Marcelo Carvalho Minhoto Teixeira Filho, and Yogesh Dashrath Naik

Abstract

Climate change presents a significant global challenge to agricultural productivity, including in vegetable crop cultivation. Traditional chemical-based methods for enhancing vegetable quality and yield often lack sustainability and adversely affect the environment. In contrast, the use of microorganisms, such as mycor-

V. K. Upadhayay (✉)
Department of Microbiology, College of Basic Sciences and Humanities, Dr. Rajendra Prasad Central Agricultural University, Pusa, Samastipur, Bihar, India

K. Gaurav
School of Agriculture, Graphic Era Hill University, Dehradun, India

R. K. Yadav
Departement of Horticulture, PGCA, Dr. Rajendra Prasad Central Agricultural University, Pusa, Samastipur, Bihar, India

A. K. Singh
Department of Vegetable Science, CHF, Acharya Narendra Deva University of Agriculture & Technology, Kumarganj, Ayodhya, UP, India

K. Kanth
College of Basic Sciences and Humanities (CBS&H), Dr Rajendra Prasad Central Agricultural University, Samastipur, Bihar, India

S. Gangola
Department of Microbiology, Graphic Era Deemed to be University, Dehradun, India

M. C. M. T. Filho
Department of Rural Engineering, Plant Health and Soils, São Paulo State University (UNESP), Ilha Solteira, Brazil

Y. D. Naik
Department of Agricultural Biotechnology and Molecular Biology, College of Basic Sciences and Humanities, Dr Rajendra Prasad Central Agricultural University, Samastipur, Bihar, India

Y. Siddiqui et al. (eds.), *Sustainable Mycorrhizal Cultivation*,
https://doi.org/10.1007/978-981-95-1778-7_5

rhizal fungi, provides an eco-friendlier approach. Mycorrhizal fungi are renowned for their ability to enhance plant growth and development, especially in vegetables, by improving nutrient uptake and resilience to stress. These fungi exhibit remarkable adaptability to various soil and climatic conditions, effectively mitigating abiotic stress, enhancing nutrient availability, and suppressing plant diseases. This chapter highlights the functional benefits of mycorrhizal fungi in boosting vegetable crop productivity and quality, positioning them as a sustainable alternative to chemical agrochemicals, thereby promoting greener and more resilient agricultural practices.

Keywords

Mycorrhizal fungi · Abiotic stress · Nutritional quality · Vegetable productivity · Disease suppression

1 Introduction

Vegetable crops are herbaceous plants primarily cultivated for human consumption, encompassing a diverse range of edible parts. These include flower buds, such as cauliflower; petioles, like celery; underground stems, such as Irish potatoes; aboveground stems, like asparagus; swollen taproots, such as carrots; lateral roots, including sweet potatoes; hypocotyls, like radishes and turnips; and leaves, such as lettuce and cabbage (Baum et al. 2015). In recent years, the production of vegetables has faced numerous challenges that threaten both sustainability and productivity (Kumari et al. 2018). Among these challenges are drastic climatic changes, environmental pollution, accelerated land degradation, and biodiversity loss. These issues are of significant concern to both growers and researchers. The indiscriminate use of chemical fertilizers and pesticides has further exacerbated these problems, leading to detrimental impacts on soil health and long-term agricultural viability (Upadhayay et al. 2023a, 2024). As the global population continues to rise, so does the demand for food. This necessitates a shift toward sustainable agricultural practices that not only enhance vegetable production but also minimize environmental impact. One promising avenue lies in the adoption of biofertilizers, particularly mycorrhizal fungi. These microorganisms play a crucial role in the rhizosphere by enhancing nutrient availability (Khan et al. 2023). The rhizosphere serves as a dynamic interface between plants and soil microorganisms, hosting numerous beneficial organisms that interact with plant roots to facilitate improved nutrient uptake and overall plant health (Upadhayay et al. 2022, 2023b; Jaggi et al. 2023). Among these beneficial organisms, arbuscular mycorrhizal fungi (AMF) stand out due to their ability to establish symbiotic relationships with the roots of 80–90% of terrestrial plants. This symbiosis significantly contributes to various stages of plant development (Zhu et al. 2022). The association with AMF not only provides essential minerals like phosphorus but also enhances the overall health and resilience of plants by improving access to nutrients and water (Alam et al. 2023). Research has

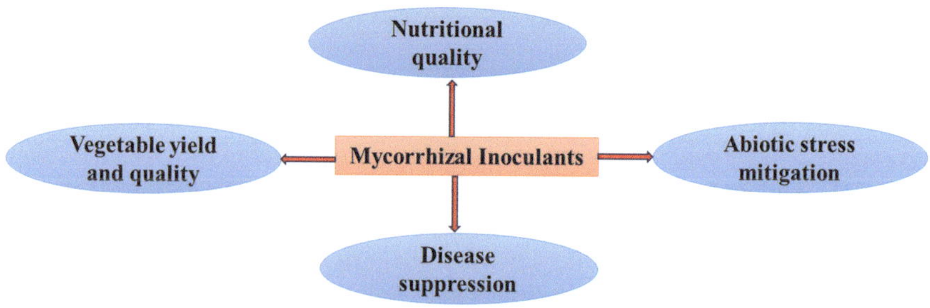

Fig. 1 Prolific impact of mycorrhizal fungi on vegetable crops

consistently demonstrated the beneficial effects of AMF on vegetable crops. For instance, they facilitate nutrient uptake—particularly phosphorus—thereby significantly enhancing the growth and yield of various vegetables such as tomatoes and onions (Baum et al., 2015; Fracasso et al. 2020; El-Sherbeny et al. 2022). Moreover, AMF have been shown to confer resistance to both abiotic and biotic stresses (Duc et al. 2018; Aljawasim et al. 2020), which are increasingly prevalent due to climate change and environmental degradation. For example, AMF improve plant resilience against drought and salinity, fostering sustainable agricultural practices (Abdel Latef and Chaoxing 2011; Ruiz-Lozano et al. 2016). In addition to enhancing nutrient uptake in vegetables—such as phosphorus and essential micronutrients—AMF also improve the nutritional value and phytochemical content in commonly cultivated crops like tomatoes (Ullah et al. 2023) and onions (Rozpądek et al. 2016). Figure 1 illustrates the significant impact of mycorrhizal fungi on vegetable crops, highlighting their roles in enhancing yield, improving nutritional quality, mitigating abiotic stresses, and suppressing diseases caused by pathogens (biotic stress mitigation). By highlighting the multifaceted advantages of mycorrhizal interventions, this chapter aims to provide a comprehensive understanding of their potential to revolutionize vegetable production sustainably. As we confront increasing environmental challenges, integrating mycorrhizal fungi into agricultural practices represents a crucial step toward achieving sustainable food production and ensuring long-term agricultural resilience.

2 Crosstalk Between Mycorrhizal Fungi and Plants

Mycorrhizal fungi are obligate biotrophs that depend on host plants for carbon. They significantly impact plant characteristics below the soil surface and facilitate communication between plants through their hyphal connections with root systems. These fungi enhance nutrient and water uptake, improve stress tolerance, and boost plant growth and productivity. The symbiotic relationship with arbuscular mycorrhizal fungi (AMF) is particularly important for crops such as carrots, which possess very short root hairs (Baum et al. 2015). This partnership helps expand the depletion zone around the roots, thereby enhancing nutrient and water absorption.

Arbuscular mycorrhizal (AM) symbiosis increases a plant's ability to absorb minerals like phosphorus and nitrogen, enhances its tolerance to various stresses, and improves overall health. In return, the fungi receive carbon compounds, such as sugars and lipids, produced during photosynthesis (Crosino and Genre 2022). Mycorrhizal fungi can be broadly categorized into four main types based on their evolution, ecology, and morphology: ectomycorrhizal, arbuscular mycorrhizal, orchid mycorrhizal, and ericoid mycorrhizal fungi (Enebe and Erasmus 2023). Among these types, arbuscular mycorrhizal fungi (a form of endomycorrhiza) are the most ancient. They form associations with both non-multicellular root plants—such as bryophytes, ferns, and lycophytes—and higher plants with multicellular root systems (Strullu-Derrien et al. 2018). These fungi penetrate the root cortex of vascular plants using their hyphae to create structures called arbuscules for nutrient exchange and vesicles for food storage; hence, they are also known as vesicular-arbuscular mycorrhizal fungi (Enebe and Erasmus 2023). Key genera in this group include *Funneliformis*, *Glomus*, *Entrophospora*, *Gigaspora*, *Archaeospora*, *Acaulospora*, *Sclerocystis*, and *Scutellospora* (Singh et al. 2021). Endomycorrhizal fungi assist plants in water absorption and survival in drought conditions by directly regulating stomatal conductance (Augé et al. 2015). Ectomycorrhizal fungi represent another category of mycorrhizal fungi distinguished by their hyphal arrangements within plant cortical tissues. These fungi belong to the basidiomycetes and ascomycetes groups and associate with approximately 10% of terrestrial plants, primarily trees. They develop a fungal mantle on feeder roots; unlike endomycorrhizal fungi, they do not penetrate root cells but instead create a Hartig net that transforms the roots into swollen and branched hyphae (Enebe and Erasmus 2023). Arbuscular mycorrhizal fungi establish symbiotic relationships with host plants through a three-step colonization process: host search, penetration, and establishment. During the initial stages of AMF symbiosis, molecular signals are exchanged between the plant and the fungus. In response to nutrient stress, the plant secretes signaling molecules such as strigolactones into the soil. AMF hyphae detect these signals and initiate a molecular response that activates genes associated with hyphal growth and colonization (Boyno et al. 2023). AMF secrete mycorrhizal factors (Myc) that facilitate communication with nitrogen-fixing bacteria (Feddermann et al. 2010). The establishment of AMF interactions involves activating specific *SYM* genes. When Myc signals are detected by host receptors, root cells respond by releasing cytosolic calcium (Ca^{2+}) and activating second membrane proteins (SYMPK) (MacLean et al. 2017; Enebe and Erasmus 2023). These proteins code for receptor-like kinases capable of recognizing AMF signals. Subsequently, SYMPK transmits these signals to the nucleus by phosphorylating an unidentified substrate, leading to repeated oscillations of Ca^{2+} ions in both the cytoplasm and nucleus (Anand et al. 2022). This oscillation is mediated by calmodulin-dependent protein kinases derived from *SYM* genes, ultimately resulting in gene regulation and root colonization (Barman et al. 2016; Anand et al. 2022). In arbuscular mycorrhizal fungi, hyphae penetrate plant root tissues to form branched structures known as arbuscules. Proteins such as RAM2 and glycerol-3-phosphate acyltransferase (GPAT) are crucial for the development of arbuscules as well as for forming hyphopodia on the root surface. The hypodermal passage cells play a significant role in facilitating the entry of mycorrhizal hyphae into host roots while also secreting strigolactones (Enebe and Erasmus 2023). A greater presence of these structures in the roots correlates with increased fungal colonization; however, their distribution and abundance are influenced by factors such as plant genotype, environmental conditions, and hormonal signals (Banasiak et al. 2021).

3 Mycorrhizae as Natural Facilitators of Phosphate Transfer

Phosphorus (P) is a critical element for plant growth and development, serving as an essential component of nucleic acids and phospholipids. Plants primarily absorb P in the form of inorganic phosphate (Pi), which often becomes a limiting factor due to its low solubility and mobility in soil. The rate of Pi absorption by plant roots frequently exceeds its diffusion rate, resulting in a depletion zone surrounding the root system (Ferrol et al. 2019). To mitigate Pi scarcity, plants have evolved various adaptive strategies aimed at enhancing Pi acquisition and utilization efficiency. These strategies include increased root branching, elongation of root hairs, and the secretion of organic acids and phosphatases that facilitate the solubilization of soil Pi. Such mechanisms are regulated by genetically controlled pathways for Pi sensing and signaling (Ha and Tran 2014). A significant adaptive strategy involves the symbiotic relationship with arbuscular mycorrhizal fungi (AMF), which enhances phosphorus availability for host plants. Through these symbioses, AMF effectively transfer inorganic phosphate (Pi) from the soil to the host plants via extensive hyphal networks (Lu et al. 2023). The partnership with AM fungi offers substantial benefits for crops cultivated in phosphorus-deficient soils, as these fungi improve P availability and uptake through multiple mechanisms. The extensive hyphal network formed by AM fungi extends the rooting zone of plants, enabling access to phosphate ions ($H_2PO_4^-$ and HPO_4^{2-}) that would otherwise be inaccessible. The absorption efficiency of phosphate ions is greater through hyphae than through root hairs due to their higher affinity for P uptake and lower absorption thresholds (Elbon and Whalen 2015). Furthermore, AM fungi alter the pH within the rhizosphere and secrete organic acids that assist in solubilizing P complexes and liberating phosphate ions chemically bound to soil minerals (Upadhayay et al. 2019; Andrino et al. 2021). Additionally, some hyphae produce phosphatase enzymes that mineralize organic P, thereby increasing the concentration of available phosphate ions in the soil solution (Jayachandran et al. 1992; Etesami et al. 2021). AMF also synthesize beneficial compounds such as glomalin—a glycoprotein that enhances nutrient uptake, particularly for hard-to-dissolve nutrients like iron and phosphorus (Shabana 2016; Emran et al. 2017). Arbuscular mycorrhizal plants employ two distinct pathways for phosphorus absorption. The first pathway, shared with non-arbuscular mycorrhizal plants, involves direct uptake of P through the root epidermis and root hairs (Jawad et al. 2023). The second pathway facilitates P entry into root cortical cells via intraradical mycelium, where arbuscules or hyphal coils create symbiotic interfaces (Smith and Smith 2011; Upadhayay et al. 2019; Etesami et al. 2021). This process enables P uptake from the interfacial apoplast through the expression of cell-specific Pi transporter genes in mycorrhizal roots (Fiorilli et al. 2013; Etesami et al. 2021). AMF induce the expression of phosphate transporters from the PHT1 family across various plant species, enhancing phosphorus uptake at the intraradical symbiotic interface (Walder et al. 2016; Upadhayay et al. 2019). Notable examples include *SlPT4* in tomato, *StPT3* in potato, *MtPT4* in barrel medic, *OsPT11* in rice, *LjPT4* in *Lotus japonicus*, and *PhPT5* in *Petunia*. These AMF-induced transporters are localized in peribranchial membranes and play a crucial role in facilitating phosphorus transport along the mycorrhizal pathway of their host plants (Rui et al. 2023). Xu et al. (2007) demonstrated that *LePT4* is essential for phosphorus uptake in tomato plants. The loss-of-function mutant (*lept4*) exhibited severe phosphorus deficiency symptoms and struggled to acquire phosphorus under low-nutrient conditions. In contrast, mycorrhizal infection resulted in increased expression of *LePT4* and enhanced phosphorus uptake. Szentpéteri et al. (2023) found that arbuscular mycorrhizal fungi (*Funneliformis mosseae*) alleviated heat and water deficit stresses in tomato plants by enhancing the expression of phosphate transporter genes—specifically *Pht1;3*, *Pht1;4*, and *Pht1;7*—indicating that increased phosphorus uptake through AM symbiosis could bolster plant resilience under abiotic stress conditions. Furthermore, AMF significantly induced the expression of a phosphate transporter gene from the *Pht1* family in cucumber seedlings, improving phosphorus uptake under both ambient and cold stress conditions (Ma et al. 2015).

4 Maximizing Vegetable Crop Production via Mycorrhizal Intervention

Currently, vegetable production is confronted with significant challenges, including severe climatic changes, environmental pollution, accelerated land degradation, and biodiversity loss. The indiscriminate and excessive use of chemical fertilizers and pesticides has negatively impacted the long-term sustainability of these practices. To mitigate these issues, it is essential to modify vegetable production systems to lessen their environmental impact while maintaining high levels of food production. The success of vegetable cultivation is closely tied to the availability of nutrients in the soil. The application of biofertilizer instead of chemical fertilizers could be the major contributor to maximizing the sustainability of vegetable production. The rhizosphere of the plant harbors a complex microbial community and serves as a hotspot for interactions between plant roots and beneficial microbes (plant growth-promoting bacteria and fungi) (Hinsinger et al. 2009; Raaijmakers et al. 2009). AMF are important root symbionts that form a symbiotic relationship with 80–90% of vascular plant species (Diagne et al. 2020). Mycorrhizal fungi form a mutualistic association with plant roots and enhance the uptake of nutrients, especially phosphorus. In return, the fungal partner obtains photosynthetically derived sugars from the plant (Upadhayay et al., 2019).

It is well known that the use of AMF is very useful in the growth and development of the vegetables due to easy access to essential nutrients, as well as providing protection against important biotic and abiotic stresses. Almost all the major vegetables are well known to host AMF to upgrade their nutrient content and be tolerant to several types of stresses. Many vegetables, such as onion, tomato, okra, beans, potato, etc., show their reliance toward AMF for increasing their growth and production (Baum et al. 2015). The AMF is applied in the field/pots in different ways depending upon the function; usually it is applied by seed dressing, in furrow application, broadcasting, root dipping, or seedling inoculation. Many species of AMF have been reported to increase the production capacity of vegetable crops in a sustainable manner without harming the ecosystem (Table 1). In tomato, the species *Claroideoglomus etunicatum, Claroideoglomus claroideum, Glomus microaggregatum, and Funneliformis mosseae* can increase the production by increasing root colonization, height, shoot dry biomass, root dry biomass, and number of fruits (Ziane et al. 2021; Almuslimawi et al. 2024). Under salinity conditions, the availability of plant mineral concentrations and fruit production, as well as the quality of chili, were improved by *Glomus intraradices* (Selvakumar and Thamizhiniyan 2011). Similarly, the use of *Funneliformis mosseae* and *Rhizophagus irregularis* together can increase the productivity and quality of sweet pepper under protected conditions (Sini et al. 2024). When the compost and biochar are joined with mycorrhizal fungi (*Glomus fasciculatum*), they enhance the growth and production of sweet potato under salinity-affected soils (Diedhiou et al. 2023). The use of AMF in vegetable production can not only increase production but also improve the quality of produce, increase tolerance to disease, and improve the growth and development of vegetables under problematic soils. By using AMF in vegetables, we can maintain the long-term productivity of vegetable crops without hampering the environment, ecosystem, and biodiversity, which are necessary for sustainable agriculture production. Various AM fungal species are presented in Table 1 in the context of their impact on improving vegetable growth and yield characteristics.

5 Mycorrhizal Strategies for Mitigating Abiotic Stress in Vegetables

Abiotic stresses, including drought, salinity, heavy metals, and temperature extremes, pose significant challenges to plant growth by disrupting physiological functions and causing oxidative stress. Drought limits water availability, leading to reduced transpiration and impaired nutrient assimilation, while salinity disrupts ion balance and promotes reactive oxygen species production. Heavy metal contamination can inhibit growth and accumulate in plant tissues, and temperature extremes can cause wilting, reduced biomass, and cellular damage, ultimately affecting crop yield and food security (Begum et al. 2019). Vegetable crops possess natural acclimation and adaptation mechanisms that enable them to survive various environmental stresses; however, these abilities may fall short in the face of rapid climate change (Dhankher and Foyer 2018). The response of plants to abiotic stresses varies based

Table 1 Use of different species of AMF to increase the production of vegetable crops

Vegetable crop	AMF species	Effect on vegetable production	References
Tomato (*Solanum lycopersicum*)	*Rhizophagus* intraradices	Enhanced fruit production by increasing chlorophyll content under water stress conditions, and net photosynthetic rate	Fracasso et al. (2020)
	Glomus microaggregatum, Claroideoglomus etunicatum, C. claroideum, Rhizophagus intraradices, Funneliformis mosseae, and *F. geosporum*	Improved root colonization, plant height, stem and root dry biomass, number of fruits, and final yield	Ziane et al. (2021)
	Funneliformis mosseae	Increases plant growth and yield	Almuslimawi et al. (2024)
	Rhizophagus clarus	Stimulated better plant development and yield	Radi et al. (2024)
Cherry tomato (*Solanum lycopersicum* L. var. *cerasiforme*)	*Rhizophagus irregularis*	Increased marketable yield and qualitative traits (fruit soluble solids content, ascorbic acid, polyphenols, and lycopene)	Vultaggio et al. (2024)
Chili (*Capsicum annuum*)	*Glomus intraradices*	Improved essential inorganic nutrients (N, P, and K) under salinity conditions	Selvakumar and Thamizhiniyan (2011)
	Glomus mosseae	Increased bio-volume index, plant biomass, phosphorus uptake, and fruit yield	Thilagar and Bagyaraj (2015)
	Glomus mosseae	Increased nutrient (Zn and P) uptake and yield	Almaca et al. (2021)
Sweet pepper (*Capsicum annuum* L. var. *grossum*)	*Funneliformis mosseae* and *Rhizophagus irregularis*	Escalated the productivity and quality of sweet pepper under greenhouse conditions	Sini et al. (2024)
Brinjal (*Solanum melongena* L.)	Endotrophic mycorrhiza	Increases the number of branches, leaf area, plant height, number of fruits, average fruit weight, yield/plant, and total yield	Alias et al. (2021)
	Glomus mosseae	Improved morphological traits and yield attributing traits	Banu et al. (2022)
	Rhizophagus irregularis	Improved the quality and yield	Sharma et al. (2022)

(continued)

Table 1 (continued)

Vegetable crop	AMF species	Effect on vegetable production	References
Okra [*Abelmoschus esculentus*]	*Glomus mosseae*	Increased nutrient uptake, plant growth, and fruit yield	Okon (2014)
	Gigaspora gigantea	Significantly improved the growth and nutrient uptake	Anozie and Okereke (2018)
Onion (*Allium cepa* L.)	*Glomus* sp., *Gigaspora* sp., *Acaulospora* sp., and *Scutellospora* sp.	Increased yield efficiently	El-Sherbeny et al. (2022)
Carrot (*Daucus carota* L.)	*Glomus mosseae* and *Gigaspora gigantea*	Enhanced carrot production under salinity stress conditions	Yadav et al. (2021)
Muskmelon (*Cucumis melo* L.)	*Funneliformis mosseae and Rhizophagus irregularis*	Enhanced the capacity of the plant to tolerate moderate water stress and improve the fruit length, sugar content, and firmness	Miceli et al. (2023)
Watermelon (*Citrullus lanatus*)	*Rhizophagus intraradices*	Increased t fruit yield under water stress conditions	Yang et al. (2022)
Sweet potato [*Ipomoea batatas* (L.) Lam.]	*Glomus fasciculatum (in combination with compost and biochar)*	Improved growth and yield in salt-affected areas	Diedhiou et al. (2023)

on species, intensity and duration of the stress, phenological stage, and the specific plant tissues involved. Such stresses can lead to both reversible and irreversible changes in plant physiology and metabolism. These factors particularly impact vegetable crops, which are generally more vulnerable to abiotic stresses (Khalid et al. 2023). To meet the increasing global demand for vegetables, it is essential to develop strategies that effectively address critical issues such as abiotic stresses, including drought, salinity, and heat stress. Mycorrhizal relationships are particularly important because they enhance nutrient and water uptake, improve soil structure, and promote plant health, all of which contribute to greater resilience and productivity under stressful conditions. Through various physiological, functional, and biochemical adaptations, mycorrhizae significantly bolster plant growth and stress tolerance, making them essential for sustainable agriculture (Malhi et al. 2021). AMF improve water and nutrient absorption for their host plants, particularly under stress conditions, by utilizing their extraradical mycelium. This mycelium enhances the absorption of water and nutrients and helps mitigate the effects of toxic elements like phytotoxic aluminum and other harmful metal(loid)s, thereby promoting plant growth and increasing crop yields (Cornejo et al. 2017). AMF-colonized plants exhibit enhanced drought tolerance through both direct and indirect mechanisms. The fine hyphae of AMF, characterized by their small diameter and high surface-to-volume ratio, penetrate soil pores inaccessible to plant roots, facilitating more efficient water extraction (Cheng et al. 2021; Leventis et al. 2021; Püschel et al. 2021).

Additionally, AMF influence root morphology and physiological processes, including the production of osmolytes, hormone production, biosynthesis of secondary metabolites (such as strigolactones that aid in establishing AMF symbiosis), and the regulation of antioxidative enzymes. These adaptations collectively contribute to increased plant resilience under water-limiting conditions (Diagne et al. 2020; Leventis et al. 2021). AMF enhance plant tolerance to salinity stress by improving water absorption, nutrient uptake, and ionic homeostasis while reducing the uptake of toxic sodium (Na^+) and chloride (Cl^-) ions (Evelin et al. 2019). They also promote the accumulation of osmoregulatory compounds like proline and sugars, enhance stomatal conductance, and mitigate oxidative damage, ultimately improving plant growth under saline conditions (Diagne et al. 2020). Under heat stress, AMF improve nutrient and water uptake while increasing antioxidant enzyme activities. They also up-regulate the expression of heat stress-related genes such as *PIPs* and *Hsp70*, further enhancing plant resilience to elevated temperatures (Tian et al. 2023). Mycorrhizal fungi also enhance cold stress tolerance in plants by reducing lipid peroxidation and maintaining membrane integrity. They boost antioxidative potential, optimize osmolyte accumulation, regulate root hydraulic conductance, and improve photosynthetic activity while promoting the expression of cold-responsive genes (Devi et al. 2019). Various strategies have been proposed to explain how AMF affect the uptake, translocation, and accumulation of heavy metals (HMs), thereby enhancing plant tolerance to HM stress. These strategies include retaining HMs within mycorrhizal roots and external hyphae, improving nutrient absorption, sequestering metals in vacuoles, adsorbing HMs onto fungal cell walls, and stimulating antioxidant responses. Additionally, AMF facilitate the chelation of HMs in their cytosol while promoting glomalin production, contributing to AMF-mediated phytoremediation efforts (Dhalaria et al. 2020). Numerous studies highlighted the role of AMF in enhancing plant resilience to abiotic stresses. For instance, the AMF *S. constrictum* has been shown to significantly improve tomato plant resilience by enhancing leaf water potential, stomatal conductance, and photosynthetic activity. This is accompanied by increased proline accumulation and elevated activities of antioxidant enzymes, alongside specific alterations in phytohormones and the expression of stress-responsive genes such as *SlAREB1*, *SlLEA*, *SlHSP70*, and *SlHSP90* (Duc et al. 2023). Similarly, AM fungi have been reported to enhance the growth and antioxidant responses of lettuce cultivars (*Lactuca sativa* cvs. 'Grand Rapids' and 'Lollo Bionda') under escalating salt stress conditions. This enhancement is evidenced by increased proline synthesis, heightened activities of antioxidant enzymes, and reduced oxidative damage (Santander et al. 2020). Additionally, the AM fungus *Funneliformis mosseae* has demonstrated its efficacy in alleviating chilling stress in cucumber seedlings. Inoculated plants exhibited higher fresh and dry weights compared to non-inoculated controls, alongside increased levels of secondary metabolites such as phenols and flavonoids. Furthermore, there was an enhancement in enzymatic activities related to secondary metabolism and antioxidant responses, coupled with upregulated expression of stress-related marker genes. This ultimately resulted in reduced hydrogen peroxide levels under low-temperature conditions (Chen et al. 2013). Research by Hashem et al. (2016) also highlights the effectiveness of various AM fungi (*Funneliformis mosseae, Rhizophagus intraradices*, and *Claroideoglomus etunicatum*) in mitigating cadmium (Cd) stress in *Solanum lycopersicum.* Their findings indicated a reduction in oxidative stress markers and an increase in antioxidant enzyme activity, along with enhanced proline and phenol content. Table 2 provides a comprehensive

Table 2 Mycorrhizal species and their role in stress mitigation for vegetable crops

Mycorrhiza name	Type of stress	Vegetable crop	Mechanism of stress mitigation and effect on plant	References
Septoglomus deserticola and *S. constrictum*	Drought and heat stress	*Solanum lycopersicum* (tomato)	Reduction in oxidative stress Improvement in antioxidant enzyme activities, stomatal conductance, leaf water potential, and biomass	Duc et al. (2018)
Rhizophagus irregularis	Drought	*Lactuca sativa* (lettuce) and *Solanum lycopersicum* (tomato)	Improvement in growth rates and photosystem II efficiency, alteration of hormonal profiles (abscisic acid and strigolactones), and enhancement in physiological responses in host plants	Ruiz-Lozano et al. (2016)
Claroideoglomus etunicatum, Rhizophagus intraradices, and *Funneliformis mosseae*	Salt	*Cucumis sativus* (cucumber)	Enhancement in biomass, pigment synthesis, and antioxidant enzyme activity (SOD, CAT, APX, and GR), mineral uptake, proline, and phenols Reduction in lipid peroxidation, electrolyte leakage, and Na^+ absorption	Hashem et al. (2018)
Paraglomus occultum	Waterlogging	*Cucumis sativus* (cucumber)	Increment in biomass (leaf, stem, root), and osmolyte levels (glucose, sucrose, betaine, proline) Upregulation of heat shock proteins (CsHsp70) and plasma membrane intrinsic proteins (CsPIP)	Xiang et al. (2024)

(continued)

Table 2 (continued)

Mycorrhiza name	Type of stress	Vegetable crop	Mechanism of stress mitigation and effect on plant	References
Funneliformis mosseae, *Claroideoglomus etunicatum*, and *Rhizophagus irregularis*	Cold stress	*Solanum melongena* (eggplant)	Improvement in photochemical reactions and antioxidant defense system Accumulation of protective molecules (e.g., proline and free phenolics) and reduction in membrane damage	Pasbani et al. (2020)
Glomus mosseae	Salt stress	(Tomato) *Lycopersicon esculentum*	Improvement in root colonization, growth, leaf area, chlorophyll content, fruit fresh weight, yield, P and K concentrations, and antioxidant enzyme activities (SOD, CAT, POD, and APX) Decrease in Na levels and MDA content	Abdel Latef and Chaoxing (2011)
Glomus mosseae (1000 spores)	Low and high root zone temperatures	Cucumber (*Cucumis sativus* cv. Super N3)	Increase in shoot and root fresh weight, enhanced antioxidant activity, and phenol content	Haghighi et al. (2015)
Funneliformis mosseae	Chilling stress	Tomato (*Solanum lycopersicum*)	Increase in fresh and dry weight, calcium precipitates in apoplast, and antioxidant enzyme activities Reduction in MDA, H_2O_2, and $O_2\cdot^-$ levels	Liu et al. (2016)
Consortium of *Rhizoglomus fasciculatum* and *Gigaspora* sp.	Salinity stress	Pea (*Pisum sativum*)	Improvement in growth, biomass production, chlorophyll synthesis, nutrient uptake, and higher accumulation of compatible osmolytes Lower cellular leakage of electrolyte	Parihar et al. (2020)
Funneliformis mosseae, *Glomus versiforme*, and *Rhizophagus intraradices*	Heavy metal (Pb, Cd)	Pakchoi (*Brassica chinensis* L. cv. Suzhou)	Increase in shoot biomass Reduction in Pb and Cd concentrations in shoots	Wu et al. (2016)

overview of the interactions between various mycorrhizal species and a range of vegetable crops. It highlights the diverse mitigation strategies employed by these symbiotic fungi to enhance plant resilience against different types of abiotic stressors, including drought, salinity, and nutrient deficiency, ultimately illustrating the crucial role that mycorrhizal associations play in promoting sustainable agricultural practices and improving crop performance under challenging environmental conditions.

6 Mycorrhizal Fungi in Disease Control for Vegetable Crops

The reliance on chemical pesticides for disease management has raised concerns due to their ineffectiveness against fungicide-resistant pathogens and their negative impact on public health and the environment. In contrast, AMF have emerged as promising biocontrol agents, enhancing plant resistance mechanisms, improving antioxidant defenses, and limiting pathogen access to roots, thereby promoting healthier plant growth and productivity (Wu et al. 2021). Research indicated that AMF alleviated biotic stress from Fusarium wilt in cucumber (*Cucumis sativus* L.), resulting in increased plant biomass and fruit yields, particularly when an AMF consortium was utilized, which significantly reduced disease incidence compared to non-inoculated controls (Hu et al. 2010). Furthermore, AMF have been shown to effectively reduce oxidative stress in watermelon plants infected by the soil-borne pathogen *Rhizoctonia solani*, as evidenced by decreased relative electrolyte leakage, hydrogen peroxide levels, and lipid peroxidation, alongside enhanced root activity and antioxidant enzyme expression (Wu et al. 2021). Specific AMF species, such as *Glomus mosseae* and *Glomus clarum*, have been reported to lessen disease severity caused by Rhizoctonia solani in cucumber plants by 21% and 25%, respectively, compared to a 65% increase in non-inoculated controls (Aljawasim et al. 2020). Additionally, *Glomus bagyarajii* has proven effective in mitigating wilt diseases in tomato caused by *Ralstonia solanacearum*, *Phytophthora capsici*, and *Meloidogyne incognita*, significantly enhancing plant growth parameters and achieving the lowest disease index (Kumar et al. 2018). In greenhouse experiments, *Glomus etunicatum* significantly suppressed disease severity from *Phytophthora infestans* in *Capsicum annuum*, demonstrating reduced pathogenic effects on seedlings (Salami 2002). Moreover, an AMF consortium comprising *Rhizophagus intraradices*, *Funneliformis mosseae*, and *Claroideoglomus etunicatum* has been shown to reduce wilt severity caused by *Fusarium oxysporum* f. sp. *lycopersici*, enhance systemic plant defenses, and improve chlorophyll content and antioxidant enzyme activities (Hashem et al. 2021). Li et al. (2021) demonstrated that the AMF species *Funneliformis caledonium* significantly alleviated *Ralstonia* wilt in tomato by improving soil quality, suppressing populations of *Ralstonia solanacearum*, and activating plant defense mechanisms, resulting in a 65.7% reduction in disease severity and a 46.5% decrease in yield loss. Ahamad et al. (2023) evaluated the synergistic effects of AMF and vermicompost in mitigating the detrimental impacts of *Meloidogyne incognita* on carrot growth and physiology, emphasizing their potential as sustainable alternatives to chemical pesticides. The ability of indigenous West African AMF to suppress root-knot nematodes in vegetables has also been highlighted by Affokpon et al. (2011), showing significant reductions in nematode multiplication and root galling while enhancing crop yields in both pot and field experiments. Additionally, Ait Rahou et al. (2021) found that soil amendment with compost and AMF significantly reduced disease severity from *Verticillium dahliae* in tomato while enhancing plant growth metrics such as stomatal conductance and chlorophyll fluorescence. The dual application of compost and AMF improved antioxidant enzyme activities and decreased malondialdehyde accumulation, leading to increased yield and fruit quality in infected plants.

7 Improving Nutritional Quality in Vegetables Through Mycorrhizal Interactions

Many commonly cultivated crops frequently exhibit essential nutrient levels that fall below the minimum recommended thresholds, which can significantly impact their nutritional quality (Rozpądek et al. 2016). Additionally, post-harvest processing and storage practices can further reduce the nutritional value of these crops (Rozpądek et al. 2016). While countries such as the United States and various European nations have made strides in addressing nutrient deficiencies through a wide range of food options and biofortification initiatives, extensive farming practices have nonetheless led to a noticeable decline in crop quality (Miret and Munné-Bosch 2014; Rozpądek et al. 2016). Sustainable agriculture presents a viable solution for enhancing nutrient uptake and crop yields by utilizing microbial symbionts like mycorrhizal fungi and nitrogen-fixing rhizobacteria, which not only improve the production of beneficial phytochemicals that may help prevent chronic diseases but also provide stress protection for plants (Rouphael et al. 2015). The levels and types of phytochemicals in plants are influenced by various factors, including plant genotype, crop management practices, harvest timing, soil quality, nutrient availability, light conditions, and water supply (Sbrana et al. 2014). Healthy soil and biological fertility are essential for ensuring the safety of plant foods. Beneficial microorganisms, particularly arbuscular mycorrhizal fungi, enhance plant nutrition and overall health. These microorganisms stimulate the synthesis of beneficial phytochemicals such as polyphenols, carotenoids, flavonoids, and phytoestrogens while boosting antioxidant enzyme activity (Sbrana et al. 2014; Agnolucci et al. 2020). Giovannetti et al. (2012) demonstrated that mycorrhizal symbiosis with *Glomus intraradices* in tomato plants improved growth, mineral nutrient uptake, and significantly increased the nutritional value of tomato fruits by elevating lycopene levels without producing genotoxic compounds. In onions, the application of a commercial arbuscular mycorrhizal fungi inoculum enhanced antioxidant activity and increased quercetin glucoside levels (Perner et al. 2008).

AMF in the soil can enhance the absorption of inorganic phosphorus from both soluble and insoluble sources, including rock phosphates. This improvement occurs through localized changes in pH and the secretion of organic acid anions that act as chelating agents. Additionally, mycorrhizal colonization can influence the uptake of other essential nutrients, such as potassium, sodium, magnesium, copper, iron, zinc, and manganese, in developing plants (Chinnathambi et al. 2024). For instance, inoculation with *Glomus mosseae* has been shown to enhance mineral nutrient uptake in tomato plants by increasing phosphorus and potassium concentrations (Abdel Latef and Chaoxing 2011). Furthermore, the application of *Glomus mosseae* significantly improved the uptake of nitrogen (N), phosphorus (P), potassium (K), calcium (Ca),

Table 3 Mycorrhizal fungal species and their impact on nutrient enhancement in host vegetable crops

Mycorrhizal fungal species	Host vegetable crop	Enhanced nutrients/ phytocompounds	References
Glomus fasciculatum	Lettuce	Cu, Fe, anthocyanins, carotenoids, and phenolics	Baslam et al. (2011)
Funneliformis mosseae	Lettuce (*Lactuca sativa* L. cv. Salinas)	N, P, Mg, carotenoids, and phenols	Cela et al. (2022)
Funneliformis mosseae	Tomato	Free amino acids, lycopene, β-carotene, and carotenoids	Ullah et al. (2023)
Glomus intraradices, G. etunicatum, G. clarum, G. caledonium, G. mosseae	Eggplant (*Solanum melongena* L.)	P and Zn	Ortas et al. (2011)
Rhizophagus irregularis, Funneliformis mosseae, Claroideoglumus etunicatum	Potato (*Solanum tuberosum* L.)	Ergothioneine	Carrara et al. (2023)
Rhizophagus prolifer, Claroideoglomus etunicatum (KS-02), *Acaulospora mellea*	Tomato (*Lycopersicon esculentum* Mill.)	Lycopene and carotenoids	Suebrasri et al. (2023)
Rhizophagus intraradices, Endogone mosseae, Claroideoglomus claroideum, Claroideoglomus etunicatum, Gigaspora margarita, and *Entrophospora* sp	Sweet pepper (*Capsicum annuum* L.)	K, Ca, and Mg	Franczuk et al. (2023)
Funneliformis mosseae, and *Rhizophagus intraradices*	Tomato (*Solanum lycopersicum* L.)	N, P, K, anthocyanins, and phenolic compounds	Singh et al. (2024)
Rhizophagus irregularis, Rhizophagus aggregatum, and *Claroideoglomus etunicatum*	Broccoli (*Brassica oleracea* L.)	Zn and Fe	Arifuzzaman et al. (2024)
Rhizophagus irregularis	Onion (*Allium cepa*)	Dietary phytochemicals, vitamin B1, and organic acids	Rozpądek et al. (2016)
Glomus fasciculatum	Green and red leaf lettuce	Carotenoids and tocopherols	Baslam et al. (2013)

iron (Fe), zinc (Zn), copper (Cu), manganese (Mn), boron (B), and molybdenum (Mo) in okra and pea crops grown in acid alfisol (Kumar et al. 2017). Table 3 further illustrates the impact of mycorrhizal interactions on enhancing nutrients and other beneficial phytocompounds in vegetable crops.

8 Conclusion

In conclusion, the significant challenges posed by climate change, environmental pollution, and the shortcomings of conventional farming practices underscore the need for a new approach to vegetable crop cultivation. Mycorrhizal fungi offer a sustainable alternative to chemical fertilizers, greatly enhancing plant growth and resilience. Their ability to improve nutrient uptake—particularly phosphorus—while providing resistance to various environmental stresses makes them essential for promoting sustainable agriculture. Incorporating mycorrhizal fungi into vegetable farming not only increases crop yields and quality but also supports soil health and biodiversity. With the rising global demand for food, it is crucial to adopt methods that reduce environmental impact while boosting productivity. By leveraging the benefits of mycorrhizal fungi, we can create a more resilient agricultural system that can cope with climate challenges and ensure food security for future generations. This chapter highlights the importance of transitioning to eco-friendly agricultural practices. Utilizing mycorrhizal fungi is a key step toward achieving sustainable food production, enhancing nutritional quality, and promoting environmentally friendly methods. As research continues to uncover the many benefits of these microorganisms, their potential to transform vegetable production becomes increasingly clear, paving the way for a more sustainable agricultural future.

References

Abdel Latef AAH, Chaoxing H (2011) Effect of arbuscular mycorrhizal fungi on growth, mineral nutrition, antioxidant enzymes activity and fruit yield of tomato grown under salinity stress. Sci Hortic 127(3):228–233. https://doi.org/10.1016/j.scienta.2010.09.020

Affokpon A, Coyne DL, Lawouin L, Tossou C, Dossou Agbèdè R, Coosemans J (2011) Effectiveness of native west African arbuscular mycorrhizal fungi in protecting vegetable crops against root-knot nematodes. Biol Fertil Soils 47(2):207–217. https://doi.org/10.1007/s00374-010-0525-1

Agnolucci M, Avio L, Palla M, Sbrana C, Turrini A, Giovannetti M (2020) Health-promoting properties of plant products: the role of mycorrhizal fungi and associated bacteria. Agronomy 10(12):1864. https://doi.org/10.3390/agronomy10121864

Ahamad L, Bhat AH, Kumar H, Rana A, Hasan Md. N, Ahmed I, Ahmed S, Machado RAR, Ameen F (2023) From soil to plant: strengthening carrot defenses against *Meloidogyne incognita* with vermicompost and arbuscular mycorrhizal fungi biofertilizers. Front Microbiol 14:1206217. https://doi.org/10.3389/fmicb.2023.1206217

Ait Rahou Y, Ait-El-Mokhtar M, Anli M, Boutasknit A, Ben-Laouane R, Douira A, Benkirane R, el Modafar C, Meddich A (2021) Use of mycorrhizal fungi and compost for improving the growth and yield of tomato and its resistance to *Verticillium dahliae*. Arch Phytopath Plant Protect 54(13–14):665–690. https://doi.org/10.1080/03235408.2020.1854938

Alam MZ, Choudhury TR, Mridha MAU (2023) Arbuscular mycorrhizal fungi enhance biomass growth, mineral content, and antioxidant activity in tomato plants under drought stress. J Food Qual 2023:1–14. https://doi.org/10.1155/2023/2581608

Alias ARN, AL-Bayati HJ, Thanoon AH (2021) Effect of endotrophic mycorrhiza and chemical (NPK) and bio-fertilizer on the growth and yield of eggplant (Solanum melongena L.). Plant Archives, 21(Suppliment-1):1094–1101. https://doi.org/10.51470/PLANTARCHIVES.2021.v21.S1.171

Aljawasim BD, Khaeim HM, Manshood MA (2020) Assessment of arbuscular mycorrhizal fungi (*Glomus* spp.) as potential biocontrol agents against damping-off disease *Rhizoctonia solani* on cucumber. J Crop Protect 9(1):141–147. https://jcp.modares.ac.ir/article-3-33473-en.html

Almaca A, Almaca ND, Senbayram M (2021) Interactive effect of arbuscular mycorrhizal fungi inoculation and phosphorus fertilizer applicatoin on yield and nutrient content of red pepper (Capsicum annuum L.) in a semi-arid region. Applied Ecology and Environmental Research, 19(4):2779–2792. https://doi.org/10.15666/aeer/1904_27792792

Almuslimawi AAA, Kuchár B, Navas SEA, Turóczi G, Posta K (2024) The effect of combined application of biocontrol microorganisms and arbuscular mycorrhizal fungi on plant growth and yield of tomato (Solanum lycopersicum L.). Agriculture 14(5):768. https://doi.org/10.3390/agriculture14050768

Anand K, Pandey GK, Kaur T, Pericak O, Olson C, Mohan R, Akansha K, Yadav A, Devi R, Kour D, Rai AK, Kumar M, Yadav AN (2022) Arbuscular mycorrhizal fungi as a potential biofertilizers for agricultural sustainability. J Appl Biol Biotech 10:90–107. https://doi.org/10.7324/JABB.2022.10s111

Andrino A, Guggenberger G, Kernchen S, Mikutta R, Sauheitl L, Boy J (2021) Production of organic acids by arbuscular mycorrhizal fungi and their contribution in the mobilization of phosphorus bound to iron oxides. Front Plant Sci 12:661842. https://doi.org/10.3389/fpls.2021.661842

Arifuzzaman M, Bagchi R, Hasan MR, Rahman MA, Rahman MM, Rahman MA, Kabir AH (2024) Arbuscular mycorrhizal fungi induce Zn uptake and antioxidant efficiency in broccoli (*Brassica oleracea* L.) exposed to Zn deficiency. Plant. Stress 11:100415. https://doi.org/10.1016/j.stress.2024.100415

Anozie H, Okereke V (2018) Response of Three Varieties of Okra (Abelmoschus esculentus L. Moench) to Arbuscular Mycorrhizal Fungi (Gigaspora gigantea) in the Humid Tropics. International Journal of Agriculture and Earth Science 4(1):16–21

Augé RM, Toler HD, Saxton AM (2015) Arbuscular mycorrhizal symbiosis alters stomatal conductance of host plants more under drought than under amply watered conditions: a meta-analysis. Mycorrhiza 25(1):13–24. https://doi.org/10.1007/s00572-014-0585-4

Banasiak J, Jamruszka T, Murray JD, Jasiński M (2021) A roadmap of plant membrane transporters in arbuscular mycorrhizal and legume–rhizobium symbioses. Plant Physiol 187(4):2071–2091. https://doi.org/10.1093/plphys/kiab280

Banu MB, Mian MH, Rahman MM, Moslehuddin AZM, Matin MQI. Assessment of the Effects of Selected AM Strain on the Growth and Yield of Brinjal. International Journal of Plant & Soil Science 34(23):1538–1547

Barman J, Samanta A, Saha B, Datta S (2016) Mycorrhiza resonance 21(12):1093–1104. https://doi.org/10.1007/s12045-016-0421-6

Baslam M, Garmendia I, Goicoechea N (2011) Arbuscular mycorrhizal fungi (AMF) improved growth and nutritional quality of greenhouse-grown lettuce. J Agric Food Chem 59(10):5504–5515. https://doi.org/10.1021/jf200501c

Baslam M, Esteban R, García-Plazaola JI, Goicoechea N (2013) Effectiveness of arbuscular mycorrhizal fungi (AMF) for inducing the accumulation of major carotenoids, chlorophylls and tocopherol in green and red leaf lettuces. Appl Microbiol Biotechnol 97(7):3119–3128. https://doi.org/10.1007/s00253-012-4526-x

Baum C, El-Tohamy W, Gruda N (2015) Increasing the productivity and product quality of vegetable crops using arbuscular mycorrhizal fungi: a review. Sci Hortic 187:131–141. https://doi.org/10.1016/j.scienta.2015.03.002

Begum N, Qin C, Ahanger MA, Raza S, Khan MI, Ashraf M, Ahmed N, Zhang L (2019) Role of arbuscular mycorrhizal fungi in plant growth regulation: implications in abiotic stress tolerance. Front Plant Sci 10. https://doi.org/10.3389/fpls.2019.01068

Boyno G, Rezaee Danesh Y, Demir S, Teniz N, Mulet JM, Porcel R (2023) The complex interplay between arbuscular mycorrhizal fungi and strigolactone: mechanisms, sinergies, applications and future directions. Int J Mol Sci 24(23):16774. https://doi.org/10.3390/ijms242316774

Brown L (2021) The impact of sowing methods on crop yield. J Agric Sci 56(2):134–145

Carrara JE, Reddivari L, Lehotay SJ, Zinati G, Heller WP (2023) Arbuscular mycorrhizal fungi increase the yield and nutritional quality of yellow and purple fleshed potatoes (*Solanum tuberosum*). Am J Potato Res 100(3):210–220. https://doi.org/10.1007/s12230-023-09910-w

Cela F, Avio L, Giordani T, Vangelisti A, Cavallini A, Turrini A, Sbrana C, Pardossi A, Incrocci L (2022) Arbuscular mycorrhizal fungi increase nutritional quality of soilless grown lettuce while overcoming low phosphorus supply. Foods 11(22):3612. https://doi.org/10.3390/foods11223612

Chen S, Jin W, Liu A, Zhang S, Liu D, Wang F, Lin X, He C (2013) Arbuscular mycorrhizal fungi (AMF) increase growth and secondary metabolism in cucumber subjected to low temperature stress. Sci Hortic 160:222–229. https://doi.org/10.1016/j.scienta.2013.05.039

Cheng S, Zou Y-N, Kuča K, Hashem A, Abd Allah EF, Wu QS (2021) Elucidating the mechanisms underlying enhanced drought tolerance in plants mediated by arbuscular mycorrhizal fungi. Front Microbiol 12. https://doi.org/10.3389/fmicb.2021.809473

Chinnathambi S, Peeran MF, Srinivasan V, Sankar SM, George P (2024) Optimizing mycorrhizal fungi application for improved nutrient uptake, growth, and disease resistance in cardamom seedlings (*Elettaria cardamomum* (L.) Maton). Heliyon 10(20):e39227. https://doi.org/10.1016/j.heliyon.2024.e39227

Cornejo P, Seguel A, Aguilera P, Meier S, Larsen J, Borie F (2017) Arbuscular mycorrhizal fungi improve tolerance of agricultural plants to cope abiotic stress conditions. In: Singh D, Singh H, Prabha R (eds) Plant-microbe interactions in agro-ecological perspectives. Springer, Singapore. https://doi.org/10.1007/978-981-10-6593-4_3

Crosino A, Genre A (2022) Peace talks: symbiotic signaling molecules in arbuscular mycorrhizas and their potential application. J Plant Interact 17(1):824–839. https://doi.org/10.1080/17429145.2022.2108150

Devi TS, Gupta S, Kapoor R (2019) Arbuscular mycorrhizal fungi in alleviation of cold stress in plants. In: Satyanarayana T, Deshmukh S, Deshpande M (eds) Advancing frontiers in mycology and mycotechnology. Springer, Singapore. https://doi.org/10.1007/978-981-13-9349-5_17

Dhalaria R, Kumar D, Kumar H, Nepovimova E, Kuča K, Torequl Islam M, Verma R (2020) Arbuscular mycorrhizal fungi as potential agents in ameliorating heavy metal stress in plants. Agronomy 10(6):815. https://doi.org/10.3390/agronomy10060815

Dhankher OP, Foyer CH (2018) Climate resilient crops for improving global food security and safety. Plant Cell Environ 41(5):877–884. https://doi.org/10.1111/pce.1320

Diagne N, Ngom M, Djighaly PI, Fall D, Hocher V, Svistoonoff S (2020) Roles of arbuscular mycorrhizal fungi on plant growth and performance: importance in biotic and abiotic stressed regulation. Diversity 12(10):370. https://doi.org/10.3390/d12100370

Diedhiou S, Diatta Y, Diedhiou L, Faye F, Goudiaby A, Ndoye I, Fall S (2023) Effects of mycorrhization and organic amendments on sweet potato (Ipomoea batatas L.) growth in a saline environment. Journal of Materials and Environmental Science 14(11):1214–1229

Duc NH, Csintalan Z, Posta K (2018) Arbuscular mycorrhizal fungi mitigate negative effects of combined drought and heat stress on tomato plants. Plant Physiol Biochem 132:297–307. https://doi.org/10.1016/j.plaphy.2018.09.011

Duc NH, Szentpéteri V, Mayer Z, Posta K (2023) Distinct impact of arbuscular mycorrhizal isolates on tomato plant tolerance to drought combined with chronic and acute heat stress. Plant Physiol Biochem 201:107892. https://doi.org/10.1016/j.plaphy.2023.107892

Elbon A, Whalen JK (2015) Phosphorus supply to vegetable crops from arbuscular mycorrhizal fungi: a review. Biol Agric Hortic 31(2):73–90. https://doi.org/10.1080/01448765.2014.966147

El-Sherbeny TMS, Mousa AM, El-Sayed E-SR (2022) Use of mycorrhizal fungi and phosphorus fertilization to improve the yield of onion (Allium cepa L.) plant. Saudi Journal of Biological Sciences 29(1):331–338. https://doi.org/10.1016/j.sjbs.2021.08.094

Emran M, Rashad M, Gispert M, Pardini G (2017) Increasing soil nutrients availability and sustainability by glomalin in alkaline soils. Int J Agric Biosyst Eng 2:74–84

Enebe MC, Erasmus M (2023) Susceptibility and plant immune control—a case of mycorrhizal strategy for plant colonization, symbiosis, and plant immune suppression. Front Microbiol 14. https://doi.org/10.3389/fmicb.2023.1178258

Etesami H, Jeong BR, Glick BR (2021) Contribution of arbuscular mycorrhizal fungi, phosphate–solubilizing bacteria, and silicon to P uptake by plant. Front Plant Sci 12:699618. https://doi.org/10.3389/fpls.2021.699618

Evelin H, Devi TS, Gupta S, Kapoor R (2019) Mitigation of salinity stress in plants by arbuscular mycorrhizal symbiosis: current understanding and new challenges. Front Plant Sci 10. https://doi.org/10.3389/fpls.2019.00470

Feddermann N, Finlay R, Boller T, Elfstrand M (2010) Functional diversity in arbuscular mycorrhiza—the role of gene expression, phosphorous nutrition and symbiotic efficiency. Fungal Ecol 3(1):1–8. https://doi.org/10.1016/j.funeco.2009.07.003

Ferrol N, Azcón-Aguilar C, Pérez-Tienda J (2019) Review: arbuscular mycorrhizas as key players in sustainable plant phosphorus acquisition: an overview on the mechanisms involved. Plant Sci 280:441–447. https://doi.org/10.1016/j.plantsci.2018.11.011

Fiorilli V, Lanfranco L, Bonfante P (2013) The expression of GintPT, the phosphate transporter of *Rhizophagus irregularis*, depends on the symbiotic status and phosphate availability. Planta 237(5):1267–1277. https://doi.org/10.1007/s00425-013-1842-z

Franczuk J, Tartanus M, Rosa R, Zaniewicz-Bajkowska A, Dębski H, Andrejiová A, Dydiv A (2023) The effect of mycorrhiza fungi and various mineral fertilizer levels on the growth, yield, and nutritional value of sweet pepper (*Capsicum annuum* L.). Agriculture 13(4):857. https://doi.org/10.3390/agriculture13040857

Fracasso A, Telò L, Lanfranco L, Bonfante P, Amaducci S (2020) Physiological beneficial effect of rhizophagus intraradices inoculation on tomato plant yield under water deficit conditions. Agronomy 10(1):71. https://doi.org/10.3390/agronomy10010071

Gallegos-Cedillo VM, Nájera C, Gruda NS, Signore A, Gallegos J, Rodríguez R, Ochoa J, Egea-Gilabert C, Fernández JA (2024) An in-depth analysis of sustainable practices in vegetable seedlings nurseries: A review. Scientia Horticulturae 334:113342. https://doi.org/10.1016/j.scienta.2024.113342

Giovannetti M, Avio L, Barale R, Ceccarelli N, Cristofani R, Iezzi A, Mignolli F, Picciarelli P, Pinto B, Reali D, Sbrana C (2012) Nutraceutical value and safety of tomato fruits produced by mycorrhizal plants. Br J Nutr 107(2):242–251. https://doi.org/10.1017/S000711451100290X

Ha S, Tran L-S (2014) Understanding plant responses to phosphorus starvation for improvement of plant tolerance to phosphorus deficiency by biotechnological approaches. Crit Rev Biotechnol 34(1):16–30. https://doi.org/10.3109/07388551.2013.783549

Haghighi M, Mozafariyan M, Abdolahipour B (2015) Effect of cucumber mycorrhiza inoculation under low and high root temperature grown on hydroponic conditions. J Crop Sci Biotechnol 18(2):89–96. https://doi.org/10.1007/s12892-014-0083-4

Hashem A, Abd Allah EF, Alqarawi AA, al Huqail AA, Egamberdieva D, Wirth S (2016) Alleviation of cadmium stress in *Solanum lycopersicum* L. by arbuscular mycorrhizal fungi via induction of acquired systemic tolerance. Saudi J Biol Sci 23(2):272–281. https://doi.org/10.1016/j.sjbs.2015.11.002

Hashem A, Alqarawi AA, Radhakrishnan R, Al-Arjani A-BF, Aldehaish HA, Egamberdieva D, Abd Allah EF (2018) Arbuscular mycorrhizal fungi regulate the oxidative system, hormones and ionic equilibrium to trigger salt stress tolerance in *Cucumis sativus* L. Saudi J Biol Sci 25(6):1102–1114. https://doi.org/10.1016/j.sjbs.2018.03.009

Hashem A, Akhter A, Alqarawi AA, Singh G, Almutairi KF, Abd Allah EF (2021) Mycorrhizal fungi induced activation of tomato defense system mitigates Fusarium wilt stress. Saudi J Biol Sci 28(10):5442–5450. https://doi.org/10.1016/j.sjbs.2021.07.025

Hinsinger P, Bengough AG, Vetterlein D, Young IM (2009) Rhizosphere: biophysics, biogeochemistry and ecological relevance. Plant and Soil 321:117–152. https://doi.org/10.1007/s11104-008-9885-9

Hu J-L, Lin X-G, Wang J-H, Shen W-S, Wu S, Peng S-P, Mao T-T (2010) Arbuscular mycorrhizal fungal inoculation enhances suppression of cucumber *Fusarium* wilt in greenhouse soils. Pedosphere 20(5):586–593. https://doi.org/10.1016/S1002-0160(10)60048-3

Jaggi V, Upadhayay VK, Joshi S, Dasila H, Sahgal M (2023) Rhizosphere engineering for sustainable agriculture. In: Advanced microbial technology for sustainable agriculture and environment. Elsevier, pp 119–136. https://doi.org/10.1016/B978-0-323-95090-9.00004-2

Jawad M, Jabeen A, Gurmani AR, Rafique M (2023) Phosphate-mobilizing mycorrhizal-based bio-fertilizers. In: Biofertilizers for sustainable soil management. CRC Press, pp 119–136

Jayachandran K, Schwab AP, Hetricic BAD (1992) Mineralization of organic phosphorus by vesicular-arbuscular mycorrhizal fungi. Soil Biol Biochem 24(9):897–903. https://doi.org/10.1016/0038-0717(92)90012-M

Khalid MF, Huda S, Yong M, Li L, Li L, Chen Z-H, Ahmed T (2023) Alleviation of drought and salt stress in vegetables: crop responses and mitigation strategies. Plant Growth Regul 99(2):177–194. https://doi.org/10.1007/s10725-022-00905-x

Khan A, Singh AV, Gautam SS, Agarwal A, Punetha A, Upadhayay VK, Kukreti B, Bundela V, Jugran AK, Goel R (2023) Microbial bioformulation: a microbial assisted biostimulating fertilization technique for sustainable agriculture. Front Plant Sci 14. https://doi.org/10.3389/fpls.2023.1270039

Kumar A, Choudhary AK, Suri VK (2017) Agronomic bio-fortification and quality enhancement in okra–pea cropping system through arbuscular mycorrhizal fungi at varying phosphorus and irrigation regimes in Himalayan acid alfisol. J Plant Nutr 40(8):1213–1229. https://doi.org/10.1080/01904167.2016.1267208

Kumar MR, Ashwin R, Bagyaraj DJ (2018) Screening arbuscular mycorrhizal fungi in order to select the best for alleviating wilt disease complex of capsicum. Proc Natl Acad Sci India Sect B Biol Sci 88(2):679–684. https://doi.org/10.1007/s40011-016-0804-1

Kumar Yadav V, Krishna Jha R, Kaushik P, Altalayan FH, al Balawi T, Alam P (2021) Traversing arbuscular mycorrhizal fungi and Pseudomonas fluorescens for carrot production under salinity. Saudi Journal of Biological Sciences 28(8):4217–4223. https://doi.org/10.1016/j.sjbs.2021.06.025

Kumari M, Verma SC, Shweta S (2018) Climate change and vegetable crops cultivation: a review. Indian J Agric Sci 88(2):167–174. https://doi.org/10.56093/ijas.v88i2.79158

Leventis G, Tsiknia M, Feka M, Ladikou EV, Papadakis IE, Chatzipavlidis I, Papadopoulou K, Ehaliotis C (2021) Arbuscular mycorrhizal fungi enhance growth of tomato under normal and drought conditions, via different water regulation mechanisms. Rhizosphere 19:100394. https://doi.org/10.1016/j.rhisph.2021.100394

Li M, Hou S, Wang J, Hu J, Lin X (2021) Arbuscular mycorrhizal fungus suppresses tomato (*Solanum lycopersicum* Mill.) *Ralstonia* wilt via establishing a soil–plant integrated defense system. J Soils Sediments 21(11):3607–3619. https://doi.org/10.1007/s11368-021-03016-8

Liu A, Chen S, Wang M, Liu D, Chang R, Wang Z, Lin X, Bai B, Ahammed GJ (2016) Arbuscular mycorrhizal fungus alleviates chilling stress by boosting redox poise and antioxidant potential of tomato seedlings. J Plant Growth Regul 35(1):109–120. https://doi.org/10.1007/s00344-015-9511-z

Lu Y, Yan Y, Qin J, Ou L, Yang X, Liu F, Xu Y (2023) Arbuscular mycorrhizal fungi enhance phosphate uptake and alter bacterial communities in maize rhizosphere soil. Front Plant Sci 14. https://doi.org/10.3389/fpls.2023.1206870

Ma J, Janoušková M, Li Y, Yu X, Yan Y, Zou Z, He C (2015) Impact of arbuscular mycorrhizal fungi (AMF) on cucumber growth and phosphorus uptake under cold stress. Funct Plant Biol 42(12):1158. https://doi.org/10.1071/FP15106

MacLean AM, Bravo A, Harrison MJ (2017) Plant signaling and metabolic pathways enabling arbuscular mycorrhizal symbiosis. Plant Cell 29(10):2319–2335. https://doi.org/10.1105/tpc.17.00555

Malhi GS, Kaur M, Kaushik P, Alyemeni MN, Alsahli AA, Ahmad P (2021) Arbuscular mycorrhiza in combating abiotic stresses in vegetables: an eco-friendly approach. Saudi J Biol Sci 28(2):1465–1476. https://doi.org/10.1016/j.sjbs.2020.12.001

Miceli A, Vetrano F, Torta L, Esposito A, Moncada A (2023) Effect of mycorrhizal inoculation on melon plants under deficit irrigation regimes. Agronomy 13(2):440. https://doi.org/10.3390/agronomy13020440

Miret JA, Munné-Bosch S (2014) Plant amino acid-derived vitamins: biosynthesis and function. Amino Acids 46(4):809–824. https://doi.org/10.1007/s00726-013-1653-3

Okon IE (2014) Growth Response of Okra (Abelmoschus Esculentus (L) Moench) to Arbuscular Mycorrhizal Fungus Inoculation in Sterile and Non-Sterile Soil. International Journal of Research in Applied, Natural and Social Sciences 2(11):185–192

Ortas I, Sari N, Akpinar C, Yetisir H (2011) Screening mycorrhizae species for increased growth and P and Zn uptake in eggplant (*Solanum melongena* L.) grown under greenhouse conditions. Eur J Hortic Sci 76(3):116–123

Parihar M, Rakshit A, Rana K, Tiwari G, Jatav SS (2020) The effect of arbuscular mycorrhizal fungi inoculation in mitigating salt stress of pea (*Pisum sativum* L.). Commun Soil Sci Plant Anal 51(11):1545–1559. https://doi.org/10.1080/00103624.2020.1784917

Pasbani B, Salimi A, Aliasgharzad N, Hajiboland R (2020) Colonization with arbuscular mycorrhizal fungi mitigates cold stress through improvement of antioxidant defense and accumulation of protecting molecules in eggplants. Sci Hortic 272:109575. https://doi.org/10.1016/j.scienta.2020.109575

Perner H, Rohn S, Driemel G, Batt N, Schwarz D, Kroh LW, George E (2008) Effect of nitrogen species supply and mycorrhizal colonization on organosulfur and phenolic compounds in onions. J Agric Food Chem 56(10):3538–3545. https://doi.org/10.1021/jf073337u

Püschel D, Bitterlich M, Rydlová J, Jansa J (2021) Drought accentuates the role of mycorrhiza in phosphorus uptake. Soil Biol Biochem 157:108243. https://doi.org/10.1016/j.soilbio.2021.108243

Raaijmakers JM, Paulitz TC, Steinberg C, Alabouvette C, Moënne-Loccoz Y (2009) The rhizosphere: a playground and battlefield for soilborne pathogens and beneficial microorganisms. Plant and Soil 321:341–361. https://doi.org/10.1007/s11104-008-9568-6

Radi AJ, Ventura MU, Barazetti AR, Andrade G, Shimizu GD (2024) Inoculation with arbuscular mycorrhizal fungus Rhizophagus clarus on tomato promotes increasing yield under organic farming inputs. Ciência Rural 54(12):e20220585. https://doi.org/10.1590/0103-8478cr20220585

Rouphael Y, Franken P, Schneider C, Schwarz D, Giovannetti M, Agnolucci M, de Pascale S, Bonini P, Colla G (2015) Arbuscular mycorrhizal fungi act as biostimulants in horticultural crops. Sci Hortic 196:91–108. https://doi.org/10.1016/j.scienta.2015.09.002

Rozpądek P, Rąpała-Kozik M, Wężowicz K, Grandin A, Karlsson S, Ważny R, Anielska T, Turnau K (2016) Arbuscular mycorrhiza improves yield and nutritional properties of onion (*Allium cepa*). Plant Physiol Biochem 107:264–272. https://doi.org/10.1016/j.plaphy.2016.06.006

Rui W, Ma J, Wei N, Zhu X, Li Z (2023) Genome-wide analysis of the PHT gene family and its response to mycorrhizal symbiosis in tomatoes under phosphate starvation conditions. Int J Mol Sci 24(12):10246. https://doi.org/10.3390/ijms241210246

Ruiz-Lozano JM, Aroca R, Zamarreño ÁM, Molina S, Andreo-Jiménez B, Porcel R, García-Mina JM, Ruyter-Spira C, López-Ráez JA (2016) Arbuscular mycorrhizal symbiosis induces strigolactone biosynthesis under drought and improves drought tolerance in lettuce and tomato. Plant Cell Environ 39(2):441–452. https://doi.org/10.1111/pce.12631

Salami AO (2002) Influence of mycorrhizal inoculation on disease severity and growth of pepper (*Capsicum annum* Linn.). Arch Agron Soil Sci 48(3):257–262. https://doi.org/10.1080/03650340213840

Santander C, Ruiz A, García S, Aroca R, Cumming J, Cornejo P (2020) Efficiency of two arbuscular mycorrhizal fungal inocula to improve saline stress tolerance in lettuce plants by changes of antioxidant defense mechanisms. J Sci Food Agric 100(4):1577–1587. https://doi.org/10.1002/jsfa.10166

Sbrana C, Avio L, Giovannetti M (2014) Beneficial mycorrhizal symbionts affecting the production of health-promoting phytochemicals. Electrophoresis 35(11):1535–1546. https://doi.org/10.1002/elps.201300568

Selvakumar G, Thamizhiniyan P (2011) The effect of the arbuscular mycorrhizal (AM) fungus Glomus intraradices on the growth and yield of chilli (Capsicum annuum L.) under salinity stress. World Applied Sciences Journal 14(8):1209–1214

Shabana A (2016) Responce of snap bean plants (Phaseolus vulgare) to mycorrhizal fungi under different levels of phosphate. J Plant Prod 7(8):851–859. https://doi.org/10.21608/jpp.2016.46196

Sharma M, Delta AK, Brar NS, Yadav A, Dhanda PS, Baslam M, Kaushik P (2022) Rhizophagus irregularis and Azotobacter chroococcum Uphold Eggplant Production and Quality under Low Fertilization. International Journal of Plant Biology 13(4):601–612. https://doi.org/10.3390/ijpb13040048

Singh U, Akhtar O, Mishra R, Zoomi I, Kehri HK, Pandey D (2021) Arbuscular mycorrhizal fungi: biodiversity, interaction with plants, and potential applications. In: Industrially important fungi for sustainable development: volume 1: biodiversity and ecological perspectives. Springer, Cham, pp 35–83. https://doi.org/10.1007/978-3-030-67561-5_2

Sini HN, Barzegar R, Mashaee, SS, Ghahsare MG, Mousavi-Fard S, Mozafarian M (2024) Effects of biofertilizer on the production of bell pepper (Capsicum annuum L.) in greenhouse. Journal of Agriculture and Food Research, 16:101060. https://doi.org/10.1016/j.jafr.2024.101060

Singh M, Chauhan A, Srivastava DK, Singh PK (2024) Arbuscular mycorrhizal fungi promote growth and enhance the accumulation of bioactive compounds in tomato (*Solanum lycopersicum* L.). Biologia Futura 75(2):251–257. https://doi.org/10.1007/s42977-024-00214-6

Smith SE, Smith FA (2011) Roles of arbuscular mycorrhizas in plant nutrition and growth: new paradigms from cellular to ecosystem scales. Ann Rev Plant Biol 62(1):227–250. https://doi.org/10.1146/annurev-arplant-042110-103846

Strullu-Derrien C, Selosse M, Kenrick P, Martin FM (2018) The origin and evolution of mycorrhizal symbioses: from palaeomycology to phylogenomics. New Phytol 220(4):1012–1030. https://doi.org/10.1111/nph.15076

Suebrasri T, Seemakram W, Lapjit C, Mongkolthanaruk W, Boonlue S (2023) Effects of arbuscular mycorrhizal fungi on enhancing growth, fruit quality, and functional substances in tomato fruits (*Lycopersicon esculentum* Mill.). J Plant Biotechnol 50:239–247. https://doi.org/10.5010/JPB.2023.50.030.239

Szentpéteri V, Mayer Z, Posta K (2023) Mycorrhizal symbiosis-induced abiotic stress mitigation through phosphate transporters in *Solanum lycopersicum* L. Plant Growth Regul 99(2):265–281. https://doi.org/10.1007/s10725-022-00906-w

Thilagar G, Bagyaraj DJ (2015) Influence of Different Arbuscular Mycorrhizal Fungi on Growth and Yield of Chilly. Proceedings of the National Academy of Sciences, India Section B: Biological Sciences 85:71–75. https://doi.org/10.1007/s40011-013-0262-y

Tian X, Liu X-Q, Liu X-R, Li Q-S, Abd Allah EF, Wu Q-S (2023) Mycorrhizal cucumber with Diversispora versiformis has active heat stress tolerance by up-regulating expression of both CsHsp70s and CsPIPs genes. Sci Hortic 319:112194. https://doi.org/10.1016/j.scienta.2023.112194

Ullah F, Ullah H, Ishfaq M, Gul SL, Kumar T, Li Z (2023) Improvement of nutritional quality of tomato fruit with *Funneliformis mosseae* inoculation under greenhouse conditions. Horticulturae 9(4):448. https://doi.org/10.3390/horticulturae9040448

Upadhayay VK, Singh J, Khan A, Lohani S, Singh AV (2019) Mycorrhizal mediated micronutrients transportation in food based plants: a biofortification strategy. In: Mycorrhizosphere and pedogenesis. Springer, Singapore, pp 1–24. https://doi.org/10.1007/978-981-13-6480-8_1

Upadhayay VK, Singh AV, Khan A (2022) Cross talk between zinc-solubilizing bacteria and plants: a short tale of bacterial-assisted zinc biofortification. Front Soil Sci 1:788170. https://doi.org/10.3389/fsoil.2021.788170

Upadhayay VK, Chitara MK, Mishra D, Jha MN, Jaiswal A, Kumari G, Ghosh S, Patel VK, Naitam MG, Singh AK, Pareek N, Taj G, Maithani D, Kumar A, Dasila H, Sharma A (2023a) Synergistic impact of nanomaterials and plant probiotics in agriculture: a tale of two-way strategy for long-term sustainability. Front Microbiol 14. https://doi.org/10.3389/fmicb.2023.1133968

Upadhayay VK, Naik YD, Ranjan N, Kastury C, Shekhar S, Kumar S, Jaggi V (2023b) Prolific microbial agents as key products for sustainable agriculture. In: Microbial bioactive compounds. Springer, pp 181–205. https://doi.org/10.1007/978-3-031-40082-7_10

Upadhayay VK, de los Santos Villalobos S, STM A, Kukreti B, Chitara MK, Jaggi V, Sharma A, Singh AV (2024) Microbial advancement in agriculture. In: Chaudhary P, Chaudhary A (eds) Microbial inoculants. Interdisciplinary biotechnological advances. Springer, Singapore. https://doi.org/10.1007/978-981-97-0633-4_5

Vultaggio L, Allevato E, Sabatino L, Ntatsi G, Rouphael Y, Torta L, la Bella S, Consentino BB (2024) Modulation of cherry tomato performances in response to molybdenum biofortification and arbuscular mycorrhizal fungi in a soilless system. Heliyon 10(13):e33498. https://doi.org/10.1016/j.heliyon.2024.e33498

Walder F, Boller T, Wiemken A, Courty P-E (2016) Regulation of plants' phosphate uptake in common mycorrhizal networks: role of intraradical fungal phosphate transporters. Plant Signal Behav 11(2):e1131372. https://doi.org/10.1080/15592324.2015.1131372

Wu Z, Wu W, Zhou S, Wu S (2016) Mycorrhizal inoculation affects Pb and Cd accumulation and translocation in Pakchoi (*Brassica chinensis* L.). Pedosphere 26(1):13–26. https://doi.org/10.1016/S1002-0160(15)60018-2

Wu M, Yan Y, Wang Y, Mao Q, Fu Y, Peng X, Yang Z, Ren J, Liu A, Chen S, Ahammed GJ (2021) Arbuscular mycorrhizal fungi for vegetable (VT) enhance resistance to *Rhizoctonia solani* in watermelon by alleviating oxidative stress. Biol Control 152:104433. https://doi.org/10.1016/j.biocontrol.2020.104433

Xiang N, Liu Z, Tian X, Wang D, Hashem A, Abd Allah EF, Wu Q-S, Zou Y-N (2024) Improved waterlogging tolerance in roots of cucumber plants after inoculation with arbuscular mycorrhizal fungi. Horticulturae 10(5):478. https://doi.org/10.3390/horticulturae10050478

Xu G-h, Chague V, Melamed-Bessudo C, Kapulnik Y, Jain A, Raghothama KG, Levy AA, Silber A (2007) Functional characterization of LePT4: a phosphate transporter in tomato with mycorrhiza-enhanced expression. J Exp Botany 58(10):2491–2501. https://doi.org/10.1093/jxb/erm096

Yang PM, He ST (2022) The effects of arbuscular mycorrhizal fungi and deficit irrigation on the yield and sugar content of watermelons (Citrullus lanatus). Horticultural Science 49(4):225–233. https://doi.org/10.17221/108/2021-HORTSCI

Zhu B, Gao T, Zhang D, Ding K, Li C, Ma F (2022) Functions of arbuscular mycorrhizal fungi in horticultural crops. Sci Hortic 303:111219. https://doi.org/10.1016/j.scienta.2022.111219

Ziane H, Hamza N, Meddad-Hamza A (2021) Arbuscular mycorrhizal fungi and fertilization rates optimize tomato (Solanum lycopersicum L.) growth and yield in a Mediterranean agro-ecosystem. Journal of the Saudi Society of Agricultural Sciences 20(7):454–458. https://doi.org/10.1016/j.jssas.2021.05.009

From Root to Fruit: Mycorrhizal Impact on Fruit Crops

Diksha Kumari, Nidhi Gupta, Nitin Sharma, and Vikas Kumar

Abstract

Arbuscular mycorrhizal fungi (AMF) are recognized for their role in eco-friendly, sustainable agriculture and help mitigate the increased input of chemicals into farmlands. Arbuscular mycorrhiza (AM) is believed to have been crucial in the evolution of land plants around 450 million years ago via increased soil phosphate uptake. In this interaction, the fungus infects the plant's root cortex, forming structures called arbuscules to support nutrient exchange. The fungus helps the plant by enhancing the uptake of phosphorus and nitrogen and can meet all the phosphate needs of the plant from its mycelium. In return, the fungus supplies the plant with carbohydrates such as hexoses and fatty acids. Symbiosis not only allows for increased nutrient acquisition by the plant but also boosts the plant's resistance to pathogens, tolerance of water and salt stress, and pollutants. Furthermore, AMF interactions resulted in earlier flowering, higher yields, and a higher number of marketable fruits. AMF have been demonstrated to increase the yield and fruit quality of crops such as strawberries, cucumbers, Washington Navel oranges, bananas, papayas, plums, peaches, guavas, pomegranates, and apples. Sustainable agriculture depends on mycorrhiza to promote plant growth, affect soil fertility, and improve absorption of phosphorus (P), ammonium (NH_4), copper (Cu), and zinc (Zn). The production of mycorrhizal inoculum can be done

D. Kumari · V. Kumar (✉)
University Institute of Biotechnology, Chandigarh University, Mohali, Punjab, India

N. Gupta
University Institute of Biotechnology, Chandigarh University, Mohali, Punjab, India

Department of Biotechnology, Chandigarh Group of Colleges, Mohali, Punjab, India

N. Sharma
Department of Biotechnology, Chandigarh Group of Colleges, Mohali, Punjab, India

Y. Siddiqui et al. (eds.), *Sustainable Mycorrhizal Cultivation*,
https://doi.org/10.1007/978-981-95-1778-7_6

in vitro, in pot cultures, aeroponics, hydroponics, and on farms. AMF contribute to the process by which resilient ecosystems are promoted by improving soil structure, cycling nutrients, and sequestering carbon. However, these effects depend on the species and the environmental context involved, and their effects on plant growth and ecological stability are highly variable. The impact of mycorrhizal fungi can be increased by exploring how they can be integrated with other sustainable agricultural practices, such as organic farming. Expanding research beyond the current traditional crops, such as corn, soybean, and wheat, and into specialty and high-value plants will allow the evaluation of the complete potential of mycorrhizal fungi in a range of agricultural systems.

Keywords

Arbuscular mycorrhizal fungi · Plant growth · Nutrient cycling · Fruit quality · Sustainable agriculture

1 Introduction

The symbiotic relationship between mycorrhizal fungi and fruit crops is one of the most important biological relationships which contributes significantly to plant health and ecosystem functioning. The term "mycorrhiza" is derived from Greek words, which means "fungus" and "root." Arbuscular mycorrhizal fungi (AMF), belonging to the phylum Glomeromycota, are one of the oldest symbiotic organisms in plants (Blackwell 2000). More than 80% of terrestrial plants form symbiotic associations with AMF, which enhances nutrient absorption, water absorption, growth, and health, and improves stress tolerance in plants (Smith and Read 2010). The coevolution of mycorrhizal fungi and plant roots, traced back to 400 million years, shows how early endophytic associations evolved into mutualistic relationships. Roots are likely developed from rhizomes to support mycorrhizal fungi, whereas different root types and mycorrhizae are adapted through selection pressures during plant evolution (Brundrett 2002). Mycorrhizae or fungal roots are involved in the colonization of plant roots by specialized fungi that extend far beyond the root zone, forming an intricate network in the soil (Brundrett 2002). This is especially important in fruit crops, where nutrient supply, water supply, and resistance to environmental stress are in high demand. In this process, mycorrhizal fungi are essential for mediating the exchange between the soil and the plant in terms of nutrients, such as phosphorus, nitrogen, and micronutrients, and provide precursors of carbohydrates from plant photosynthesis (Ruytinx et al. 2020).

Fruit crops, which are the cornerstone of global agriculture, face challenges such as nutrient depletion, water scarcity, and increasing uncertainty of climate change. These challenges result in a decrease in fruit quantity and quality. Many of these challenges can be overcome using mycorrhizal fungi, which improve root function, increase soil structure, and enhance the plant's ability to resist biotic and abiotic stressors (Shi et al. 2023). Mycorrhizal fungi play an important role that extends beyond their facilitation of better nutrient uptake, and they influence the life cycle of fruit crops from root establishment to fruit maturity. The mycorrhizal relationship in many fruit crop species, including apples, grapes, citrus, and berries, significantly determines crop yield and yield quality (Berdeni et al. 2018; Naik et al. 2018; Mikiciuk et al. 2019). Fruit yield and quality are severely affected by abiotic stresses such as extreme temperatures, salinity, drought, and heavy metals, as well as biotic stresses caused by bacteria, fungi, viruses, nematodes, and herbivorous insects. Plants inoculated with AMF show increased biomass, root and shoot length, chlorophyll content, and less proline accumulation, indicating enhanced drought tolerance (Tallapragada et al. 2016). Mycorrhizal association is involved in the immobilization of soil salinity and heavy metal toxicity, and resistance against pathogen attacks indicates a complex protective system (Sabra et al. 2018; Wahab et al. 2023). Mikiciuk et al. (2019) found that AMF application also increases root colonization, chlorophyll content, CO_2 assimilation and ultimately results in increased yield and fruit size.

The increased use of chemical fertilizers has resulted in an increase in crop yield but has also led to environmental pollution, resource depletion, higher costs, and loss of biodiversity (Jote 2023; Bauer et al. 2016). To balance agricultural productivity and environmental sustainability and ensure future food security, it is important to find alternatives to chemical fertilizers. Sustainable farming is spreading, as are mycorrhizae for their benefits in growing fruit crops (Singh et al. 2024). The detrimental effects of agrochemicals on soil health and the environment can be reduced by utilizing mycorrhizal fungi, which is an eco-friendly method to improve crop yield and quality (Thomas and Singh 2019; Olaniyan et al. 2022). Recent research has shown that systemic resistance to the necrotrophic pathogen *Dickeya fangzhongdai* in terrestrial orchids, *Bletilla striata*, is induced by root colonization by the compatible orchid mycorrhizal fungus, *Tulasnella calospora* (Pujasatria et al. 2024). The use of these beneficial fungi by farmers can decrease reliance on chemical inputs, resulting in healthier soils, more resilient plants, and better fruits. This chapter investigates the effect of mycorrhizal fungi on fruit crops while exploring their role in soil nourishment, nutrient acquisition, stress tolerance, and general plant health. In addition, the diversity of mycorrhizae that benefit fruit crop productivity and quality will also be investigated, as well as the possible benefits for agricultural sustainability. Finally, case studies of several fruit crops where mycorrhizal associations have been applied to enhance growth, yield, and resistance to biotic and abiotic stresses are presented in the chapter.

2 Understanding Mycorrhizal Fungi

As defined by Brundrett (2002), mutualistic relationships occur between the specialized absorptive structures of mycorrhizae, which primarily comprise plant roots (photobionts) and fungal hyphae (mycobionts). Mycorrhizae are categorized into four types based on their structure and function: arbuscular mycorrhizal, ectomycorrhizal, ericoid mycorrhizal, and orchid mycorrhizal (van Der Heijden et al. 2015). Mycorrhizae can also be classified into two main categories based on their

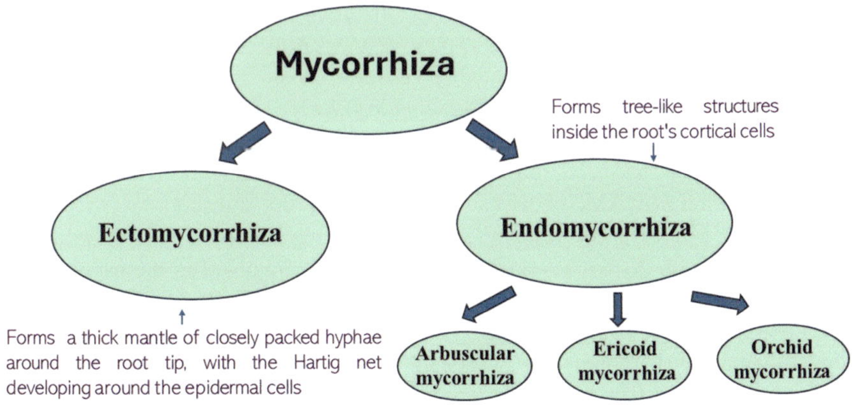

Fig. 1 Types of mycorrhizae

anatomical characteristics: ectomycorrhizae (EMs) and endomycorrhizas. This difference was based on the way the fungus affected the roots of the plant. EMs live between the spaces of cells in the root, whereas endomycorrhizas live inside root cells. Endomycorrhizas are further subdivided into three types: orchid mycorrhiza, ericoid mycorrhiza, and arbuscular mycorrhiza (AMs) (Fig. 1), of which AMs form symbiotic relationships with a range of plant species (Smith and Read 2010).

2.1 Ectomycorrhiza (EM)

Ectomycorrhiza (EM), including Basidiomycota and Ascomycota, form external fungal sheaths around the roots of many tree species, such as pines and oaks. Ectomycorrhiza create an extensive mycelial network in plants (Fig. 2), which enhances water and nutrient uptake, provides protection against biotic stress, and improves plant resilience to environmental stresses. Ectomycorrhizal fungi form a thick mantle of closely packed hyphae around the root tip, with the Hartig net developing around epidermal cells (Bonfante and Genre 2010).

2.2 Ericoid Mycorrhizae

Ericoid Mycorrhizae, primarily associated with Ascomycota and some Basidiomycota, form a unique relationship with ericaceous plants that grow in acidic, low-nutrient environments. These fungi colonize the fine root hairs, helping in nutrient acquisition, particularly nitrogen and phosphorus (Perotto et al. 2002).

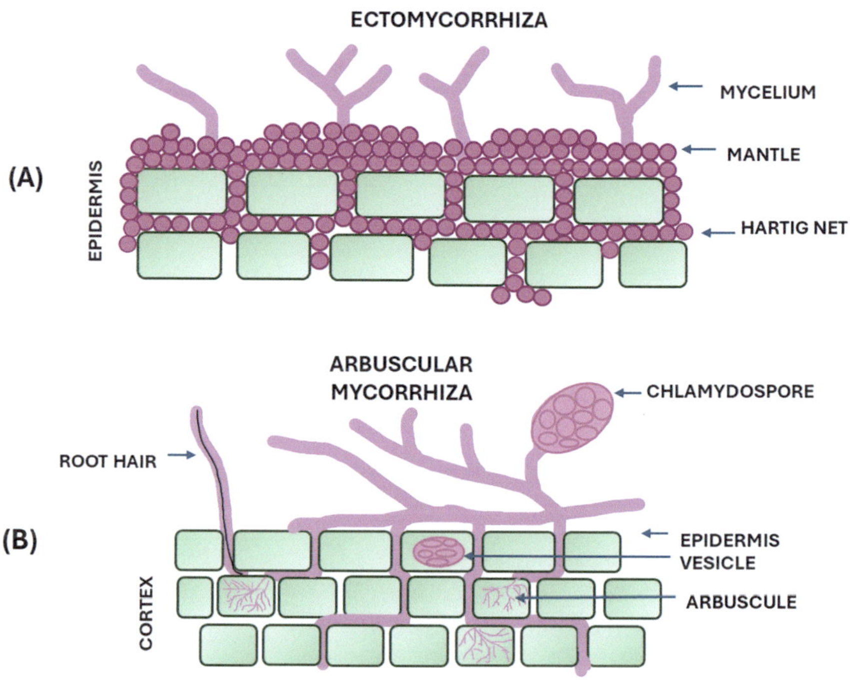

Fig. 2 Ectomycorrhizal (**a**) and endomycorrhizal (**b**) colonization of plant roots. (Huey et al. 2020, license number 6044930630615)

2.3 Arbuscular Mycorrhiza (AM)

Arbuscular Mycorrhiza (AM), formed by fungi in the phylum Glomeromycota, is the most widespread, establishing intracellular associations with over 80% of land plants (Brundrett 2002). AM differs from the other types by producing large amounts of fungal hyphae that grow along the endodermis within the root cortex. They are named after "arbuscules," which are tree-like structures formed by the fungus inside the root's cortical cells (Fig. 2). AM fungi enhance the uptake of vital nutrients, such as phosphorus and water, especially in nutrient-poor soils (Hart and Forsythe 2012).

2.4 Orchid Mycorrhiza

Basidiomycota fungi associated with orchid mycorrhiza are essential for orchid seed germination and early development, as orchid seeds are nutritionally deficient and do not germinate without mycorrhiza. Favre-Godal et al. (2020) found that fungi provide essential carbohydrates and other nutrients to orchids necessary for establishment and growth. Mycorrhizas of each type fulfill a different role in the plant's survival, adaptation, and ecosystem functioning.

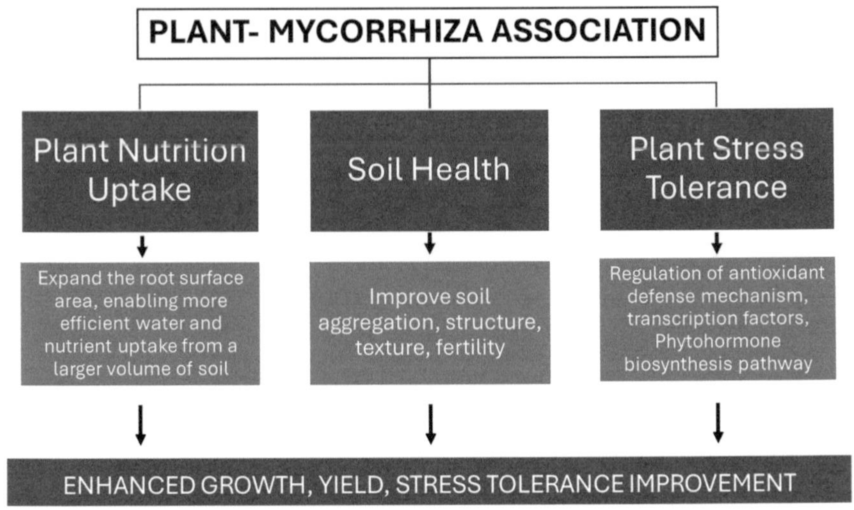

Fig. 3 Mycorrhizal functions in ecosystem regulation and plant growth promotion

3 Mycorrhizal Effects on Plant Growth

Roots colonized by mycorrhizae have two nutrient uptake pathways: the mycorrhizal pathway (MP) and plant pathway (PP). Plant pathway: nutrients are directly absorbed into plant roots by the epidermis of the root and root hairs. In both mycorrhizal pathways, nutrients are transferred by fungal transporters from the extraradical mycelium (ERM) in ectomycorrhizal fungi or intraradical mycelium (IRM) in arbuscular mycorrhizae to the cortex through the Hartig net in the ECM or to the apoplast in AM. Transporters induced by mycorrhizae allow the uptake of nutrients at the plant-fungal interface. The ability of roots to colonize multiple fungal species also varies with respect to their nutrient transfer efficiency (Bücking et al. 2012). AMF can improve plant resistance to environmental stresses such as drought, temperature, and soil salinity, thereby improving plant growth and survival. These technologies can help increase crop yields and maintain the sustainability of smallholder farming practices for future generations (Tsiknia et al. 2021). Figure 3 briefly summarizes the mycorrhizal functions in ecosystem regulation and plant growth promotion.

3.1 Nutrient Uptake

The leaves of plants are used as feed by fungi in association with plants by transporting sugars. Mycorrhizae provide plants with a vast root surface area, aiding the uptake of water and nutrients from a greater volume of soil (Nadeem et al. 2014; Bowles et al. 2016). Results from one study revealed that mycorrhizal inoculation of *Glomus mosseae, G. intraradices,* and moderate levels of phosphorus increased nutrition, increased shoot concentrations of N, P, and Ca, and improved growth and ornamental traits in statice plants. Mycorrhizal treatment delays the flowering of statice plants and improves the yield and quality of statice when inoculum and phosphorus levels are important (Sheikh-Assadi et al. 2023). AM symbiosis enhances crop yield and plant phosphorus acquisition (Hou et al. 2021). Orta and Ustuner (2014) showed that different mycorrhizal species respond uniquely to changes in the growth media with mycorrhizal inoculation, resulting in higher shoot and root dry weights than those of non-inoculated plants. AMF can aid plants in growing more efficiently in low phosphorus (P) and low nitrogen (N) soils by better absorption of nutrients (Liu et al. 2018). Many studies have shown that AMF symbiosis increases nitrogen (N) nutrition in plants by taking up N from the soil (Koegel et al. 2017). Plants can survive better when inoculated with mycorrhiza, yield more fruit, and have higher phosphorus and zinc concentrations in the shoots (Ortas 2010).

3.2 Soil Nourishment

Mycorrhizae form a network of filaments around the roots of plants, increasing the surface area and soil access. Essentially, this symbiotic relationship helps plants increase the uptake of certain essential minerals found in soil, such as zinc, copper, and phosphorus, which are often not mobile in the soil (Huey et al. 2020). Indeed, AMF improve plant health by affecting soil quality through soil structure and texture (Zou et al. 2016; Thirkell et al. 2017). Although small, they have an important role in soil carbon storage by secreting an N-linked glycoprotein, glomalin, present in their spores and hyphae. Glomalin is important in the conservation of soil organic matter; it helps to enhance the aggregation of soil and protects hyphae, which are essential for nutrient cycling, water retention, and soil erosion prevention. Glomalin also binds the soil particles together, thereby stabilizing the soil structure and increasing the capacity of plants to absorb water and nutrients (Atakan and Özkaya 2021). The mycelial network continuously regenerates, and even dead mycelia contribute to the conservation of soil structure until they decompose and contribute to stabilizing the soil and improving its health (Gianinazzi et al. 2010). Dead mycelia also contribute to organic matter stocks and serve as physical binders (i.e., they are a major agent of soil aggregation) (Hamel and Plenchette 2017). Fungal hyphae break down complex organic matter to accelerate the decomposition of soil organic matter and the availability of nutrients for plants (Paterson et al. 2016). AMF inoculation in clay soil has been shown to improve plant growth, nutrient uptake, and soil aggregation, and increase soil organic carbon and microbial biomass, which makes it a potential strategy to enhance plant productivity and soil health in clay soils (Akter et al. 2024). AM fungi also improve the physicochemical properties of barren soils, such as nutrient availability and soil structure. This finding highlights the importance of mycorrhiza in semi-arid environments, where the soil is often of low quality (Pal and Pandey 2016).

3.3 Plant Stress Tolerance

AM fungal inoculation (*Rhizophagus intraradices* and *R. fasciculatum*) and *Burkholderia seminalis* in plants resulted in a significant biomass increase with increased root and shoot length and chlorophyll content, with reduced proline accumulation, indicating alleviation of drought stress while implying sustainable agriculture (Tallapragada et al. 2016). The presence of AMF prevents heavy metal stress. AM inoculation increases plant growth (leaf count, shoot height, dry biomass, and phosphorus concentration in both shoots and roots) and can alleviate Cu toxicity in contaminated soils and improve plant growth (Quan et al. 2023). AMF enhance growth, increase chlorophyll content, increase osmolyte levels, and upregulate CsPIP1;6, CsPIP2;1, and CsPIP2;6 in plants under heat stress (Liu et al. 2023). AMF inoculation in *Elymus nutans* improves plant resistance to the pathogen *Locusta migratoria* by increasing biomass, nutrients, and defense enzyme activities, particularly through the jasmonic acid (JA) signaling pathway. Insect-resistant compounds have also been detected, suggesting AMF's potential for insect control in forage grasses (Zhang et al. 2022). Alkalinity stress increases LOX activity in the leaves and roots of plants, whereas mycorrhizal inoculation generally reduces LOX activity in both stressed and unstressed plants (Farghaly et al. 2022). *G. intraradices* (Biofertilizante INIFAP®) provides salinity tolerance in plants by promoting growth and improving stress-related metabolites, including chlorophyll, carotenoids, proline, glycine betaine, soluble phenols, and antioxidants (Baltazar-Bernal et al. 2022). AM fungi influence phytohormones (ABA, SA, JA, and IAA) and stress-responsive genes (SlAREB1, SlLEA, SlHSP70, and SlHSP90) under combined drought and chronic heat, as well as under drought and heat shock, resulting in enhanced leaf water potential, stomatal conductance, photosynthetic activity, and antioxidant enzyme activity, indicating species-specific benefits in alleviating the effects of combined stress (Duc et al. 2023).

4 Mycorrhizal Fungal Diversity in Fruit Crops

The composition, species, and quantity of mycorrhizal fungi can vary depending on the species of the fruit crop, geographical conditions, environmental factors, soil type, plant age, and agricultural management practices, which influence plant growth, nutrient and water uptake, plant disease resistance, and yield and quality of fruit. Several AMF species, such as *Gigaspora albida, G. macrocarpum, G. mosseae, G. fasciculatum, G. clarum, G. constrictum, Acaulospora scrobiculata, A. bireticulata*, and *Scutellospora* spp., have been found to enhance nutrient uptake, improve growth, and increase resilience against various environmental stresses in temperate agroecosystems of sweet cherry (*Prunus avium*) plants (Kirti et al. 2016). In sweet cherry (*P. avium*) grown in Himachal Pradesh, India, *Glomus* species were the most prevalent (80%), including *G. mosseae, G. macrocarpum, G. clarum, G. fasciculatum, G. epigenum*, and *G. constrictum,* followed by *Acaulospora* (10%) (*A. scrobiculata* and *A. bireticulata*), *Gigaspora* (5%) (*G. heterogama* and

G. albida), and *Scutellospora* (5%) (*S. pellucida*) (Sharma et al. 2016). AMF species found predominantly in apples are *G. fasciculatum*, *G. mosseae*, and *G. macrocarpum,* although younger apple orchards have fewer mycorrhizal associations than mature orchards (Mehta and Bharat 2013).

AMF colonization is affected by altitude, geographical location, and management practices of orchards, although the composition of species varies, as observed in apples (Turrini et al. 2017). Fruit quality is also affected by geographical variation and environmental factors, which significantly influence the AMF composition. *Claroideoglomus*, *Dominikia*, and *Funneliformis* genera were correlated with environmental factors, such as soil pH, and influenced the total sugar and flavonoid content of *Lycium barbarum* fruit (Chen et al. 2022b). AMF community also varied according to seasonal variations in hilly areas. In mandarin orchards (*Citrus reticulata*, *C. limonia*, *C. medica*), *Acaulospora* is found throughout the year, and *Gigaspora* species dominate during the monsoon and winter, whereas *Glomus* species (*G. mosseae*, *G. fasciculatum*, and *G. aggregatum*) prevail in summer (Allay et al. 2021). *R. intraradices*, *G. aggregatum*, and *Funneliformis mosseae* are the predominant species present in hillside red tangerine (*C. reticulata* Blanco) orchards. The number of propagules, root colonization, and diversity of red tangerines are determined by environmental factors (Wang et al. 2013). Co-cultivation of grass in tangerine orchards enhances the stress resistance and stability of AMF species present in the soil (Xiao et al. 2022). The use of cover crops, such as faba bean, enhances the colonization of walnuts (*Juglans regia*) by mycorrhiza (Thioye et al. 2021). In bananas, *Glomus* species dominate the AMF community, comprising 91.1% of the spores, while *Acaulospora* and *Scutellospora* species were also found (Sumathi and Thangavelu 2016). Grape wine (*Vitis vinifera* L.) orchards at the foot of the mountains contain *Glomus* as the dominant genus, with *G. melanosporum* as the dominant AMF, and soil pH and organic matter are key factors influencing AMF abundance, diversity, and community composition (Massa et al. 2020; Zhang et al. 2024). In the Kashmir Himalayas, the mycorrhizal fungi isolated from apples (*Malus domestica* Borkh) included the genera *Glomus, Scutellospora*, *Gigaspora*, and *Acaulospora,* with *Glomus* being the most predominant and *Scutellospora* being the least abundant (Ajaz et al. 2017). Orchard management practices also affect the root colonization rate of fruit crops. A study revealed that organically managed almond orchards have higher root colonization (73.2%) than the conventional technique (62.1%) (Vasilikiotis et al. 2020).

5 Impact on Fruit Crops

Studying AM symbiosis in relation to various biotic and abiotic stresses, such as drought, salinity, alkalinity, and temperature, can inform the practical uses of arbuscular mycorrhizal fungi (AMF) in sustainable farming. AMF applications can improve the growth and uptake of nutrients such as phosphorus and nitrogen, thereby reducing dependency on chemical fertilizers and promoting both higher crop yields and environmental protection (Wang et al. 2024a, b).

5.1 Nutrient Absorption and Resistance to Replant Diseases

A recent pot and field study showed that the inoculation of apple rootstocks (*M. hupehensis*) with AMF (*Paraglomus* sp. SW1) can help manage apple replant disease (ARD). AMF improve plant growth, elongate branches, alter soil fungal communities, increase beneficial fungi, reduce harmful *Fusarium* species (*Fusarium moniliforme*, *F. solani*, and *F. proliferatum*), and enhance soil enzyme activity and nutrient levels. The study also found that the levels of nitrogen (NH_4^+ and NO_3^-), available potassium (AK), available phosphorus (AP), organic matter, and soil pH in the rhizosphere significantly influence the structure of fungal and AMF communities (Wang et al. 2024a). Inoculating peach seedlings (*P. persica* L. Batsch) with *A. scrobiculata* (AMF) effectively alleviated soil replant disease. AMF treatment in replant soil can enhance concentrations of calcium, copper, phosphorus, zinc, iron, boron, and potassium in root and shoot biomass, increase soil bacteria and total microbial populations, but decrease soil fungi. It also elevates soil urease and acid phosphatase activities while lowering sucrase and catalase activities. With changes in the richness of root exudates, the magnitude of these allelochemical effects can be reduced to improve overall soil quality for further plant growth (Lǚ et al. 2019). AMF *F. mosseae* inoculation in red tangerines (*C. tangerine* ex. Tanaka), fragrant citrus (*C. junos* Sieb. ex. Tanaka), and kumquat (*Fortunella margarita* L. Swingle) enhance phosphorus content and indole-3-acetic acid (IAA) concentrations, which improve root morphology with gene expression analysis showing upregulation of the P transporter gene PT5 and downregulation of the phosphatase gene PAP1, auxin synthesis gene YUC2, and transporter gene LAX2 (Liu et al. 2024).

5.2 Water Absorption

AMF can extend the root system of plants by continuously moving water from the soil to the air. This water flow is driven by plant transpiration, and water travels along the surface of the AMF hyphae (Kakouridis et al. 2022). AMF (*A. scrobiculata*) treatment of peach (*P. persica* L. Batsch) plants increases the percentage of water-stable aggregates (WSAs) in the 2.00–4.00 mm and 1.00–2.00 mm size ranges, which leads to higher soil aggregate stability in replant soil than in non-replant soil. The presence of mycorrhizal hyphae and glomalin secreted by AMF contributed to this stability by acting as a "glue" for macroaggregate formation. AMF colonization stimulates soil organic carbon (SOC) accumulation in replant soil, further stabilizing WSAs (Lǚ et al. 2019).

5.3 Photosynthesis

Inoculating plants with *G. margarita* and *G. albida* results in enhanced transpiration rates, net CO_2 assimilation, and stomatal conductance compared to control plants, as demonstrated in blackberry plants (*Rubus fruticosus*), resulting in increased fresh and dry shoot biomass (Dewir et al. 2023). Inoculation with mycorrhizal preparations Mykoflor (*R. irregularis, F. mosseae,* and *Claroideoglomus etunicatum*) and MYC 800 (*R. intraradices*), along with the bacterial preparation Rhizocell C (*Bacillus amyloliquefaciens* IT45), enhanced chlorophyll concentrations, transpiration rates, and CO_2 assimilation in strawberry plants, with Rhizocell C and MYC 800 resulting in higher CO_2 assimilation rates than in control plants (Mikiciuk et al. 2024). AMF *Diversispora spurca, A. scrobiculata, G. etunicatum, and G. mosseae* showed similar effects of increased leaf photosynthesis rates in walnut plants (Huang et al. 2020a). AMF also lower oxidative stress markers (H_2O_2, methylglyoxal, and malondialdehyde) and improve chlorophyll content in pistachio (Fattahi et al. 2021).

5.4 Flowering and Pollination

R. intraradices inoculum (MYKOS® Xtreme Gardening, Canada) enhanced the yield of *R. idaeus* (raspberry) by increasing flower production (33%) and fruit number (35%), leading to a 43% yield enhancement compared with non-inoculated plants. This increase could be linked to improved nutrient uptake (P and K) and hormone levels. AMF also increase pollinator visits at moderate fertilizer levels, thereby increasing yield benefits. With maximum fertilizer and pollination, yield rises by 135%, suggesting that ecosystem services, combined with conventional farming, can synergistically improve crop productivity (Chen et al. 2022a, b). Mycorrhizal symbiosis can encourage early flowering in cherry tomatoes (*Solanum lycopersicum* var. *cerasiforme*), enhance the quality, number, weight, and germination rates of seeds, and enhance female functions (Wang et al. 2022).

5.5 Plant Yield and Fruit Quality

AMF (*R. irregularis*) colonization enhances fruit quality by increasing sugar content (higher BRIX values) and levels of lycopene, β-carotene, and amino acids in the red fruits of tomatoes, thus increasing the fruit's nutritional and health benefits (Schubert et al. 2020). Inoculating tomato plants with the AMF *C. etunicatum* can enhance plant growth, increase fruit yield, and increase production of bioactive compounds (Suebrasri et al. 2023). AMF inoculation combined with 50% NPK enhanced root colonization and significantly increased mineral content, BRIX values, and carotenoids in plants compared to fully fertilized, non-inoculated tomato plants (Chafai et al. 2023). AMF (*D. versiformis, F. mosseae,* and *R. intraradices*) enhances fruit quality by improving size, weight, pulp and peel quality, coloration, and soluble solids content, along with boosting root health in *C. reticulata* (Ponkan mandarin) (Cao et al. 2021). The use of *R. intraradices* leads to an increase in fruit production in tomato (*S. lycopersicum*) plants by increasing the chlorophyll content, electron transport rate (ETR), efficiency of photosystem II (PSII), and Fv/Fm ratio, along with an improved net photosynthetic rate. Whole canopy water use efficiency

(WUE) is also enhanced, supporting better growth and fruit yield under limited water availability (Fracasso et al. 2020). In strawberry plants, AMF and plant growth-promoting bacteria (PGPB) inoculation improves fruit quality by increasing anthocyanins, vitamins, sugar content, size, and yield, as well as low levels of phosphorus and nitrogen (Bona et al. 2015). *A. scrobiculata* and *D. spurca* enhance the quality of Lane late navel orange (*C. sinensis*) fruits by upregulating CsSWEET, CsVINV1, and CsA/NINV4 expression, increasing the concentrations of trehalose, fructose, and glucose, reducing sucrose concentrations, and improving external fruit quality (Li et al. 2023).

5.6 Seed Germination and Growth

AMF can be applied early in the developmental process because, during the early phase, plants are more prone to environmental pressures and in their growth and developmental phases, and they have a higher requirement for phosphorus. In a pot study on *Carica papaya* (var. Gujarat Junagadh Papaya-1), *Trichoderma harzianum* proves to be most effective for increasing seed germination (89%) and early emergence (6 days), while also enhancing overall seedling growth, including tap root length and diameter. The combined use of *G. intraradices*, *Pseudomonas fluorescens*, *T. viride*, and *T. harzianum* further improves growth (Desai et al. 2022). Treatment of purple passion fruit (*Passiflora edulis* f. edulis Sims) with AMF *G. fasciculatum* can also increase the growth of seedlings when soluble phosphorus concentration is around 0.02 mg L^{-1} in soil (Ramírez Gil et al. 2015).

5.7 Abiotic Stress Tolerance

5.7.1 Drought Tolerance

Drought stress poses a serious threat to crop production worldwide, and AMF application is an effective means by which crop productivity can be improved. Drought stress was effectively mitigated when tomato plants were inoculated with *Rhizophagus* spp. and *Burkholderia seminalis,* where marked changes were observed in the biomass, root and shoot length, total chlorophyll content, and antioxidant enzyme activity, including catalase and guaiacol peroxidase, together with a reduction in proline content compared to the non-inoculated control (Tallapragada et al. 2016). The effects of *R. irregularis* in apple plants, *M. hupehensis*, increased drought tolerance through improvement in gas exchange, an increase in chlorophyll fluorescence, osmotic adjustment, and scavenging of active reactive oxygen species, in addition to mitogen-activated protein kinase (MAPK) signaling in mycorrhizal fungi–apple plant interaction. The GH3 gene family is present in apples (*M. domestica*), with 14 genes, of which MdGH3-2 and MdGH3-12 are important for AM symbiosis. Mycorrhization upregulated these genes, silenced AM colonization, and downregulated five strigolactone synthesis genes, leading to changes in root strigolactone levels (Huang et al. 2021a). AMF (*D. spurca*) can enhance drought tolerance by activating antioxidant defense systems and upregulating stress-related genes (JrHsf03, JrHsf22, and JrHsf24) to alleviate oxidative damage in the walnut variety 'Liaohe No. 1.' AMF decreases the levels of MDA, H_2O_2, and O_2^- and increases the activities of superoxide dismutase, peroxidase, and catalase (Ma et al. 2022a, b). *D. spurca* also enhances growth in drought-stressed walnut (*J. regia* cv. Qingxiang) by increasing the levels of osmolytes such as betaine, proline, fructose, glucose, sucrose, and soluble protein, and reducing H_2O_2 levels (Wen et al. 2024). A recent study showed that arbuscular mycorrhization of black walnut (*J. nigra* L.) seedlings by *F. mosseae* and *R. irregularis* enhances plant performance under salt stress by enhancing chlorophyll content, increasing proline content, decreasing H_2O_2 levels, and promoting growth regulatory hormones (Li et al. 2024a, b). AM fungi and plant growth-promoting bacteria (PGPB), such as *Azotobacter chroococcum* and *Azospirillum lipoforum,* enhance walnut growth by increasing the levels of metabolites, such as phenolics, proline, sugars, and starch, and boosting peroxidase activity under drought stress. *G. etunicatum* was more effective than *G. mosseae* or a mixed AM fungal inoculum (Behrooz et al. 2019). Biotization of red raspberry (*R. idaeus* L.) with a combined inoculum of AMF and *Paenibacillus*-associated AMF spores enhances plant growth, photosynthetic rate, chlorophyll content, carbon assimilation, stomatal conductance, and transpiration under water stress (Ważny et al. 2024).

5.7.2 Soil Alkalinity Tolerance

The combinations of *G. etunicatum, G. chlorodum, G. deserticola, and G. etunicatum* significantly enhanced the number and weight of stems and leaves, and leaf water content in strawberry plants compared to the control. This indicates that they have the potential to stimulate growth in areas with calcareous soils. Furthermore, nutrient analysis demonstrated that the leaf concentrations of Mg, Fe, N, Mn, and B were higher, which suggests that both mycorrhizal combinations are efficient for improving plant nutrient status (Arıkan et al. 2023).

5.7.3 Salinity Stress Tolerance

Inoculation of salt-stressed *Pistacia vera* seedlings with *R. irregularis* improved biomass, reduced oxidative damage, and enhanced antioxidant enzyme activity and gene expression. AMF play a positive role in reducing the effect of salinity stress on fruit crops through the stimulation of antioxidant enzymes and concentrations of bioactive compounds (Afshar and Abbaspour 2023). A study evaluated the impact of AMF on four pistachio rootstocks under drought and salinity stress for over 60 days. AMF-inoculated plants showed a significant increase in root and shoot dry weights, improved mineral accumulation (N, P, Mg^{2+}, and Ca^{2+}), and reduced sodium (Na^+) and chloride (Cl^-) concentrations compared to non-inoculated plants. AMF *R. irregularis* alleviates salinity stress in pistachio plants grown under NaCl conditions in a greenhouse. After 4 months of symbiosis, root colonization significantly enhances plant growth, indicated by increased root (48.6%) and shoot (53.2%) dry weights, nutrient content (P by 71.6% and Fe by 60.2%), and favorable Ca/Na (80.1%) and K/Na (87.5%) ratios compared to non-mycorrhizal (NMF) plants. In addition, the levels of toxic ions were reduced. Gene expression analysis revealed that salinity stress upregulated genes related to ion transport (SOS1, CCX2, SKOR) and that AMF modulated their expression (Abbaspour et al. 2021). *G. versiforme* has been used successfully in reducing the impact of salt stress on pistachio seedlings (Pistacia vera L.) (Soleimanian et al. 2016). Under salt stress, AM seedlings of Chinese jujube (*Ziziphus jujuba*) exhibited elevated phytohormone levels, enhanced ion transport, and sustained K+/Na + balance. Fatty acid levels increased in both the leaves and roots. Gene analysis revealed 267 core genes associated with fatty acid metabolism and phytohormone signaling. The results revealed that AM symbiosis helps to reprogram fatty acid metabolism, helping plants adapt to salt stress (Ma et al. 2022a, b). The exogenous application of dopamine (100 μM) can enhance the colonization of AMF *F. mosseae* in apple trees (*M. domestica* Borkh.) under salt stress and work synergistically to improve salt tolerance by enhancing photosynthesis, chlorophyll content, cell membrane stability, and root biomass. Overexpression of the dopamine synthetase tyrosine decarboxylase (MdTYDC) gene can enhance AMF colonization (Gao et al. 2020).

5.7.4 Temperature Stress Tolerance

The AM fungus *F. mosseae* improved chilling resistance in tomato seedlings by promoting growth, lowering oxidative stress markers (MDA, H_2O_2, and O_2^-), enhancing antioxidant enzyme activity, and maintaining redox balance and calcium levels in root cell apoplasts and vacuoles under low temperatures (Liu et al. 2016). *F. mosseae* has also been found to be effective in protecting tomato seedlings from severe chilling stress by reducing electrolyte leakage and improving photosystem II (PSII) efficiency (Caradonia et al. 2019).

5.7.5 Heavy Metal Stress Tolerance

Cadmium (Cd) stress causes plants to initiate physiological and transcriptional changes to reduce oxidative damage caused by excessive reactive oxygen species (ROS). Huang et al. (2021b) demonstrated the role of MdGH3-2/12 genes during Cd stress tolerance in *M. domestica* by regulating colonization of AME. In addition, co-inoculation of tomato (*S. lycopersicum*) roots with *R. irregularis* (AMF) and metal-tolerant fungi, *Lecythophora* sp. and *Hormonema viticola*, isolated from metallophyte species, reduces the expression of metal-tolerant genes (MT, HSP, PCS, and NRAMP), thus supporting the growth of plants in Zn- and Cu-contaminated soils (Fuentes-Quiroz et al. 2022).

5.8 Biotic Stress

5.8.1 Bacterial Stress Tolerance

Ralstonia solanacearum is a plant pathogen that causes bacterial wilt disease in many crops, causing global economic losses. *R. irregularis* MUCL 41833 can delay the onset of bacterial wilt symptoms in tomato plantlets, which may be due to the induction of systemic resistance and exerting its biocontrol effects within the root tissues rather than in the rhizosphere (Chave et al. 2017).

5.8.2 Fungal Stress Tolerance

Inoculating apple trees with AMF, *F. mosseae,* and *R. irregularis* significantly enhanced resistance to *Neonectria ditissima*, a major fungal pathogen affecting apple orchards worldwide. AMF treatment reduces the extent of infection by approximately 18% compared to that in trees without AMF inoculation (Berdeni et al. 2018). Both AMF (*F. mosseae* and *G. margarita*) reduce the severity of fungal diseases in strawberry crops caused by *Rhizoctonia fragariae*, *Alternaria alternata,* and *F. oxysporum,* as well as increase growth, total phenolic and phosphorus content, and antioxidant activity, where *F. mosseae* was found to be more efficient in both AMF (Demir et al. 2023). The synergistic action of both endophytic *Bacillus* strains (*B. velezensis* and *Bacillus* sp.) and AMF species (*F. mosseae* and *G. fasciculatum*) reduced Fusarium wilt in tomatoes. AMF inoculation may reduce the level of fusaric acid, which limits wilt within vascular tissues, and increase the chlorophyll content and activity of phosphate-metabolizing enzymes (alkaline and acid phosphatase) in infected plants (Devi et al. 2022). Compost and AMF (Rhizolive comprising 26 microbial species) combined improve the resistance of tomato (*Lycopersicon esculentum*) to the pathogen *Verticillium dahliae* and enhance growth, fruit quality, chlorophyll fluorescence, stomatal conductance, antioxidant enzyme activity, and lower malonyldialdehyde levels (Ait Rahou et al. 2021). AMF can also enhance plant growth and improve the symptoms of *Phytophthora cactorum* crown rot in strawberries (Hautsalo et al. 2016). Combined application of rhizobacteria (*P. putida*) and AMF species (*G. intraradices, G. etunicatum, G. mosseae,* and *G. albida*) reduces *F. oxysporum* colonization in papaya (*Carica papaya*) seedlings, thus enhancing both disease resistance and growth (Hernández-Montiel et al. 2013). Co-inoculation of the fungal endophyte *Serendipita herbamans* and AMF *F. mosseae* effectively controlled Fusarium wilt caused by *F. oxysporum* in tomato plants (Hallasgo et al. 2022). Inoculation of *L. barbarum with R. intraradices* improves plant growth, enhances resistance to root rot by boosting enzyme activities, and increases lignin and flavonoid production (Li et al. 2024a, b). Co-inoculation with AMF *G. intraradices* and antifungal *Bacillus* strains effectively controls the fungal pathogen *Sclerotinia sclerotiorum* in infected strawberries and increases plant biomass (Delgado et al. 2023).

5.8.3 Viruses' Stress Tolerance

Tomato mosaic virus (ToMV) causes tomato mosaic disease in tomato plants, leading to loss in quality, size, and yield of tomato fruits. AMF inoculum consisting of *F. mosseae*, *R. aggregatus,* and *R. clarum* in equal proportions potentially mitigates the effects of ToMV infection in tomatoes by upregulating the genes, phenylalanine ammonia lyase 1 (PAL1) and hydroxycinnamoyl-CoA:quinate hydroxycinnamoyl transferase (HQT), enhancing growth, flavonoid content, and photosynthetic pigments in infected plants, thus reducing the severity of infection (Aseel et al. 2019).

5.8.4 Nematode and Other Pest Tolerance

The root rot nematode *Meloidogyne incognita* is a plant parasite which affects cherry tomatoes as well. The treatment of nematode-infected cherry tomatoes (*S. lycopersicum* var. cerasiforme) with AMF (*F. mosseae*, *R. intraradices*, *G. versiforme*, and a mixture of these) can reduce the density of pathogens and enhance growth, uptake of nutrients (phosphorus and nitrogen), flower and fruit weight, and reproductive capacity (Wang et al. 2023). Treatment of guava (*Psidium guajava* L.) seedlings with AMF species (*G. albida*, *C. etunicatum*, and *A. longula*) helps to control *M. enterolobii* infection by reducing nematode numbers in the roots. It enhances plant growth, height, leaf number, stem diameter, and shoot fresh and dry weights (De Sá and Campos 2020). *R. intraradices* and *F. mosseae*, individually and in combination, reduced the penetration of the nematode *Nacobbus aberrans* in the roots of tomato plants (Marro et al. 2018). In a recent study, AMF (*F. mosseae*, *R. intraradices*, *R. irregularis*, and *G. iranicus*) treatment increased the resistance of tomato against the whitefly *Bemisia tabaci* by reducing the survival rates of whiteflies compared to non-treated plants, thereby improving plant growth and increasing defense enzyme activities (peroxidase, phenylalanine ammonia lyase, and polyphenol oxidase) (Shafiei et al. 2024).

6 Conclusion and Future Prospects

Mycorrhizal associations promote plant growth, nutrient and water uptake, and biotic and abiotic stress resistance, thereby promoting sustainable horticulture. With climate change and environmental problems arising in global agriculture, using mycorrhizal symbiosis can help decrease reliance on agrochemical products such as chemical fertilizers and pesticides. Nevertheless, the effectiveness of mycorrhizal inoculants should be researched further to improve the mycorrhizal fungi formulations tailored to each plant species and the specific site conditions, and to determine the balance of processes concerning fungi, host plants, and the environment. Current knowledge of AMF needs to be combined with AMF-specific inoculants for various types of soils and crops, as well as unambiguous instructions on agricultural practices. However, while some of these AM fungi might provide good results under numerous circumstances, the local types show better results.

References

Abbaspour H, Pour FS, Abdel-Wahhab MA (2021) Arbuscular mycorrhizal symbiosis regulates the physiological responses, ion distribution and relevant gene expression to trigger salt stress tolerance in pistachio. Physiol Mol Biol Plants 27:1765–1778

Afshar AS, Abbaspour H (2023) Mycorrhizal symbiosis alleviates salinity stress in pistachio plants by altering gene expression and antioxidant pathways. Physiol Mol Biol Plants 29:263–276

Ait Rahou Y, Ait-El-Mokhtar M, Anli M, Boutasknit A, Ben-Laouane R, Douira A, Benkirane R, El Modafar C, Meddich A (2021) Use of mycorrhizal fungi and compost for improving the growth and yield of tomato and its resistance to *Verticillium dahliae*. Arch Phytopathol Plant Prot 54:665–690

Ajaz M, Mohammad YZ, Jagana CS (2017) Isolation, identification, and characterization of arbuscular mycorrhizal fungi in apple (*Malus domestica* Borkh) growing area of Kashmir Himalaya. Int J Curr Microbiol Appl Sci 6:25–37

Akter S, Kamruzzaman M, Sarder MP, Amin MS, Joardar JC, Islam MS, Nasrin S, Islam MU, Islam F, Rabbi S, Halder M (2024) Mycorrhizal fungi increase plant nutrient uptake, aggregate stability, and microbial biomass in clay soil. Symbiosis 93:1–14

Allay S, Thapa K, Chakraborty B (2021) Mycorrhizal association and root colonization of *Citrus reticulata*, *Citrus medica* and *Citrus limonia* grown in Darjeeling hills and foothills. J Mycopathol Res 59:375–384

Arıkan Ş, İpek M, Pırlak L, Eşitken A, Şahin M (2023) The effect of mycorrhiza applications on growth and yield in some strawberry cultivars under calcareous soil conditions. Selcuk J Agric Food Sci 37:326–336

Aseel DG, Rashad YM, Hammad SM (2019) Arbuscular mycorrhizal fungi trigger transcriptional expression of flavonoid and chlorogenic acid biosynthetic pathways genes in tomato against tomato mosaic virus. Sci Rep 9:9692

Atakan A, Özkaya HÖ (2021) Arbuscular mycorrhizal fungi and glomalin. Turk J Agric Food Sci Technol 9:2371–2375

Baltazar-Bernal O, Spinoso-Castillo JL, Mancilla-Álvarez E, Bello-Bello JJ (2022) Arbuscular mycorrhizal fungi induce tolerance to salinity stress in *Taro* plantlets (*Colocasia esculenta* L. Schott) during acclimatization. Plants 11:1780

Bauer SE, Tsigaridis K, Miller R (2016) Significant atmospheric aerosol pollution caused by world food cultivation. Geophys Res Lett 43:5394–5400

Behrooz A, Vahdati K, Rejali F, Lotfi M, Sarikhani S, Leslie C (2019) Arbuscular mycorrhiza and plant growth-promoting bacteria alleviate drought stress in walnut. HortScience 54:1087–1092

Berdeni D, Cotton TEA, Daniell TJ, Bidartondo MI, Cameron DD, Evans KL (2018) The effects of arbuscular mycorrhizal fungal colonization on nutrient status, growth, productivity, and canker resistance of apple (*Malus pumila*). Front Microbiol 9:1461

Blackwell M (2000) Terrestrial life--fungal from the start? Science 289:1884–1885

Bona E, Lingua G, Manassero P, Cantamessa S, Marsano F, Todeschini V, Copetta A, D'Agostino G, Massa N, Avidano L, Berta G (2015) AM fungi and PGP pseudomonads increase flowering, fruit production, and vitamin content in strawberry grown at low nitrogen and phosphorus levels. Mycorrhiza 25:181–193

Bonfante P, Genre A (2010) Mechanisms underlying beneficial plant–fungus interactions in mycorrhizal symbiosis. Nat Commun 1:48

Bowles TM, Barrios-Masias FH, Carlisle EA, Cavagnaro TR, Jackson LE (2016) Effects of arbuscular mycorrhizae on tomato yield, nutrient uptake, water relations, and soil carbon dynamics under deficit irrigation in field conditions. Sci Total Environ 566:1223–1234

Brundrett MC (2002) Coevolution of roots and mycorrhizas of land plants. New Phytol 154:275–304

Bücking H, Liepold E, Ambilwade P (2012) The role of the mycorrhizal symbiosis in nutrient uptake of plants and the regulatory mechanisms underlying these transport processes. Plant Sci 4:108–132

Cao MA, Wang P, Hashem A, Wirth S, Abd Allah EF, Wu QS (2021) Field inoculation of arbuscular mycorrhizal fungi improves fruit quality and root physiological activity of citrus. Agriculture 11:1297

Caradonia F, Francia E, Morcia C, Ghizzoni R, Moulin L, Terzi V, Ronga D (2019) Arbuscular mycorrhizal fungi and plant growth-promoting rhizobacteria avoid processing tomato leaf damage during chilling stress. Agronomy 9:299

Chafai W, Haddioui K, Serghini-Caid H, Labazi H, AlZain MN, Noman O, Parvez MK, Addi M, Khalid A (2023) Impact of Arbuscular mycorrhizal fungal strains isolated from soil on the growth, yield, and fruit quality of tomato plants under different fertilization regimens. Horticulturae 9:973

Chave M, Crozilhac P, Deberdt P, Plouznikoff K, Declerck S (2017) *Rhizophagus irregularis* MUCL 41833 transitorily reduces tomato bacterial wilt incidence caused by *Ralstonia solanacearum* under in vitro conditions. Mycorrhiza 27:719–723

Chen K, Huang G, Li Y, Zhang X, Lei Y, Li Y, Xiong J, Sun Y (2022a) Illumina MiSeq sequencing reveals correlations among fruit ingredients, environmental factors, and AMF communities in three *Lycium barbarum* producing regions of China. Microbiol Spectr 10:e02293–e02221

Chen K, Kleijn D, Scheper J, Fijen TP (2022b) Additive and synergistic effects of arbuscular mycorrhizal fungi, insect pollination and nutrient availability in a perennial fruit crop. Agric Ecosyst Environ 325:107742

De Sá CSB, Campos MAS (2020) Arbuscular mycorrhizal fungi decrease *Meloidogyne enterolobii* infection of guava seedlings. J Helminthol 94:e183

Delgado A, Toro M, Memenza-Zegarra M, Zúñiga-Dávila D (2023) Control of white rot caused by *Sclerotinia sclerotiorum* in strawberry using arbuscular mycorrhizae and plant-growth-promoting bacteria. Sustainability 15:2901

Demir S, Durak ED, Güneş H, Boyno G, Mulet JM, Rezaee Danesh Y, Porcel R (2023) Biological control of three fungal diseases in strawberry (*Fragaria*× *ananassa*) with arbuscular mycorrhizal fungi. Agronomy 13:2439

Desai DH, Tank RV, Desai KD, Desai HS, Mehta VS, Champaneri DD (2022) Effect of arbuscular mycorrhizal fungi and bio-inoculants on germination and seedling growth of *Carica papaya* L. Var. Gujarat Junagadh papaya-1. Int J Plant Soil Sci 34:1–10

Devi NO, Tombisana Devi RK, Debbarma M, Hajong M, Thokchom S (2022) Effect of endophytic *bacillus* and arbuscular mycorrhiza fungi (AMF) against *Fusarium* wilt of tomato caused by *Fusarium oxysporum* f. sp. *lycopersici*. Egypt J Biol Pest Contr 32:1–14

Dewir YH, Al-Qarawi AA, Alshahrani T, Bansal Y, Mujib A, Murthy HN, Alebidi AI, Almutairi KF, Al-Saif AM (2023) Influence of arbuscular mycorrhizal fungi on the growth and development of micropropagated *Rubus fruticosus* 'P45' plants during acclimatization. HortScience 58:871–876

Duc NH, Szentpéteri V, Mayer Z, Posta K (2023) Distinct impact of arbuscular mycorrhizal isolates on tomato plant tolerance to drought combined with chronic and acute heat stress. Plant Physiol Biochem 201:107892

Farghaly FA, Nafady NA, Abdel-Wahab DA (2022) The efficiency of arbuscular mycorrhiza in increasing tolerance of *Triticum aestivum* L. to alkaline stress. BMC Plant Biol 22:490

Fattahi M, Mohammadkhani A, Shiran B, Baninasab B, Ravash R, Gogorcena Y (2021) Beneficial effect of mycorrhiza on nutritional uptake and oxidative balance in pistachio (*Pistacia* spp.) rootstocks submitted to drought and salinity stress. Sci Hortic 281:109937

Favre-Godal Q, Gourguillon L, Lordel-Madeleine S, Gindro K, Choisy P (2020) Orchids and their mycorrhizal fungi: an insufficiently explored relationship. Mycorrhiza 30:5–22

Fracasso A, Telò L, Lanfranco L, Bonfante P, Amaducci S (2020) Physiological beneficial effect of *Rhizophagus intraradices* inoculation on tomato plant yield under water deficit conditions. Agronomy 10:71

Fuentes-Quiroz A, Herrera H, Soto J, Campos-Vargas R, Ortiz J, Arriagada C (2022) Rhizosphere fungi regulate the expression of metal tolerance genes in *Solanum lycopersicum* L. (*Solanaceae*) growing in a metal(loid)-contaminated soil. Rhizosphere 24:100599

Gao T, Liu X, Shan L, Wu Q, Liu Y, Zhang Z, Ma F, Li C (2020) Dopamine and arbuscular mycorrhizal fungi act synergistically to promote apple growth under salt stress. Environ Exp Bot 178:104159

Gianinazzi S, Gollotte A, Binet MN, van Tuinen D, Redecker D, Wipf D (2010) Agroecology: the key role of arbuscular mycorrhizas in ecosystem services. Mycorrhiza 20:519–530

Hallasgo AM, Hauser C, Steinkellner S, Hage-Ahmed K (2022) Single and coinoculation of *Serendipita herbamans* with arbuscular mycorrhizal fungi reduces *Fusarium* wilt in tomato and slows disease progression in the long term. Biol Control 168:104876

Hamel C, Plenchette C (2017) Implications of past, current, and future agricultural practices for mycorrhiza-mediated nutrient flux. In: *Mycorrhizal mediation of soil*. Elsevier, pp 175–186

Hart MM, Forsythe JA (2012) Using arbuscular mycorrhizal fungi to improve the nutrient quality of crops; nutritional benefits in addition to phosphorus. Sci Hortic 148:206–214

Hautsalo J, Vestberg M, Parikka P, Kukkonen S, Karhu S, Tahvonen R (2016) Biological control of strawberry crown rot is substrate-dependent phenomenon. J Berry Res 6:65–79

Hernández-Montiel LG, Rueda-Puente EO, Cordoba-Matson MV, Holguín-Peña JR, Zulueta-Rodríguez R (2013) Mutualistic interaction of rhizobacteria with arbuscular mycorrhizal fungi and its antagonistic effect on *Fusarium oxysporum* in *Carica papaya* seedlings. Crop Prot 47:61–66

Hou L, Zhang X, Feng G, Li Z, Zhang Y, Cao N (2021) Arbuscular mycorrhizal enhancement of phosphorus uptake and yields of maize under high planting density in the black soil region of China. Sci Rep 11:1100

Huang D, Ma M, Wang Q, Zhang M, Jing G, Li C, Ma F (2020a) Arbuscular mycorrhizal fungi enhanced drought resistance in apple by regulating genes in the MAPK pathway. Plant Physiol Biochem 149:245–255

Huang D, Wang Q, Zhang Z, Jing G, Ma M, Ma F, Li C (2021a) Silencing *MdGH3-2/12* in apple reduces drought resistance by regulating AM colonization. Hortic Res 8:84

Huang D, Wang Q, Zou Y, Ma M, Jing G, Ma F, Li C (2021b) Silencing *MdGH3-2/12* in apple reduces cadmium resistance via the regulation of AM colonization. Chemosphere 269:129407

Huey CJ, Gopinath SC, Uda MNA, Zulhaimi HI, Jaafar MN, Kasim FH, Yaakub ARW (2020) Mycorrhiza: a natural resource assists plant growth under varied soil conditions. 3 Biotech 10:1–9

Jote CA (2023) The impacts of using inorganic chemical fertilizers on the environment and human health. Org Med Chem Int J 13:555864

Kakouridis A, Hagen JA, Kan MP, Mambelli S, Feldman LJ, Herman DJ, Weber PK, Pett-Ridge J, Firestone MK (2022) Routes to roots: direct evidence of water transport by arbuscular mycorrhizal fungi to host plants. New Phytol 236:210–221

Kirti S, Sharma R, Dipta B, Pawar R, Kaushal R, Bharat NK (2016) Status of arbuscular mycorrhizae associated with rhizosphere of sweet cherry (*Prunus avium*) in high hills temperate zone of Himachal Pradesh. Indian Phytopathol 69:286

Koegel S, Mieulet D, Baday S, Chatagnier O, Lehmann MF, Wiemken A, Boller T, Wipf D, Bernèche S, Guiderdoni E, Courty PE (2017) Phylogenetic, structural, and functional characterization of AMT3;1, an ammonium transporter induced by mycorrhization among model grasses. Mycorrhiza 27:695–708

Li QS, Srivastava AK, Zou YN, Wu QS (2023) Field inoculation responses of arbuscular mycorrhizal fungi versus endophytic fungi on sugar metabolism associated changes in fruit quality of lane late navel orange. Sci Hortic 308:111587

Li N, Chen W, Wang B, Zhang C, Wang Y, Li R, Yan Y, He J (2024a) Arbuscular mycorrhizal fungi improve the disease resistance of *Lycium barbarum* to root rot by activating phenylpropane metabolism. Front Plant Sci 15:1459651

Li A, Wu C, Zheng X, Nie R, Tang J, Ji X, Zhang J (2024b) Physiological and biochemical responses of arbuscular mycorrhizal fungi in symbiosis with *Juglans nigra* L. seedlings to alleviate salt stress. Rhizosphere 31:100928

Liu A, Chen S, Wang M, Liu D, Chang R, Wang Z, Lin X, Bai B, Ahammed GJ (2016) Arbuscular mycorrhizal fungus alleviates chilling stress by boosting redox poise and antioxidant potential of tomato seedlings. J Plant Growth Regul 35:109–120

Liu C, Ravnskov S, Liu F, Rubæk GH, Andersen MN (2018) Arbuscular mycorrhizal fungi alleviate abiotic stresses in potato plants caused by low phosphorus and deficit irrigation/partial root-zone drying. J Agric Sci 156:46–58

Liu XR, Rong ZY, Tian X, Hashem A, Abd Allah EF, Zou YN, Wu QS (2023) Mycorrhizal fungal effects on plant growth, osmolytes, and *CsHsp70s* and *CsPIPs* expression in leaves of cucumber under a short-term heat stress. Plants 12:2917

Liu CY, Guo XN, Dai FJ, Wu QS (2024) Mycorrhizal symbiosis enhances P uptake and indole-3-acetic acid accumulation to improve root morphology in different citrus genotypes. Horticulturae 10:339

Lŭ LH, Zou YN, Wu QS (2019) Mycorrhizas mitigate soil replant disease of peach through regulating root exudates, soil microbial population, and soil aggregate stability. Commun Soil Sci Plant Anal 50:909–921

Ma WY, Qin QY, Zou YN, Kuča K, Giri B, Wu QS, Hashem A, Al-Arjani ABF, Almutairi KF, Abd Allah EF, Xu YJ (2022a) Arbuscular mycorrhiza induces low oxidative burst in drought-stressed walnut through activating antioxidant defense systems and heat shock transcription factor expression. Front Plant Sci 13:1089420

Ma Z, Zhao X, He A, Cao Y, Han Q, Lu Y, Yong JWH, Huang J (2022b) Mycorrhizal symbiosis reprograms ion fluxes and fatty acid metabolism in wild jujube during salt stress. Plant Physiol 189:2481–2499

Marro N, Caccia M, Doucet ME, Cabello M, Becerra A, Lax P (2018) Mycorrhizas reduce tomato root penetration by false root-knot nematode *Nacobbus aberrans*. Appl Soil Ecol 124:262–265

Massa N, Bona E, Novello G, Todeschini V, Boatti L, Mignone F, Gamalero E, Lingua G, Berta G, Cesaro P (2020) AMF communities associated with *Vitis vinifera* in an Italian vineyard subjected to integrated pest management at two different phenological stages. Sci Rep 10:9197

Mehta P, Bharat NK (2013) Arbuscular-mycorrhizal associations with apple plantations of different age groups in Himachal Pradesh. Indian Phytopathol 66:88–92

Mikiciuk G, Sas-Paszt L, Mikiciuk M, Derkowska E, Trzciński P, Głuszek S, Lisek A, Wera-Bryl S, Rudnicka J (2019) Mycorrhizal frequency, physiological parameters, and yield of strawberry plants inoculated with endomycorrhizal fungi and rhizosphere bacteria. Mycorrhiza 29:489–501

Mikiciuk G, Miller T, Kisiel A, Cembrowska-Lech D, Mikiciuk M, Łobodzińska A, Bokszczanin K (2024) Harnessing Beneficial Microbes for Drought Tolerance: A Review of Ecological and Agricultural Innovations. Agriculture 14(12):2228. https://doi.org/10.3390/agriculture14122228

Nadeem SM, Ahmad M, Zahir ZA, Javaid A, Ashraf M (2014) The role of mycorrhizae and plant growth promoting rhizobacteria (PGPR) in improving crop productivity under stressful environments. Biotechnol Adv 32:429–448

Naik SR, Nandini MLN, Jameel MA, Venkataramana KT, Mukundalakshmi L (2018) Role of arbuscular mycorrhiza in fruit crops production. Int J Pure Appl Biosci 6:1126–1133

Olaniyan FT, Alori ET, Adekiya AO, Ayorinde BB, Daramola FY, Osemwegie OO, Babalola OO (2022) The use of soil microbial potassium solubilizers in potassium nutrient availability in soil and its dynamics. Ann Microbiol 72:45

Ortas I (2010) Effect of mycorrhiza application on plant growth and nutrient uptake in cucumber production under field conditions. Span J Agric Res 8:116–122

Ortas I, Ustuner O (2014) Determination of different growth media and various mycorrhizae species on citrus growth and nutrient uptake. Sci Hortic 166:84–90

Pal A, Pandey S (2016) Role of arbuscular mycorrhizal fungi on plant growth and reclamation of barren soil with wheat (*Triticum aestivum* L.) crop. Int J Soil Sci 12:25–31

Paterson E, Sim A, Davidson J, Daniell TJ (2016) Arbuscular mycorrhizal hyphae promote priming of native soil organic matter mineralization. Plant Soil 408:243–254

Perotto S, Girlanda M, Martino E (2002) Ericoid mycorrhizal fungi: some new perspectives on old acquaintances. In: Diversity and integration in mycorrhizas: proceedings of the 3rd international conference on mycorrhizas (ICOM3) Adelaide, Australia, 8–13 July 2001. Springer, pp 41–53

Pujasatria GC, Miura C, Yamaguchi K, Shigenobu S, Kaminaka H (2024) Colonization by orchid mycorrhizal fungi primes induced systemic resistance against necrotrophic pathogen. Front Plant Sci 15:1447050

Quan L, Duan K, Wei Z, Li W, Chen Y, Duan W, Xia Y (2023) Beneficial effects of arbuscular mycorrhizae on cu detoxification in *Mimosa pudica* L. grown in cu-polluted soils. Environ Sci Pollut Res 30:25755–25763

Ramírez Gil JG, Agudelo MM, Bedoya LO, Osorio NW, Osorio JGM (2015) Germination and growth of purple passion fruit seedlings under pre-germination treatments and mycorrhizal inoculation. Agric Res Trop/Pesquisa Agropecuária Trop 45:3

Ruytinx J, Kafle A, Usman M, Coninx L, Zimmermann SD, Garcia K (2020) Micronutrient transport in mycorrhizal symbiosis; zinc steals the show. Fungal Biol Rev 34:1–9

Sabra M, Aboulnasr A, Franken P, Perreca E, Wright LP, Camehl I (2018) Beneficial root endophytic fungi increase growth and quality parameters of sweet basil in heavy metal contaminated soil. Front Plant Sci 9:1726

Schubert R, Werner S, Cirka H, Rödel P, Tandron Moya Y, Mock HP, Hutter I, Kunze G, Hause B (2020) Effects of arbuscular mycorrhization on fruit quality in industrialized tomato production. Int J Mol Sci 21:7029

Shafiei F, Shahidi-Noghabi S, Sedaghati E, Smagghe G (2024) Arbuscular mycorrhizal fungi inducing tomato plant resistance and its role in control of *Bemisia tabaci* under greenhouse conditions. Neotrop Entomol 53:424–438

Sharma R, Kirti S, Dipta B, Kaushal R, Bharat NK (2016) Biodiversity of arbuscular mycorrhizal fungi in sweet cherry (*Prunus avium* L.) of Kullu district of Himachal Pradesh. Bioscan 11:1555–1558

Sheikh-Assadi M, Khandan-Mirkohi A, Taheri MR, Babalar M, Sheikhi H, Nicola S (2023) Arbuscular mycorrhizae contribute to growth, nutrient uptake, and ornamental characteristics of statice (*Limonium sinuatum* [L.] mill.) subject to appropriate inoculum and optimal phosphorus. Horticulturae 9:564

Shi J, Wang X, Wang E (2023) Mycorrhizal symbiosis in plant growth and stress adaptation: from genes to ecosystems. Annu Rev Plant Biol 74:569–607

Singh S, Singh R, Singh K, Katoch K, Zaeen AA, Birhan DA, Singh A, Sandhu HS, Singh H, Sahrma LK (2024) Smart fertilizer technologies: an environmental impact assessment for sustainable agriculture. Smart Agric Technol 8:100504

Smith SE, Read DJ (2010) Mycorrhizal symbiosis. Academic

Soleimanian S, Abbaspour H, Mohammadi NA (2016) Contributions of arbuscular mycorrhizal fungi to growth, biomass, and nutrient status of pistachio seedlings under saline conditions. J Nuts 7:67–74

Suebrasri T, Seemakram W, Lapjit C, Mongkolthanaruk W, Boonlue S (2023) Effects of arbuscular mycorrhizal fungi on enhancing growth, fruit quality, and functional substances in tomato fruits (*Lycopersicon esculentum* mill.). J Plant Biotechnol 50:239–247

Sumathi S, Thangavelu R (2016) Biodiversity of arbuscular mycorrhizal fungi (AMF) in banana plantations in India. Plant Arch 16:210–216

Tallapragada P, Dikshit R, Seshagiri S (2016) Influence of *Rhizophagus* spp. and *Burkholderia seminalis* on the growth of tomato (*Lycopersicon esculatum*) and bell pepper (*Capsicum annuum*) under drought stress. Commun Soil Sci Plant Anal 47:1975–1984

Thioye B, Legras M, Castel L, Hirissou F, Chaftar N, Trinsoutrot-Gattin I (2021) Understanding arbuscular mycorrhizal colonization in walnut plantations: the contribution of cover crops and soil microbial communities. Agriculture 12:1

Thirkell TJ, Charters MD, Elliott AJ, Sait SM, Field KJ (2017) Are mycorrhizal fungi our sustainable saviours? Considerations for achieving food security. J Ecol 105:921–929

Thomas L, Singh I (2019) Microbial biofertilizers: types and applications. In: Biofertilizers for sustainable agriculture and environment, pp 1–19

Tsiknia M, Tsikou D, Papadopoulou KK, Ehaliotis C (2021) Multi-species relationships in legume roots: from pairwise legume-symbiont interactions to the plant–microbiome–soil continuum. FEMS Microbiol Ecol 97:fiaa222

Turrini A, Agnolucci M, Palla M, Tomé E, Tagliavini M, Scandellari F, Giovannetti M (2017) Species diversity and community composition of native arbuscular mycorrhizal fungi in apple roots are affected by site and orchard management. Appl Soil Ecol 116:42–54

Van Der Heijden MG, Martin FM, Selosse MA, Sanders IR (2015) Mycorrhizal ecology and evolution: the past, the present, and the future. New Phytol 205:1406–1423

Vasilikiotis C, Li M, Schmidt JE, Azimi A, Garcia J, Volder A, Lampinen B, Gaudin AC (2020) Orchard management practices affect arbuscular mycorrhizal fungal root colonisation of almond. Biol Agric Hortic 36:230–248

Wahab A, Muhammad M, Munir A, Abdi G, Zaman W, Ayaz A, Khizar C, Reddy SPP (2023) Role of arbuscular mycorrhizal fungi in regulating growth, enhancing productivity, and potentially influencing ecosystems under abiotic and biotic stresses. Plants 12:3102

Wang P, Shu B, Wang Y, Zhang DJ, Liu JF, Xia RX (2013) Diversity of arbuscular mycorrhizal fungi in red tangerine (*Citrus reticulata* Blanco) rootstock rhizospheric soils from hillside citrus orchards. Pedobiologia 56:161–167

Wang L, Chen X, Wang S, Du Y, Zhang D, Tang Z (2022) Effects of arbuscular mycorrhizal symbiosis on the growth and reproduction of cherry tomato can be persistent to the next generation. Eur J Soil Biol 112:103429

Wang L, Chen X, Tang Z (2023) Arbuscular mycorrhizal symbioses improved biomass allocation and reproductive investment of cherry tomato after root-knot nematodes infection. Plant Soil 482:513–527

Wang Q, Liu M, Wang Z, Li J, Liu K, Huang D (2024a) The role of arbuscular mycorrhizal symbiosis in plant abiotic stress. Front Microbiol 14:1323881

Wang M, Xiang L, Tang W, Chen X, Li C, Yin C, Mao Z (2024b) Improving apple orchard health: the role of arbuscular mycorrhizal fungi in alleviating replant disease and strengthening soil microbial communities. Appl Soil Ecol 196:105278

Ważny R, Jędrzejczyk RJ, Rozpądek P, Domka A, Tokarz KM, Janicka M, Turnau K (2024) Bacteria associated with spores of arbuscular mycorrhizal fungi improve the effectiveness of fungal inocula for red raspberry biotization. Microb Ecol 87:50

Wen Y, Zhou LJ, Xu YJ, Hashem A, Abd Allah EF, Wu QS (2024) Growth performance and osmolyte regulation of drought-stressed walnut plants are improved by mycorrhiza. Agriculture 14:367

Xiao L, Lai S, Chen M, Long X, Fu X, Yang H (2022) Effects of grass cultivation on soil arbuscular mycorrhizal fungi community in a tangerine orchard. Rhizosphere 24:100583

Zhang W, Yu L, Han B, Liu K, Shao X (2022) Mycorrhizal inoculation enhances nutrient absorption and induces insect-resistant defense of *Elymus nutans*. Front Plant Sci 13:898969

Zhang Q, Wang R, Li L, Ju M, Zhang Q, Gu P (2024) The diversity of AMF in the rhizosphere soil of wine grape and its correlation with soil factors in the eastern foot of Helan Mountain in Ningxia. SSRN 4892725

Zou YN, Srivastava AK, Wu QS (2016) Glomalin: a potential soil conditioner for perennial fruits. Int J Agric Biol 18:293–297

Symbiotic Relationships: The Role of Mycorrhizae in Optimizing Flower Crop Productivity

Rajni Dhalaria, Harsh Kumar, Ajay Katoch, Elsayed Fathi Abd-Allah, Yasmeen Siddiqui, Swarnendra Banerjee, M. A. Elinson, and Vishnu D. Rajput

Abstract

Mycorrhizal fungi, crucial components of soil ecosystems, form symbiotic relationships with plant roots and significantly enhance plant growth and resilience. This chapter explores the effects of mycorrhizal associations on flower crops, focusing on nutrient uptake, soil health, and plant stress tolerance. Mycorrhizal fungi improve the absorption of essential nutrients such as phosphorus, nitrogen, and trace elements, promoting more vigorous flowering and increased yield in

R. Dhalaria
Department of Biosciences, Minerva P.G. College of Arts, Science and Commerce, Kangra, Himachal Pradesh, India

H. Kumar
Centre of Advanced Technologies, Faculty of Science, University of Hradec Kralove, Hradec Králové, Czech Republic

A. Katoch
Department of Biosciences, MCM DAV College Kangra (Himachal Pradesh), Kangra, Himachal Pradesh, India

E. F. Abd-Allah
Plant Production Department, College of Food and Agricultural Sciences, King Saud University, Riyadh, Saudi Arabia

Y. Siddiqui
Department of Biological Sciences, College of Science, King Faisal University, Al Ahsa, Saudi Arabia

S. Banerjee · V. D. Rajput (✉)
Academy of Biology and Biotechnology, Southern Federal University, Rostov-on-Don, Russia
e-mail: rvishnu@sfedu.ru

M. A. Elinson
Ufa University of Science and Technology, Ufa, Russia

Y. Siddiqui et al. (eds.), *Sustainable Mycorrhizal Cultivation*,
https://doi.org/10.1007/978-981-95-1778-7_7

ornamental species. Additionally, these symbionts enhance plant resistance to drought, pathogens, and salinity, which are critical factors for the sustainability of flower production. Mycorrhizal associations offer an environmentally friendly approach to floriculture by reducing the need for chemical fertilizers and pesticides. This review synthesizes recent research on the mechanisms through which mycorrhizae influence flower crops and discusses potential applications for improving crop productivity and sustainability in horticulture.

Keywords

Mycorrhizal fungi · Flower crops · Interactions · Growth · Environmental stress

1 Introduction

Floriculture is a specialized sector of horticulture that involves the cultivation of flowers, ornamental plants, and foliage for aesthetic, emotional, and economic purposes. The global floriculture industry is a significant contributor to the economy, with growing demand for cut flowers, potted plants, and other ornamental products (Datta and Gupta 2022). In addition to their aesthetic value, many flowers and ornamental plants also possess medicinal properties, making them an important part of traditional and modern medicine (Arya et al. 2014; Benvenuti and Mazzoncini 2021). For example, rose petals are used in herbal teas and skincare products because of their antioxidant and anti-inflammatory properties (Takahashi 2025), whereas lavender flowers are used for their calming effects and essential oils (Crişan et al. 2023). However, the industry faces challenges, such as climate change, water scarcity, and the need for sustainable production practices (De et al. 2024). Moreover, the excessive use of chemical fertilizers and pesticides in floriculture has been linked to environmental degradation, soil pollution, and biodiversity loss, ultimately affecting the sustainability and profitability of the industry (Gelaye 2023; Zhang et al. 2024). To address these challenges, floriculturists have adopted innovative strategies to promote plant growth and sustainability.

The application of arbuscular mycorrhizal fungi (AMF) has gained attention as a promising option to improve crop productivity and sustainability in floriculture. AMF form symbiotic associations with a wide range of plant species, including many ornamental plants, and can potentially improve plant growth by enhancing nutrient uptake and rhizospheric soil structure (van der Heijden et al. 2008; Smith and Read 2010; Wenjing et al. 2023). The fungal mycelial network extends under the plant roots, facilitating the uptake of essential nutrients, such as phosphorus, and promoting healthy plant development (Garg and Chandel 2011). By acting as a natural biofertilizer, AMF can help reduce the use of agrochemicals in floricultural production, leading to more sustainable and environmentally friendly practices (Vosátka and Albrechtová 2008; Madawala 2021). AMF inoculation has been shown to enhance the quality and yield of various ornamental crops, including those with medicinal properties (Sohn et al. 2003; Popescu and Popescu 2022) by improving the production of bioactive compounds in these plants. AMF contribute to the development of high-quality medicinal flowers and ornamental plants used in traditional medicine, cosmetics, and other applications.

Thus, this chapter aims to provide an overview of the potential benefits and applications of AMF in floriculture, with a focus on improving crop productivity, quality, and resilience, while reducing environmental impacts.

2 Significance of Mycorrhizal Diversity

Mycorrhizal fungi play a crucial role in flower crop productivity by forming symbiotic relationships with plant roots (Hart and Reader 2002). The diversity of mycorrhizal fungi is essential for optimizing flower crop productivity, as different species and isolates of fungi can have varying effects on plant growth and development (Powell and Rillig 2018). Ecosystem functionality, including nutrient uptake and transfer, plant interactions, and soil environment modification, is critical for the maintenance of healthy flower crops (Wagg et al. 2014). Soil management practices significantly affect mycorrhizal diversity, which in turn affects ecosystem services and flower crop productivity (Helgason and Fitter 2009).

Variability in fungal growth strategies among different families of mycorrhizal fungi affects their ability to colonize plant roots and provide ecosystem services (Hart and Reader 2002). For example, the Gigasporaceae family colonizes more slowly than the Glomaceae and Acaulosporaceae families. Understanding the significance of mycorrhizal diversity can enhance strategies to improve flower crop production. By promoting mycorrhizal diversity and selecting the right fungal partners, flower crop producers can improve plant growth, reduce environmental impact, and enhance ecosystem services (Dowarah et al. 2022).

Mycorrhizal fungi enhance nutrient uptake and transfer to flower crops, thereby promoting healthy plant growth and development (Powell and Rillig 2018). They also alter plant interactions with other biota, including pests and pathogens, which affect flower crop productivity (Wagg et al. 2014). Additionally, mycorrhizal fungi modify the soil environment and improve soil structure and fertility, which benefit flower crops (Helgason and Fitter 2009) and lead to improved flower quality, increased yields, and reduced fertilizer requirements.

Mycorrhizal fungi play a critical role in sustainable agricultural practices. By promoting soil health and biodiversity, mycorrhizal fungi help reduce the environmental impacts on flower crop production (Powell and Rillig 2018). This includes reducing the use of chemical fertilizers and pesticides, which can harm the environment and human health.

3 Mechanisms of Mycorrhizal Benefits in Flower Crops

3.1 AMF in Stress Tolerance

3.1.1 Abiotic Stress

Plants often encounter a variety of environmental stresses that cause adverse changes in their growth and metabolism, ultimately affecting plant production (Gonzalez Guzman et al. 2022). Drought, salinity, temperature, and heavy metals are among the major abiotic stress factors that reduce crop yields and productivity. Drought is a challenging stress caused by climate change that impedes plant growth and productivity by reducing photosynthesis and disrupting the nutritional equilibrium (Cao et al. 2023). Soil salinity affects several lands globally, especially in arid areas, disrupting plant growth owing to cell toxicity caused by ionic stress (Arif et al. 2020). High temperature stress is a primary consequence of global warming, which affects agricultural productivity (Hassan et al. 2021). This results in the generation of reactive oxygen species (ROS) that harm plant growth (Zandalinas et al. 2018). Low temperatures impede plant development and agricultural productivity by slowing plant metabolism, altering the activity of macromolecules, imposing mechanical constraints, lowering osmotic potential, and decreasing productivity (Devireddy et al. 2021). Heavy metal contamination is another abiotic stressor that can affect environmental safety. Moreover, it hinders ecosystem development and causes high levels of environmental pollution (Ghori et al. 2019).

AMF are beneficial fungal organisms that have symbiotic relationships with several terrestrial plants (Thangavel et al. 2022; Dhalaria et al. 2024a, b). Colonization with AMF boosts plant growth and vitality, while altering plant morphology, nutrition, and physiology to enhance tolerance to various abiotic stressors (Iqbal et al. 2023; Negi et al. 2023). It enhances soil properties, leading to increased plant development under both normal and stress conditions (Mitra et al. 2021). Different types of AMF have been investigated for their potential roles in the management of various abiotic stresses (Table 1). An overview of the mechanisms evolved by AMF to assist plants in surviving stressful conditions is provided below.

3.1.1.1 Improved Water Relations

AMF can enhance the root surface area by effectively producing extraradical and intraradical mycelia for better water utilization and transportation (Abdalla et al. 2023). A large network of extraradical hyphae forms a link between the soil and the plant. Extraradical mycelia have been reported to transfer approximately 4% of the water from the hyphal compartment to the root compartment, owing to their small diameter (2–20 μm), which enables them to easily penetrate the fine pores of the soil (Ganugi et al. 2019; Zhao et al. 2022). Moreover, AMF hyphal tips are hydrophilic, aiding water flow from the soil to the hyphal tip and its subsequent transport into the plant cells. This vast mycorrhizal network prevents water depletion around the roots (Tyagi et al. 2017). A study on *Rosa damascena* found that inoculation with mycorrhizal fungi enhanced water absorption by increasing the effectiveness of root hydraulic conductivity and decreased root vulnerability to drought compared with

Table 1 Representative studies highlighting the impact of various AMF species on different flower crops under different abiotic stress conditions

Abiotic stress	Flower crop type Scientific/common name	Mycorrhizal species	Growing conditions	Positive effects on flower crops	References
Drought	*Tagetes erecta* L./marigold	*Glomus constrictum* Trappe	Greenhouse	Improved plant growth, flower quality, pigments, and P content	Asrar and Elhindi (2011)
	Helianthus annuus L./sunflower	*Funneliformis mosseae* *Glomus hoi*	Greenhouse	Increased dry matter, heavier seeds, greater seed, oil yields, N, P, and K content	Gholamhoseini et al. (2013)
	Rosa damascena Mill./d amask rose	n.d.	Greenhouse	Enhanced plant growth, flower quality, P_n, g_s, E, RWC, N, P, K, Ca, Mg, Chl and carotenoid content	Abdel-Salam et al. (2018)
	Zinnia elegans L./zinnia	*Rhizophagus irregularis* *Rhizophagus intraradices* *F. mosseae* Mixture	Greenhouse	Improved growth and nutrient uptake	Bagheri et al. (2019)
	Lavandula angustifolia (Mill.)/lavender	Mixture of AMF (*R. irregularis, Claroideoglomus claroideum, F. mosseae, Funneliformis geosporum, Claroideoglomus etunicatum, and Glomus microaggregatum)*	Open field	Improved plant growth, total plant biomass, photosynthetic pigments, $P_{n,}$ and ornamental value	Popescu and Popescu (2022)
	Satureja hortensis L./summer savory	*F. mosseae*	Greenhouse	Higher plant height, shoot mass (fresh and dry weight), RWC, MSI, photosynthesis pigments, and total soluble proteins	Rasouli et al. (2023)

(continued)

Table 1 (continued)

Abiotic stress	Flower crop type Scientific/common name	Mycorrhizal species	Growing conditions	Positive effects on flower crops	References
Salinity	*Gerbera jamesonii* cvs. Beaudine and Palm Beach/Gerbera	*F. mosseae*	Greenhouse	Increased flower yield and quality	Othman et al. (2022)
	Chrysanthemum morifolium (Ramat.) Tzvel/hangbaiju	*F. mosseae* *Diversispora versiformis* *Mixture*	Greenhouse	Improved root length, shoot and root dry weight, total dry weight, and root N concentration	Wang et al. (2018)
	Tagetes minuta L./marigold	*R. intraradices* *F. mosseae*	Greenhouse	Increased absorption of nutrients such as P and N	Iraji Mareshk and Moghaddam (2020)
Heavy metal	*Calendula officinalis* L./marigold	*Claroideoglomus claroideum* *F. mosseae*	Glasshouse	Promoted the accumulation of important secondary metabolites (phenols, flavonoids, and carotenoids)	Hristozkova et al. (2016)
	Tagetes erecta L./marigold	*Glomus constrictum*	Greenhouse	Increased yield, relative Chl content, leaf area, flower quality, and element (P, N, K, and Mg) content	Elhindi et al. (2018)
	Helianthus annuus L./sunflower	*F. mosseae* *Funneliformis caledonium*	Greenhouse	Increased biomass, P content in shoots and roots	Zhang et al. (2018)
	Monarda citriodora Cerv. ex lag/lemon beebalm	*Glomus fasciculatum*	Greenhouse	Improved the overall growth and development of plants	Luxmi et al. (2023)
	Helianthus annuus L./sunflower	Mixture of *F mosseae, R. intraradices, and Claroideoglomus etunicatum*	Greenhouse	Higher activities of acid and alkaline phosphatases	Abd-Allah et al. (2015

Ca: calcium; Chl: chlorophyll; *E*: transpiration rate; g_s: stomatal conductance; K: potassium; Mg: magnesium; MSI: membrane stability index; N: nitrogen; n.d. not defined; P: phosphorus; P_n: leaf net photosynthetic rate; RWC: relative water content non-inoculated plants (Abdel-Salam et al. 2018). Asrar et al. (2012) reported that mycorrhizal plants exhibited higher relative water content in leaves during drought conditions in *Antirrhinum majus*, suggesting that AMF hyphae penetrated deeply into the soil and provided moisture to the plants. Moreover, plants growing in high-salinity soils experience physiological drought, as the water they need is bound by ions such as Na^+ and Cl^-, making it unavailable for uptake (Kapoor et al. 2012). Several studies have shown that plants treated with AMF retain a higher water content, even under stress conditions (Othman et al. 2022; Janah et al. 2021). This process is supported by increased root hydraulic conductivity under low water potential conditions (Abdalla and Ahmed 2021). AMF-inoculated plants have been shown to exhibit higher stomatal conductance, resulting in increased transpiration demand (Abdalla et al. 2023). Studies have also indicated that these plants maintain a lower osmotic potential facilitated by solutes accumulated by the fungi, thereby enhancing the osmotic adjustment of the plant (Kumar et al. 2019). Enhanced turgor potential and reduced water saturation deficits in AMF plants contribute to improved plant water status (Asrar et al. 2014). All of these improved parameters resulting from AMF colonization promote more efficient water use by the host plants, thereby enabling them to maintain a lower concentration of intercellular carbon dioxide.

Aquaporins are a type of major intrinsic protein (MIP) family that functions as selective water channels across the plasma membrane (by *PIPs*) and tonoplast (by *TIPs*), while restricting the movement of ions and other small molecules. Abiotic factors, including drought and salt stress, are known to affect aquaporin expression through phytohormones such as abscisic acid (ABA) and gibberellic acid (Evelin et al. 2009). Research on aquaporin gene expression in AMF-inoculated plants under salt stress has shown divergent results. However, AMF colonization was found to have a more pronounced effect on aquaporin transcript levels than salinity stress alone. In a study by Zou et al. (2024), AMF colonization upregulates the expression of *LjNIP1;5* in *Lotus japonicus* under drought stress conditions. These results suggest that AMF may differentially regulate the expression of aquaporin genes, indicating a gene-specific response within symbiotic interactions. Moreover, the distinct expression patterns observed imply that each PIP gene may play a unique role in the AMF-mediated response to environmental stress, reflecting a complex regulatory network involved in water transport and stress adaptation in host plants.

3.1.1.2 Osmotic Adjustment

AMF confers tolerance to host plants by adjusting the osmotic imbalance induced by stress (Borde et al. 2017; Mitra et al. 2021). Osmotic adjustment occurs through the accumulation of metabolites that function as compatible solutes. Compatible solutes, also known as osmolytes, do not disrupt normal metabolic reactions but instead substitute for water in these reactions. Commonly studied osmolytes are proline, glycine betaine, and sugars (Yadav et al. 2017; Evelin et al. 2019). Proline accumulation is frequently altered by water and salt stress in plants. Under abiotic conditions, plants accumulate proline as a non-toxic osmolyte, which helps maintain osmotic balance during low water availability. It also serves as a source of energy and nitrogen during stressful conditions (Evelin et al. 2019). Glycine betaine, a non-toxic cellular osmolyte, improves salt stress tolerance through several mechanisms, such as (1) maintaining membrane integrity against salt damage, (2) activating specific stress tolerance genes, (3) safeguarding the photosynthetic machinery, (4) stabilizing enzyme and protein complex structures and activities, (5) regulating ion-channel protein activities directly or by protecting the plasma membrane, and (6) decreasing ROS levels under stress (Kapoor et al. 2012). Several studies have reported that mycorrhizal plants show higher proline levels than non-mycorrhizal plants, although some findings have indicated reduced proline accumulation in mycorrhizal associations (Ali and Hassan 2014). The increased proline content in mycorrhizal plants is attributed to the enhanced activity of the proline 5 carboxylate synthetase (P5CS) enzyme, upregulation of the gene encoding P5CS, and suppression of proline dehydrogenase (ProDH) activity, which catalyzes proline degradation, and elevated activity of the glutamate dehydrogenase enzyme, which is involved in glutamate synthesis (Chun et al. 2018). In addition to its role as an osmoprotectant, proline has been recognized as a stress marker (Sannazzaro et al. 2007; Evelin et al. 2019). A study by Abdel-Salam et al. (2018) reported lower proline content in AMF plants than in non AMF plants, and this accumulation decreased with decreasing drought stress levels. This suggests that AMF colonization enhances drought stress tolerance in the host plants.

3.1.1.3 Nutrient Uptake

AMF have been demonstrated to positively affect the mineral nutrient composition of plants under stress conditions, particularly by enhancing and/or selectively improving the uptake of phosphorus (P) (Bhantana et al. 2021). The extraradical mycelium of AMF extends from the host root to the soil, increasing the effective surface area for nutrient absorption. This network facilitates the uptake of inaccessible minerals, thereby improving the nutrient supply to the host plant and alleviating nutrient deficiency (Nouri et al. 2015; Huang et al. 2021). Various studies have documented the role of AMF in soil nutrient uptake, particularly P and N, which can effectively stimulate host plant growth (Bhupenchandra et al. 2024). Studies have revealed that AMF can boost plant phosphate levels primarily through three pathways: by improving root architecture to enhance the soil phosphate uptake area, by stimulating the breakdown of organic phosphate in soil into inorganic phosphate (Pi) suitable for plant uptake, by enhancing the secretion and activity of phosphatase, and by increasing the expression level of mycorrhizal-specific phosphate transporter (PT) genes that improve the efficiency of P transport in hosts (Huang et al. 2021). In addition to facilitating Pi uptake by plants, AMF are responsible for regulating the growth of arbuscular structures and shapes, as well as sustaining symbiotic associations (Igiehon and Babalola 2017). The ensuing symbiotic interface boosts photosynthate generation by allowing the transfer of nutrients, such as P and N (Denison and Kiers 2011). According to the available data, arbuscular mycorrhiza (AM) symbiosis can provide up to 90% of a plant's soil-based Pi requirements (Van der Heijden et al. 2008).

Furthermore, AMF promote phosphatase and organic acid secretion in plant roots, improve the soil environment, and accelerate insoluble P conversion (Huang et al. 2021). Javot et al. (2007) observed that AMF-colonized plants increased plant phosphatase secretion and thus improved plant P uptake. Following the acquisition of Pi by AMF, several Pi transporters facilitate root Pi uptake and homeostasis. High-affinity Pi transporters from the PHOSPHATE TRANSPORTER1 (PHT1) gene family aid in Pi remobilization (Chiou and Lin 2011). The increased uptake of P by AMF in plants under salt stress may mitigate the harmful effects of Na^+ and Cl^-. This is achieved by maintaining the integrity of vacuolar membranes, which aids in compartmentalization within vacuoles and regulates the intake of specific ions, thus safeguarding ions from interfering with metabolic processes essential for plant growth (Evelin et al. 2009).

Similarly, AMF contribute to the uptake of N in the soil, where it exists in mineral and organic forms (nitrites, nitrates, and ammonium ions). During stress conditions, the availability of N is reduced, leading to a decreased uptake of N and lower assimilation rates (Dubey et al. 2021). The extraradical mycelium absorbs N from the soil as inorganic N in the form of nitrate, which is assimilated by nitrate reductase found in arbuscule-containing cells and then converted to arginine through the GS-GOGAT (glutamine synthetase-glutamine: 2-oxoglutarate aminotransferase) cycle. Arginine is degraded to urea in the hyphae and eventually transmitted to the plant as NH_4^+ (Fig. 1), with the resultant C skeletons from arginine degradation being re-incorporated into the fungal C pools (Kapoor et al. 2012). Various studies

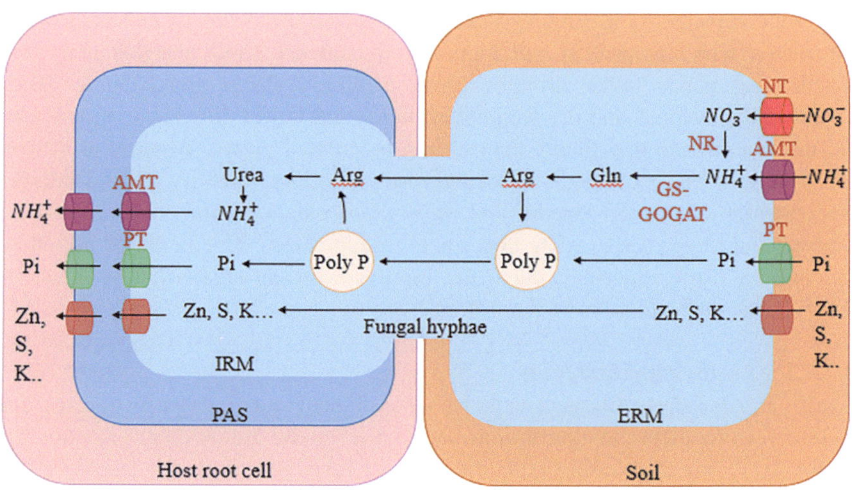

Fig. 1 Mycorrhizal pathways for the uptake of nutrients. (Adopted from Wang et al. 2017, license number 6032030766105). Ammonium transporter (AMT), arginine (Arg), extraradical mycelium (ERM), glutamine (Gln), glutamine synthetase-glutamine, 2-oxoglutarate aminotransferase (GS-GOGAT), intraradical mycelium (IRM), potassium (K), ammonium (NH_4^+), nitrate (NO_3^-), nitrate reductase (NR), nitrate transporter (NT), peri-arbuscular space (PAS), inorganic phosphate (Pi), polyphosphate (Poly P), phosphate transporter (PT), sulfur (S), zinc (Zn)

have shown that AMF can transfer 20–75% of the total N absorption of AMF plants to their hosts (Ahanger et al. 2013; Begum et al. 2019). Increased N in AMF-treated plants leads to increased chlorophyll concentrations because chlorophyll molecules can easily trap N (Gholamhoseini et al. 2013). Asrar et al. (2012) observed that the specified fungal association increased the macronutrient concentration, including K, P, N, Mg, and Ca, in drought-stressed *Antirrhinum majus*.

The introduction of AMF improves the yield and nutrient content of several horticultural crops (Zhu et al. 2022). Gholamhoseini et al. (2013) reported an increased percentage of mineral nutrients (N, P, and K) in the leaves and seeds of sunflower plants inoculated with mycorrhiza under drought stress. AMF inoculation enhances the uptake of various micronutrients and macronutrients, leading to increased photosynthate synthesis and improved biomass accumulation (Sohn et al. 2003). Dhalaria et al. (2024a, b) observed elevated concentrations of mineral nutrients in the leaves and flowers of the *Gomphrena globosa* plant following AMF inoculation, which might be due to the expanded absorptive surface provided by the extraradical mycelial networks. Furthermore, AMF interactions alter the physiology and rhizophore environment of the host, leading to higher nutrient uptake and biomass (Wahab et al. 2023, Cruz et al. 2004).

3.1.1.4 Plant Growth and Biomass

Mycorrhization improves the fitness of host plants by increasing their growth and biomass (Mitra et al. 2021). Several studies have reported that plants inoculated with arbuscular mycorrhizal fungi (AMF) exhibit better growth than non-inoculated plants under stressful conditions (Bagheri et al. 2019; Popescu and Popescu 2022). A study conducted by Rasouli et al. (2023) reported that *Satureja hortensis* plants inoculated with mycorrhiza had higher fresh and dry shoot weights than non-mycorrhizal plants under drought stress. Similarly, Asrar and Elhindi (2011) observed higher fresh and dry weights of shoots and flower diameters in mycorrhizal-inoculated marigold plants than in non-inoculated plants. Wang et al. (2018) reported improved root length, shoot and root dry weight, total dry weight, and root N concentration in *Chrysanthemum morifolium* plants colonized by arbuscular mycorrhizal fungi (AMF) when exposed to salinity stress. The increased growth of mycorrhizal plants is partially due to the improved nutrient acquisition facilitated by AMF, particularly in terms of P nutrition.

3.1.1.5 Antioxidant Enzymes

AMF-mediated stress tolerance in host plants is linked to their capacity to neutralize ROS and to regulate the equilibrium between ROS and antioxidants (Kapoor and Singh 2017; Zou et al. 2021). ROS are natural byproducts of plant metabolism and are produced during photorespiration, photosynthesis, and respiration (Bhattacharjee 2019). However, abiotic stress also promotes ROS overproduction. At low concentrations, ROS play a role in signaling, development, and behavior; however, high concentrations of ROS cause harm to cells by inducing membrane lipid peroxidation, protein oxidation, enzyme inhibition, and damage to DNA and RNA (Sharma et al. 2019). Multiple studies have indicated that AMF symbiosis assists plants in reducing abiotic stress by enhancing the activities of antioxidant enzymes, such as catalase (CAT), peroxidase (POD), superoxide dismutase (SOD), glutathione reductase (GR), ascorbate peroxidase (APX), and antioxidant compounds, such as tocopherols, glutathione, ascorbic acid, and carotenoids, while preventing the overproduction of ROS (Kheyri et al. 2022; Rasouli et al. 2023).

Increased SOD activity induced by AMF facilitates the conversion of superoxide anions ($O^{2\cdot-}$) into oxygen and H_2O_2. The generated H_2O_2 was further broken down into water (H_2O) by CAT and POD (Dhalaria et al. 2020). APX is also involved in the detoxification of generated H_2O_2 in the chloroplasts of stressed host plants (Choudhary et al. 2017). Increased GR activity ensures an adequate supply of $NADP^+$, which accepts electrons via the photosynthetic electron transport chain. This process helps divert electrons away from oxygen, thereby limiting the production of superoxide radicals (O_2^-) (Foyer and Harbinson 2019). Thus, AMF quickly eliminates surplus ROS to maintain the optimum ROS concentration for its physiological function and prevents the shift toward a destructive state. Amiri et al. (2015) reported that mycorrhizal plants exhibit reduced malondialdehyde (MDA), hydrogen peroxide (H_2O_2) content, and increased activity of SOD, CAT, APX, and GPX when compared with non-mycorrhizal plants. Similarly, Rasouli et al. (2023) reported higher antioxidant enzyme levels in *Satureja hortensis* colonized by *F. mosseae* than in non-colonized plants.

3.1.1.6 Reduction of HM Ion Uptake

AMF colonization of higher plant roots allows the plant to store heavy metals in its roots and external mycelia (Agarwal et al. 2017). AMF reduce HM toxicity in plants by sequestering HMs in specialized structures called vesicles and fungal mycelium, where high concentrations of HMs accumulate, which hinders their transfer to aboveground plant parts (Riaz et al. 2021; Karunanantham et al. 2022). Fungal vesicles in AMF-colonized plants are similar to plant vacuoles and are involved in the storage of toxic substances (Dhalaria et al. 2020). The external mycelium has a greater impact on the immobilization of HMs in stressed soil because of its larger surface area, making it more efficient than plant roots in absorbing metals (Hashem et al. 2022). This retention mechanism by AMF reduces the movement of HMs in roots and shields leaf tissues from damage, thereby aiding phytostabilization (Riaz et al. 2021).

The fungal cell wall also contributes to the absorption of HM ions. The cell wall consists of polysaccharides and chitin, which serve as barriers to prevent HM ions and other solutes from entering the cells, thus controlling their absorption (Dhalaria et al. 2020; Ayele et al. 2021). Multiple studies have documented that fungal cell walls serve as the first line of defense against increasing metal concentrations by preventing the entry of HMs. Organic acids, amino acids, and other compounds in mycelium cells interact with HMs to produce precipitates or crystals, thereby reducing their availability and mobility (Dhalaria et al. 2020). In addition, AMF can release certain compounds, such as glomalin, that bind to HMs and decrease their harmful effects on plants (Boorboori and Zhang 2022).

Chelation of HMs with ligands that have a strong affinity for HM ions in the cytosol is another strategy used by AMF (Ferrol et al. 2016). When the concentration of harmful metals inside cells surpasses a certain threshold, chelating compounds are produced. Chelation reduces the availability of HMs in the cytoplasm by decreasing their reactivity and solubility, thereby preventing their harmful effects (Dhalaria et al. 2020). Phytochelatins, metallothioneins, amino acids, and organic acids are among the various chelating compounds found in AMF and plants and play an important role in plant HM tolerance by chelating these substances. A study conducted by Elhindi et al. (2018) found that inoculation with *Glomus constrictum* reduced HM toxicity in marigold by retaining HMs in glomalin produced by AMF and enhancing HM uptake by the AM mycelium. Abd-Allah et al. (2015) found that mycorrhizal inoculation in sunflower plants reduces the Cd concentration, which may be attributed to the ability of fungal hyphae to bind heavy metals outside and inside the roots and restrict their uptake to the upper parts.

3.1.2 Biotic Stress

Plants are constantly threatened by biotic agents such as fungi, bacteria, viruses, parasites, and insect herbivores, with severe economic and environmental consequences (Bilgin et al. 2010; Umar et al. 2021). Early transcriptome studies have demonstrated that several photosynthetic genes are downregulated after biotic assault (Bilgin et al. 2010). AMFs are essential in mitigating biotic stress as they strengthen the plant's defense system and are involved in pathogen resistance (Diagne et al. 2020). AMF protect against different biotic stresses by acting alone or collaborating with other related microorganisms living in plant-influenced zones. Plants colonized with AMF exhibit increased growth, survival, and resistance to plant pathogens (El Omari and El Ghachtouli 2021). Research findings inferred that host defense induction, rhizosphere alteration, and competition for nutrients, photosynthates, and space are the prevalent mechanisms shown by AMF (Diagne et al. 2020; Dowarah et al. 2022). Table 2 lists a variety of phytopathogens for which AMF conferred tolerance.

Generally, AMF colonization in plant roots causes relay reactions that activate plant defense systems. Most plant pathogenic and defense-activating microbes, such as AMF, have small molecules known as microbe-associated molecular patterns (MAMP) on their surface, along with effector proteins (EP) secreted to the interior or exterior of plant cells. MAMP recognition by plant receptor protein complexes, namely pattern recognition receptors (PRR), produces salicylic acid (SA)-dependent defense responses, which activate MAMP-triggered immunity (MTI) (McDowell 2019; Dowarah et al. 2022). Some effector molecules inhibit MTI by targeting host proteins, whereas others are identified as plant defense proteins. The recognition of effector molecules by plant resistance proteins activates a defense response called effector-triggered immunity (ETI), which serves as the plant's first line of defense to prevent further pathogen invasion (Dangl et al. 2013). In addition, ETI responses generate a hypersensitivity response (HR) that principally involves rapid localized necrosis at the site of the first pathogen entrance.

Table 2 Representative studies highlighting the impact of various AMF species on different flower crops under different biotic stress conditions

Biotic stressors	Host crop	AMF species	Positive effects of AMF	Mechanism used by AMF	References
Fusarium oxysporum f. sp. *Lini* and *Oidium lini* bond	*Linum usitatissimum* L.	*Glomus intraradices* Schenck and Smith	Increased shoot biomass, CO_2 assimilation, and sucrose content	Increased respiration rate, ethylene production, and DNA demethylation	Dugassa et al. (1996)
Botrytis cinerea	Rose (*Rosa* L. (*polyantha-hybrid*), cv. Karina)	*Glomus mosseae*	*G. mosseae* significantly reduced non-specific leaf wilt incidence; increased the total number of leaves per plant	Changes in the root exudate composition	Møller et al. (2009)
Candidatus Phytoplasma asteris strain CY	*Chrysanthemum carinatum*	*Glomus mosseae* BEG12	Improved plant growth and altered root morphology	Activation of metabolic pathways	D'Amelio et al. (2011)
Armillaria mellea	*Lavandula angustifolia* L.	*Rhizophagus irregularis*	Decreased disease symptoms and increased flowering potential	n.d.	Calvet et al. (2015)
Fusarium oxysporum f. sp. *ranunculi*	*Ranunculus asiaticus*	Commercial formulation (*Rhizophagus irregularis*, *Funelliformis mosseae*, and *Glomus* spp.)	Decreased disease symptoms	Production of root exudates	Cangelosi et al. (2017)

(continued)

Table 2 (continued)

Biotic stressors	Host crop	AMF species	Positive effects of AMF	Mechanism used by AMF	References
Sclerotium rolfsii	*Helianthus tuberosus*	*Glomus* spp.	Decreased disease incidence and severity; improved shoot biomass and tuber yield	Improved SOD, CAT, APX activities	Eid et al. (2019)
Lymantria dispar	*Rosa rugosa* Thunb.	*Funneliformis mosseae*, *Rhizophagus intraradices*, and their mixture	Improved plant growth	Improved accumulation of jasmonic acid and abscisic acid in rose leaves	Lv et al. (2022)

APX: ascorbate peroxidase; CAT: catalase; n.d.: not defined; SOD: superoxide dismutase

Furthermore, AMF can potentially decrease the impact of plant diseases via an induced systemic resistance (ISR) mechanism activated by rhizospheric microorganisms (Pineda et al. 2010), as shown in Fig. 2.

Although the AM symbiotic connection is restricted to the roots, MIR is a mechanism that defends plants from aboveground diseases. MIR occurs via jasmonic acid (JA)-dependent defense pathways. D'Amelio et al. (2011) stated that MIR-mediated bioprotection is effective against *Candidatus* Phytoplasma asteris strain CY in *Chrysanthemum carinatum*. Higher activities of superoxide dismutase, catalase, and ascorbate peroxidase were reported in *Helianthus tuberosus* plants when treated with *Glomus* spp. upon infection with *Sclerotium rolfsii* (Eid et al. 2019). Møller et al. (2009) observed changes in the root exudate composition of rose plants inoculated with *Glomus mosseae* upon infection with *Botrytis cinerea*. In addition, activation of metabolic pathways has been observed upon pathogen attack compared to that in non-AMF plants (D'Amelio et al. 2011).

AMF have also been shown to provide pathogen resistance by increasing the competition for host photosynthates (Trouvelot et al. 2015). Pathogens infect plants to receive the photosynthates produced by their hosts. Soil-borne pathogens enter the roots, which is a key factor for pathogens to spread further. However, in AM symbiosis, the fungal partner receives photosynthetic assimilates from the companion plant in exchange for its benefits. Increased carbon requirement is expected to have a deleterious impact on pathogens when AMF have access to photosynthates. Furthermore, AMF can control infections through nutritional competition (Pozo et al. 2013). Since AMF considerably increases the nutritional intake of the linked plant roots, there is strong competition for nutrients between AMF-infected plants and soil-borne plant diseases in the rhizosphere. AM symbiosis increases the surface area of plant roots by more than 100 times compared to plants that do not have

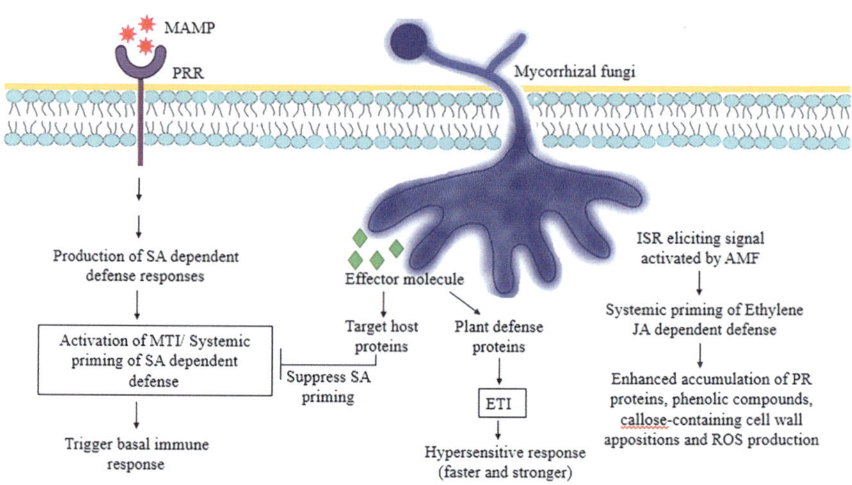

Fig. 2 Host defense induction by AMF. (Modified from Dowarah et al. 2022 with license number 6032040057127)

AMF (Smith and Read 2010). Moreover, AMF colonization alters the root morphology of the host plant, making it difficult for soil-borne phytopathogens to reach the root system.

4 Conclusion

As the global demand for high-quality flowers continues to rise, the floriculture industry must adopt sustainable and efficient practices to meet this demand, while promoting eco-friendly agriculture. The use of AMF has emerged as a promising strategy to enhance flower crop productivity, improve soil health, and promote ecosystem services. By harnessing the benefits of AMF, the floriculture industry can reduce its reliance on chemical fertilizers and pesticides, mitigate environmental pollution, and promote sustainable agricultural practices. Further research is needed to fully explore the potential of AMF in floriculture, including the identification of genes that enable host plants to survive under stressful conditions and the determination of interaction mechanisms in natural field conditions. Advancing knowledge in these areas will pave the way for innovative, resilient, and environmentally responsible practices in flower cultivation, ultimately supporting both industry growth and ecological balance.

Acknowledgments VDR acknowledges the support by the Priority 2030 program implemented at Southern Federal University, Russia.

References

Abd-Allah EF, Abeer H, Alqarawi AA, Hend AA (2015) Alleviation of adverse impact of cadmium stress in sunflower (*Helianthus annuus* L.) by arbuscular mycorrhizal fungi. Pak J Bot 47(2):785–795

Abdalla M, Ahmed MA (2021) Arbuscular mycorrhiza symbiosis enhances water status and soil-plant hydraulic conductance under drought. Front Plant Sci 12:722954

Abdalla M, Bitterlich M, Jansa J, Püschel D, Ahmed MA (2023) The role of arbuscular mycorrhizal symbiosis in improving plant water status under drought. J Exp Bot 74(16):4808–4824

Abdel-Salam E, Alatar A, El-Sheikh MA (2018) Inoculation with arbuscular mycorrhizal fungi alleviates harmful effects of drought stress on damask rose. Saudi J Biol Sci 25(8):1772–1780

Agarwal AN, Singh JA, Singh AP (2017) Arbuscular mycorrhizal fungi and its role in sequestration of heavy metals. Trends Biosci 10(21):4068–4077

Ahanger MA, Tyagi SR, Wani MR, Ahmad P (2013) Drought tolerance: role of organic osmolytes, growth regulators, and mineral nutrients. In: Physiological mechanisms and adaptation strategies in plants under changing environment: Volume 1. Springer, New York, pp 25–55

Ali EF, Hassan FA (2014) Alleviatory effects of salt stress by mycorrhizal fungi and gibberellic acid on chamomile plant. Int J Sci Res 3(11):109–118

Amiri R, Nikbakht A, Etemadi N (2015) Alleviation of drought stress on rose geranium [*Pelargonium graveolens* (L.) Herit.] in terms of antioxidant activity and secondary metabolites by mycorrhizal inoculation. Sci Hortic 197:373–380

Arif Y, Singh P, Siddiqui H, Bajguz A, Hayat S (2020) Salinity induced physiological and biochemical changes in plants: an omic approach towards salt stress tolerance. Plant Physiol Biochem 156:64–77

Arya V, Kumar D, Gautam M (2014) Phytopharmacological review on flowers: source of inspiration for drug discovery. Biomed Prev Nutr 4(1):45–51

Asrar AW, Elhindi KM (2011) Alleviation of drought stress of marigold (*Tagetes erecta*) plants by using arbuscular mycorrhizal fungi. Saudi J Biol Sci 18(1):93–98

Asrar AA, Abdel-Fattah GM, Elhindi KM (2012) Improving growth, flower yield, and water relations of snapdragon (*Antirhinum majus* L.) plants grown under well-watered and water-stress conditions using arbuscular mycorrhizal fungi. Photosynthetica 50:305–316

Asrar AA, Abdel-Fattah GM, Elhindi KM, Abdel-Salam EM (2014) The impact of arbuscular mychorrhizal fungi in improving growth, flower yield and tolerance of kalanchoe (*Kalanchoe blossfeldiana* Poelin) plants grown in NaCl-stress conditions. J Food Agric Environ 12(1):105–112

Ayele A, Haile S, Alemu D, Tesfaye T, Kamaraj M (2021) Mycoremediation: fungal-based technology for biosorption of heavy metals–a review. In: Strategies and tools for pollutant mitigation: avenues to a cleaner environment, pp 355–373

Bagheri V, Shamshiri MH, Alaei H, Salehi H (2019) The role of inoculum identity for growth, photosynthesis, and chlorophyll fluorescence of zinnia plants by arbuscular mycorrhizal fungi under varying water regimes. Photosynthetica 57(2):409

Begum N, Qin C, Ahanger MA, Raza S, Khan MI, Ashraf M, Ahmed N, Zhang L (2019) Role of arbuscular mycorrhizal fungi in plant growth regulation: implications in abiotic stress tolerance. Front Plant Sci 10:1068

Benvenuti S, Mazzoncini M (2021) The biodiversity of edible flowers: discovering new tastes and new health benefits. Front Plant Sci 11:569499

Bhantana P, Rana MS, Sun XC, Moussa MG, Saleem MH, Syaifudin M, Shah A, Poudel A, Pun AB, Bhat MA, Mandal DL (2021) Arbuscular mycorrhizal fungi and its major role in plant growth, zinc nutrition, phosphorous regulation and phytoremediation. Symbiosis 84:19–37

Bhattacharjee S (2019) Reactive oxygen species in plant biology. New Delhi, Springer

Bhupenchandra I, Chongtham SK, Devi AG, Dutta P, Sahoo MR, Mohanty S, Kumar S, Choudhary AK, Devi EL, Sinyorita S, Devi SH (2024) Unlocking the potential of arbuscular mycorrhizal fungi: exploring role in plant growth promotion, nutrient uptake mechanisms, biotic stress alleviation, and sustaining agricultural production systems. J Plant Growth Regul:1–39

Bilgin DD, Zavala JA, Zhu JI, Clough SJ, Ort DR, DeLUCIA EH (2010) Biotic stress globally downregulates photosynthesis genes. Plant Cell Environ 33(10):1597–1613

Boorboori MR, Zhang HY (2022) Arbuscular mycorrhizal fungi are an influential factor in improving the phytoremediation of arsenic, cadmium, lead, and chromium. J Fungi 8(2):176

Borde M, Dudhane M, Kulkarni M (2017) Role of arbuscular mycorrhizal fungi (AMF) in salinity tolerance and growth response in plants under salt stress conditions. In: Mycorrhiza-eco-physiology, secondary metabolites, nanomaterials, pp 71–86

Calvet C, Garcia-Figueres F, Lovato P, Camprubi A (2015) Role of the arbuscular mycorrhizal symbiosis in tolerance response against *Armillaria mellea* in lavender. Span J Agric Res 13(3):e1008

Cangelosi B, Curir P, Beruto M, Monroy F, Borriello R, Beruto M (2017) Protective effects of Arbuscular mycorrhizae against *Fusarium oxysporum* f. sp. *ranunculi* in *Ranunculus asiaticus* cultivations for flower crop. J Hortic Sci Res 1(2):36–41

Cao H, Ding R, Kang S, Du T, Tong L, Zhang Y, Chen J, Shukla MK (2023) Drought, salt, and combined stresses in plants: effects, tolerance mechanisms, and strategies. Adv Agron 178:107–163

Chiou TJ, Lin SI (2011) Signaling network in sensing phosphate availability in plants. Annu Rev of Plant Biol 62(1):185–206

Choudhary KK, Chaudhary N, Agrawal SB, Agrawal M (2017) Reactive oxygen species: generation, damage, and quenching in plants during stress. In: Reactive oxygen species in plants: boon or bane-revisiting the role of ROS, pp 89–115

Chun SC, Paramasivan M, Chandrasekaran M (2018) Proline accumulation influenced by osmotic stress in arbuscular mycorrhizal symbiotic plants. Front Microbiol 9:2525

Crişan I, Ona A, Vârban D, Muntean L, Vârban R, Stoie A, Mihăiescu T, Morea A (2023) Current trends for lavender (*Lavandula angustifolia* mill.) crops and products with emphasis on essential oil quality. Plants 12(2):357

Cruz C, Green JJ, Watson CA, Wilson F, Martins-Loução MA (2004) Functional aspects of root architecture and mycorrhizal inoculation with respect to nutrient uptake capacity. Mycorrhiza 14:177–184

D'Amelio R, Berta G, Gamalero E, Massa N, Avidano L, Cantamessa S, D'agostino G, Bosco D, Marzachì C (2011) Increased plant tolerance against chrysanthemum yellows phytoplasma ('*Candidatus* Phytoplasma asteris') following double inoculation with *glomus mosseae* BEG12 and *pseudomonas putida* S1Pf1Rif. Plant Pathol 60(6):1014–1022

Dangl JL, Horvath DM, Staskawicz BJ (2013) Pivoting the plant immune system from dissection to deployment. Science 341(6147):746–751

Datta SK, Gupta YC (eds) (2022) Floriculture and ornamental plants. Springer

De LC, Anumala NV, Kumar A, Biswas SS, Ns K, Natta S, Gowda C, Das SP (2024) Impact of climate change on floriculture industry of India. Medicon Agric Environ Sci 7(4):26–33

Denison RF, Kiers ET (2011) Life histories of symbiotic rhizobia and mycorrhizal fungi. Curr Biol 21(18):R775–R785

Devireddy AR, Tschaplinski TJ, Tuskan GA, Muchero W, Chen JG (2021) Role of reactive oxygen species and hormones in plant responses to temperature changes. Int J Mol Sci 22(16):8843

Dhalaria R, Kumar D, Kumar H, Nepovimova E, Kuča K, Torequl Islam M, Verma R (2020) Arbuscular mycorrhizal fungi as potential agents in ameliorating heavy metal stress in plants. Agronomy 10(6):815

Dhalaria R, Verma R, Kumar D, Upadhyay NK, Alomar S, Kuca K (2024a) Impact assessment of beneficial mycorrhizal fungi on phytochemical constituents and nutrient uptake in *Gomphrena globosa*. Sci Hortic 325:112646

Dhalaria R, Verma R, Sharma R, Jomova K, Nepovimova E, Kumar H, Kuca K (2024b) Assessing the potential role of arbuscular mycorrhizal fungi in improving the phytochemical content and antioxidant properties in *Gomphrena globosa*. Sci Rep 14(1):22830

Diagne N, Ngom M, Djighaly PI, Fall D, Hocher V, Svistoonoff S (2020) Roles of arbuscular mycorrhizal fungi on plant growth and performance: importance in biotic and abiotic stressed regulation. Diversity 12(10):370

Dowarah B, Gill SS, Agarwala N (2022) Arbuscular mycorrhizal fungi in conferring tolerance to biotic stresses in plants. J Plant Growth Regul 41(4):1429–1444

Dubey RS, Srivastava RK, Pessarakli M (2021) Physiological mechanisms of nitrogen absorption and assimilation in plants under stressful conditions. In: Pessarakli M (ed) Handbook of plant and crop physiology. CRC Press, pp 579–616

Dugassa GD, Von Alten H, Schönbeck F (1996) Effects of arbuscular mycorrhiza (AM) on health of *Linum usitatissimum* L. infected by fungal pathogens. Plant Soil 185:173–182

Eid KE, Abbas MH, Mekawi EM, ElNagar MM, Abdelhafez AA, Amin BH, Mohamed I, Ali MM (2019) Arbuscular mycorrhiza and environmentally biochemicals enhance the nutritional status of *Helianthus tuberosus* and induce its resistance against *Sclerotium rolfsii*. Ecotoxicol Environ Saf 186:109783

El Omari B, El Ghachtouli N (2021) Arbuscular mycorrhizal fungi-weeds interaction in cropping and unmanaged ecosystems: a review. Symbiosis 83(3):279–292

Elhindi KM, Al-Mana FA, El-Hendawy S, Al-Selwey WA, Elgorban AM (2018) Arbuscular mycorrhizal fungi mitigates heavy metal toxicity adverse effects in sewage water contaminated soil on *Tagetes erecta* L. Soil Sci Plant Nutr 64(5):662–668

Evelin H, Kapoor R, Giri B (2009) Arbuscular mycorrhizal fungi in alleviation of salt stress: a review. Ann Bot 104(7):1263–1280

Evelin H, Devi TS, Gupta S, Kapoor R (2019) Mitigation of salinity stress in plants by arbuscular mycorrhizal symbiosis: current understanding and new challenges. Front Plant Sci 10:470

Ferrol N, Tamayo E, Vargas P (2016) The heavy metal paradox in arbuscular mycorrhizas: from mechanisms to biotechnological applications. J Exp Bot 67:6253

Foyer CH, Harbinson J (2019) Oxygen metabolism and the regulation of photosynthetic electron transport. In: Causes of photooxidative stress and amelioration of defense systems in plants. CRC Press, pp 1–42

Ganugi P, Masoni A, Pietramellara G, Benedettelli S (2019) A review of studies from the last twenty years on plant–arbuscular mycorrhizal fungi associations and their uses for wheat crops. Agronomy 9(12):840

Garg N, Chandel S (2011) Arbuscular mycorrhizal networks: process and functions. Sustain Agric 2:907–930

Gelaye Y (2023) The status and natural impact of floriculture production in Ethiopia: a systematic review. Environ Sci Pollut Res 30(4):9066–9081

Gholamhoseini M, Ghalavand A, Dolatabadian A, Jamshidi E, Khodaei-Joghan A (2013) Effects of arbuscular mycorrhizal inoculation on growth, yield, nutrient uptake and irrigation water productivity of sunflowers grown under drought stress. Agric Water Manag 117:106–114

Ghori NH, Ghori T, Hayat MQ, Imadi SR, Gul A, Altay V, Ozturk M (2019) Heavy metal stress and responses in plants. Int J Environ Sci Technol 16:1807–1828

Gonzalez Guzman M, Cellini F, Fotopoulos V, Balestrini R, Arbona V (2022) New approaches to improve crop tolerance to biotic and abiotic stresses. Physiol Plant 174(1):e13547

Hart MM, Reader RJ (2002) Taxonomic basis for variation in the colonization strategy of arbuscular mycorrhizal fungi. New Phytol 53(2):335–344

Hashem A, Shameem N, Parray JA, Abd-Allah EF (2022) Mycorrhizal strategy for the management of hazardous chromium contaminants. In: Core microbiome: improving crop quality and productivity, pp 298–314

Hassan MU, Chattha MU, Khan I, Chattha MB, Barbanti L, Aamer M, Iqbal MM, Nawaz M, Mahmood A, Ali A, Aslam MT (2021) Heat stress in cultivated plants: nature, impact, mechanisms, and mitigation strategies—a review. Plant Biosyst 155(2):211–234

Helgason T, Fitter AH (2009) Natural selection and the evolutionary ecology of the arbuscular mycorrhizal fungi (phylum glomeromycota). J Exp Bot 60(9):2465–2480

Hristozkova M, Geneva M, Stancheva I, Boychinova M, Djonova E (2016) Contribution of arbuscular mycorrhizal fungi in attenuation of heavy metal impact on *Calendula officinalis* development. Appl Soil Ecol 101:57–63

Huang GM, Xu YJ, Wu QS (2021) Mycorrhiza-improved P acquisition of host plants: a mini-review. GSC Biol Pharm Sci 14(3):62–67

Igiehon NO, Babalola OO (2017) Biofertilizers and sustainable agriculture: exploring arbuscular mycorrhizal fungi. Appl Microbiol Biotechnol 101:4871–4881

Iqbal B, Li G, Alabbosh KF, Hussain H, Khan I, Tariq M, Javed Q, Naeem M, Ahmad N (2023) Advancing environmental sustainability through microbial reprogramming in growth improvement, stress alleviation, and phytoremediation. Plant Stress 10:100283

Iraji Mareshk M, Moghaddam M (2020) Effect of mycorrhizal fungi on growth and absorption of nutrients of Mexican marigold (*Tagetes minuta* L.) under salinity stress. Environ Stresses Crop Sci 13(3):969–982

Janah I, Meddich A, Elhasnaoui A, Khayat S, Anli M, Boutasknit A, Aissam S, Loutfi K (2021) Arbuscular mycorrhizal fungi mitigates salt stress toxicity in *Stevia rebaudiana* Bertoni through the modulation of physiological and biochemical responses. J Soil Sci Plant Nutr 1:152

Javot H, Pumplin N, Harrison MJ (2007) Phosphate in the arbuscular mycorrhizal symbiosis: transport properties and regulatory roles. Plant Cell Environ 30(3):310–322

Kapoor R, Singh N (2017) Arbuscular mycorrhiza and reactive oxygen species. In: Arbuscular mycorrhizas and stress tolerance of plants, pp 225–243

Kapoor R, Evelin H, Mathur P, Giri B (2012) Arbuscular mycorrhiza: approaches for abiotic stress tolerance in crop plants for sustainable agriculture. In: Plant acclimation to environmental stress. Springer, New York, pp 359–401

Karunanantham K, Lakshminarayanan SP, Ganesamurthi AK, Ramasamy K, Rajamony VR (2022) Arbuscular mycorrhiza—a health engineer for abiotic stress alleviation. In: Rhizosphere engineering. Academic, pp 171–198

Kheyri Z, Moghaddam M, Farhadi N (2022) Inoculation efficiency of different mycorrhizal species on growth, nutrient uptake, and antioxidant capacity of *Calendula officinalis* L.: a comparative study. J Soil Sci Plant Nutr 22:1–3

Kumar D, Priyanka P, Yadav P, Yadav A, Yadav K (2019) Arbuscular mycorrhizal fungi-mediated mycoremediation of saline soil: current knowledge and future prospects. In: Recent advancement in white biotechnology through fungi, Volume 3: Perspective for sustainable environments, pp 319–348

Luxmi S, Singh R, Ahmed S, Gandhi SG, Bhanwaria R (2023) *Glomus fasciculatum* an arbuscular mycorrhizal fungus alleviate the adverse effects of lead, arsenic, nickel, and improves growth parameters of *Monarda citriodora* Cerv. ex Lag (lemon beebalm). Rhizosphere 27:100753

Lv Y, Wang F, Chen H, Zhang T, Yan J, Hu Y (2022) Unraveling arbuscular mycorrhizal fungi-induced resistance of purple branch rose (*Rosa rugosa* 'Zizhi') to *Lymantria dispar* based on metabolomics. Biol Control 172:104971

Madawala HM (2021) Arbuscular mycorrhizal fungi as biofertilizers: current trends, challenges, and future prospects, Biofertilizers, pp 83–93

McDowell JM (2019) Focus on activation, regulation, and evolution of MTI and ETI. Mol Plant-Microbe Interact 32(1):5

Mitra D, Djebaili R, Pellegrini M, Mahakur B, Sarker A, Chaudhary P, Khoshru B, Gallo MD, Kitouni M, Barik DP, Panneerselvam P (2021) Arbuscular mycorrhizal symbiosis: plant growth improvement and induction of resistance under stressful conditions. J Plant Nutr 44(13):1993–2028

Møller K, Kristensen K, Yohalem D, Larsen J (2009) Biological management of gray mold in pot roses by co-inoculation of the biocontrol agent *Ulocladium atrum* and the mycorrhizal fungus *Glomus mosseae*. Biol Control 49(2):120–125

Negi NP, Narwal P, Sharma A (2023) Plant–microbe interaction in alleviating drought stress. In: Sharma V, Salwan R, Moliszewska E, Ruano-Rosa D, Jędryczka M (eds) The chemical dialogue between plants and beneficial microorganisms. Elsevier, Academic, pp 49–65

Nouri E, Breuillin-Sessoms F, Feller U, Reinhardt D (2015) Correction: phosphorus and nitrogen regulate arbuscular mycorrhizal symbiosis in *Petunia hybrida*. PLoS One 10(4):e0127472

Othman YA, Tahat M, Alananbeh KM, Al-Ajlouni M (2022) Arbuscular mycorrhizal fungi inoculation improves flower yield and postharvest quality component of Gerbera grown under different salinity levels. Agriculture 12(7):978

Pineda A, Zheng SJ, van Loon JJ, Pieterse CM, Dicke M (2010) Helping plants to deal with insects: the role of beneficial soil-borne microbes. Trends Plant Sci 15(9):507–514

Popescu GC, Popescu M (2022) Role of combined inoculation with arbuscular mycorrhizal fungi, as a sustainable tool, for stimulating the growth, physiological processes, and flowering performance of lavender. Sustainability 14(2):951

Powell JR, Rillig MC (2018) Biodiversity of arbuscular mycorrhizal fungi and ecosystem function. New Phytol 220(4):1059–1075

Pozo MJ, Jung SC, Martínez-Medina A, López-Ráez JA, Azcón-Aguilar C, Barea JM (2013) Root allies: arbuscular mycorrhizal fungi help plants to cope with biotic stresses. In: Symbiotic endophytes, pp 289–307

Rasouli F, Amini T, Skrovankova S, Asadi M, Hassanpouraghdam MB, Ercisli S, Buckova M, Mrazkova M, Mlcek J (2023) Influence of drought stress and mycorrhizal (*Funneliformis mosseae*) symbiosis on growth parameters, chlorophyll fluorescence, antioxidant activity, and essential oil composition of summer savory (*Satureja hortensis* L.) plants. Front Plant Sci 14:1151467

Riaz M, Kamran M, Fang Y, Wang Q, Cao H, Yang G, Deng L, Wang Y, Zhou Y, Anastopoulos I, Wang X (2021) Arbuscular mycorrhizal fungi-induced mitigation of heavy metal phytotoxicity in metal contaminated soils: a critical review. J Hazard Mater 402:123919

Sannazzaro AI, Echeverría M, Albertó EO, Ruiz OA, Menéndez AB (2007) Modulation of polyamine balance in *Lotus glaber* by salinity and arbuscular mycorrhiza. Plant Physiol Biochem 45(1):39–46

Sharma P, Sharma P, Arora P, Verma V, Khanna K, Saini P, Bhardwaj R (2019) Role and regulation of ROS and antioxidants as signaling molecules in response to abiotic stresses. In: Khan MIR, Reddy PS, Ferrante A, Khan NA (eds) Plant signaling molecules. Elsevier/Woodhead Publishing, pp 141–156

Smith SE, Read DJ (2010) Mycorrhizal symbiosis, 3rd edn. Academic
Sohn BK, Kim KY, Chung SJ, Kim WS, Park SM, Kang JG, Rim YS, Cho JS, Kim TH, Lee JH (2003) Effect of the different timing of AMF inoculation on plant growth and flower quality of chrysanthemum. Sci Hortic 98(2):173–183
Takahashi N (2025) Rose (Rosa sp.) more than just beautiful: exploring the therapeutic properties of the rose species. In: Advances in medicinal and aromatic plants. Apple Academic Press, p vol2-263
Thangavel P, Anjum NA, Muthukumar T, Sridevi G, Vasudhevan P, Maruthupandian A (2022) Arbuscular mycorrhizae: natural modulators of plant–nutrient relation and growth in stressful environments. Arch Microbiol 204(5):264
Trouvelot S, Bonneau L, Redecker D, Van Tuinen D, Adrian M, Wipf D (2015) Arbuscular mycorrhiza symbiosis in viticulture: a review. Agron Sustain Dev 35:1449–1467
Tyagi J, Sultan E, Mishra A, Kumari M, Pudake RN (2017) The impact of AMF symbiosis in alleviating drought tolerance in field crops. In: Mycorrhiza-nutrient uptake, biocontrol, ecorestoration, pp 211–234
Umar OB, Ranti LA, Abdulbaki AS, Bola AU, Abdulhamid AK, Biola MR, Victor KO (2021) Stresses in plants: biotic and abiotic. In: Current trends in wheat research. IntechOpen
Van Der Heijden MG, Bardgett RD, Van Straalen NM (2008) The unseen majority: soil microbes as drivers of plant diversity and productivity in terrestrial ecosystems. Ecol Lett 11(3):296–310
Vosátka M, Albrechtová J (2008) Theoretical aspects and practical uses of mycorrhizal technology in floriculture and horticulture. In: Floriculture, ornamental and plant biotechnology: advances and topical, vol 5, pp 466–479
Wagg C, Bender SF, Widmer F, Van Der Heijden MG (2014) Soil biodiversity and soil community composition determine ecosystem multifunctionality. Proc Natl Acad Sci 111(14):5266–5270
Wahab A, Muhammad M, Munir A, Abdi G, Zaman W, Ayaz A, Khizar C, Reddy SP (2023) Role of arbuscular mycorrhizal fungi in regulating growth, enhancing productivity, and potentially influencing ecosystems under abiotic and biotic stresses. Plants 12(17):3102
Wang W, Shi J, Xie Q, Jiang Y, Yu N, Wang E (2017) Nutrient exchange and regulation in arbuscular mycorrhizal symbiosis. Mol Plant 10(9):1147–1158
Wang Y, Wang M, Li Y, Wu A, Huang J (2018) Effects of arbuscular mycorrhizal fungi on growth and nitrogen uptake of *Chrysanthemum morifolium* under salt stress. PLoS One 13(4):e0196408
Wenjing HU, Shuhua YA, Hong GE, Yaping KO, Xin ZH, Ruidong JI (2023) Research of AMF on the growth and development of ornamental plants. Chin Agric Sci Bull 39(7):55–63
Yadav RS, Mahatma MK, Thirumalaisamy PP, Meena HN, Bhaduri D, Arora S, Panwar J (2017) Arbuscular mycorrhizal fungi (AMF) for sustainable soil and plant health in salt-affected soils. In: Bioremediation of salt affected soils: an Indian perspective, pp 133–156
Zandalinas SI, Mittler R, Balfagón D, Arbona V, Gómez-Cadenas A (2018) Plant adaptations to the combination of drought and high temperatures. Physiol Plant 162(1):2–12
Zhang Y, Hu J, Bai J, Wang J, Yin R, Wang J, Lin X (2018) Arbuscular mycorrhizal fungi alleviate the heavy metal toxicity on sunflower (*Helianthus annuus* L.) plants cultivated on a heavily contaminated field soil at a WEEE-recycling site. Sci Total Environ 628:282–290
Zhang P, Zhou J, He D, Yang Y, Lu Z, Yang C, Zhang D, Li F, Wang J (2024) From flourish to nourish: cultivating soil health for sustainable floriculture. Plants 13(21):3055
Zhao Y, Cartabia A, Lalaymia I, Declerck S (2022) Arbuscular mycorrhizal fungi and production of secondary metabolites in medicinal plants. Mycorrhiza 32(3):221–256
Zhu B, Gao T, Zhang D, Ding K, Li C, Ma F (2022) Functions of arbuscular mycorrhizal fungi in horticultural crops. Sci Hortic 303:111219
Zou YN, Wu QS, Kuča K (2021) Unravelling the role of arbuscular mycorrhizal fungi in mitigating the oxidative burst of plants under drought stress. Plant Biol 23:50–57
Zou R, Zhou J, Cheng B, Wang G, Fan J, Li X (2024) Aquaporin LjNIP1; 5 positively modulates drought tolerance by promoting arbuscular mycorrhizal symbiosis in *Lotus japonicus*. Plant Sci 342:112036

Harnessing Mycorrhizal Symbiosis: Exploring Its Impact on Medicinal Crops

Rajdeep Mohanta, Soumik Dey Roy, Parijat De, Payal Maiti, Subhadwip Ghorai, and Amit Samanta

Abstract

Medicinal crops are a critical resource for humans and play an essential role in countering disease-causing pathogens. An effective approach to enhance the yield and quality of medicinal crops is through the application of mycorrhizae. These fungi form symbiotic partnerships with plant roots, enhancing nutrient and water absorption, bolstering plant disease resistance, and fostering overall soil health. By facilitating the mobilization of vital nutrients, such as phosphorus, which is often lacking in soils, mycorrhizae support better plant growth and help plants withstand various diseases and environmental stresses. In turn, this leads to healthier plants that produce higher quantities of bioactive compounds with medicinal potential. The use of mycorrhizal associations in medicinal crop cultivation provides a sustainable means to enhance both plant health and efficacy of medicinal products. Medicinal crops inoculated with mycorrhizal fungi generally exhibit increased biomass, superior secondary metabolite profiles, and higher overall yields than non-inoculated plants. By harnessing mycorrhizal symbiosis, it is possible to cultivate medicinal plants of higher quality, which holds significant promise in both the pharmaceutical and agricultural sectors. Despite this, our understanding of the potential application of mycorrhizal symbiosis in enhancing medicinal crop production remains limited. This review highlights the need for further investigation to elucidate the effects of mycorrhizal fungi on medicinal crops and devise effective strategies for their utilization in various agricultural settings.

R. Mohanta (✉) · S. D. Roy · P. De · S. Ghorai · A. Samanta
Department of Agriculture, Brainware University, Kolkata, West Bengal, India

P. Maiti
Department of Post-Harvest Management, Bidhan Chandra Krishi Viswavidyalaya, Nadia, West Bengal, India

Y. Siddiqui et al. (eds.), *Sustainable Mycorrhizal Cultivation*,
https://doi.org/10.1007/978-981-95-1778-7_8

Keywords

Medicinal crops · Mycorrhizae · Resilience · Secondary metabolites · Symbiosis

1 Introduction

Medicinal plants have long been at the heart of traditional and modern medicine, yet many face the challenges of declining yields, nutrient-poor soils, and environmental stress. A promising, nature-based solution lies beneath our feet in mycorrhizal fungi. These ancient symbionts form intricate partnerships with plant roots, significantly boosting nutrient uptake, water absorption, and resilience against stress. Recent studies, including work on *Angelica glauca*, reveal that inoculation with arbuscular mycorrhizal fungi (AMF) not only enhances plant growth and biomass but also amplifies the production of valuable phytochemicals (Tapwal 2024). Beyond supporting endangered medicinal species, this fungal alliance strengthens the very compounds used in herbal remedies, making crops both more productive and pharmacologically potent. With AMF improving soil structure, stimulating metabolic pathways, and enabling plants to thrive in challenging conditions, harnessing this symbiosis offers a sustainable path forward. As we face the dual pressures of ecological degradation and rising demand for plant-based therapies, unlocking the full potential of mycorrhizal associations could transform medicinal agriculture into a more resilient, regenerative system.

Currently, between 320,000 and 340,000 ventricular and inactive plant species can develop mycorrhizal relationships (Tedersoo et al. 2020). Angiosperms make up 85–90% of them, making them the most numerous and varied group of associated organisms. Vegetation, foliage, herbal remedies, and the majority of staple crops, rice, maize, wheat, sweet potatoes, potatoes, tomatoes, and cassava are included in this group (Fernie and Yan 2019)**.** Approximately 50,000 kinds of fungi belonging to the Ascomycota, Basidiomycota, and Mucoromycota families are thought to be important for the formation of mycorrhizal relationships (Van der Heijden et al. 2015). The emergence of four main mycorrhizal types—arbuscular mycorrhizas (AM), ectomycorrhizas (EM), orchid mycorrhizas (ORM), and ericoid mycorrhizas (ErM) is a result of this extraordinary diversity, which has been shaped by more than 400 million years of co-evolution between plants and fungi. Although the majority of plants develop microbial interactions, their distribution is not uniform; AMF accounts for 72% of the total, followed by ORM (10%), EMF (2%), and ErM (1.4%) (Tedersoo et al. 2020). Different mycorrhizal species coexist in many habitats, and some plant and fungal species exhibit multiple mycorrhizal relationships. Furthermore, a "typical rhizome connection," which is an underground network of hyphae connecting various individual plants, can be formed when a single fungus colonizes multiple host plants (Simard et al. 2012a, b). Mycorrhizal fungi are recognized as key players within the plant microbiota, which is a diverse ecosystem of microbial communities that coexist closely with multicellular organisms. These fungi significantly influence the health of plants and humans (Compant et al. 2019; Elinav et al. 2019).

1.1 Mycorrhizal Symbiosis

Symbiosis, particularly in its mutualistic form, has received strong support from Frank's research on mycorrhizas. He proposed that the relationships between phylogenetically distinct organisms represent a general phenomenon, varying from loose to intimate and essential connections and from antagonistic to mutually beneficial interactions. Frank introduced the term "Symbiotismus" as a neutral descriptor that avoids labeling these relationships as purely parasitic. He emphasized the importance of considering all instances where different species coexist, irrespective of their roles, advocating the use of the term symbiosis to encompass this concept (Frank 1877). Frank introduced the term "mycorrhiza" to describe the symbiotic relationship between fungi and roots, referring to it as "Pilzsymbiosis" or "Wurzelsymbiosis" (Frank 1885). The German State Forestry Department's interest in expanding truffle growth in Prussia served as an impetus for mycorrhizal research. Although he did not succeed in advancing truffle culture, his work highlighted the widespread presence of mycorrhizas. He made a crucial distinction between "ectotrophic" mycorrhizas, where fungi form a mycelial sheath around the outside of the roots, and "endotrophic" mycorrhizas, where fungi penetrate the root cells.

Mycorrhizal fungi are broadly classified into two main categories: endomycorrhizal and ectomycorrhizal. Recently, mycorrhizal fungi have developed symbiotic connections with plants, especially in tree crops such as eucalyptus trees and needle-leaf forests. The mycelial systems of these fungi extend both inside and outside the plant roots system. The outside mycelium extends through the soil, allowing it to infiltrate tiny pores that are usually inaccessible to plant roots, thereby facilitating nutrient absorption and enhancing plant growth (Smith and Read 2008).

1.1.1 Endomycorrhizae

An endomycorrhizal fungus derives its name from its ability to penetrate the root cells of the host plants. The development of specific structures called receptacles and arbuscules within the roots of these fungi sets them apart. Although vesicles typically appear during the mid to late stages of vegetative growth, not all endomycorrhizal fungi, such as those from the genera *Gigaspora* and *Scutellospora*, produce vesicles. Arbuscules, on the other hand, serve as the primary sites for metabolic exchanges between the fungus and the plant, typically forming within the inner layers of the root epidermis. As the fungal hyphae invade the plant root system, they develop bifurcated branches that become increasingly fine, creating a network resembling small shrubs. This extensive network enhances the contact surface area with the host cells, facilitating the efficient exchange of metabolites (Johansson et al. 2004).

1.1.2 Ectomycorrhizae

An ectomycorrhizal fungus, belonging primarily to the classes Basidiomycetes and Ascomycetes, and includes some imperfect fungi and Zygomycetes like *Andogon*, does not penetrate the root cells; hence, their designation as "ecto" (external). These fungi establish a dense network known as the Hartig network within the intercellular spaces of the root epidermis, which facilitates metabolite exchange with the host plant. Additionally, ectomycorrhizae form a thick sheath around the root tips, often altering the color and shape of the roots, which can exhibit frequent branching. Morphological changes in the root sheath provide a clear indication of ectomycorrhizal association (Ramanankierana et al. 2007).

1.2 Arbuscular Mycorrhizal Fungi

AMF derive their name from the Latin words "arbusculum" (little tree), "mycos" (fungus), and "rhiza" (root) (Leake et al. 2004). AM fungi are particularly effective because of their ability to colonize a broad range of crop plants, thereby enhancing both plant productivity (Rocha et al. 2019) and crop quality (Torres et al. 2018). However, as noted in a meta-analysis, there is some increase in yield, especially in cereals, which is not always significant. Additionally, AM fungi play essential roles in ecosystem functions, such as improving soil structure through aggregation (Rillig et al. 2016) and reducing nutrient losses (Cavagnaro et al. 2015).

However, AMF implementation in agro-ecosystems faces several challenges. Environmental factors can greatly influence the efficacy of AM colonization. Additionally, the obligate biotic nature of AMF complicates the large-scale inoculum production. Lipid feeding is an innovation that makes it easier to produce pure AMF inocula in vitro (Sugiura et al. 2020). Additionally, breeding programs have mostly ignored the significance of beneficial plant-fungus interactions, producing crop varieties that do well in fertilized soils but are less susceptible to AMF invasion (Sosa-Hernández et al. 2019). According to recent studies, exogenous treatment with mycorrhizal compounds (including short-chain chito-oligosaccharides) may improve AMF facilities, creating new avenues for mycorrhizal inoculation-based agricultural practice improvement (Volpe et al. 2020).

2 Types of Mycorrhizal Fungi and Their Functions

2.1 Arbuscular Mycorrhizal Fungi

AMF form symbiotic relationships with plant roots, enhancing nutrient uptake and plant health and quality. These benefits are supported by bacterial communities within AMF structures (spores and hyphae), which aid in nitrogen fixation, nutrient mobilization, and the production of plant hormones and antibiotics, collectively promoting growth (Ujvári et al. 2021). In biodynamic agriculture, AMF improves root development, water and nutrient acquisition, and resilience to environmental stressors (Malik et al. 2011; Basu et al. 2018; Begum et al. 2019).

AMF form symbiotic relationships with over 80% of terrestrial plants, although colonization rates depend on fungal and plant species and environmental factors, such as soil properties (Kim et al. 2017). In fruit, vegetable, and medicinal plant cultivation, AMF enhance macro- and micronutrient uptake (Chen et al. 2017), whereas variations in AMF diversity affect plant growth and agricultural systems (Marro et al. 2020; Rillig et al. 2019). AMF also improve soil nutrient and moisture retention; increase plant resilience to drought, salinity, and heavy metals; and enhance crop quality by stimulating secondary metabolite production, supporting sustainable agriculture (Bhantana et al. 2021).

The mutualistic association between AMF and plants supports essential physical and biochemical processes, helping plants manage abiotic stresses that impair their growth (Begum et al. 2019). As a biofertilizer, AMF offer an eco-friendly alternative to fertilizers and pesticides, enhancing plant resilience to heat, salinity, drought, heavy metals, and temperature extremes, while sustaining crucial metabolic pathways (Bhantana et al. 2021).

AMF-plant symbiosis boosts nutrient uptake by enhancing transport proteins, such as aquaporins, phosphate, ammonium, nitrate, sulfur, zinc, carbon, and protein transporters (Burleigh and Bechmann 2002; Allen and Shachar-Hill 2009; Sieh et al. 2013; Giovannetti et al. 2014; Nguyen et al. 2019). In wheat, F. mosseae and R. intraradices inoculation raised zinc levels by 1.13–2.76 times (Ma et al. 2019a). AMF also form common mycorrhizal networks that link plants to facilitate nutrient exchange and adaptive interactions across species (Gorzelak et al. 2015; Ma et al. 2019b; Diagne et al. 2020).

AMF inoculation mitigates pathogen damage, reduces soybean charcoal root-rot severity (Spagnoletti et al. 2020), and increases shoot dry weight in *Fusarium*-infected plants (Vos et al. 2012). This mycorrhiza-induced resistance (MIR) resembles ISR and SAR, which combat pathogens via rhizobacteria (Pozo and Azcón-Aguilar 2007; Cameron et al. 2013; Nguvo and Gao 2019). AMF also reduce nematode infections (e.g., *M. incognita* and *P. penetrans*) by 45–87%, although outcomes vary by environment and host (Schüßler et al. 2001).

Research on AMF-insect interactions is limited, although *R. intraradices* has shown potential in reducing the impact of leaf-chewing insects (Koricheva et al. 2009; Diagne et al. 2020). In *Plantago lanceolata*, AMF increased resistance to *Arctia caja* but enhanced *the performance of Myzus persicae* (Gange and West 1994; Koricheva et al. 2009).

2.2 Ectomycorrhizal Fungi

Belowground mutualisms, such as those with ectomycorrhizal fungi (EMF), are crucial for global forest stability (Steidinger et al. 2019). EMF forms extensive hyphal networks that improve soil exploration and nutrient uptake (Finlay 2008). EMF contains more labile and recalcitrant compounds than AMF, which are rich in acid-hydrolysable compounds (Huang et al. 2022). These symbioses enhance water relations, drought resistance (Smith and Read 2008; Osonubi et al. 1991), and pathogen defense (Vishwanathan et al. 2020; Cahanovitc et al. 2022). EMF also create CMNs that foster support systems among diverse trees (Simard et al. 2012a, b; Molina and Horton 2015; van der Heijden et al. 2015).

Tropical forests with EMF associations host unique plant-fungal communities, highlighting specific ecological interactions (Corrales et al. 2018; Corrales et al. 2022). EMF facilitates belowground carbon (C) transfer; however, our understanding of carbon distribution in these complex soil systems remains limited. EMF extends partnerships beyond single trees through common mycorrhizal networks (CMNs), although knowledge of the specific EMF species that transfer carbon and their impact on carbon dynamics is sparse. While EMF is vital to forest stability, gaps remain in our understanding of how these relationships contribute to broader network structures (Cahanovitc et al. 2022).

The impact of EMF on soil organic matter (SOM) varies widely across their 85+ evolutionary lineages, many of which are derived from saprotrophic ancestors (Tedersoo and Smith 2013). Genomic studies have shown differences in SOM-modifying genes (Kohler et al. 2015; Shah et al. 2016; Pellitier and Zak 2018). EMF influences SOM across soil layers, shaping ecological feedback that affects plant growth and the carbon and nitrogen cycles (Lindahl et al. 2007; Craig et al. 2018; Zak et al. 2019).

2.3 Other Types

2.3.1 Orchid Mycorrhizae

The Orchidaceae family is highly diverse, and many species possess medicinal properties. These plants form mycorrhizal partnerships with fungi from the Serendipitaceae and Tulasnellaceae, known as orchid mycorrhiza (ORM) (Leng et al. 2024). ORM has significant potential for ecosystem conservation and restoration, involving the development of intracellular fungal pelotons in the roots of orchids (Dai 2011).

Recent studies on orchid mycorrhiza (ORM) have identified various mycorrhizal fungi that aid in seed germination and seedling development in medicinal orchids, such as *Dendrobium* spp. (Zhang et al. 2020a, b; Tang et al. 2022), *Bletilla striata* (Xi et al. 2020; Shi et al. 2023), *Gymnadenia conopsea* (Gao et al. 2020), and *Cremastra appendiculata* (Gao et al. 2022). ORM enhances host survival under environmental stressors by improving photosynthetic efficiency, water use efficiency, osmoprotectant production, antioxidant activity, and gene expression (Santander et al. 2017; Zhang et al. 2024). Additionally, the inoculation of *Coelogyne viscosa* with endophytes and mycorrhizal fungi boosts seedling biomass and drought tolerance (Liu et al. 2022).

Orchid medicinal plants face biological stresses from pathogens and pests, which affect development and product quality (Jain et al. 2021). Mycorrhizal fungi enhance disease resistance by activating defense mechanisms and increasing phytohormones, signaling compounds, gene expression, and protein production (Pu et al. 2022; Weng et al. 2022).

2.3.2 Ericoid Mycorrhizae

Ericaceous mycorrhizal (ErM) fungi, mainly from the phyla Ascomycota and Basidiomycota (Perotto et al. 2018; Fehrer et al. 2019; Vohnik 2020), including *Hyaloscypha hepaticicola* aggregates. They form loose hyphal networks on ericaceous plant roots, leading to intracellular coils that enable bidirectional exchange of nutrients and carbohydrates (Vohnik et al. 2012; Wei et al. 2022).

ErM fungi play a vital role in enhancing seed germination and root development in ericaceous plant cuttings, significantly improving rooting success in both stems and microcuttings (Eccher et al. 2010; Wei et al. 2020). For instance, blueberry plants exhibit better root development and growth when inoculated with ErM fungi (Scagel et al. 2005a; Scagel et al. 2005b). These fungi secrete bioactive compounds, including iron-chelating siderophores (Ahmed and Holmström 2014) and cell-wall-degrading enzymes that enhance nutrient availability (Kohler et al. 2015; Martino et al. 2018). They also exhibit antimicrobial properties (Hosoe et al. 1999; Ouyang 2021) and antioxidant capabilities, such as rutin production (Lin et al. 2021). Although natural colonization rates are often below 15%, inoculation can boost rates to approximately 30%, promoting growth in blueberry cultivars (Scagel 2005).

Ericoid mycorrhizal (ErM) fungi demonstrate a significant tolerance to heavy metal toxicity (Meharg and Cairney 1999). For instance, *Chamaedaphne vulgaris* thrives in mine spoil areas because of its symbiotic relationship with ErM fungi, which helps it to tolerate high Cu^{2+} and Zn^{2+} levels. Additionally, inoculation with *Mycena variabilis* or *Oidiodendron maius* enhanced salt tolerance in velvetleaf blueberries, labrador tea, and lingonberries under NaCl stress, with *Oidiodendron maius* showing greater effects at 30 mM NaCl (Fadaei et al. 2020). Limited studies have indicated that ErM fungi can inhibit pathogenic mycelium growth and reduce infection rates from *Pythium* spp. and *Phytophthora cinnamomic* (Grunewaldt-Stöcker et al. 2013; Wei et al. 2022).

3 Mechanisms of Symbiosis

AMF exist as spores in soil, which under suitable environmental conditions, germinate and develop into hyphae under suitable environmental conditions (Hetrick 2018). In the absence of a plant host, AMF can survive for approximately 2–4 weeks by relying on internal energy reserves such as triacylglycerides (Cooper 2018). During this period, they retract the hyphal cytoplasm back into the spores as a conservation strategy (Lee 2011). Upon successful invasion of a host plant, AMF predominantly colonize the root cortex, intercellular spaces, and root surface. Following staining, structures such as hyphae, vesicles, clusters, and spores were visible (Vohnik 2020).

The fungal mycelium produces glomalin-related soil proteins (GRSPs), a group of glycoproteins noted for their water insolubility, remarkable stability in natural environments, and association with metal ions (Agnihotri et al. 2022). GRSPs adhere to the mycelial surface and fall into the soil, where they act as a source of carbon (Zou et al. 2014), functioning as an effective soil-structuring agent. Their ability to bind soil particles has been shown to be 3–10 times more potent than that of other soil carbohydrates (Morales-García et al. 2019). By coating extraradical hyphae within the rhizosphere, GRSP helps plants harness hyphae for moisture uptake and shields the root zone from microbial invasions (Agnihotri et al. 2021). As hyphae degrade, GRSP is gradually incorporated into the soil, enhancing both its microenvironment and its structural integrity (Wang et al. 2022).

In medicinal plants, AMF synthesize and secrete GRSP, which improves soil structure, regulates microbial communities, boosts enzyme activity, enriches soil fertility, and enhances soil matrix quality.

3.1 Nutrient Exchange

Mycorrhizal fungi supply up to 90% of essential nutrients to their host plants, which, in return, allocate approximately 30% of their photosynthetic carbon to the fungi. This exchange supports the life cycle of fungi and promotes efficient nutrient redistribution within ecosystems (Wahab et al. 2023). Under conditions of phosphorus (P) scarcity, AMF fulfill 50–75% of the plant's P needs (Bhantana et al. 2021). Additionally, external fungal hyphae facilitate nitrogen (N) uptake, contributing up to 30% of the total N acquired by host plants (Miransari 2011). This symbiotic interaction enhances the uptake of ammonium, nitrate, complex organic nitrogen, and other substances, converting them into arginine in extraradical hyphae. Arginine is then transported to the intraradical hyphae, where it is converted to NH_4^+, thus regulating the N metabolism of the host plant (Jin et al. 2012).

AMF also enhance potassium (K) uptake, promoting its transport over long distances, as well as its movement within the xylem and phloem of plants. These fungi help maintain the Na^+ and K^+ balance within the plant (Dominguez-Nuñez et al. 2016). By extending their hyphal networks, AMF facilitate the absorption of a broad range of nutrients, thereby improving the chlorophyll levels, photosynthetic rate, and overall nutrient uptake in plants. The fungi significantly increase the availability of micronutrients and low-mobility minerals, such as iron (Fe), zinc (Zn), and copper (Cu) (Bhantana et al. 2021). For instance, in greenhouse-grown lettuce with optimal irrigation, mycorrhizal associations boosted Cu and Fe levels, although the effects varied based on the lettuce cultivar and the source of applied P (Baslam et al. 2011). The extensive hyphal network of these fungi is highly effective at extracting both nutrients and water from the soil; however, only a few specific fungal transporters are known to be involved in this uptake process.

3.2 Improvement of Soil Structure and Health

AM hyphae are critical for facilitating water absorption in plants. These hyphae penetrate plant root cortical cells, forming complex branched arbuscular structures that resemble dendrites or cauliflower heads, serving as efficient channels for water transport and boosting soil water retention capacity. By extending outside the root, these extra-root hyphae increase root length and form an extensive hyphal network within the soil, thereby improving root hydraulic conductivity and enhancing water uptake. AMF dynamically adjust water transport, redistribution, and host plant leaf transpiration rates based on water availability and plant demand, thereby promoting a balanced dry/wet cycle (Yu et al. 2022).

Beyond the water absorbed through AM fungal hyphae, plants utilize aquaporins to control water movement across cell membranes. During drought stress, AMF influence the expression of host plant aquaporin genes, including those encoding plasma membrane intrinsic proteins and vacuolar intrinsic proteins, to mediate water intake and release. This regulation enhances swelling pressure, osmotic balance, and root water conductivity (Nehls and Dietz 2014), supports water uptake and retention, optimizes water metabolism, and preserves cellular water balance.

3.3 Stress Resistance

Drought resistance in plants is closely linked to proline and soluble sugar levels in plant cells and tissues. Under drought conditions, plants produce large amounts of proline to support osmotic adjustment and stress tolerance (Nahar et al. 2016). However, plants inoculated with AMF maintained lower proline levels to safeguard enzymatic function. These fungi decrease the activity of Delta-1-pyrroline-5-carboxylate synthetase, enhance proline dehydrogenase activity, and limit proline buildup in host plants, such as trifoliate orange (Wu et al. 2017).

AMF are a dominant group of soil microorganisms that colonize the roots of most terrestrial plants. They provide numerous advantages to their host plants, including improved uptake of minerals and water as well as enhanced resilience to various environmental stresses (Singh et al. 2011). AMF enhance host plant resistance to salinity stress through physiological and biochemical mechanisms, such as increased water use efficiency, greater photosynthetic capacity, maintenance of ionic balance, osmoprotection, preservation of cell structures, and strengthened antioxidant metabolism (Singh et al. 2015). For instance, in lettuce exposed to salt stress, AMF inoculation increased proline levels and the activity of antioxidants, such as catalase, superoxide dismutase, and ascorbate peroxidase, while reducing the synthesis of phenolic compounds and oxidative stress, leading to robust plant growth (Hashem et al. 2018).

Diseases in medicinal plants can be either infectious or non-infectious depending on the causative factors (Nazarov et al. 2020). Non-infectious diseases result from abiotic factors, such as toxic pollutants, drought, or flooding due to moisture imbalance, temperature extremes, nutrient deficiencies or excesses, and suboptimal cultivation practices, such as improper planting density or timing (Bal and Minhas 2017). AMF help mitigate non-infectious diseases by extending the reach of plant roots through their hyphal network, enhancing nutrient uptake, and boosting plant resilience and survival under stress conditions (Sati et al. 2020). Conversely, infectious diseases arise from biotic pathogens, including fungi, bacteria, viruses, and nematodes (Singh 2018). AMF combat infectious diseases by enhancing the activity of plant defense enzymes, upregulating defense-related genes, and increasing the host-pathogen sensitivity within its defense system (Dowarah et al. 2022).

4 Impact on Medicinal Crop Quality and Yield

4.1 Enhancement of Secondary Metabolites

4.1.1 What Are Secondary Metabolites?

Although they are not seen as energy sources by cells, secondary metabolites are crucial for how plants interact with their surroundings. They shield plants from biotic (bacteria, fungi, insects, and nematodes) and abiotic (high temperatures and droughts) challenges. Conversely, secondary metabolites can serve as markers and aid in the systematic determination of plant categorization. Low molecular weight substances and certain organs are involved in the manufacture of secondary metabolites. Similarly, primary metabolites help plants grow and develop through a variety of metabolic activities (Pagare et al. 2015). Secondary plant metabolites play various ecological roles in the natural world. They are essential to a plant's defense and reproductive systems and are engaged in all interactions with their surroundings (Abbas et al. 2017). Compounds derived from the isoprene units are terpenes and isomeric terpenoids. They are vital for ecological communication and frequently erratic. They are components of a plant's defense mechanisms and are frequently eluted or absorbed by the plant in response to injury from herbivores or machinery (Bustos-Segura and Foley 2018).

4.1.2 Secondary Metabolites Found in Therapeutic Plants

Secondary metabolites (SMs) of plants with medicinal properties are the foundation of their therapeutic properties. They are also important markers for assessing the quality of pharmaceutical products. Plants produce a wide variety of secondary metabolic products (SMs), which are living chemical factories. Such SMs serve as the foundation for a large number of commercially available pharmaceutical drugs, as well as herbal remedies derived from therapeutic plants. The different biological activities of the chemical components of medicinal plants may improve human health and the food industry through the nutrition and medicinal sectors; however, they also have significant applications in agrochemical, beauty products, and fragrance businesses (Hassan 2012). It is crucial to quickly discuss the novel methods which are being developed and required to improve the research of botanical medicines before delving into how plants control their secondary metabolites. Now used to describe all tiny molecules found in a cell, the term "metabolome" is the fourth element of an integrated biology strategy, following transcriptomics (RNA), the field of proteomics, and genomics (DNA). Secondary metabolites can be considered an element of an organism's metabolome in such circumstances and are generally thought to be separate from the essential metabolites required for life in all organisms, such as those engaged in glycolysis or the mitochondrial TCA cycle. Furthermore, recent studies have demonstrated the numerous and potent uses of mass chromatography in the research and measurement of metabolites, particularly secondary metabolites (Wang et al. 2019).

Semi-terpenes (such as 2-methyl-3-buten-2-ol), monoterpenes (such as linalool and thymol), sesquiterpenes (such as bisabolol and chamazulene), and a few diterpenoids (such as sclareol) are the main types of chemicals in essential oils (Sicari and Poiana 2017). In addition to terpenoids, volatile phenylpropanoids (apiol and eugenol), fatty acid derivatives (hexenal), and chemical amino acid compounds (2-phenylacetaldehyde) are also present in essential oils (Rout et al. 2007). A naturally occurring heterocyclic, bicyclic monoterpenoid, eucalyptol (also called 1, 8-cineole) is frequently found in a variety of medicinal plants, including sage (*Salvia officinalis*), rosemary (*Rosmarinus officinalis*), and eucalyptus species.

Therapeutic and fragrant vegetation usually contains a wide variety of hydrocarbon-rich and fragrant scents produced by a combination of approximately 30 terpenes, whether liberated and replenished with oxygen or glycosidically connected. This terpene profile can vary significantly depending on environmental factors, stages of growth and ripening, and genetic differences among plant varieties. Different climatic conditions can influence the activation of transcription factors, which in turn may initiate various enzymatic processes that lead to the accumulation of these terpenes.

4.1.3 Enhancement of Plant Secondary Metabolites by Microorganisms

A wide variety of microbes, including yeast, mold, and algal cells, are part of the soil microflora and are essential for almost all the chemical reactions that occur in the soil. These microbes perform vital tasks that are critical for plant survival and growth, including nutrient intake, atmospheric nitrogen and carbon cycling, and improving resistance to different challenges. However, plants have a significant impact on terrestrial microorganisms, especially the rhizosphere. The impact results from the soil's increased carbon availability, mostly via the breakdown of exudates from roots and decaying plant tissues. These materials promote the growth and activity of microbes by acting as sources of nutrition, architectural encouragement, and signaling chemicals (Barea et al. 2005). Many plant species form symbiotic relationships with specific soil fungi through their mycorrhizal structures. These mycorrhizal relationships are essential for sustainable management of agricultural ecosystems (Smith and Read 2008). In symbiosis, nutrients are exchanged in two ways: inorganic nutrients are sent from the fungus to the plant, and carbon is transported through the host to the mold. This creates a vital connection between the plant, its roots, and soil. Higher plants associated with AMF develop more quickly and accumulate additional substances, such as terpenoids, flavonoids, and alkaloids, which have significant medicinal qualities.

4.1.3.1 Arbuscular Mycorrhizal Fungi

More than 80% of vascular plants worldwide have symbiotic interactions with AMF. Regardless of the type of soil, vegetation, or growing conditions, they flourish across all seasons and environments. In agriculture, the introduction of AMF into plants can improve soil quality, increase nutrient uptake, and increase resistance to environmental stressors. Additionally, AMF may boost the output of therapeutic plants by altering the concentration of bioactive chemicals. The presence of hyphae, vesicles, and arbuscular mycorrhizal structures created by the Glomeromycete fungus within the root cortex defines arbuscular mycorrhizas. Colonization of root cortical cells by AMF leads to various cytological and metabolic changes, including significant proliferation of plastids during the formation of arbuscules (Lohse et al. 2005). By establishing a network between a plant's roots and the surrounding soil, mycorrhizae can allow fungi to take up nutrients from larger regions. This improved water availability and increased the outer region of the roots, fostering better plant growth (Sylvia et al. 2005). This mutually beneficial relationship is advantageous for both fungi and plants. The fungus helps the plant by absorbing water and nutrients, while the plant provides the fungus with 10–20% of the carbon it fixes during photosynthesis. Apart from providing plants with a better nutritional source, AMF also help plants become more resilient to dehydration and saltwater. They also work in concert with additional positive soil microbes, such as phosphorus-solubilizers and nitrogen-fixers (Auge et al. 2015), and help boost resistance to a variety of illnesses (Pozo and Azcón-Aguilar 2007). These fungal organisms have recently been shown to increase the amount of secondary metabolites in a number of significant plants used for medicinal purposes (Varma et al. 2017).

Through morphological and physiological changes, it has been demonstrated that inoculating medicinal plants with AMF reduces the infections caused by bacterial and fungal pathogens. In terms of morphology, AMF can result in vascular bundle strengthening and cell wall thickening. These fungi increase the amount of sulfur compounds containing amino acids and phenolics in the body. Studies have shown that AMF and a number of medicinal plants, such as *Ocimum basilicum* (Prasad et al. 2011), Panax ginseng (Zhang et al. 2013), and *Salvia miltiorrhiza* and *S. officinalis* (Nell et al. 2010), have symbiotic associations. It has been shown that AMF symbiosis changes the makeup of bioactive plant chemicals. For example, mycorrhizal relationships in Salvia officinalis led to higher concentrations of essential oil constituents, such as 1,8-cineole, α-thujones, β-thujones, and bornyl acetate (Geneva et al. 2010). Similarly, mycorrhization altered the essential oil makeup of *Mentha viridis* and *Origanum onites* (Karagiannidis et al. 2011). Kapoor et al. (2017) revealed that mycorrhizal plants' higher phosphorus uptake greatly boosts the formation of terpenoid compounds, which are prized for their therapeutic qualities (Gianinazzi et al. 2010).

4.1.3.2 AMF Application to Medicinal Plants

Mycorrhizal fungi induce significant changes in enzymatic activities, including superoxide dismutase and catalase, in addition to physiological processes that result in the buildup of secondary metabolites such as polyphenols and carotenoids (Marulanda et al. 2007). The production of additional antioxidative enzymes and carotenoids by mycorrhizal plants in response to oxidative stress is indicated by elevated levels of reactive oxygen species (ROS) in their roots. The generation of secondary metabolites has been associated with increased levels of reactive oxygen species (ROS), which are a consequence of both abiotic and biotic stresses (Chaudhary et al. 2008). These stresses can disrupt the balance between primary and secondary metabolism in plants, diverting resources from growth to defense and leading to overproduction of secondary metabolites (Szakiel et al. 2010). Accumulation of ROS has also been observed in the roots of mycorrhizal plants (Fester and Hause 2005). One possible defense mechanism against AMF invasion is artemisinin concentration.

4.1.3.3 Reference Studies

Table 1 summarizes recent findings on the effects of AMF on the secondary metabolite profiles of medicinal plants. Secondary metabolites, including alkaloids, phenolics, and terpenes, are bioactive compounds that contribute to the therapeutic properties of these plants. AMF symbiosis often enhances the production of these compounds and improves their medicinal efficacy. The table outlines the specific plant parts, AMF strains involved, and observed changes in metabolite concentrations (Walid et al. 2021).

5 Applications and Techniques in Medicinal Crop Cultivation

5.1 Mycorrhizal Inoculation Methods

A range of biological agro-inoculants, including biostimulants, biological pesticides, biocontrol agents, and fertilizers, are available as alternatives to synthetic chemicals in sustainable agriculture. These products contain active components, such as amino acids, humic and fulvic acids, herb or algal extracts, living microbes, and microbial metabolites. This study reviews the benefits, mechanisms, formulations, and application techniques of microbial-based agro-inoculants, which can be classified based on their active components, intended uses, or application methods. Notable microbial metabolites include Bt endotoxin from *Bacillus thuringiensis*, which is known for its insecticidal properties (Azizoğlu 2019). Nitrogen-fixing microbes, which are essential for plant growth, contribute significantly to crop productivity by forming amino acids, proteins, and nucleic acids (Thilakarathna et al. 2016). Members of the Rhizobiaceae family, including *Azorhizobium* and *Rhizobium*, have been well studied for their positive effects on soil moisture, leguminous plant growth, and yields (Fagorzi et al. 2020).

In addition to the legume-rhizobia symbiosis, Frankia actinobacteria are recognized as nitrogen-fixers that induce nodule formation in woody species across Fagales, Rosales, and Cucurbitales. *Frankia* spp. are gram-positive, filamentous, spore-forming bacteria with diazotrophic traits. Free-living nitrogen-fixing bacteria such as Azotobacter and Azospirillum are commonly used as biological fertilizers for grains and vegetables (Vurukonda et al. 2016). These bacteria establish beneficial rhizospheric relationships with plants and enhance plant growth and nutrient availability (Gangwar et al. 2018). Cyanobacteria, or blue-green algae such as

Table 1 The impact of AMF on secondary metabolites in medicinal plants (Zhao et al. 2022)

Secondary metabolites		Plant name	Plant parts	Medicinal value	AMF	Change in secondary metabolite
Alkaloid	Alliin	*Allium sativum*	Leaves, flowers, and cloves	To treat fevers, diabetes, rheumatism, and intestinal worms	*Glomus fasciculatum*	Significant increase
	Vinblastine and vincristine	*Catharanthus roseus*	Aerial part	Diabetes, malaria, Hodgkin's lymphoma, and leukemia	*Glomus* species	Significant increase
	Forskolin	*Coleus forskohlii*	Root	For angina, anti-inflammatory activity, antihypertensive, and a cure placebo	*Acaluspora laevis*	Significant increase
	Colchicine	*Gloriosa superba*	Tuber	Urinary pains, colic, chronic ulcers, snake bites, scorpion stings, parasite skin conditions, and piles	*Gigaspora margarita, Gigaspora gilmorei*	1.9% increasing
	Scopolamine	*Eclipta prostrata*	Leaves	Diabetes type II, dizziness, haemoptysis, and liver diseases	*Rhizophagus irregularis, Funneliformis mosseae*	C.34% increasing

Phenols	Caffeic acid, rosmarinic acid, and luteolin	*Mentha spicata*	Leaves	Stomach issues, headaches, and fevers. The herb has antispasmodic, antiemetic, and carminative properties	*G. etunicatum, F. mosseae*	Significant increase
	Gallic acid	*Libidibia ferrea*	Fruit	Antianemic, antidiabetic, and for the treatment of pulmonary haemolytic disorders, and preventing cancer	*Claroideoglomus etunicatum*	21% increasing
	Diobulbinone	*Dioscorea bulbifera*	Bulbils	Treatment of cancer, leprosy, diabetes, asthma, ulcers as well as syphilis, diarrhea, and piles	*G. clarum,G. etunicatum*	Significant increase
	Catechin, hydroxyl benzoic acid, gallic acid, and chlorogenic acid	*Valeriana jatamansi*	Rhizome and root	Insect-repelling and antihelmintic properties, antidepressant, antioxidant, and antimicrobial activities	*G. intraradices*	Significant increase
	Caffeic acid	*Viola tricolor*	Aerial part	Epilepsy, skin conditions like dermatitis, and respiratory issues including asthma and bronchitis	*Rhizophagus irregularis*	Significant increase
	Citronellal and neral	*Melissa officinalis*	Leaves	For its diuretic, tonic, analgesic, sedative, digestive, carminative, and antispasmodic qualities	*Claroideoglomus etunicatum, Claroideoglomus claroideum, Rhizophagus intraradices*	11–17% increasing
	Salvianolic acid	*Salvia miltiorrhiza*	Root	Coronary heart disease and cerebrovascular diseases	G. mosseae, G. aggregatum	Significant increase

(continued)

Table 1 (continued)

Secondary metabolites		Plant name	Plant parts	Medicinal value	AMF	Change in secondary metabolite
Terpenoids	Sesquiterpene lactones	*Arnica montana*	Fresh or dried flowers	Joint pain, superficial phlebitis, muscular aches, sprains, and bruises	*G. geosporum, G. constrictum*	Significant increase
	Linalool and geraniol	*Ocimum basilicum*	Seed	Headaches, warts, worms, coughing, diarrhea, and constipation	*Gigaspora margarita, Gigaspora rosea*	Significant increase
	Valerenic acid	*Valeriana officinalis*	Root	Insomnia/sleep disorders, attention deficit-hyperactivity disorder, anxiety disorders	*G. Intraradices*	Relative increasing
	Artemisinin	*Artemisia annua*	Dried above-ground parts	Fever, inflammation, malaria, coughing, and disturbed stomach and intestines	*G. macrospermum, G. fasciculatum, G. mosseae*	15.2–32.8% increasing
	Patchoulol	*Pogostemon cablin*	Leaves	Antimicrobial and antioxidant qualities are used to cure fever, headaches, and colds	*Acaulospora laevis, G. mosseae, Scutellospora calaspora*	Significant improvement
	Stevioside, rebaudioside-A	*Stevia rebaudiana*	Leaves	Depression, hypertension, dental cavities, diabetes, and skin abrasions	*Rhizophagus fasciculatus*	Significant increase
	β caryophyllene, p-cymene, geraniol	*Coriandrum sativum*	Seeds or leaves	Upset stomach, loss of appetite, hernia, nausea, diarrhea, bowel spasms, and intestinal gas	*Glomus hoi*	Significant improvement

Nostoc and Anabaena, thrive in aquatic environments and convert atmospheric nitrogen into ammonium using specialized cells called heterocysts (Kumar et al. 2010; Singh et al. 2016).

Phosphorus-solubilizing microbes are also essential for agricultural growth because phosphorus is vital for enzymatic processes in plants. Plant growth-promoting microorganisms (PGPM), including microalgae, fungi, and bacteria, enhance plant development through bioactive substances such as phytohormones and nutrient-solubilizing enzymes (Gouda et al. 2018). Among the most studied PGPM are plant growth-promoting rhizobacteria (PGPR) and actinomycetes, which are known for their sporulation and bioactive compound production (Anwar et al. 2016). Microbial endophytes, both fungi (Murphy et al. 2018) and bacteria (Mburu et al. 2021), are gaining attention because of their close relationship with host plants. PGPM promote plant growth by enhancing nutrient acquisition, priming defense responses, and fostering resistance to abiotic and biotic stresses (Umesha et al. 2018).

The formulation of agro-inoculants is critical to their success. This process combines specific microbial strains with suitable carriers, thereby affecting microbial performance and product stability (Bargaz et al. 2018). Additives may be included to enhance the stability during manufacturing and storage (Namasivayam et al. 2014). Over time, formulations have evolved from concentrated suspensions to diverse types including granules, pellets, and liquid dispersions. Innovations, such as the encapsulation of microbial cells, improve survivability and controlled release, particularly for endophyte-based formulations (Aswani et al. 2020). Effervescent pills are another innovative method that rapidly dissolve in water for accurate dosing (Nair and Brahmaprakash 2017).

5.1.1 Microbial Inoculant Production

Effective microorganisms selected for production must promote plant growth, multiply rapidly, compete within root systems, connect with native soil microbes, and have a minimal environmental impact. They are cultivated in affordable, nutrient-rich media using fermentation techniques (liquid, semisolid, solid-state) before being impregnated into carriers based on the microbe type and intended product form (Stamenkovic et al. 2018). Before commercialization, the bioproduct must undergo extensive greenhouse and field testing to confirm that (i) it lacks eco-toxicological effects, (ii) it delivers the anticipated plant benefits, and (iii) it enhances crop yields. Regulatory approval and product registration are mandatory before entering a market (Backer et al. 2018).

Packaging should include essential information, such as product name, microbial strain(s), microbial load, target crops, manufacturer details, production and expiry dates, and application guidelines and suggestions (Garcia-Fraile et al. 2015).

5.1.2 Environmental Impact of Microbial Inoculants

The application of microbial inoculants significantly affects soil microbial communities as well as soil physical and chemical properties (Javoreková et al. 2015). The impact varies based on the local microbial populations, with various methods employed to study these changes, including denaturing gradient gel electrophoresis (DGGE), amplified ribosomal DNA restriction analysis (ARDRA), terminal restriction fragment length polymorphism (t-RFLP), single-strand conformation polymorphism (SSCP), and community-level physiological profiling (CLPP) using the BIOLOG method (Rastogi and Sani 2011).

Agro-inoculants can be applied directly to soil, seeds, or plantlets, with seed treatment being the most common method because of its ease of use and low product requirements (Asif et al. 2018). Techniques include seed coating, slurry, and dusting, although dusting is the least effective because of poor microbial adherence (Malusa and Ciesielska 2012). The slurry involves mixing inoculants with wetted seeds or water, and overnight soaking is recommended. Adhesives, such as Arabic cellulose and sugar solutions, help ensure proper microbial adherence (Bashan et al. 2014). For larger inoculations, soil inoculation is advised, often using granular carriers such as peat, perlite, or talc, whereas liquid formulations can be sprayed in seed furrows (Macik et al. 2020).

5.1.3 Microbial Inoculants in the Market

The industrial use of bacterial inoculants began over 120 years ago with the registration of "Nitragin" (O'Callaghan 2016). Currently, biological fertilizers constitute approximately 5% of the global fertilizer market, comprising over 150 approved bacteria-based products. Rhizobium inoculants dominate, accounting for approximately 79% of sales, followed by phosphate-solubilizing biofertilizers at 15% and pathogenic fungi at 7% (Verma et al. 2019). In the European Union, 52.7% of the approved microbial strains are used for bactericidal or fungicidal purposes, 29.1% have insecticidal effects, and 18.2% serve as viral inoculants or nematicides (Cornea et al. 2019). Different species of mycorrhiza are associated with their host plants, and their effects are summarized in Table 2.

5.2 Challenges and Prospects

Species from the Glomeraceae family are the primary source of arbuscular mycorrhizal inoculum production, though other families like Gigasporaceae, Scutellosporaceae, and Acaulosporaceae are increasingly utilized for their symbiotic relationships with various host plants worldwide. These species show potential for efficient inoculum production, comparable to that of Glomeraceae. However, despite claims from manufacturers regarding their broad applicability across different plant species and conditions, field performance often falls short, raising doubts about the reliability of commercial AM inoculants. The specific plant species that thrive with a particular AM inoculum across various climates and soil types remains uncertain (Berruti et al. 2014).

Manufacturers face critical challenges in maintaining AM inoculant standards to comply with legal requirements. Key requirements include (1) ensuring that products effectively establish relationships with plants post-inoculation, (2) guaranteeing the absence of contaminants that could hinder plant growth, (3) reducing

Table 2 Integration of mycorrhizal associations into farming practices (Kumar et al. 2017)

Mycorrhizal species	Host plant	Effect(s) on host plants
F. mosseae and *R. intraradices*	*Pistacia vera*	Increased P, K, Zn, and Mn concentrations in the leaf in comparison with a placebo
F. mosseae, *G. versiforme*, *R. intraradices*	*Cucumis melo*	Increased growth metrics and physiological investigations showed that *G. mosseae* inoculation improved drought resistance
F. mosseae, *G. versiforme*	*Solanum lycopersicum*	Enhanced development and outcomes (19–32%) in various drought situations
F. mosseae, *F. geosporus*, and *mixed inoculation*	*Fragaria* × *ananassa*	Improved water use efficiency (WUE), growth, and production of strawberries during water stress
R. intraradices	*Tagetes erecta*	Increased biomass production results from improved antioxidant enzyme activity and decreased Cd translocation to shoots
F. mosseae	*Chrysanthemum maximum*	Pb and Cu accumulated less in the shoot; however, there was no exclusion effect observed for Zn
R. intraradices	*Ocimum basilicum*	High inoculum concentration decreased metal concentration in shoots, which improved yield; increased concentration of heavy metals (Cd, Pb, and Ni) in shoots decreased yield
G. gigaspora and *Scutellospora* spp.	*Petunia* hybrid, *Callistephus chinensis*, *Impatiens balsamina*	Enhanced three ornamentals' vegetative and reproductive characteristics; phosphorus fertilization costs might be lowered by up to 70%

production costs while ensuring adequate shelf life, and (4) avoiding chemical additives. Controlling contamination levels and maintaining a minimum viable cell count for each unit are vital; however, many countries lack regulatory standards for acceptable contaminant levels in bio-inoculants.

To address these challenges, inoculum manufacturers must ensure that their products effectively establish mycorrhizal relationships with plants when applied, guaranteeing that consumers see a return on their investment. Upholding product quality and raising standards of excellence are crucial for safeguarding the reputation of the bio-inoculant industry. Recent advances in the understanding of arbuscular mycorrhizal plant associations have highlighted their potential applications in agriculture, horticulture, landscaping, and phytoremediation. These advancements include exploring coexistence mechanisms within root frameworks, developing molecular methods to identify fungal inoculants in field settings, and understanding the interactions between fungi and other microbes such as saprophytic fungi and plant growth-promoting rhizobacteria (PGPR). Notable discoveries, including strigolactones that promote mycorrhizal activity and insights into the physiology of AM fungi grown in vitro, further support these applications (Vosatka et al. 2012).

The global shift toward reducing phosphorus-based chemical inputs has intensified interest in mycorrhiza-based inoculants as a sustainable agricultural practice.

Mycorrhizal fungi show promise for phytoremediation, enhancing phytoextraction and stabilization of contaminated soils. Research indicates that certain AM fungi can minimize pollutants reaching plant aerial parts, thereby improving the quality of edible produce (Rivera-Becerril et al. 2002). Additionally, some fungal species promote phytoextraction, aiding in soil decontamination (Cabral et al. 2015).

Another significant area of research is the enhancement of crop nutritional value through mycorrhizal treatments. Studies have shown that AM fungi can increase the levels of sugars, essential minerals, such as magnesium and zinc, antioxidants, and other beneficial compounds in crop plants. As demand for sustainable agricultural practices increases, manufacturers must continue to enhance the efficacy and value of mycorrhiza-based inoculants. Despite the promise of natural biotechnologies, challenges, such as limited funding for applied research, threaten their effective implementation. Therefore, collaboration between scientific researchers and the business sector is vital for driving further development in this market, ensuring that mycorrhizal products are optimized for broader commercial use.

6 Future Directions and Research Needs

Medicinal plants are valued for compounds such as paclitaxel, camptothecin, and vincristine, which are driving global demand; however, unsustainable harvesting and environmental degradation threaten their cultivation (Rui-Ting et al. 2021). AMF form beneficial root symbioses that enhance plant adaptation and influence the root-associated endophytic bacterial microbiota (Ujvári et al. 2021). These bacteria may be recruited through physiological changes in AMF-colonized plants or increased root tissue access via AMF hyphae (Smith and Read 2008; Toljander et al. 2006; de Novais et al. 2020). Further research is needed to explore the potential of AMF-associated bacteria as plant growth-promoting (PGP) endophytes and how AMF diversity affects root endophyte communities.

Ectomycorrhizal fungi (EMF) play key roles in forest nutrient cycling, primarily through carbon and nitrogen cycling (Liu et al. 2020). Future research should integrate field and lab studies to understand EMF nutrient-cycling mechanisms, develop models to predict EMF community dynamics, and create soil biological sample banks to track EMF evolution over time.

Ericoid mycorrhizal (ErM) fungi, which support Ericaceae plants, degrade soil organic matter (SOM) and improve nutrient availability, growth, and resilience (Wei et al. 2022). To enhance soil fertility and plant productivity, studies should target ErM genes related to SOM biodegradation, evaluate hormone-producing ErM strains in bioreactors, and identify metal-tolerant strains for phytoremediation. Developing ErM biofertilizers that combine specific growth-promoting and stress-tolerant traits could significantly benefit ericaceous plants.

7 Conclusion

Mycorrhizal fungi are vital for cultivating medicinal plants and enhancing crop resilience, quality, and yield through symbiotic relationships with plant roots. These fungi improve nutrient and water uptake, particularly phosphorus and nitrogen, which are critical for plant growth and bioactive compound production and are essential for medicinal crops. Key types, such as AMF and ectomycorrhizal fungi (EMF), play specialized roles; AMF penetrate root cells, boosting nutrient exchange and soil structure, whereas EMF forms external networks around roots for enhanced water and nutrient absorption. Mycorrhizal fungi also strengthen plant resistance to environmental stresses, including drought, salinity, and heavy metal exposure, and mitigate pathogen and pest damage, thereby offering a sustainable alternative to synthetic fertilizers and pesticides. However, challenges associated with commercial inoculant efficacy, large-scale production, and environmental adaptability require further research. The future aims to advance biofertilizer formulations, establish regulatory standards, and enhance mycorrhizal compatibility across diverse plant species to unlock the full potential of fungi. Optimizing these symbiotic interactions supports medicinal crop cultivation to meet rising pharmaceutical needs sustainably while promoting environmental health by reducing chemical dependency.

References

Abbas F, Ke Y, Yu R, Yue Y, Amanullah S, Jahangir M, Fan Y (2017) Volatile terpenoids: multiple functions, modulation and manipulation by genetic engineering. Planta 246:803–816

Agnihotri R, Bharti A, Ramesh A, Prakash A, Sharma MP (2021) Glomalin related protein and C16: 1ω5 PLFA associated with AM fungi as potential signatures for assessing the soil C sequestration under contrasting soil management practices. Eur J Soil Biol 103:103286

Agnihotri R, Sharma MP, Prakash A, Ramesh BS, Patra AK, Manna MC, Kurganova I, Kuzyakov Y (2022) Glycoproteins of arbuscular mycorrhiza for soil carbon sequestration: review of mechanisms and controls. Sci Total Environ 806:150571

Ahmed E, Holmström SJ (2014) Siderophores in environmental research: roles and applications. Microb Biotechnol 7(3):196–208

Allen JW, Shachar-Hill Y (2009) Sulfur transfer through an arbuscular mycorrhiza. Plant Physiol 149(1):549–560

Anwar S, Ali B, Sajid I (2016) Screening of rhizospheric actinomycetes for various in-vitro and in-vivo plant growth promoting (PGP) traits and for agroactive compounds. Front Micobiol 1334(7):1–11

Asif M, Mughal AH, Bisma R, Mehdi Z, Saima S, Ajaz M, Masood A, Ma M, Sidique S (2018) Application of different strains of biofertilizers for raising quality forest nursery. Int J Curr Microbiol Appl Sci 7:3680–3686

Aswani R, Jishma P, Radhakrishnan EK (2020) Endophytic bacteria from the medicinal plants and their potential applications. In: Kumar A, Singh VK (eds) Woodhead publishing series in food science, technology and nutrition, microb endophytes. Woodhead Publishing, Sawston

Auge RM, Toler HD, Saxton AM (2015) Arbuscular mycorrhizal symbiosis alters stomatal conductance of host plants more under drought than under amply watered conditions: a metaanalysis. Mycorrhiza 25(1):13–24

Azizoğlu U (2019) Bacillus thuringiensis as a biofertilizer and biostimulator: a mini-review of the little-known plant growth-promoting properties of Bt. Curr Microbiol 76(1):1379–1385
Backer R, Js R, Ilangumaran G, Lamont J, Praslickova D, Ricci E, Sowmyalakshmi S, Smith DL (2018) Plant growth-promoting rhizobacteria: context, mechanisms of action, and roadmap to commercialization of biostimulants for sustainable agriculture. Front Plant Sci 9(1473):1–17
Bal SK, Minhas PS (2017) Atmospheric stressors: challenges and coping strategies. In: Minhas PS, Rane J, Pasala RK (eds) Abiotic stress management for resilient agriculture. Springer, Singapore
Barea JM, Pozo MJ, Azcon R, Azcon-Aguilar C (2005) Microbial co-operation in the rhizosphere. J Exp Bot 56:1761–1778
Bargaz A, Lyamlouli K, Chtouki M, Zeroual Y, Dhiba D (2018) Soil microbial resources for improving fertilizers efficiency in an integrated plant nutrient management system. Front Micobiol 1606(9):1–25
Bashan Y, De-Bashan LE, Prabhu SR, Hernandez JP (2014) Advances in plant growth-promoting bacterial inoculant technology: formulations and practical perspectives (1998–2013). Plant Soil 378:1–33
Baslam M, Pascual I, Sánchez-Díaz M, Erro J, García-Mina JM, Goicoechea N (2011) Improvement of nutritional quality of greenhouse-grown lettuce by arbuscular mycorrhizal fungi is conditioned by the source of phosphorus nutrition. J Agric Food Chem 59(20):11129–11140
Basu S, Rabara RC, Negi S (2018) AMF: the future prospect for sustainable agriculture. Physiol Mol Plant Pathol 102:36–45
Begum N, Qin C, Ahanger MA, Raza S, Khan MI, Ashraf M, Ahmed N, Zhang L (2019) Role of arbuscular mycorrhizal fungi in plant growth regulation: implications in abiotic stress tolerance. Front Plant Sci 10:1068
Berruti A, Borriello R, Orgiazzi A, Barbera AC, Lumini E, Bianciotto V (2014) Arbuscular mycorrhizal fungi and their value for ecosystem management. In: Oscar G (ed) Biodiversity-the dynamic balance of the planet. InTech, Rijeka
Bhantana P, Rana MS, Sun XC, Moussa MG, Saleem MH, Syaifudin M, Shah A, Poudel A, Pun AB, Bhat MA, Mandal DL (2021) Arbuscular mycorrhizal fungi and its major role in plant growth, zinc nutrition, phosphorous regulation and phytoremediation. Symbiosis 84:19–37
Burleigh SH, Bechmann IE (2002) Plant nutrient transporter regulation in arbuscular mycorrhizas. Plant Soil 244:247–251
Bustos-Segura C, Foley W (2018) Foliar terpene chemotypes and herbivory determine variation in plant volatile emissions. J Chem Ecol 44:51–61
Cabral L, de Sousa Soares CRF, Giachini AJ, Siqueira O (2015) Arbuscular mycorrhizal fungi in phytoremediation of contaminated areas by trace elements: mechanisms and major benefits of their applications. World J Microbiol Biotechnol 31:1655
Cahanovitc R, Livne-Luzon S, Angel R, Klein T (2022) Ectomycorrhizal fungi mediate below-ground carbon transfer between pines and oaks. ISME J 16(5):1420–1429
Cameron DD, Neal AL, van Wees SC, Ton J (2013) Mycorrhiza-induced resistance: more than the sum of its parts? Trends Plant Sci 18(10):539–545
Cavagnaro TR, Bender SF, Asghari HR, van der Heijden MGA (2015) The role of arbuscular mycorrhizas in reducing soil nutrient loss. Trends Plant Sci 20:283–290
Chaudhary V, Kapoor R, Bhatnagar AK (2008) Effectiveness of two arbuscular mycorrhizal fungi on concentrations of essential oil and artemisinin in three accessions of *Artemisia annua* L. Appl Soil Ecol 40:174–181
Chen S, Zhao H, Zou C, Li Y, Chen Y, Wang Z, Jiang Y, Liu A, Zhao P, Wang M, Ahammed GJ (2017) Combined inoculation with multiple arbuscular mycorrhizal fungi improves growth, nutrient uptake and photosynthesis in cucumber seedlings. Front Microbiol 8:2516
Compant S, Samad A, Faist H, Sessitsch AA (2019) Review on the plant microbiome: ecology, functions, and emerging trends in microbial application. J Adv Res 19:29–37
Cooper KM (2018) Physiology of VA mycorrhizal associations. In: Conway L, Powell D, Bagyaraj P (eds) VA mycorrhiza. CRC Press, Florida

Cornea CP, Voaideş C, Boiu-Sicuia OA, Matei F, Băbeanu N (2019) Creating products and services in environmental biotechnology. In: Matei F, Zirra D (eds) Introduction to biotech entrepreneurship: from idea to business. Springer, Romania

Corrales A, Henkel TW, Smith ME (2018) Ectomycorrhizal associations in the tropics–biogeography, diversity patterns and ecosystem roles. New Phytol 220(4):1076–1091

Corrales A, Koch RA, Vasco-Palacios AM, Smith ME, Ge ZW, Henkel TW (2022) Diversity and distribution of tropical ectomycorrhizal fungi. Mycologia 114(6):919–933

Craig ME, Turner BL, Liang C, Clay K, Johnson DJ, Phillips RP (2018) Tree mycorrhizal type predicts within-site variability in the storage and distribution of soil organic matter. Glob Change Biol 24(8):3317–3330

Dai XY (2011) The endophytic fungi from wild cymbidium plants and effects on seed germination and seeding growth (doctoral dissertation, master thesis, Beijing Forestry University, Beijing. 12–13 (in Chinese))

de Novais CB, Sbrana C, da Conceição JE, Rouws LFM, Giovannetti M, Avio L, Siqueira JO, Saggin Junior OJ, da Silva EMR, de Faria SM (2020) Mycorrhizal networks facilitate the colonization of legume roots by a symbiotic nitrogen-fixing bacterium. Mycorrhiza 30(2):389–396

Diagne N, Ngom M, Djighaly PI, Fall D, Hocher V, Svistoonoff S (2020) Roles of arbuscular mycorrhizal fungi on plant growth and performance: importance in biotic and abiotic stressed regulation. Diversity 12(10):370

Dominguez-Nuñez JA, Benito B, Berrocal-Lobo M, Albanesi A (2016) Mycorrhizal fungi: role in the solubilization of potassium. In: Mena VS, Maurya BR, Verma JP, Meena RS (eds) Potassium solubilizing microorganisms sustainable agriculture. Springer, New Delhi

Dowarah B, Gill SS, Agarwala N (2022) Arbuscular mycorrhizal fungi in conferring tolerance to biotic stresses in plants. J Plant Growth Regul 41(4):1429–1444

Eccher T, Piagnani MC, Zordan C (2010) Influence of thirteen different strains of ericoid endomycorrhizae on rooting and growth of micropropagated *Azalea mollis*. Acta Hortic 865:321–326

Elinav E, Garrett WS, Trinchieri G, Wargo J (2019) The cancer microbiome. Nat Rev Cancer 19:371–376

Fadaei S, Vaziriyeganeh M, Young M, Sherr I, Zwiazek JJ (2020) Regulation of the European parliament and of the council laying down rules on the making available on the market of EU fertilising products and amending regulations (EC) No 1069/2009 and (EC) No 1107/2009 and repealing regulation (EC) No 2003/2003

Fagorzi C, Ilie A, Decorosi F, Cangioli L, Viti C, Mengoni M, Dicenzo GC (2020) Symbiotic and nonsymbiotic members of the genus Ensifer (syn. Sinorhizobium) are separated into two clades based on comparative genomics and high-throughput phenotyping. Genome Biol Evol 12(12):2521–2534

Fehrer J, Réblová M, Bambasová V, Vohník M (2019) The root-symbiotic Rhizoscyphus ericae aggregate and Hyaloscypha (Leotiomycetes) are congeneric: phylogenetic and experimental evidence. Stud Mycol 92(1):195–225

Fernie AR, Yan J (2019) De novo domestication: an alternative route toward new crops for the future. Mol Plant 12:615–631

Fester T, Hause G (2005) Accumulation of reactive oxygen species in arbuscular mycorrhizal roots. Mycorrhiza 15:373–373

Finlay RD (2008) Ecological aspects of mycorrhizal symbiosis: with special emphasis on the functional diversity of interactions involving the extraradical mycelium. J Exp Bot 59(5):1115–1126

Frank AB (1877) Uber die biologischen Verhältnisse des Thallus eineger Krustenflechnten Beitr Biol Pflanz 2:123–200. [In German]

Frank AB (1885) Uber die auf Wurzelsymbiose beruhende Ernährung gewisser Bäumedurch unterirdische Pilze. German

Gange AC, West HM (1994) Interactions between arbuscular mycorrhizal fungi and foliar-feeding insects in *Plantago lanceolata* L. New Phytol 128(1):79–87

Gangwar M, Saini P, Nikhanj P, Kaur S (2018) Plant growth-promoting microbes (PGPM) as potential microbial bio-agents for eco-friendly agriculture. In: Adhya TK (ed) Advances in soil microbiology: recent trends future prospects, microorganisms for sustainability 4. Springer, Singapore

Gao Y, Zhao Z, Li J, Liu N, Jacquemyn H, Guo S, Xing X (2020) Do fungal associates of co-occurring orchids promote seed germination of the widespread orchid species *Gymnadenia conopsea*? Mycorrhiza 30:221–228

Gao Y, Ji J, Zhang Y, Yang N, Zhang M (2022) Biochemical and transcriptomic analyses of the symbiotic interaction between *Cremastra appendiculata* and the mycorrhizal fungus *Coprinellus disseminatus*. BMC Plant Biol 22(1):15

Garcia-Fraile P, Menendez E, Rivas R (2015) Role of bacterial biofertilizers in agriculture and forestry. AIMS Bioeng 2:183–205

Geneva MP, Stancheva IV, Boychinova MM, Mincheva NH, Yonova PA (2010) Effects of foliar fertilization and arbuscular mycorrhizal colonization on *Salvia officinalis* L. growth, antioxidant capacity, and essential oil composition. J Sci Food Agric 90:696–702

Gianinazzi S, Gollotte A, Binet MN, van Tuinen D, Redecker D, Wipf D (2010) Agroecology: the key role of arbuscular mycorrhizas in ecosystem services. Mycorrhiza 20(8):519–530

Giovannetti M, Tolosano M, Volpe V, Kopriva S, Bonfante P (2014) Identification and functional characterization of a sulfate transporter induced by both sulfur starvation and mycorrhiza formation in *Lotus japonicus*. New Phytol 204(3):609–619

Gorzelak MA, Asay AK, Pickles BJ, Simard SW (2015) Inter-plant communication through mycorrhizal networks mediates complex adaptive behaviour in plant communities. AoB Plants 7:plv050

Gouda S, Kerry RG, Das G, Paramithiotis S, Shin HS, Patra JK (2018) Revitalization of plant growth promoting rhizobacteria for sustainable development in agriculture. Microbiol Res 206:131–140

Grunewaldt-Stöcker G, von den Berg C, Knopp J, von Alten H (2013) Interactions of ericoid mycorrhizal fungi and root pathogens in rhododendron: in vitro tests with plantlets in sterile liquid culture. Plant Root 7:33–48

Hashem A, Alqarawi AA, Radhakrishnan R, Al-Arjani ABF, Aldehaish HA, Egamberdieva D, Abd Allah EF (2018) Arbuscular mycorrhizal fungi regulate the oxidative system, hormones and ionic equilibrium to trigger saltstress tolerance in Cucumis sativus L. Saudi J Biol Sci 25(6):1102–1114

Hassan ARB (2012) Medicinal plants (importance and uses). Pharm Anal Acta 3:e139

Hetrick BAD (2018) Ecology of VA mycorrhizal fungi. In: Conway L, Powell D, Bagyaraj P (eds) VA mycorrhiza. CRC Press, Florida

Hosoe T, Nozawa K, Lumley TC, Currah RS, Fukushima K, Takizawa K, Miyaji M, Kawai KI (1999) Tetranorditerpene lactones, potent antifungal antibiotics for human pathogenic yeasts, from a unique species of Oidiodendron. Chem Pharm Bull 47(11):1591–1597

Huang W, van Bodegom PM, Declerck S, Heinonsalo J, Cosme M, Viskari T, Liski J, Soudzilovskaia NA (2022) Mycelium chemistry differs markedly between ectomycorrhizal and arbuscular mycorrhizal fungi. Commun Biol 5(1):398

Jain A, Sarsaiya S, Chen J, Wu Q, Lu Y, Shi J (2021) Changes in global Orchidaceae disease geographical research trends: recent incidences, distributions, treatment, and challenges. Bioengineered 12(1):13–29

Javoreková S, Maková J, Medo J, Kovácsová S, Charousová I, Horák J (2015) Effect of bio-fertilizers application on microbial diversity and physiological profiling of microorganisms in arable soil. Eurasian J Soil Sci 4:54–61

Jin H, Liu J, Liu J, Huang X (2012) Forms of nitrogen uptake, translocation, and transfer via arbuscular mycorrhizal fungi: a review. Sci China Life Sci 55:474–482

Johansson JF, Paul LR, Finlay RD (2004) Microbial interactions in the mycorrhizosphere and their significance for sustainable agriculture. FEMS Microbiol Ecol 48:1–13

Kapoor R, Anand G, Gupta P, Mandal S (2017) Insight into the mechanisms of enhanced production of valuable terpenoids by arbuscular mycorrhiza. Phytochem Rev 16(4):677–692

Karagiannidis N, Thomidis T, Lazari D, Panou-Filotheou E, Karagiannidou C (2011) Effect of three Greek arbuscular mycorrhizal fungi in improving the growth, nutrient concentration, and production of essential oils of oregano and mint plants. Sci Hortic 129:329–334

Kim SJ, Eo JK, Lee EH, Park H, Eom AH (2017) Effects of arbuscular mycorrhizal fungi and soil conditions on crop plant growth. Mycobiology 45(1):20–24

Kohler A, Kuo A, Nagy LG, Morin E, Barry KW, Buscot F, Canbäck B, Choi C, Cichocki N, Clum A, Colpaert J (2015) Convergent losses of decay mechanisms and rapid turnover of symbiosis genes in mycorrhizal mutualists. Nat Genet 47(4):410–415

Koricheva J, Gange AC, Jones T (2009) Effects of mycorrhizal fungi on insect herbivores: a meta-analysis. Ecology 90(8):2088–2097

Kumar K, Mella-Herrera RA, Golden JW (2010) Cyanobacterial heterocysts. Cold Spring Harb. perspect. biol 2(4):a000315

Kumar A, Singh R, Adholeya A (2017) Biotechnological advancements in industrial production of arbuscular mycorrhizal fungi: achievements, challenges, and future prospects. In: Developments in fungal biology and applied mycology, pp 413–431

Leake J, Johnson D, Donnely D, Muckle G, Boddy L, Read D (2004) Networks of power and influence: the role of mycorrhizal mycelium in controlling plant communities and agroecosystem functioning. Can J Bot 82:1016–1045

Lee J (2011) The distribution of cytoplasm and nuclei within the extra-radical mycelia in Glomus intraradices, a species of arbuscular mycorrhizal fungi. Mycobiology 39(2):9–84

Leng C, Hou M, Xing Y, Chen J (2024) Perspective and challenges of mycorrhizal symbiosis in orchid medicinal plants. Chin Herb Med 16:172–179

Lin LC, Lin CY, Lin WR, Tung YT, Wu JH (2021) Effects of ericoid Mycorrhizal fungi or dark septate endophytic fungi on the secondary metabolite of *Rhododendron pseudochrysanthum* (R. *Morii*) seedlings. Appl Ecol Environ Res 19:1221–1232

Lindahl BD, Ihrmark K, Boberg J, Trumbore SE, Högberg P, Stenlid J, Finlay RD (2007) Spatial separation of litter decomposition and mycorrhizal nitrogen uptake in a boreal forest. New Phytol 173(3):611–620

Liu Y, Li X, Kou Y (2020) Ectomycorrhizal fungi: participation in nutrient turnover and community assembly pattern in forest ecosystems. Forests 11(4):453

Liu N, Jacquemyn H, Liu Q, Shao SC, Ding G, Xing X (2022) Effects of a dark septate fungal endophyte on the growth and physiological response of seedlings to drought in an epiphytic orchid. Front Microbiol 13:961172

Lohse S, Schliemann W, Ammer C, Kopka J, Strack D, Fester T (2005) Organisation and metabolism of plastids and mitochondria in arbuscular mycorrhizal roots of *Medicago truncatula*. Plant Physiol 139:329–340

Ma J, Janoušková M, Ye L, Bai LQ, Dong RR, Yan Y, Yu XC, Zou ZR, Li YS, He CX (2019a) Role of arbuscular mycorrhiza in alleviating the effect of cold on the photosynthesis of cucumber seedlings. Photosynthetica 57(1):86–95

Ma X, Luo W, Li J, Wu F (2019b) Arbuscular mycorrhizal fungi increase both concentrations and bioavilability of Zn in wheat (*Triticum aestivum* L) grain on Zn-spiked soils. Appl Soil Ecol 135:91–97

Macik M, Gryta A, Frac M (2020) Biofertilizer in agriculture: an overview on concepts, strategies and effects on soli microorganism. Adv Agron 16:31–76

Malik AA, Sanghmitra Sryapani SS, Javed Ahmad JA (2011) Chemical vs organic cultivation of medicinal and aromatic plants: the choice is clear 5–13

Malusa M, Ciesielska J (2012) Biofertilizers: a source for sustainable plant nutrition. Synthetis 1:282–320

Marro N, Cofre N, Grilli G, Alvarez C, Labuckas D, Maestri D, Urcelay C (2020) Soybean yield, protein content and oil quality in response to interaction of arbuscular mycorrhizal fungi and native microbial populations from mono-and rotation-cropped soils. Appl Soil Ecol 152:103575

Martino E, Morin E, Grelet GA, Kuo A, Kohler A, Daghino S, Barry KW, Cichocki N, Clum A, Dockter RB, Hainaut M (2018) Comparative genomics and transcriptomics depict ericoid mycorrhizal fungi as versatile saprotrophs and plant mutualists. New Phytol 217(3):1213–1229

Marulanda A, Porcel R, Barea JM, Azcon R (2007) Drought tolerance and antioxidant activities in lavander plants colonized by native drought-tolerant of drought-sensitive Glomus species. Microb Ecol 54:543–552

Mburu SW, Koskey G, Njeru EM, Maingi JM (2021) Revitalization of bacterial endophytes and rhizobacteria for nut rients bioavailability in degraded soils to promote crop production. AIMS Agric Food 6(2):496–524

Meharg AA, Cairney JW (1999) Co-evolution of mycorrhizal symbionts and their hosts to metal-contaminated environments. In: Advances in ecological research, vol 30. Academic, New York, pp 69–112

Miransari M (2011) Arbuscular mycorrhizal fungi and nitrogen uptake. Arch Microbiol 193(2):77–81

Molina R, Horton TR (2015) Mycorrhiza specificity: its role in the development and function of common mycelial networks. In: Horton TR (ed) Mycorrhizal networks. Springer, pp 1–39

Morales-García AL, Bailey RG, Jana S, Burgess JG (2019) The role of polymers in cross-kingdom bioadhesion. Philos Trans R Soc B Biol Sci 374(1784):20190192

Murphy BR, Doohan FM, Hodkinson TR (2018) From concept to commerce: developing a successful` fungal endophyte inoculant for agricultural crops. J Fungi 4:24

Nahar K, Hasanuzzaman M, Fujita M (2016) Roles of osmolytes in plant adaptation to drought and salinity. In: Iqbal N, Nazar R, Khan NA (eds) Osmolytes plants acclimation changing environment: emerging omics technologies. Springer, New Delhi

Nair SS, Brahmaprakash GP (2017) Effect of effervescent biofertilizer consortial tablets on growth of tomato (*Lycopersicon esculentum* Mill.). Int J Curr Microbiol Appl Sci 6(9):615–623

Namasivayam SKR, Saikia SL, Bharani RSA (2014) Evaluation of persistence and plant growth promoting effect of bioencapsulated formulation of suitable bacterial biofertilizers. Biosci Biotechnol Res Asia 11:407–415

Nazarov PA, Baleev DN, Ivanova MI, Sokolova LM, Karakozova MV (2020) Infectious plant diseases: etiology, current status, problems and prospects in plant protection. Acta Nat 12(3):46

Nehls U, Dietz S (2014) Fungal aquaporins: cellular functions and ecophysiological perspectives. Appl Microbiol Biotechnol 98:8835–8851

Nell M, Wawrosch C, Steinkellner S, Vierheilig H, Kopp B, Lössl A et al (2010) Root colonization by symbiotic arbuscular mycorrhizal fungi increases sesquiterpenic acid concentrations in *Valeriana officinalis* L. Planta Med 76(04):393–398

Nguvo KJ, Gao X (2019) Weapons hidden underneath: bio-control agents and their potentials to activate plant induced systemic resistance in controlling crop Fusarium diseases. J Plant Dis Prot 126:177–190

Nguyen TD, Cavagnaro TR, Watts-Williams SJ (2019) The effects of soil phosphorus and zinc availability on plant responses to mycorrhizal fungi: a physiological and molecular assessment. Sci Rep 9(1):14880

O'Callaghan M (2016) Microbial inoculation of seed for improved crop performance: issues and opportunities. Appl Microbiol Biotechnol 100:5729–5746

Osonubi O, Mulongoy K, Awotoye OO, Atayese MO, Okali DUU (1991) Effects of ectomycorrhizal and vesicular-arbuscular mycorrhizal fungi on drought tolerance of four leguminous woody seedlings. Plant Soil 136:131–143

Ouyang X (2021) Witnessing the process of bacterial cell death: novel antimicrobials and their mechanisms of action doctoral dissertation, Leiden University

Pagare S, Manila Bhatia M, Tripathi N et al (2015) Secondary metabolites of plants and their role: overview. Curr Trends Biotechnol Pharm 9:294–305

Pellitier PT, Zak DR (2018) Ectomycorrhizal fungi and the enzymatic liberation of nitrogen from soil organic matter: why evolutionary history matters. New Phytol 217(1):68–73

Perotto S, Daghino S, Martino E (2018) Ericoid mycorrhizal fungi and their genomes: another side to the mycorrhizal symbiosis? New Phytol 220(4):1141–1147

Pozo MJ, Azcón-Aguilar C (2007) Unraveling mycorrhizainduced resistance. Curr Opin Plant Biol 10(4):393–398

Prasad A, Kumar S, Khaliq A, Pandey A (2011) Heavy metals and arbuscular mycorrhizal (AM) fungi can alter the yield and chemical composition of volatile oil of sweet basil (Ocimum basilicum L.). Biol. Fertil. Soils 47(8):853–861

Pu C, Ge Y, Yang G, Zheng H, Guan W, Chao Z, Shen Y, Liu S, Chen M, Huang L (2022) Arbuscular mycorrhizal fungi enhance disease resistance of *Salvia miltiorrhiza* to Fusarium wilt. Front Plant Sci 13:975558

Ramanankierana N, Ducousso M, Rakotoarimanga N, Prin Y, Thioulouse J, Randrianjohany E, Ramaroson L, Kisa M, Galiana A, Duponnois R (2007) Arbuscular mycorrhizas and ectomycorrhizas of Uapaca bojeri L. (Euphorbiaceae): sporophore diversity, patterns of root colonization, and effects on seedling growth and soil microbial catabolic diversity. Mycorrhiza 17(3):195–208

Rastogi G, Sani RK (2011) Molecular techniques to assess microbial community structure, function, and dynamics in the environment. In: Ahmad I (ed) Microbes and microbial technology: agricultural and environmental applications. Springer

Rillig MC et al (2016) Towards an integrated mycorrhizal technology: harnessing mycorrhizae for sustainable intensification in agriculture. Front Plant Sci 7:1625

Rillig MC, Aguilar-Trigueros CA, Camenzind T, Cavagnaro TR, Degrune F, Hohmann P, Lammel DR, Mansour I, Roy J, van Der Heijden MG, Yang G (2019) Why farmers should manage the arbuscular mycorrhizal symbiosis. New Phytol 222(3):1171–1175

Rivera-Becerril F, Calantzis C, Turnau K, Caussanel JP, Belimov AA, Gianinazzi S, Strasse RJ, Gianinazzi-Pearson V (2002) Cadmium accumulation and buffering of cadmium-induced stress by arbuscular mycorrhiza in three *Pisum sativum* L. genotypes. J Exp Bot 53:1177–1185

Rocha I et al (2019) Seed coating with arbuscular mycorrhizal fungi for improved field production of chickpea. Agronomy 9:471–482

Rout P, Rao Y, Sree A, Naik S (2007) Composition of essential oil, concrete, absolute, wax and headspace volatiles of *Murraya panicuata* (Linn.) Jack fowers. Flavour Fragr 22(5):352–357

Rui-Ting SUN, Zhang ZZ, Nong ZHOU, Srivastava A, Kuča K, Abd-Allah EF, Hashem A, Qiang-Sheng WU (2021) A review of the interaction of medicinal plants and arbuscular arbuscular mycorrhizal fungi in the rhizosphere. Not Bot Horti Agrobot Cluj-Napoca 49(3):12454

Santander C, Aroca R, Ruiz-Lozano JM, Olave J, Cartes P, Borie F, Cornejo P (2017) Arbuscular mycorrhiza effects on plant performance under osmotic stress. Mycorrhiza 27:639–657

Sati D, Pandey SC, Pande V, Upreti S, Gouri V, Joshi T, Gangola S, Debbarma P, Pandey A, Samant M (2020) Toward an enhanced understanding of plant growth promoting microbes for sustainable agriculture. In: Recent advancements in microbial diversity. Academic, Cambridge, MA

Scagel CF (2005) Inoculation with ericoid mycorrhizal fungi alters fertilizer use of highbush blueberry cultivars. HortScience 40(3):786–794

Scagel CF, Wagner A, Winiarski P (2005a) Frequency and intensity of root colonization by ericoid mycorrhizal fungi in nursery production of blueberry plants. Small Fruits Rev 4(4):95–112

Scagel CF, Wagner A, Winiarski P (2005b) Inoculation with ericoid mycorrhizal fungi alters root colonization and growth in nursery production of blueberry plants from tissue culture and cuttings. Small Fruits Rev 4(4):113–135

Schüßler A, Schwarzott D, Walker C (2001) A new fungal phylum, the Glomeromycota: phylogeny and evolution. Mycol Res 105(12):1413–1421

Shah F, Nicolás C, Bentzer J, Ellström M, Smits M, Rineau F, Canbäck B, Floudas D, Carleer R, Lackner G, Braesel J (2016) Ectomycorrhizal fungi decompose soil organic matter using oxidative mechanisms adapted from saprotrophic ancestors. New Phytol 209(4):1705–1719

Shi HF, Dong ZY, Shi Y, Hu MJ, Cong K, Yang LY et al (2023) Effect of symbiotic germinating fungi on seed germination of *Dendrobium officinale* Yunnan agricultural. Sci Technol 4:8–10

Sicari V, Poiana M (2017) Comparison of the volatile component of the essential oil of kumquat (*Fortunella margarita swingle*) extracted by supercritical carbon dioxide, hydrodestillation and conventional solvent extraction. J Essent Oil-Bear Plants 20:87–94

Sieh D, Watanabe M, Devers EA, Brueckner F, Hoefgen R, Krajinski F (2013) The arbuscular mycorrhizal symbiosis influences sulfur starvation responses of *Medicago truncatula*. New Phytol 197(2):606–616

Simard SW et al (2012a) Mycorrhizal networks: mechanisms, ecology and modelling. Fungal Biol Rev 26:39–60

Simard SW, Beiler KJ, Bingham MA, Deslippe JR, Philip LJ, Teste FP (2012b) Mycorrhizal networks: mechanisms, ecology and modelling. Fungal Biol Rev 26(1):39–60

Singh RS (2018) Plant diseases. Oxford and IBH Publishing, New Delhi

Singh LP, Gill SS, Tuteja N (2011) Unraveling the role of fungal symbionts in plant abiotic stress tolerance. Plant Signal Behav 6(2):175–191

Singh M, Kumar J, Singh S, Singh VP, Prasad SM (2015) Roles of osmoprotectants in improving salinity and drought tolerance in plants: a review. Rev Environ Sci Biotechnol 14:407–426

Singh JS, Kumar A, Rai AN, Singh DP (2016) Cyanobacteria: a precious bio-resource in agriculture, ecosystem, and environmental sustainability. Front Microbiol 7:529

Smith SE, Read D (2008) Nitrogen mobilization and nutrition in ectomycorrhizal plants. Mycorrhizal symbiosis 3:321–348

Sosa-Hernández MA, Leifheit EF, Ingraffia R, Rillig MC (2019) Subsoil arbuscular mycorrhizal fungi for sustainability and climate-smart agriculture: a solution right under our feet? Front Microbiol 10:744

Spagnoletti FN, Cornero M, Chiocchio V, Lavado RS, Roberts IN (2020) Arbuscular mycorrhiza protects soybean plants against *Macrophomina phaseolina* even under nitrogen fertilization. Eur J Plant Pathol 156:839–849

Stamenkovic S, Besˇkoski V, Karabegovic I, Lazic M, Nikolic N (2018) Microbial fertilizers: a comprehensive review of current findings and future perspectives. Span J Agric Res 16:1–18

Steidinger BS, Crowther TW, Liang J, Van Nuland ME, Werner GD, Reich PB, Nabuurs GJ, De-Miguel S, Zhou M, Picard N, Hérault B (2019) Climatic controls of decomposition drive the global biogeography of forest-tree symbioses. Nature 569(7756):404–408

Sugiura Y, Akiyama R, Tanaka S, Yano K, Kameoka H, Marui S, Saito M, Kawaguchi M, Akiyama K, Saito K (2020) Myristate can be used as a carbon and energy source for the asymbiotic growth of arbuscular mycorrhizal fungi. Proc Natl Acad Sci 117(41):25779–25788

Sylvia D, Fuhrmann J, Hartel P, Zuberer D (2005) Principles and applications of soil microbiology. Pearson, Upper Saddle River

Szakiel A, Paczkowski C, Henry M (2010) Influence of environmental biotic factors on the content of saponins in plants. Phytochem Rev 10:493–502

Tang YJ, Zhou DY, Dai J, Li Y, Xing YM, Guo SX, Chen J (2022) Potential specificity between mycorrhizal fungi isolated from widespread *Dendrobium* spp. and rare *D. huoshanense* seeds. Curr Microbiol 79(9):264

Tapwal A (2024) Harnessing Native Arbuscular Mycorrhizal Fungi in Enhancing Biomass and Phytochemicals Profiles in Angelica Glauca Edgew Available at SSRN 5037018

Tedersoo L, Smith ME (2013) Lineages of ectomycorrhizal fungi revisited: foraging strategies and novel lineages revealed by sequences from belowground. Fungal Biol Rev 27(3–4):83–99

Tedersoo L, Bahram M, Zob M (2020) How mycorrhizal associations drive plant population and community. Science 367:eaba1223

Thilakarathna MS, Mcelroy MS, Chapagain T, Papadopoulos YA, Raizada MN (2016) Belowground nitrogen transfer from legumes to non-legumes under managed herbaceous cropping systems: a review. Agron Sustain Dev 36:1–16

Toljander JF, Artursson V, Paul LR, Jansson JK, Finlay RD (2006) Attachment of different soil bacteria to arbuscular mycorrhizal fungal extraradical hyphae is determined by hyphal vitality and fungal species. FEMS Microbiol Lett 254(1):34–40

Torres N, Antolin MC, Goicoechea N (2018) Arbuscular mycorrhizal symbiosis as a promising resource for improving berry quality in grapevines under changing environments. Front Plant Sci 9:18

Ujvári G, Turrini A, Avio L, Agnolucci M (2021) Possible role of arbuscular mycorrhizal fungi and associated bacteria in the recruitment of endophytic bacterial communities by plant roots. Mycorrhiza 31(5):527–544

Umesha S, Singh PK, Singh RP (2018) Microbial biotechnology and sustainable agriculture. In: Singh RL, Mondal S (eds) Biotechnology for Sustainable Agriculture. Woodhead Publishing, Cambridge

Van Der Heijden MG, Martin FM, Selosse MA, Sanders IR (2015) Mycorrhizal ecology and evolution: the past, the present, and the future. New Phytol 205(4):1406–1423

Varma A, Prasad R, Tuteja N (eds) (2017) Mycorrhiza-eco-physiology, secondary metabolites, nanomaterials. Springer, Cham, pp 183–207

Verma M, Mishra J, Arora, NK (2019) Plant growth-promoting rhizobacteria: diversity and applications. In: Sobti RC, Arora NK, ·Kothari R (Eds.), Environmental biotechnology: for sustainable future. Springer, Singapore

Vishwanathan K, Zienkiewicz K, Liu Y, Janz D, Feussner I, Polle A, Haney CH (2020) Ectomycorrhizal fungi induce systemic resistance against insects on a nonmycorrhizal plant in a CERK1-dependent manner. New Phytol 228(2):728–740

Vohnik M (2020) Ericoid mycorrhizal symbiosis: theoretical background and methods for its comprehensive investigation. Mycorrhiza 30(6):671–695

Vohnik M, Sadowsky JJ, Kohout P, Lhotakova Z, Nestby R, Kolařík M (2012) Novel root-fungus symbiosis in Ericaceae: sheathed ericoid mycorrhiza formed by a hitherto undescribed basidiomycete with affinities to Trechisporales. PLoS One 7(6):e39524

Volpe V et al (2020) Short chain chito-oligosaccharides promote arbuscular mycorrhizal colonization in *Medicago truncatula*. Carbohydr Polym 229:115–505

Vos CM, Tesfahun AN, Panis B, De Waele D, Elsen A (2012) Arbuscular mycorrhizal fungi induce systemic resistance in tomato against the sedentary nematode *Meloidogyne incognita* and the migratory nematode *Pratylenchus penetrans*. Appl Soil Ecol 61:1–6

Vosatka M, Latr A, Gianinazzi S, Albrechtova J (2012) Development of arbuscular mycorrhizal biotechnology and industry: current achievements and bottlenecks. Symbiosis 58:29–37

Vurukonda SSKP, Vardharajula S, Shrivastava M, Skz A (2016) Enhancement of drought stress tolerance in crops by plant growth promoting rhizobacteria. Microbiol Res 184:13–24

Wahab A, Muhammad M, Munir A, Abdi G, Zaman W, Ayaz A, Khizar C, Reddy SP (2023) Role of arbuscular mycorrhizal fungi in regulating growth, enhancing productivity, and potentially influencing ecosystems under abiotic and biotic stresses. Plants 12(17):3102

Walid N, Kasaam R, Ashfaq MA (2021) Interaction between mycorrhizal and medicinal plants towards enhancement of secondary metabolites. Int J Chem Stud 9(2):140–152

Wang XJ, Ren JL, Zhang AH, Sun H, Yan GL, Han Y, Liu L (2019) Novel applications of mass spectrometry-based metabolomics in herbal medicines and its active ingredients: current evidence. Mass Spectrom Rev 38(4–5):380–402

Wang F, Zhang L, Zhou J, Rengel Z, George TS, Feng G (2022) Exploring the secrets of hyphosphere of arbuscular mycorrhizal fungi: processes and ecological functions. Plant Soil 481(1):1–22

Wei X, Chen J, Zhang C, Liu H, Zheng X, Mu J (2020) Ericoid mycorrhizal fungus enhances microcutting rooting of Rhododendron fortunei and subsequent growth. Hortic Res 7:140

Wei X, Zhang W, Zulfiqar F, Zhang C, Chen J (2022) Ericoid mycorrhizal fungi as biostimulants for improving propagation and production of ericaceous plants. Front Plant Sci 13:1027390

Weng W, Yan J, Zhou M, Yao X, Gao A, Ma C, Cheng J, Ruan J (2022) Roles of arbuscular mycorrhizal fungi as a biocontrol agent in the control of plant diseases. Microorganisms 10(7):1266

Wu HH, Zou YN, Rahman MM, Ni QD, Wu QS (2017) Mycorrhizas alter sucrose and proline metabolism in trifoliate orange exposed to drought stress. Sci Rep 7(1):42389

Xi G, Shi J, Li J, Han Z (2020) Isolation and identification of beneficial orchid mycorrhizal fungi in *Paphiopedilum barbigerum* (*Orchidaceae*). Plant Signal Behav 15(12):1816644

Yu L, Zhang Z, Zhou L (2022) Advances in the studies on symbiotic arbuscular mycorrhizal fungi of traditional Chinese medicinal plants. Biocell 46(12):2559–2573

Zak DR, Pellitier PT, Argiroff W, Castillo B, James TY, Nave LE, Averill C, Beidler KV, Bhatnagar J, Blesh J, Classen AT (2019) Exploring the role of ectomycorrhizal fungi in soil carbon dynamics. New Phytol 223(1):33–39

Zhang SF Pan SY, Zhou SF, Gao SH, Yu ZL, Tang MK et al (2013) New perspectives on how to discover drugs from herbal medicines: CAM's outstanding contribution to modern therapeutics. Evid-based Complement Altern Med

Zhang Y, Li Y, Chen X, Meng Z, Guo S (2020a) Combined metabolome and transcriptome analyses reveal the effects of mycorrhizal fungus *Ceratobasidium* sp. AR2 on the flavonoid accumulation in *Anoectochilus roxburghii* during different growth stages. Int J Mol Sci 21(2):564

Zhang Y, Li YY, Chen XM, Guo SX, Lee YI (2020b) Effect of different mycobionts on symbiotic germination and seedling growth of Dendrobium officinale, an important medicinal orchid. Bot Stud 61:1–10

Zhang E, Wang Y, Chen S, Zhou D, Shangguan Z, Huang J, He JS, Wang Y, Sheng J, Tang L, Li X (2024) Mycorrhizal symbiosis increases plant phylogenetic diversity and regulates community assembly in grasslands. Ecol Lett 27(9):e14516

Zhao Y, Cartabia A, Lalaymia I, Declerck S (2022) Arbuscular mycorrhizal fungi and production of secondary metabolites in medicinal plants. Mycorrhiza 32(3):221–256

Zou YN, Srivastava AK, Wu QS, Huang YM (2014) Glomalin-related soil protein and water relations in mycorrhizal citrus (*citrus tangerina*) during soil water deficit. Arch Agron Soil Sci 60(8):1103–1114

Rooting for Growth: Mycorrhizal Impact on Forestry Ecosystems

Prasann Kumar, Amit Raj, Jagadeesh Bathula, Purumandla Vennela Reddy, Appani Laxman Kumar, and Indu Kale

Abstract

Forestry ecosystems rely heavily on the intricate relationship between trees and soil microorganisms, with mycorrhizal fungi playing a critical role in promoting sustainable forest growth. This chapter examines the symbiotic relationships between mycorrhizae and forest trees, highlighting their substantial contributions to nutrient cycling, water uptake, and overall plant health. By forming a dynamic partnership with tree roots, mycorrhizal fungi enhance nutrient absorption, particularly in phosphorus-deficient soils, and improve tree resilience to environmental stresses, such as drought and soil degradation. The chapter discusses the various types of mycorrhizal associations, including ectomycorrhizae and arbuscular mycorrhizae, as well as their distinct roles in diverse forest ecosystems. It highlights the ecological significance of these associations in sustaining forest productivity and biodiversity, particularly in hilly regions where soil fertility and water availability are limiting factors. Furthermore, the chapter explores the practical applications of mycorrhizae in forestry practices, including

P. Kumar · A. Raj
Department of Agronomy, School of Agriculture, Lovely Professional University, Phagwara, Punjab, India

J. Bathula (✉)
Department of Forest Resource Management, Forest College and Research Institute, Mulugu, Siddipet, Telangana, India

P. V. Reddy
Department of Agroforestry, Bundelkhand University, Jhansi, Uttar Pradesh, India

A. L. Kumar
Sri Konda Laxman Telangana Horticultural University, Siddipet, Telangana, India

I. Kale
Department of Forest Biology and Tree Improvement, College of Forestry, Navsari Agriculture University, Navsari, Gujarat, India

Y. Siddiqui et al. (eds.), *Sustainable Mycorrhizal Cultivation*,
https://doi.org/10.1007/978-981-95-1778-7_9

their potential in reforestation efforts, the restoration of degraded lands, and the enhancement of growth in commercially valuable tree species. By integrating recent research on the influence of mycorrhizae on soil health and plant growth, this chapter provides valuable insights for forest managers, researchers, and policymakers seeking to leverage the benefits of mycorrhizal associations for sustainable forestry practices. The use of mycorrhizae in forestry not only supports the growth and health of forest trees but also contributes to the broader goal of environmental conservation and carbon sequestration, aligning with global initiatives to combat climate change and land degradation.

Keywords

Mycorrhiza · Symbiosis · Forestry · Sustainability · Nutrient acquisition · Soil health

1 Introduction

1.1 Importance of Mycorrhizae in Forestry Ecosystems

Mycorrhizae, which are symbiotic associations between fungi and plant roots, play a crucial role in enhancing the health and resilience of forest ecosystems. These associations facilitate nutrient uptake, particularly of phosphorus and nitrogen, which are often limiting factors in forest soils (Zhao and Riaz 2024). Mycorrhizal fungi significantly contribute to soil structure and fertility, which in turn support biodiversity and ecosystem functioning (Zhang et al. 2023). Furthermore, these fungi help trees withstand environmental stresses such as drought and soil contamination, underscoring their importance in sustainable forestry practices (Wahab et al. 2023). Mycorrhizae are essential components of forest ecosystems, playing a critical role in enhancing plant health, soil fertility, and overall ecosystem stability. These symbiotic relationships between fungi and plant roots facilitate the uptake of essential nutrients, particularly phosphorus and nitrogen, which are often limited in forest soils (Johnson et al. 2023). The presence of mycorrhizal fungi not only improves nutrient availability but also enhances water absorption, thus increasing plant resilience to drought and other abiotic stresses (Wahab et al. 2023). Moreover, mycorrhizal associations contribute to soil structure by promoting the formation of stable aggregates, which improve aeration and water retention in the soil (Imran and Ortas 2025). This is particularly important in forestry, where healthy soil conditions are vital for the growth and sustainability of tree species. The ecological benefits of mycorrhizae extend beyond individual plants; they enhance biodiversity by supporting a diverse range of plant species, which in turn fosters a more resilient ecosystem (Nizamani et al. 2024).

1.2 Overview of Mycorrhizal Associations and Tree Roots

Mycorrhizal associations can be classified primarily into two types: arbuscular mycorrhizae (AM) and ectomycorrhizae (ECM). AM fungi are predominantly associated with herbaceous plants and many tree species, forming a network that enhances nutrient exchange (Yang et al. 2022a, b). In contrast, ECM fungi form sheaths around tree roots, particularly common in temperate and boreal forests (Henriksson et al. 2023). According to a comprehensive review by Saha et al. (2023), these mycorrhizal relationships are crucial for forest ecosystem health, significantly influencing plant community dynamics and playing a vital role in the carbon cycle. Mycorrhizal associations can be broadly categorized into two main types: arbuscular mycorrhizal fungi (AMF) and ectomycorrhizal fungi (EMF). AMF form symbiotic relationships primarily with herbaceous plants and many tree species, penetrating root cells and facilitating nutrient exchange (Kashyap et al. 2024). In contrast, EMF is typically associated with woody plants, enveloping root tips, and forming a protective mantle that enhances nutrient uptake and provides defense against pathogens (Javeed et al. 2023). These associations are characterized by a complex network of fungal hyphae that extend far beyond the root zone, effectively increasing root surface area and enabling trees to access nutrients and water from a larger volume of soil (Wahab et al. 2023). This symbiotic relationship is beneficial for nutrient acquisition and plays a crucial role in establishing new plant communities, particularly in disturbed or degraded environments (Pedrinho et al. 2024). Understanding the dynamics of these mycorrhizal associations is essential for developing effective forestry management strategies.

1.3 Role of Mycorrhizae in Sustainable Forest Growth

Sustainable forest growth is increasingly recognized as a vital component of environmental stewardship and climate change mitigation. Mycorrhizal fungi can enhance forest productivity by improving nutrient cycling and soil health (Fahad et al. 2022). Sachsenmaier et al. (2024) demonstrated that forests with diverse mycorrhizal communities show greater resilience to environmental changes. Additionally, these fungi play a vital role in carbon sequestration, as they facilitate the transfer of carbon from trees to the soil, promoting long-term soil health and stability (Dou et al. 2025). Therefore, understanding and utilizing the potential of mycorrhizal associations are essential for achieving sustainable forestry practices amid global environmental challenges.

The role of mycorrhizae in promoting sustainable forest growth cannot be overstated. As forests face increasing pressure from climate change, land-use changes, and pollution, the resilience provided by mycorrhizal fungi has become ever more critical. These fungi not only enhance nutrient and water uptake but also improve the overall health of trees, making them more resilient to environmental stresses (Wahab et al. 2023). Furthermore, mycorrhizal fungi contribute to carbon sequestration by facilitating the storage of carbon in both plant biomass and soil organic

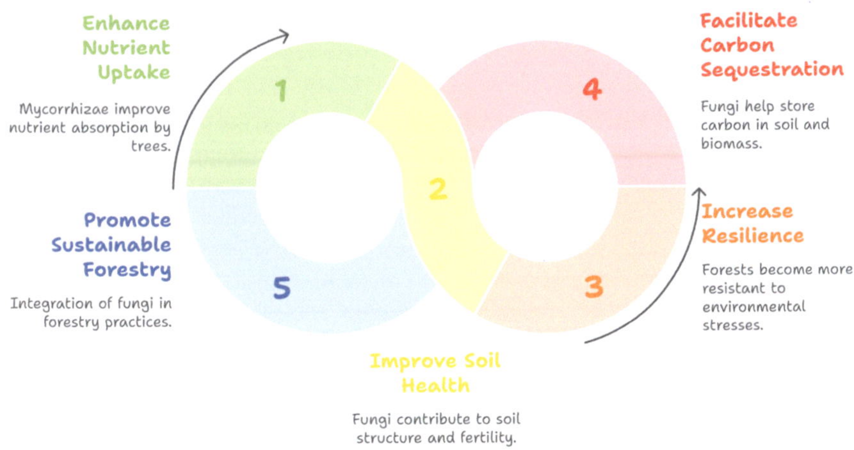

Fig. 1 Mycorrhizal fungi's role in forest sustainability

matter (Basiru and Hijri 2024). This process is vital for mitigating the effects of climate change and enhancing the sustainability of forest ecosystems. By integrating mycorrhizal management into forestry practices, such as reforestation and afforestation, we can promote the establishment and growth of trees, leading to healthier and more productive forests (Urgoiti et al. 2022). In conclusion, mycorrhizal fungi are indispensable allies in the quest for sustainable forestry. Their contribution to nutrient cycling, soil health, and ecosystem resilience underscores the importance of incorporating them into forestry management practices. As we move forward, understanding and harnessing the power of mycorrhizae will be crucial for achieving sustainable cultivation in hilly and diverse agricultural landscapes (Fig. 1).

2 Types of Mycorrhizal Associations in Forestry

2.1 Ectomycorrhizae (EM)

Ectomycorrhizae (EM) represent a type of mycorrhizal association characterized by the formation of a fungal sheath around the roots of host trees. This sheath, or mantle, is vital for the interaction between the fungus and the tree, facilitating nutrient exchange and enhancing the tree's ability to absorb water and minerals from the soil (Sudharsan et al. 2023). The formation of EM is essential in temperate and boreal forests, where these associations are predominant.

2.1.1 Formation and Function

The formation of ectomycorrhizal associations begins when fungal spores germinate in the rhizosphere of host trees, resulting in the colonization of root tips. The fungi penetrate the root cortex without invading the plant cells; instead, they create a network of hyphae that extends into the soil (Cargill et al. 2025). This structure not only increases the surface area for nutrient absorption but also helps protect roots from pathogens and environmental stressors (Wahab et al. 2023).

2.1.2 Host Tree Species in Forestry Ecosystems

Ectomycorrhizae are often associated with various tree species, particularly conifers like pine, spruce, and fir, as well as hardwoods such as oak and beech (Adnan et al. 2022). These associations are crucial for the growth and health of these trees, as they enhance nutrient uptake and improve resistance to diseases and environmental stresses (Wahab et al. 2023).

Recent studies have revealed that EMFs show both host specificity and flexibility depending on environmental and biological contexts. For example, *Hebeloma cylindrosporum* has been shown to form ectomycorrhizal associations not only with *Pinus pinaster* but also with other species, such as *Larix laricina*, *Dryas integrifolia*, and *Quercus acutissima*, under controlled conditions (Dyshko et al. 2024). Similarly, *Suillus spraguei* exhibits strong specificity to five-needle pines in the subgenus Strobus, though it can establish mycorrhizal relationships with other pines under experimental conditions (Pérez-Pazos et al. 2021).

Environmental variables such as soil pH, climate, and stand age also play significant roles in shaping EMF community structure. For example, research by Liu et al. (2023) in semi-arid forests of Northern China showed that the composition and function of EMF communities in *Pinus sylvestris* plantations varied significantly with soil properties and plantation age, directly impacting tree performance. Similarly, Zheng et al. (2023a, b) reported that soil pH is a key determinant of EMF richness and abundance across different forest types.

These findings highlight the significance of understanding EMF-host specificity and adaptability for sustainable forestry. The intentional inoculation of seedlings with selected EMF strains can enhance reforestation, boost tree growth in degraded soils, and increase resilience to climate-induced stressors. Therefore, incorporating mycorrhizal symbioses into forest management strategies offers both ecological and practical benefits.

2.1.3 Role in Nutrient Acquisition and Soil Structure

Ectomycorrhizal fungi play a crucial role in nutrient acquisition, particularly in the uptake of nitrogen and phosphorus, which are often scarce in forest soils. The hyphal networks formed by these fungi can access nutrients beyond the reach of tree roots, effectively increasing the nutrient pool available to the host (Sudharsan et al. 2023). These extraradical hyphae not only enhance the surface area for nutrient absorption but also contain specialized enzymes that mobilize organic forms of N and P, facilitating their uptake by the host plant (Wang et al. 2022a, b).

Within the root cortex, the Hartig net—a network of hyphae that penetrate root cells—serves as the leading site for nutrient exchange between the fungus and the host plant. This structure facilitates the efficient transfer of nutrients from the fungus to the plant and carbohydrates from the plant to the fungus, thereby maintaining a mutualistic relationship (Mishra et al. 2024). Additionally, EM fungi enhance soil structure by binding soil particles, which improves soil aeration and water retention.

2.2 Arbuscular Mycorrhizae (AM)

Arbuscular mycorrhizae (AM) represent a prevalent type of mycorrhizal association, distinguished by the formation of arbuscules within the root cells of host plants. These structures promote nutrient exchange between the fungus and the plant, making arbuscular mycorrhizal (AM) associations especially significant in various ecosystems, including forests.

2.2.1 Symbiotic Mechanisms with Tree Roots

The symbiotic relationship in AM involves the penetration of fungal hyphae into the root cells, where they form arbuscules to facilitate the transfer of nutrients (Wahab et al. 2023). This symbiosis is established through a series of molecular and physical interactions, starting with root exudates—chemical signals released by tree roots—that attract compatible fungal partners in the rhizosphere (Ahmed et al. 2025). Unlike arbuscular mycorrhizal fungi, EMF do not invade the host cell cytoplasm but remain in the apoplastic space between root cortical cells, allowing for the controlled exchange of nutrients and signaling molecules. In this partnership, the fungus provides enhanced access to soil-derived nutrients, such as nitrogen and phosphorus, through its extraradical mycelium. At the same time, the tree supplies the fungus with photosynthetically derived carbohydrates, primarily in the form of glucose and sucrose (Sudharsan et al. 2023). This intimate association enables efficient exchange of carbohydrates from the plant and nutrients, particularly phosphorus, from the fungus.

Molecular studies have demonstrated that this interaction is mediated by a complex network of signaling pathways, involving hormones such as auxins, cytokinins, and jasmonates, which regulate root architecture and fungal colonization (Selwal et al. 2023). Furthermore, both partners undergo transcriptional reprogramming during symbiosis, with EMF expressing a suite of symbiosis-induced genes, known as mycorrhiza-induced small secreted proteins (MiSSPs), which suppress host immune responses and facilitate colonization (Genre et al. 2020).

The functional integration of symbiosis contributes to several ecological benefits. EMF enhances the tree's tolerance to drought, heavy metals, and pathogenic attacks by modulating root exudation profiles and stimulating systemic defense mechanisms. Additionally, EMF networks can connect multiple trees of the same or different species within a forest, forming common mycorrhizal networks (CMNs) that facilitate the redistribution of resources and interplant communication (Chaudhury et al. 2024).

2.2.2 Role in Phosphorus and Water Uptake

AM fungi play a crucial role in enhancing phosphorus and water uptake in forest ecosystems. Phosphorus is a vital macronutrient for plant growth; however, it is often poorly mobile and sparsely available in forest soils due to its tendency to bind with soil particles or form insoluble compounds (Likhanova et al. 2022). Mycorrhizal fungi overcome this limitation by extending their hyphal networks far beyond the root depletion zones, allowing them to access and solubilize phosphorus from organic and inorganic pools that are otherwise unavailable to the plant. Studies have shown that trees associated with arbuscular mycorrhizal (AM) fungi can absorb significantly more phosphorus than their non-mycorrhizal counterparts (Luo et al. 2024).

These hyphae secrete organic acids (e.g., oxalate and citrate) and phosphatases that mobilize bound phosphates, enhancing their bioavailability to the host plant. Ectomycorrhizal fungi (EMF), in particular, have been shown to effectively mobilize organic phosphorus compounds through enzymatic degradation in nutrient-poor forest soils, thereby facilitating sustainable tree nutrition (Alnaass et al. 2025).

In addition to phosphorus uptake, mycorrhizal fungi significantly enhance plant water acquisition, particularly during drought conditions. The extensive hyphal network increases the surface area for water absorption, enabling plants to draw moisture from micropores that roots alone cannot reach. Moreover, mycorrhizal symbiosis affects plant water regulation by improving osmotic adjustment and root hydraulic conductivity (Abbaspour et al. 2021). AMF colonization has been linked to the regulation of aquaporin expression in plant roots, thereby facilitating water transport under conditions of limited water availability (Sharma et al. 2021).

The combined enhancement of phosphorus and water uptake by mycorrhizal fungi contributes to improved plant vigor, biomass accumulation, and resilience against environmental stressors in forest ecosystems. Thus, these symbiotic relationships are essential for the sustainability of natural forests and for optimizing forest productivity under climate variability and soil degradation (Fig. 2).

2.2.3 Host Range and Ecological Adaptation

Arbuscular mycorrhizae have a broad host range and are associated with approximately 80% of terrestrial plant species, including several tree species. This adaptability enables AM fungi to thrive in diverse ecological conditions, making them essential for maintaining forest health and resilience in changing environments (Wahab et al. 2023).

2.3 Comparison of Ectomycorrhizae and Arbuscular Mycorrhizae

Both ectomycorrhizae and arbuscular mycorrhizae enhance nutrient uptake and support tree health, yet they differ in structure, host range, function, and ecological niche. Ectomycorrhizae primarily form a protective mantle around the root tips and develop a Hartig net, which penetrates the cortical cells but not the endodermis (Priyadarshani et al. 2025). In contrast, arbuscular mycorrhizae create intracellular structures, such as arbuscules and vesicles, within the root cortex, enabling a more direct interface for nutrient exchange. Arbuscular mycorrhizal fungi (AM fungi), mainly from the phylum Glomeromycota, are associated with approximately 80% of terrestrial plant species, including most herbaceous and tropical trees (Duchicela et al. 2022). Ectomycorrhizal (EM) fungi, however, are predominantly associated with woody plants, especially in temperate and boreal forests, including genera

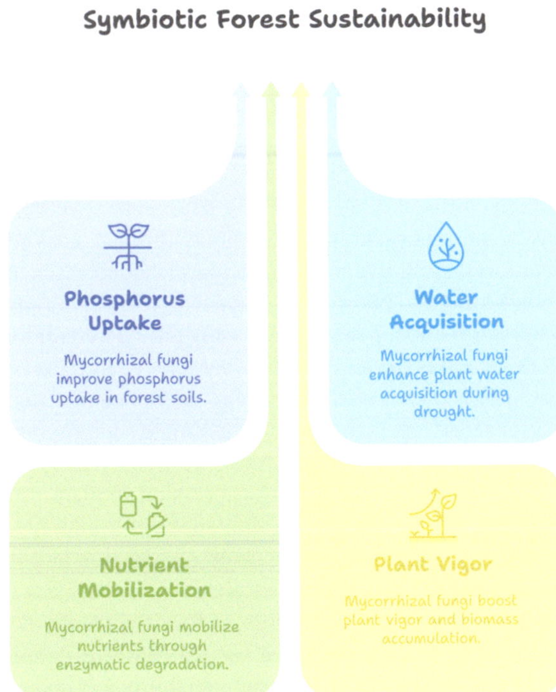

Fig. 2 Symbiotic forest sustainability

such as Pinus, Fagus, and Eucalyptus. Both types of mycorrhizal associations are vital for forest ecosystems, contributing to nutrient cycling, soil structure, and overall plant health. However, EM fungi possess additional enzymatic capabilities that enable them to access organic forms of nitrogen and phosphorus, which are more prevalent in forest soils (Tunlid et al. 2022). AM fungi primarily facilitate the uptake of inorganic phosphate and enhance the availability of micronutrients. Understanding these differences is crucial for developing effective forest management and conservation strategies that bolster forest resilience and productivity in the face of environmental challenges.

3 Ecological Contributions of Mycorrhizae in Forestry3.1. Enhancing Nutrient Cycling and Availability

Mycorrhizae play a pivotal role in enhancing nutrient cycling and availability in forest ecosystems. These symbiotic relationships between fungi and plant roots significantly increase the surface area for nutrient absorption, particularly for essential elements like phosphorus and nitrogen. Mycorrhizal fungi can mobilize nutrients from organic matter and soil minerals, making them accessible to trees. This is particularly important in nutrient-poor soils, where the presence of mycorrhizae can boost tree growth and health. Mohan and Joshi (2025) emphasized that mycorrhizal networks facilitate nutrient exchange not only between individual plants but also across different species, promoting a more interconnected and resilient forest ecosystem. Furthermore, the ability of mycorrhizae to break down organic matter contributes to the overall nutrient pool in the soil, enhancing the fertility of the forest floor (Rathna et al. 2025).

3.1 Water Absorption and Drought Tolerance in Trees

Mycorrhizal associations significantly enhance the water absorption capabilities of trees, thereby improving their resilience to drought conditions. Mycorrhizal fungi extend the root systems of trees through their hyphal networks, enabling access to moisture in deeper soil layers (Henriksson et al. 2023). According to a study, trees with mycorrhizal associations exhibit improved water uptake, which is critical during periods of low water availability. Mycorrhizae enhance the ability of trees to tolerate drought stress by mitigating the impact of water scarcity on physiological processes (Tang et al. 2022). This increased drought tolerance is crucial for maintaining forest health in changing climates, where fluctuations in precipitation are becoming more common. The capacity of mycorrhizal fungi to enhance tree-water relations is essential for sustainable forestry practices, particularly in hilly agricultural systems where water availability is often limited.

3.2 Influence on Soil Structure and Health

Mycorrhizae significantly contribute to soil structure and overall health, enhancing soil quality and fertility. Mycorrhizal fungi produce extracellular compounds that help bind soil particles together, forming stable aggregates which improve soil texture and porosity. These aggregates enhance soil aeration and water infiltration, promoting a healthier soil environment for roots and microorganisms (Ali et al. 2024). Mycorrhizal activity enhances soil organic matter content, vital for maintaining soil health and fertility. The hyphal networks created by mycorrhizal fungi also play a crucial role in preventing soil erosion, particularly in hilly terrains where soil stability is a significant concern (Wahab et al. 2023). By fostering a healthy soil ecosystem, mycorrhizae support the growth of diverse plant communities, contributing to the sustainability of forestry practices.

3.3 Mycorrhizal Impact on Forest Biodiversity and Productivity

Mycorrhizae have a profound effect on forest biodiversity and productivity. The presence of diverse mycorrhizal fungi in forest ecosystems can enhance plant species richness by facilitating the coexistence of different plant species through shared nutrient networks. Biodiversity is crucial for ecosystem resilience, allowing forests to adapt to environmental changes and disturbances (Zheng et al. 2025). Studies have shown that forests with abundant mycorrhizal communities tend to be more productive, as these fungi enhance nutrient and water availability, leading to increased tree growth and biomass accumulation. Moreover, mycorrhizal associations can influence plant community dynamics by affecting resource competition (Wang et al. 2022a, b). This interaction promotes a stable and productive ecosystem where diverse plant species can thrive, supporting a wide range of animal life and contributing to the overall ecological balance.

4 Mycorrhizae in Hilly Agriculture and Forestry

4.1 Soil Fertility Challenges in Hilly Regions

Hilly regions often face significant challenges to soil fertility, which can hinder both agricultural productivity and forest health. The steep slopes in these areas contribute to soil erosion, leading to the loss of topsoil and essential nutrients. Additionally, hilly terrains frequently exhibit poor soil structure, low organic matter content, and limited nutrient availability due to leaching and runoff. These conditions create a challenging environment for both crops and forest ecosystems, necessitating effective management strategies to enhance soil fertility and sustain productivity (Antwi et al. 2024). Research has shown that hilly regions often have acidic soils, which can further limit the availability of nutrients, particularly phosphorus. The complex terrain also makes conventional farming practices difficult, resulting in lower yields and reduced forest growth (Wahab et al. 2023). Therefore, understanding the role of mycorrhizae in addressing these challenges is crucial for promoting sustainable cultivation practices in hilly agriculture and forestry (Fig. 3).

4.2 Role of Mycorrhizae in Improving Soil Conditions

Mycorrhizae play a significant role in improving soil conditions by enhancing fertility and promoting healthy root systems. These symbiotic fungi boost nutrient uptake, particularly phosphorus, which is often limited in hilly soils. The mycorrhizal networks formed between fungi and plant roots increase the surface area for absorption, allowing trees and crops to access nutrients beyond their immediate root systems (Zhao and Riaz 2024). Furthermore, mycorrhizal fungi contribute to soil aggregation, enhancing soil structure and aeration. This aggregation helps retain moisture and reduces erosion, which are crucial factors in hilly regions prone to runoff. Mycorrhizal associations can increase soil organic matter content, thereby enhancing overall health and fertility. By promoting beneficial soil conditions, mycorrhizae support the growth of diverse plant species, which are essential for maintaining ecosystem balance in hilly forestry (Yang et al. 2022a, b).

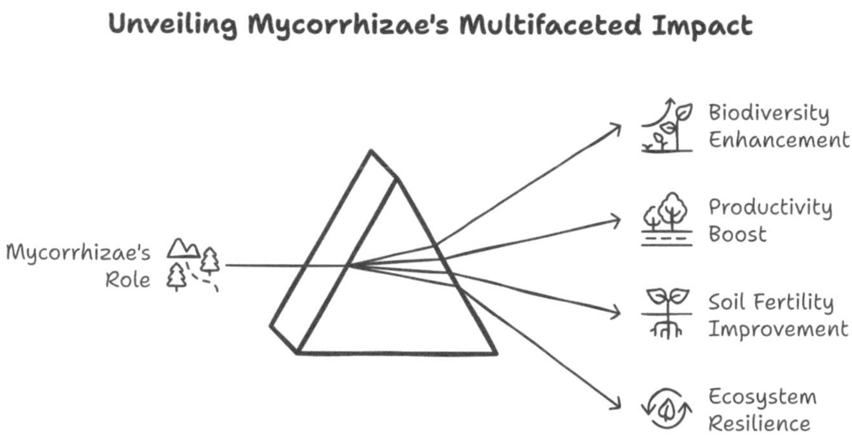

Fig. 3 Unveiling the mycorrhizae's multifaceted impact

4.3 Water Management in Hilly Forests Through Mycorrhizal Symbiosis

Efficient water management is crucial in hilly forests, where water availability can fluctuate significantly due to topography and climate. Mycorrhizal fungi enhance the water absorption capabilities of trees, enabling them to access moisture from deeper soil layers. This is particularly important during dry periods when surface water may be scarce (Wu et al. 2024a, b). Studies suggest that mycorrhizal associations increase tree resilience to drought stress by improving their water relations. The extensive hyphal networks formed by mycorrhizal fungi act as bridges between the soil and tree roots, facilitating the movement and retention of water in the soil. This enhanced water management capability is vital for sustaining forest ecosystems in hilly regions where trees face the dual challenges of steep slopes and variable moisture availability (Singh et al. 2022).

4.4 Case Studies: Mycorrhizal Impact in Hill-Based Forestry Ecosystems

Numerous case studies have highlighted the positive impacts of mycorrhizae on hill-based forestry ecosystems. For instance, a survey conducted in the Appalachian Mountains demonstrated that afforestation efforts utilizing mycorrhizal inoculation significantly improved tree survival and growth rates on eroded slopes (Gunathunga et al. 2023). This finding suggests that promoting mycorrhizal associations can be an effective strategy for restoring degraded hilly landscapes. Another case study in the Western Ghats of India found that native tree species with established mycorrhizal relationships exhibited better nutrient uptake and overall growth compared to non-mycorrhizal counterparts. This research underscores the importance of

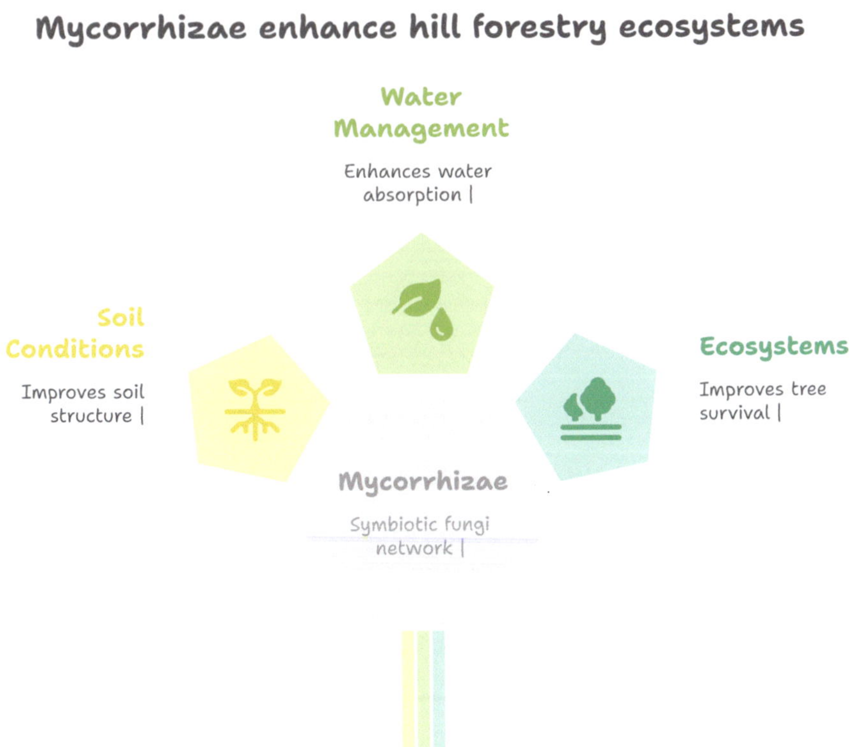

Fig. 4 Mycorrhiza enhances hill forestry ecosystems

preserving native mycorrhizal communities in forest management practices, particularly in biodiverse hilly regions. Additionally, a meta-analysis by Thanni et al. (2022) confirmed that forests with higher mycorrhizal diversity tend to exhibit greater resilience and productivity. This emphasizes the need for sustainable forestry practices that support and enhance mycorrhizal associations to foster ecological balance and productivity in hilly agriculture (Fig. 4).

5 Mycorrhizae and Tree Resilience to Environmental Stresses

5.1 Drought Tolerance and Soil Degradation Resistance

Mycorrhizal fungi significantly enhance tree drought tolerance and resistance to soil degradation, which are critical for maintaining forest health in hilly regions. These fungi extend the root systems of trees through their hyphal networks, allowing greater access to water and nutrients, particularly during periods of water scarcity. Research indicates that trees with mycorrhizal associations can better withstand drought conditions than non-mycorrhizal plants due to their enhanced water retention and nutrient absorption capabilities. Furthermore, mycorrhizal fungi contribute to soil structure by promoting soil aggregation and improving soil health (Wahab et al. 2023). This enhanced soil structure reduces erosion and degradation, ensuring that the soil remains fertile and capable of supporting tree growth. Mycorrhizal associations can significantly increase the root biomass and surface area of trees, which not only aid in water uptake but also stabilize the soil, thereby resisting degradation (Wahab et al. 2023).

5.2 Managing Soil Salinity and Heavy Metal Stress

Soil salinity and heavy metal contamination pose significant environmental stresses that can adversely affect tree health and productivity, particularly in hilly agricultural systems. Mycorrhizal fungi have been shown to enhance resilience against these stresses. They help mitigate the effects of soil salinity by improving nutrient uptake and osmotic balance within plants (Ahmed et al. 2025). Mycorrhizal fungi can also modify the composition of root exudates, which assists in mobilizing nutrients and reducing the toxic effects of saline conditions. Regarding heavy metal stress, mycorrhizal fungi can aid in detoxifying metals in the soil (Wahab et al. 2023). Studies have demonstrated that mycorrhizal associations boost the tolerance of trees to heavy metals by facilitating their uptake and sequestration in fungal tissues, thereby decreasing the metal concentration in the host plant (Zheng et al. 2023a, b). This detoxification mechanism is crucial in contaminated sites, enabling trees to survive and thrive in conditions that would otherwise threaten their health.

5.3 Mitigating the Impact of Climate Change on Forest Ecosystems

Climate change poses significant threats to forest ecosystems, including rising temperatures, altered precipitation patterns, and an increased frequency of extreme weather events. Mycorrhizae play a crucial role in enhancing tree resilience to climate change. By improving nutrient and water uptake, mycorrhizal fungi enable trees to adapt to changing environmental conditions (Chaudhury et al. 2024). Research indicates that trees associated with mycorrhizal fungi demonstrate better growth and survival rates under climate stress compared to those without such associations. Additionally, mycorrhizal networks can enhance carbon sequestration in forest soils, which is essential for mitigating the effects of climate change. By promoting soil health and supporting diverse plant communities, mycorrhizae contribute to forest stability and resilience, enabling ecosystems to better cope with the challenges posed by climate change (Wahab et al. 2023).

6 Practical Applications of Mycorrhizae in Forestry Management

6.1 Use of Mycorrhizae in Reforestation Programs

Mycorrhizal fungi play a crucial role in the success of reforestation programs by enhancing the survival and growth of seedlings. The incorporation of mycorrhizal inoculants in reforestation efforts has been shown to improve nutrient uptake, particularly phosphorus uptake, which is often limited in degraded soils (Islam et al. 2024). Reforestation projects utilizing mycorrhizal inoculation yielded significantly higher survival rates and seedling growth compared to those without inoculation. Moreover, mycorrhizal associations can enhance the resilience of newly planted trees to environmental stresses such as drought and soil degradation. For instance, seedlings inoculated with mycorrhizal fungi exhibit improved drought tolerance and nutrient availability, which are critical factors for successful reforestation in challenging environments (Branco et al. 2022).

6.2 Restoration of Degraded Lands with Mycorrhizal Inoculation

Restoring degraded lands through mycorrhizal inoculation has gained attention as a sustainable management practice. Mycorrhizal fungi can facilitate soil rehabilitation by improving soil structure, enhancing nutrient cycling, and increasing organic matter content. This is particularly important in areas affected by mining, agriculture, or deforestation, where soil fertility is severely compromised (Kumar et al. 2022). Research has shown that applying mycorrhizal inoculants can accelerate the recovery of degraded soils by promoting the establishment of vegetation. Mycorrhizal inoculation not only enhances plant growth but also improves soil microbial diversity, which is crucial for maintaining long-term soil health (Chauhan et al. 2023). Therefore, incorporating mycorrhizal fungi into restoration strategies is a promising approach for enhancing ecosystem recovery in degraded landscapes.

6.3 Mycorrhizae in the Growth of Commercially Important Tree Species

Mycorrhizal fungi are essential for the successful growth of commercially important tree species such as eucalyptus, pine, and oak. These species often rely on mycorrhizal associations to access nutrients and water in soils that are nutrient-poor. Studies have demonstrated that mycorrhizal inoculation can significantly enhance growth rates and biomass accumulation in these species (Xue et al. 2022). For example, the eucalyptus seedlings inoculated with mycorrhizal fungi exhibited increased growth and improved nutrient uptake compared to non-inoculated controls. This is particularly beneficial in plantation forestry, where maximizing growth rates is critical for economic returns. Therefore, the application of mycorrhizal inoculants can be integrated into forestry practices to ensure healthy growth of commercially valuable tree species (Sudheer et al. 2023).

6.4 Integration into Plantation Forestry: Challenges and Solutions

Although the benefits of mycorrhizal fungi in forestry management are well documented, challenges also arise from their integration into plantation forestry. One primary challenge is the variability of mycorrhizal communities across different soil types and environmental conditions, which can affect inoculation effectiveness (Wahab et al. 2023). Furthermore, the successful establishment of mycorrhizal associations can be hindered by soil disturbances, such as tillage and compaction, which disrupt fungal networks. To overcome these challenges, it is essential to conduct site-specific assessments to understand the existing mycorrhizal communities and tailor inoculation strategies accordingly (Pedrinho et al. 2024). Moreover, educating forest managers and stakeholders about the significance of mycorrhizae in plantation forestry can encourage sustainable practices that enhance mycorrhizal associations (Fig. 5). Developing best management practices that minimize soil disturbance and promote healthy mycorrhizal communities will be crucial for successfully integrating mycorrhizae into forestry management.

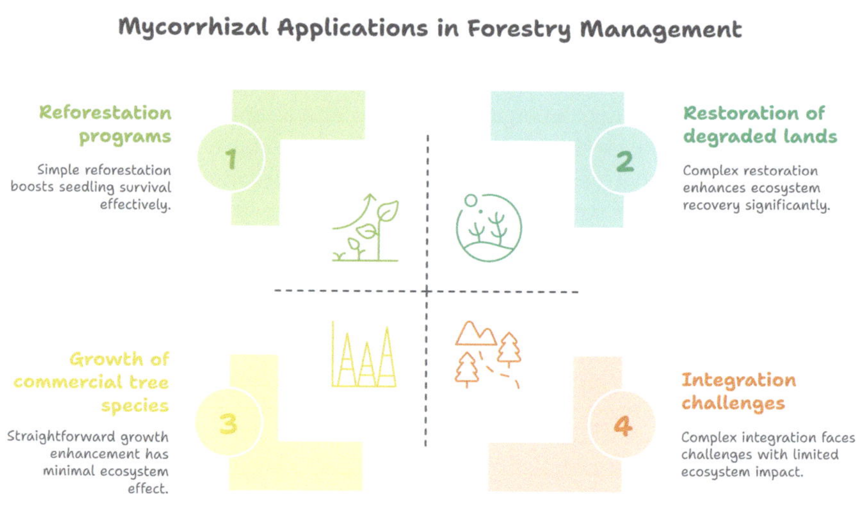

Fig. 5 Mycorrhizal applications in forestry management

7 Mechanisms of Mycorrhizal Influence on Soil Health

7.1 Fungal-Root Interactions and Soil Carbon Sequestration

Mycorrhizal fungi play a significant role in enhancing soil carbon sequestration through their interactions with plant roots. These fungi form extensive networks of hyphae that connect plant roots to the soil, facilitating the transfer of carbon compounds from plants to the soil (Wang et al. 2022a, b). As plants photosynthesize, they allocate a portion of the carbon fixed to their roots, which is subsequently transferred to mycorrhizal fungi. In return, fungi provide essential nutrients to plants, creating a mutualistic relationship that enhances both carbon storage and nutrient availability. Research indicates that mycorrhizal fungi contribute to the formation of stable soil organic matter through the production of glomalin, a glycoprotein that binds soil particles. This process not only increases soil carbon stocks but also enhances soil structure and fertility. Bhattacharyya et al. 2022 highlighted that ecosystems with diverse mycorrhizal communities exhibit greater soil carbon storage potential, emphasizing the importance of mycorrhizal fungi in long-term carbon sequestration strategies.

7.2 Mycorrhizal Contributions to Soil Microbial Diversity

Mycorrhizal fungi significantly enhance soil microbial diversity, which is crucial for maintaining soil health and ecosystem functionality. The presence of mycorrhizae creates a favorable environment for various soil microorganisms, including bacteria, archaea, and other fungi. This is because mycorrhizal networks provide organic carbon and other nutrients that support the growth of diverse microbial communities. Additionally, the interactions between mycorrhizal fungi and soil microbes can lead to increased microbial biomass and activity, which in turn contributes to nutrient cycling and organic matter decomposition (Wu et al. 2024a, b). Studies have shown that soils enriched with mycorrhizal fungi exhibit higher microbial diversity and activity, leading to enhanced soil health and resilience. Microbial diversity is essential for ecosystem stability, allowing the soil to adapt to environmental changes and stresses (Pedrinho et al. 2024).

7.3 Mycorrhizal Mediation in Organic Matter Decomposition

Mycorrhizal fungi play a crucial role in the decomposition of organic matter, which is essential for nutrient cycling and maintaining soil fertility. These fungi interact with decomposing plant materials, facilitating the breakdown of complex organic compounds into simpler forms that can be utilized by plants and other soil organisms (Tunlid et al. 2022). Research has shown that mycorrhizal associations can enhance the rate of organic matter decomposition by increasing the efficiency of nutrient uptake and promoting the activity of soil microorganisms involved in the decomposition processes. Mycorrhizal fungi not only aid in the breakdown of organic substrates but also enhance the availability of nutrients released during decomposition, thereby improving soil health and fertility.

7.4 Impact on Nutrient Transformation in Forest Soils

Mycorrhizal fungi have a significant influence on nutrient transformation in forest soils, particularly with respect to essential nutrients such as nitrogen and phosphorus. Through their extensive hyphal networks, these fungi enhance nutrient uptake by extending the root system's reach, enabling plants to access nutrients that would otherwise remain unavailable (Wahab et al. 2023). Research indicates that mycorrhizal fungi facilitate nitrogen transformation processes, including nitrogen mineralization and immobilization, which are critical for maintaining soil fertility. Additionally, mycorrhizal associations are particularly beneficial in phosphorus-poor soils, where they facilitate the mobilization of phosphorus from soil particles and organic matter. Mycorrhizal inoculation significantly enhances nutrient availability and uptake in forest ecosystems, underscoring the crucial role of these fungi in nutrient cycling and forest productivity (Pena and Tibbett 2024).

8 Research Advances on Mycorrhizal Symbiosis in Forestry

8.1 Insights from Recent Studies on Mycorrhizal Fungi

Recent studies have significantly advanced our understanding of mycorrhizal fungi and their ecological role in forestry. Research has shown that mycorrhizal fungi play a crucial role in nutrient uptake, enhancing soil structure, and promoting plant health. These fungi form symbiotic relationships with a wide range of plant species, facilitating access to essential nutrients such as phosphorus and nitrogen, which are often limiting in forest ecosystems (Wahab et al. 2023). Moreover, previous studies have highlighted the significance of mycorrhizal diversity in influencing plant community dynamics and ecosystem resilience. For instance, the presence of diverse mycorrhizal fungi can enhance plant growth and survival under various environmental stresses, including drought and nutrient deficiency. This adaptability is crucial in the context of climate change, where forest ecosystems are experiencing increasing pressure (Falk et al. 2022). Innovative research has focused on the molecular mechanisms underlying mycorrhizal interactions. Recent findings suggest that specific signaling pathways and gene expression play a crucial role in establishing and maintaining mycorrhizal symbiosis, offering valuable insights into how plants and fungi communicate and cooperate (Wahab et al. 2023).

8.2 Innovative Techniques for Mycorrhizal Cultivation and Application

The cultivation and application of mycorrhizal fungi have led to significant innovations in recent years. Advances in biotechnology have enabled the development of effective mycorrhizal inoculants that can be used in forestry to enhance tree growth and resilience. These inoculants were designed to strengthen the establishment of mycorrhizal associations in various soil types, thereby optimizing nutrient uptake and improving plant health. Researchers are currently exploring the use of bioengineering techniques to enhance the efficacy of mycorrhizal fungi in forestry (Joshi et al. 2025). For instance, the application of microbial consortia that include mycorrhizal fungi and beneficial bacteria has been shown to improve plant growth and soil health in a synergistic manner. This approach not only enhances nutrient availability but also promotes soil microbial diversity, which is essential for maintaining ecosystem stability (Pedrinho et al. 2024). The integration of mycorrhizal fungi into precision agriculture practices is gaining traction. By utilizing soil health assessments and targeted inoculation strategies, forest managers can optimize the benefits of mycorrhizal associations, leading to improved forest productivity and sustainability.

8.3 Role of Genetic Engineering in Enhancing Mycorrhizal Efficiency

Genetic engineering is emerging as a promising tool to enhance the efficiency of mycorrhizal symbiosis in forestry. Recent studies have focused on identifying and manipulating specific genes involved in mycorrhizal interactions to improve plant responses to mycorrhizal colonization. For instance, researchers have successfully enhanced the expression of genes related to nutrient transport and signaling pathways, resulting in improved nutrient uptake and enhanced plant growth in mycorrhizal plants (Mitra et al. 2023). Moreover, genetic engineering can be used to develop tree species that are more compatible with beneficial mycorrhizal fungi, thereby optimizing the benefits of these symbiotic relationships. This approach has the potential to revolutionize forestry practices by creating tree varieties that are more resilient to environmental stresses and better suited for specific soil conditions. Additionally, the application of CRISPR technology in forestry research paves the way for precise modifications that can enhance mycorrhizal efficiency. By targeting genes associated with mycorrhizal interactions, researchers can develop tree species that exhibit improved growth rates and stress tolerance, ultimately contributing to more sustainable forestry practices (Cao et al. 2024).

9 Policy and Management Implications

9.1 Role of Mycorrhizae in Forestry Policy Development

The integration of mycorrhizal research into forestry policy development is crucial for enhancing sustainable forest management practices. Policymakers are increasingly recognizing the significance of mycorrhizal fungi in promoting forest health, resilience, and productivity. Scientific evidence suggests that mycorrhizae play a crucial role in nutrient cycling, soil health, and plant stress tolerance, which can inform policies aimed at conserving forest ecosystems and promoting sustainable practices (Wahab et al. 2023). Recent studies have shown that mycorrhizal fungi can significantly enhance the effectiveness of reforestation and afforestation programs, thereby supporting carbon sequestration efforts. Policies that encourage the use of mycorrhizal inoculants in these programs can lead to more successful tree establishment and growth, thereby contributing to climate change mitigation goals (Ghorui et al. 2024). Furthermore, incorporating mycorrhizae into land-use planning and forest restoration strategies can enhance biodiversity and ecosystem services.

9.2 Integration into National and Regional Forestry Programs

The integration of mycorrhizal management into national and regional forestry programs is essential for realizing the full potential of these symbiotic relationships. Governments and forestry agencies can develop guidelines and frameworks to promote the use of mycorrhizal fungi in forestry practices, including afforestation, reforestation, and agroforestry initiatives (Mishenin et al. 2024). For instance, national forestry programs can include training for forest managers and practitioners on the benefits and application of mycorrhizal inoculants. Additionally, funding and support for research on local mycorrhizal communities can help identify the most effective fungal species for different forest types and conditions. By integrating mycorrhizal considerations into regional forestry policies, stakeholders can enhance forest resilience, improve soil health, and increase forest productivity (Ghorui et al. 2024).

9.3 Strategies for Promoting Mycorrhizal Use in Sustainable Forestry

Several strategies can be employed to encourage the use of mycorrhizae in sustainable forestry effectively.

9.3.1 Awareness and Education

Raising awareness of the benefits of mycorrhizal fungi among forest managers, landowners, and the general public is essential. Educational programs can highlight the ecological and economic benefits of mycorrhizal associations (Purahong et al. 2022).

9.3.2 Research and Development

Continued investment in research is essential to explore the various roles of mycorrhizae in forest ecosystems. This includes understanding species-specific interactions, the effects of mycorrhizae on tree health and productivity, and their potential for biotechnological applications (Dyshko et al. 2024).

9.3.3 Policy Incentives

Governments can create incentives for landowners and forest managers to adopt practices that enhance mycorrhizal associations, such as subsidies for mycorrhizal inoculants or funding for sustainable land management practices (Sudheer et al. 2023).

9.3.4 Collaborative Approaches

Engaging multiple stakeholders, including researchers, practitioners, and policymakers, can foster collaborative approaches for integrating mycorrhizal management into forestry practices. Partnerships between the public and private sectors can facilitate knowledge sharing and resource allocation (Mishenina and Dvorak 2022).

9.3.5 Monitoring and Evaluation

Establishing monitoring frameworks to assess the impact of mycorrhizal management practices on forest health and productivity is essential. These data can inform adaptive management strategies and policy adjustments (Wahab et al. 2023).

By implementing these strategies, the forestry sector can capitalize on the benefits of mycorrhizal fungi, thereby contributing to sustainable forest management and the overall health of forest ecosystems.

10 Conclusion

10.1 Summary of Mycorrhizal Contributions to Forestry Ecosystems

Mycorrhizal fungi play a pivotal role in enhancing the health and productivity of forest ecosystems. The symbiotic relationships between fungi and plant roots significantly contribute to soil health, nutrient cycling, and overall forest resilience. Mycorrhizae enhance tree growth by expanding the root system's capacity to access water and essential nutrients, particularly phosphorus and nitrogen, which are often limited in forest soils. Furthermore, mycorrhizal fungi improve soil structure by forming stable aggregates that promote aeration and water retention, ultimately leading to healthier soil ecosystems. In addition to their direct benefits to tree growth, mycorrhizal associations support biodiversity and ecosystem stability. By facilitating interactions among various plant species, mycorrhizae help maintain diverse plant communities that are more resilient to environmental changes and stress. Moreover, these fungi play a crucial role in carbon sequestration, helping mitigate the impacts of climate change by storing carbon in both plant biomass and soil organic matter.

10.2 Future Directions for Mycorrhizal Research in Forestry

As our understanding of mycorrhizal symbiosis continues to evolve, several future research directions are essential to harness the full potential of mycorrhizae in forestry. First, extensive studies on the functional diversity of mycorrhizal fungi in various forest ecosystems are needed. Understanding the specific roles of different mycorrhizal species can help identify the most effective inoculants for particular forestry applications. Second, incorporating molecular tools and genomic studies into mycorrhizal research can offer insights into the genetic mechanisms behind mycorrhizal interactions. This knowledge will facilitate the development of tree varieties that are more compatible with beneficial mycorrhizal fungi, thus optimizing their growth and resilience. Ultimately, further research is necessary to understand the impact of environmental changes, including climate change, land use alterations, and pollution, on mycorrhizal communities and their associated functions. This understanding is crucial for developing adaptive management strategies in forestry that utilize the benefits of mycorrhizal fungi.

10.3 Call for Sustainable Forestry Practices Based on Mycorrhizal Symbiosis

In conclusion, there is a pressing need to adopt sustainable forestry practices that prioritize mycorrhizal symbiosis. Policymakers, forest managers, and practitioners must acknowledge the vital role of mycorrhizal fungi in maintaining forest health and productivity. Implementing practices such as mycorrhizal inoculation during reforestation and afforestation efforts, minimizing soil disturbance, and promoting biodiversity can significantly enhance the resilience and sustainability of forest ecosystems. By fostering a deeper understanding of mycorrhizal relationships and their ecological significance, we can create more effective management strategies that support the long-term health of forests.

References

Abbaspour H, Pour FS, Abdel-Wahhab MA (2021) Arbuscular mycorrhizal symbiosis regulates the physiological responses, ion distribution, and relevant gene expression to trigger salt stress tolerance in pistachio. Physiol Mol Biol Plants 27(8):1765–1778. https://doi.org/10.1007/s12298-021-01043-w

Adnan M, Islam W, Gang L, Chen HY (2022) Advanced research tools for fungal diversity and its impact on the forest ecosystem. Environ Sci Pollut Res 29(30):45044–45062. https://doi.org/10.1007/s11356-022-20317-8

Ahmed N, Li J, Li Y, Deng L, Deng L, Chachar M, Chachar Z, Chachar S, Hayat F, Raza A, Umrani JH (2025) Symbiotic synergy: how arbuscular mycorrhizal fungi enhance nutrient uptake, stress tolerance, and soil health through molecular mechanisms and hormonal regulation. IMA Fungus 16:e144989. https://doi.org/10.3897/imafungus.16.144989

Ali N, Abbas SAAA, Sharif L, Shafiq M, Kamran Z, Haseeb M, Shahid MA (2024) Microbial extracellular polymeric substance and impacts on soil aggregation. In: Bacterial secondary metabolites. Elsevier, pp 221–237. https://doi.org/10.1016/B978-0-323-95251-4.00021-1

Alnaass NS, Trabalsiy ZBK, Abubaira MH, Alyaseer NA (2025) Fungi as nature's recyclers optimizing nutrient dynamics in agroecosystems. Afr J Adv Pure App Sci 4(1):110–126

Antwi RA, Chappell C, Twumasi YA, Anning DK, Oppong J, Annan JB, Ahoma G, Atayi J, Ferchaud V (2024) Fertility and management strategies of soils in rural and urban forest ecosystems: a review of selected rural and urban forests in Ghana and USA. Geol Ecol Landsc 1–12. https://doi.org/10.1080/24749508.2024.2334512

Basiru S, Hijri M (2024) Trade-off between soil organic carbon sequestration and plant nutrient uptake in arbuscular mycorrhizal symbiosis. Fungal Biol Rev 49:100381. https://doi.org/10.1016/j.fbr.2024.100381

Bhattacharyya SS, Ros GH, Furtak K, Iqbal HM, Parra-Saldívar R (2022) Soil carbon sequestration–an interplay between soil microbial community and soil organic matter dynamics. Sci Total Environ 815:152928. https://doi.org/10.1016/j.scitotenv.2022.152928

Branco S, Schauster A, Liao HL, Ruytinx J (2022) Mechanisms of stress tolerance and their effects on the ecology and evolution of mycorrhizal fungi. New Phytol 235(6):2158–2175. https://doi.org/10.1111/nph.18308

Cao HX, Michels D, Vu GTH, Gailing O (2024) Applications of CRISPR technologies in forestry and molecular wood biotechnology. Int J Mol Sci 25(21):11792. https://doi.org/10.3390/ijms252111792

Cargill RI, Shimizu TS, Kiers ET, Kokkoris V (2025) Cellular anatomy of arbuscular mycorrhizal fungi. Curr Biol 35(11):R545–R562. https://doi.org/10.1016/j.cub.2025.03.053

Chaudhury R, Chakraborty A, Rahaman F, Sarkar T, Dey S, Das M (2024) Mycorrhization in trees: ecology, physiology, emerging technologies and beyond. Plant Biol 26(2):145–156. https://doi.org/10.1111/plb.13613

Chauhan P, Sharma N, Tapwal A, Kumar A, Verma GS, Meena M, Seth CS, Swapnil P (2023) Soil microbiome: diversity, benefits and interactions with plants. Sustainability 15(19):14643. https://doi.org/10.3390/su151914643

Dou T, Zhang K, Shi X, Liu W, Yu F, Liu D (2025) Crop–mushroom rotation: a comprehensive review of its multifaceted impacts on soil quality, agricultural sustainability, and ecosystem health. Agronomy 15(3):563. https://doi.org/10.3390/agronomy15030563

Duchicela J, Valdivieso A, Prado-Vivar B, Arévalo-Granda V, Hickey-Darquea A, Hof PVT (2022) Diversity of arbuscular mycorrhizal fungi in the Ecuadorian Amazon region. In: Mycorrhizal fungi in South America: biodiversity, conservation, and sustainable food production. Springer, Cham, pp 141–170. https://doi.org/10.1007/978-3-031-12994-0_7

Dyshko V, Hilszczańska D, Davydenko K, Matić S, Moser WK, Borowik P, Oszako T (2024) An overview of mycorrhiza in pines: research, species, and applications. Plants 13(4):506. https://doi.org/10.3390/plants13040506

Fahad S, Chavan SB, Chichaghare AR, Uthappa AR, Kumar M, Kakade V, Pradhan A, Jinger D, Rawale G, Yadav DK, Kumar V (2022) Agroforestry systems for soil health improvement and maintenance. Sustainability 14(22):14877. https://doi.org/10.3390/su142214877

Falk DA, van Mantgem PJ, Keeley JE, Gregg RM, Guiterman CH, Tepley AJ, Young DJ, Marshall LA (2022) Mechanisms of forest resilience. For Ecol Manag 512:120129. https://doi.org/10.1016/j.foreco.2022.120129

Genre A, Lanfranco L, Perotto S, Bonfante P (2020) Unique and common traits in mycorrhizal symbioses. Nat Rev Microbiol 18(11):649–660. https://doi.org/10.1038/s41579-020-0402-3

Ghorui M, Chowdhury S, Balu P, Burla S (2024) Arbuscular Mycorrhizal inoculants and its regulatory landscape. Heliyon 10(9). https://doi.org/10.1016/j.heliyon.2024.e30359

Gunathunga SU, Gagen EJ, Evans PN, Erskine PD, Southam G (2023) Anthropedogenesis in coal mine overburden; the need for a comprehensive, fundamental biogeochemical approach. Sci Total Environ 892:164515. https://doi.org/10.1016/j.scitotenv.2023.164515

Henriksson N, Marshall J, Högberg MN, Högberg P, Polle A, Franklin O, Näsholm T (2023) Re-examining the evidence for the mother tree hypothesis–resource sharing among trees via ectomycorrhizal networks. New Phytol 239(1):19–28. https://doi.org/10.1111/nph.18935

Imran, Ortas I (2025) Mechanism in soil health improvement with soil microbial actions and its role in potential agriculture. Commun Soil Sci Plant Anal 56(13):1–22. https://doi.org/10.1080/00103624.2025.2486597

Islam W, Zeng F, Alotaibi MO, Khan KA (2024) Unlocking the potential of soil microbes for sustainable desertification management. Earth Sci Rev 104738. https://doi.org/10.1016/j.earscirev.2024.104738

Javeed HM, Ali M, Zamir MS, Qamar R, Mubeen M, Shahzad M, Khalid S, Sabagh AE (2023) Ectomycorrhizal fungi: role as bio-fertilizers in forestry. In: Biofertilizers for sustainable soil management. CRC Press, pp 167–177

Johnson D, Liu X, Burslem DF (2023) Symbiotic control of canopy dominance in subtropical and tropical forests. Trends Plant Sci 28(9):995–1003. https://doi.org/10.1016/j.tplants.2023.03.027

Joshi D, Kaushik A, Kumar R, Arya A, Santoyo G, Singh VK, Kashyap N, Solanki MK, Kumari M, Bhardwaj N, Kumar A (2025) Improving plant performance through microbiome manipulation: the potential role of current bioengineering approaches. Bacteria 4(1):12. https://doi.org/10.3390/bacteria4010012

Kashyap P, Sharma I, Kashyap S, Agarwala N (2024) Arbuscular mycorrhizal fungi (AMF)-mediated control of foliar fungal diseases. In: Arbuscular mycorrhizal fungi and higher plants: fundamentals and applications. Springer, Singapore, pp 193–223. https://doi.org/10.1007/978-981-99-8220-2_9

Kumar R, Kumar A, Saikia P (2022) Deforestation and forests degradation impacts on the environment. In: Environmental degradation: challenges and strategies for mitigation. Springer, Cham, pp 19–46. https://doi.org/10.1007/978-3-030-95542-7_2

Likhanova IA, Deneva SV, Kholopov YV, Kuznetsova EG, Shakhtarova OV, Lapteva EM (2022) The effect of hydromorphism on soils and soil organic matter during the primary succession processes of forest vegetation on ancient alluvial sands of the European north-east of Russia. Forests 13(2):230. https://doi.org/10.3390/f13020230

Liu X, Zhao PS, Gao GL, Ren Y, Ding GD, Zhang Y (2023) Growing season stage determines the stability of root symbiotic and pathogenic fungi associated with *Pinus sylvestris* var. *mongolica* in a semi-arid desert. Appl Soil Ecol 190:104993. https://doi.org/10.1016/j.apsoil.2023.104993

Luo Y, Ma L, Feng Q, Luo H, Chen C, Wang S, Yuan Y, Liu C, Cao X, Li N (2024) Influence and role of fungi, bacteria, and mixed microbial populations on phosphorus acquisition in plants. Agriculture 14(3):358. https://doi.org/10.3390/agriculture14030358

Mishenin Y, Yarova I, Koblianska I (2024) Sustainable spatial agroforestry in the context of forestry globalization: strategic guidelines. In: Agroforestry, pp 155–198. https://doi.org/10.1002/9781394231164.ch6

Mishenina H, Dvorak J (2022) Public–private partnership as a form of ensuring sustainable development of the forest management sphere. Adm Sci 12(4):156. https://doi.org/10.3390/admsci12040156

Mishra S, Srivastava A, Singh A, Pandey GC, Srivastava G (2024) An overview of symbiotic and pathogenic interactions at the fungi-plant interface under environmental constraints. Front Fungal Biol 5:1363460. https://doi.org/10.3389/ffunb.2024.1363460

Mitra D, Nayeri FD, Sansinenea E, Ortiz A, Bhatta BB, Adeyemi NO, Janeeshma E, Tawfeeq Al-Ani LK, Sharma SB, Boutaj H, Priyadarshini A (2023) Unraveling arbuscular mycorrhizal fungi interaction in rice for plant growth development and enhancing phosphorus use efficiency through recent development of regulatory genes. J Plant Nutr 46(13):3184–3220. https://doi.org/10.1080/01904167.2023.2191638

Mohan VK, Joshi SR (2025) Geobiomineralization at the interface of plant–microbe interaction and the role of mycorrhizal fungi. In: Mineral transformation and bioremediation by geomicrobes. Springer, Singapore, pp 369–391. https://doi.org/10.1007/978-981-96-3033-2_14

Nizamani MM, Hughes AC, Qureshi S, Zhang Q, Tarafder E, Das D, Acharya K, Wang Y, Zhang ZG (2024) Microbial biodiversity and plant functional trait interactions in multifunctional ecosystems. Appl Soil Ecol 201:105515. https://doi.org/10.1016/j.apsoil.2024.105515

Pedrinho A, Mendes LW, de Araujo Pereira AP, Araujo ASF, Vaishnav A, Karpouzas DG, Singh BK (2024) Soil microbial diversity plays an important role in resisting and restoring degraded ecosystems. Plant Soil 500(1):325–349. https://doi.org/10.1007/s11104-024-06489-x

Pena R, Tibbett M (2024) Mycorrhizal symbiosis and the nitrogen nutrition of forest trees. Appl Microbiol Biotechnol 108(1):461. https://doi.org/10.1007/s00253-024-13298-w

Pérez-Pazos E, Certano A, Gagne J, Lebeuf R, Siegel N, Nguyen N, Kennedy PG (2021) The slippery nature of ectomycorrhizal host specificity: Suillus fungi associated with novel pinoid (Picea) and abietoid (Abies) hosts. Mycologia 113(5):891–901. https://doi.org/10.1080/00275514.2021.1921525

Priyadarshani TDC, Weerasinghe PA, Law S, Zakeel MCM (2025) Hidden allies: exploring the cutting-edge insights into the role of mycorrhizal fungi in plant health. In: Fungal endophytes volume II: applications in agroecosystems and plant protection. Springer, Singapore, pp 551–585. https://doi.org/10.1007/978-981-97-8804-0_19

Purahong W, Günther A, Gminder A, Tanunchai B, Gossner MM, Buscot F, Schulze ED (2022) City life of mycorrhizal and wood-inhabiting macrofungi: importance of urban areas for maintaining fungal biodiversity. Landsc Urban Plan 221:104360. https://doi.org/10.1016/j.landurbplan.2022.104360

Rathna V, Prasanna SL, Anandakumar J, Swathi B, Patil SS, Kujur RS (2025) Forest alchemists: microbe-mediated nutrient recycling. In: Tree biology and biotechnology. Springer, Singapore, pp 67–86. https://doi.org/10.1007/978-981-96-0002-1_5

Sachsenmaier L, Schnabel F, Dietrich P, Eisenhauer N, Ferlian O, Quosh J, Richter R, Wirth C (2024) Forest growth resistance and resilience to the 2018–2020 drought depend on tree diversity and mycorrhizal type. J Ecol 112(8):1787–1803. https://doi.org/10.1111/1365-2745.14360

Saha S, Huang L, Khoso MA, Wu H, Han D, Ma X, Poudel TR, Li B, Zhu M, Lan Q, Sakib N (2023) Fine root decomposition in forest ecosystems: an ecological perspective. Front Plant Sci 14:1277510. https://doi.org/10.3389/fpls.2023.1277510

Selwal N, Wani AK, Akhtar N, Kaur M, Jassal PS (2023) Molecular insights of strigolactone biosynthesis, signalling pathways, regulatory roles, and hormonal crosstalks in plant systems. S Afr J Bot 160:9–22. https://doi.org/10.1016/j.sajb.2023.06.046

Sharma K, Gupta S, Thokchom SD, Jangir P, Kapoor R (2021) Arbuscular mycorrhiza-mediated regulation of polyamines and aquaporins during abiotic stress: deep insights on the recondite players. Front Plant Sci 12:642101. https://doi.org/10.3389/fpls.2021.642101

Singh M, Sinha B, Bisaria J, Thomas T, Srivastava P (2022) Understanding synergies and tradeoffs between forests, water, and climate change. Wiley Interdiscip Rev Water 9(6):e1614. https://doi.org/10.1002/wat2.1614

Sudharsan MS, Rajagopal K, Banu N (2023) An insight into fungi in forest ecosystems. In: Plant mycobiome: diversity, interactions and uses. Springer, Cham, pp 291–318

Sudheer S, Johny L, Srivastava S, Adholeya A (2023) The trade-in-trade: multifunctionalities, current market and challenges for arbuscular mycorrhizal fungal inoculants. Symbiosis 89(3):259–272. https://doi.org/10.1007/s13199-023-00905-z

Tang H, Hassan MU, Feng L, Nawaz M, Shah AN, Qari SH, Liu Y, Miao J (2022) The critical role of arbuscular mycorrhizal fungi to improve drought tolerance and nitrogen use efficiency in crops. Front Plant Sci 13:919166. https://doi.org/10.3389/fpls.2022.919166

Thanni B, Merckx R, De Bauw P, Boeraeve M, Peeters G, Hauser S, Honnay O (2022) Spatial variability and environmental drivers of cassava—arbuscular mycorrhiza fungi (AMF) associations across Southern Nigeria. Mycorrhiza 1–13. https://doi.org/10.1007/s00572-021-01058-x

Tunlid A, Floudas D, Op De Beeck M, Wang T, Persson P (2022) Decomposition of soil organic matter by ectomycorrhizal fungi: mechanisms and consequences for organic nitrogen uptake and soil carbon stabilization. Front For Glob Change 5:934409. https://doi.org/10.3389/ffgc.2022.934409

Urgoiti J, Messier C, Keeton WS, Reich PB, Gravel D, Paquette A (2022) No complementarity no gain—net diversity effects on tree productivity occur once complementarity emerges during early stand development. Ecol Lett 25(4):851–862. https://doi.org/10.1111/ele.13959

Wahab A, Muhammad M, Munir A, Abdi G, Zaman W, Ayaz A, Khizar C, Reddy SP (2023) Role of arbuscular mycorrhizal fungi in regulating growth, enhancing productivity, and potentially influencing ecosystems under abiotic and biotic stresses. Plants 12(17):3102. https://doi.org/10.3390/plants12173102

Wang G, Koziol L, Foster BL, Bever JD (2022a) Microbial mediators of plant community response to long-term N and P fertilization: evidence of a role of plant responsiveness to mycorrhizal fungi. Glob Chang Biol 28(8):2721–2735. https://doi.org/10.1111/gcb.16091

Wang F, Zhang L, Zhou J, Rengel Z, George TS, Feng G (2022b) Exploring the secrets of hyphosphere of arbuscular mycorrhizal fungi: processes and ecological functions. Plant Soil 481(1):1–22. https://doi.org/10.1007/s11104-022-05621-z

Wu C, Bi Y, Zhu W, Xue C (2024a) Optimizing water use strategies in arid coal mining areas: the synergistic effects of layered soil profiles and arbuscular mycorrhizal fungi on plant growth and water use efficiency. Environ Exp Bot 221:105722. https://doi.org/10.1016/j.envexpbot.2024.105722

Wu S, Fu W, Rillig MC, Chen B, Zhu YG, Huang L (2024b) Soil organic matter dynamics mediated by arbuscular mycorrhizal fungi–an updated conceptual framework. New Phytol 242(4):1417–1425. https://doi.org/10.1111/nph.19178

Xue J, Bakker MR, Milin S, Graham D (2022) Enhancement in soil fertility, early plant growth and nutrition and mycorrhizal colonization by vermicompost application varies with native and exotic tree species. J Soils Sediments 22(6):1662–1676. https://doi.org/10.1007/s11368-022-03180-5

Yang Y, Zhang X, Hartley IP, Dungait JA, Wen X, Li D, Guo Z, Quine TA (2022a) Contrasting rhizosphere soil nutrient economy of plants associated with arbuscular mycorrhizal and ectomycorrhizal fungi in karst forests. Plant Soil 470(1):81–93. https://doi.org/10.1007/s11104-021-04950-9

Yang N, Zhang Y, Li J, Li X, Ruan H, Bhople P, Keiblinger K, Mao L, Liu D (2022b) Interaction among soil nutrients, plant diversity and hypogeal fungal trophic guild modifies root-associated fungal diversity in coniferous forests of Chinese southern Himalayas. Plant Soil 481(1):395–408. https://doi.org/10.1007/s11104-022-05646-4

Zhang H, Dong L, Yao X, Wang W (2023) Soil fertility shifts the relative importance of saprotrophic and mycorrhizal fungi for maintaining ecosystem stability. Glob Chang Biol 29(4):1206–1216. https://doi.org/10.1111/gcb.16540

Zhao S, Riaz M (2024) Plant–soil interactions and nutrient cycling dynamics in tropical rainforests. In: Environment, climate, plant and vegetation growth. Springer, Cham, pp 229–264. https://doi.org/10.1007/978-3-031-69417-2_8

Zheng J, Xie X, Li C, Wang H, Yu Y, Huang B (2023a) Regulation mechanism of plant response to heavy metal stress mediated by endophytic fungi. Int J Phytoremediation 25(12):1596–1613. https://doi.org/10.1080/15226514.2023.2176466

Zheng J, Zhang F, Zhang B, Chen D, Li S, Zhao T, Wang Q, Han G, Zhao M (2023b) Biodiversity and soil pH regulate the recovery of ecosystem multifunctionality during secondary succession of abandoned croplands in northern China. J Environ Manag 327:116882. https://doi.org/10.1016/j.jenvman.2022.116882

Zheng X, Hu Y, Ge X, Zhou X, Li Y, Zhao R, Fang Y, Ding H (2025) Fluctuations in species diversity in evergreen broad-leaved forests and changes in their co-occurrence network. Forests 16(4):594. https://doi.org/10.3390/f16040594

Impact of Mycorrhizal Application on Major Crops Such as Cereals, Pulses, and Oilseeds

Khizar Razzaq, Luis Del Rio Mendoza, Bita Babakhani, and Hasnain Razzaq

Abstract

The fungus-root association known as "mycorrhiza" is an ancient symbiotic relationship, estimated to be 400–500 million years old, in which fungi form mutually beneficial interactions with plant roots. Mycorrhizae play a key role in sustainable agriculture by improving soil health and nutrient acquisition, ultimately reducing dependence on chemical fertilizers and pesticides. Two primary types of mycorrhizal fungi are recognized: arbuscular mycorrhizal fungi (AMF), which associate with the majority of agricultural crops and are especially important for phosphorus uptake, and ectomycorrhizal fungi (EMF), which are less common in crop systems but crucial to forest ecosystems. Both AMF and EMF contribute to increased plant resilience against abiotic stresses, many of which are intensified by climate change. Studies have shown that AMF can enhance yield in pulses and increase disease resistance in cereals, while also benefiting oilseed crops. Beyond individual plant benefits, mycorrhizal associations play a broader ecological role by supporting biodiversity and maintaining soil structure and fertility, thereby contributing to the stability of agroecosystems. Given their ecological and agronomic advantages, mycorrhizal fungi represent an environmentally friendly alternative to synthetic agrochemicals. This chapter explores current research on the physiological and agronomic impacts of mycorrhizal application in major crops, with a particular focus on AMF. Emphasis is placed on the potential of integrating mycorrhizal technologies into modern agricultural

K. Razzaq · L. D. R. Mendoza (✉) · B. Babakhani
Department of Plant Pathology, North Dakota State University, Fargo, ND, USA
e-mail: luis.delriomendoza@ndsu.edu

H. Razzaq
Department of Plant Pathology, University of Agriculture Faisalabad, Faisalabad, Punjab, Pakistan

Y. Siddiqui et al. (eds.), *Sustainable Mycorrhizal Cultivation*,
https://doi.org/10.1007/978-981-95-1778-7_10

practices to enhance crop productivity and sustainability in the face of environmental challenges.

Keywords

Mycorrhiza · Sustainable crop production · Cereals · Biodiversity

Abbreviations

AMF	Arbuscular mycorrhizal fungi
EMF	Ectomycorrhizal fungi
N	Nitrogen
P	Phosphorus
Zn	Zinc
S	Sulfur
Ca	Calcium (C)
Fe	Iron
Mn	Manganese
Cu	Copper

1 Introduction

The global population is at the edge of food security risks. A growing population can increase food demand by 50% if it reaches an estimated ten billion by 2050. In addition to population pressure on agriculture, changing climate can be a major threat to meeting global food demand. Current climatic changes can affect arable lands in many countries worldwide, including Asia and Africa (Zhang and Cai 2011), and the expansion of arable lands is necessary to increase food production. Even the current cultivation area is facing issues such as salinization and low fertility due to land mismanagement (Shrivastava and Kumar 2015). High crop productivity is directly linked to inputs in the form of fertilizers, which in the long term can deteriorate soil health and reduce the soil ecosystem by 60% (Lal 2015). The only way to alleviate soil degradation is through integrated nutrient management involving soil microorganisms, to improve the future of agriculture.

Soil ecosystems consist of different kinds of microorganisms (beneficial bacteria and mycorrhizal fungi) that can colonize plants through symbiotic relationships, resulting in improved plant performance. The most common and non-pathogenic symbiotic association was mycorrhizal. This association has been observed in a broad range of plants, including crop lands, grasslands, and forests (Bonfante and Genre 2010; Smith and Read 2010). In addition to improving the availability of major nutrients to plants, the symbiotic association also protects plants against disease and toxicity of heavy metals (Al-Karaki 2000; Armada et al. 2015; Birhane et al. 2012; Liu et al. 2015). DNA sequencing confirmed the evolution of arbuscular mycorrhizal fungi into living fossils in 460 million years, benefiting the plants (Redecker et al. 2000). Among the seven types of mycorrhizae, arbuscular mycorrhizal fungi (AMF) and ectomycorrhizal fungi (EMF) are the most abundant. In most crops, mycorrhizal association results in the transportation of photosynthesized carbon to AMF and nutrients to plants, which facilitates fast growth (Ortas et al. 2001). Most of the immobile nutrients, including P, Cu, Zn, and partly NH4, are transported to the plant by the extended hyphae of AMF that grow beyond the root system, and an estimated 20% of the carbon fixed by plants is transported to the fungus (Marschner 2011; Ortas and Sari 2003; Parniske 2008).

Furthermore, AMF are the major sources of nutrient transport and uptake in soils with poor texture (Kayama and Yamanaka 2014). The wide hyphal diameter with the ability to absorb nutrients from an area of 25 cm around the roots makes the AMF fungi adaptable to different climatic conditions. AMF lack the ability to absorb carbohydrates from sources other than plants, which makes them obligate biotrophs (Jansa et al. 2003; Munkvold et al. 2004). AMF plays a major role in increasing the sustainability of agricultural systems through diverse and active biological communities (Gianinazzi and Schüepp 1994). In agriculture, 50% of the soil microbial biomass constitutes AMF, highlighting its abundance and profound growth (Olsson et al. 1999). The natural association of AMF exists in most of the major crops. Symbiotic associations improve growth and health and protect crops against pathogens (Khan et al. 2021a, b). In different types of cereal crops, including maize (*Zea mays*), wheat (*Triticum aestivum*), and rice (*Oryza sativa*), symbiotic relationships have been reported to be effective in increasing growth and productivity (Bernaola et al. 2018; Parniske 2008). The effect of AMF depends on the crop stage and type of fungi; for example, 255 AMF species have been identified and used in advanced agriculture systems to enhance crop productivity (Douds Jr et al. 2012; Khan et al. 2021a, b). In pulses and oil seeds, the use of AMF resulted in higher dry weight and nitrogen and phosphorus content, contributing to higher production (Choudhury and Azad 2004; Laad et al. 2019). Ectomycorrhiza are mostly associated with forest trees which stabilize the ecosystem by acting as a driver of global carbon (Agarwal and Sah 2009). The aim of this chapter is to explore AMF for sustainable agriculture. In modern agricultural practices, the integration of mycorrhiza can increase yield, particularly in major cereal, pulse, and oil seed crops, which reduces the need for pesticides and fertilizers while allowing wheat, rice, and canola to grow in areas with poor soil. This can improve the quality and quantity of oil produced by canola and sunflowers. In cereals, it can improve grain quality and yield, and nitrogen fixation increases in pulses, which reduces the use of nitrogen fertilizers. Overall, mycorrhizae create more sustainable food production systems for generation and offer a promising avenue for addressing global challenges, including environmental degradation, climate change, and food security (Fig. 1).

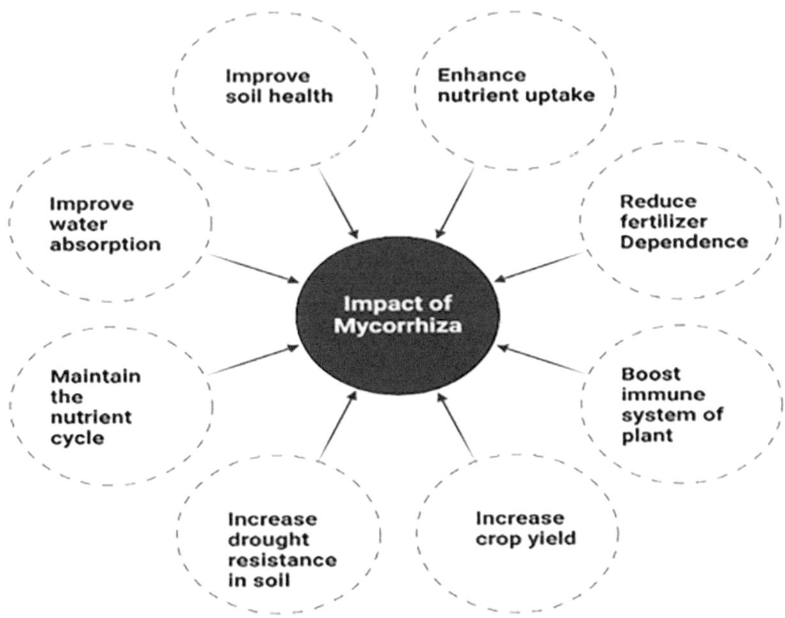

Fig. 1 Overall impact of mycorrhiza on soil and plant health. (Zhang et al. 2024)

2 Mycorrhiza

Mycorrhizal fungi are used in various crops, particularly in cereals, to enhance nutrient uptake and physiological characteristics such as plant growth and production. Arbuscular mycorrhiza are the most commonly used fungi in major crops as compared to ectomycorrhiza, so our aim was to summarize the importance of the most efficient and commonly used mycorrhiza (AMF) in cereals. The overall interaction of mycorrhiza with cereals, oilseeds, and pulses is shown in Fig. 2, which explains the general mechanism of host and mycorrhizal interactions. Some recent key references are provided in Table 1 to obtain a better understanding of recent advancements in mycorrhizal studies.

3 The Role of AMF in Optimizing Nutrient Uptake, Photosynthetic Efficiency, and Metabolic Processes in Cereal Crops

Environment-friendly agriculture is only possible through AMF fungi because of their positive effects on plant systems (Battini et al. 2017), and most AMF enhance nutrient uptake, including macro and micro nutrients such as nitrogen (N), phosphorus (P), zinc (Z), sulfur (S), calcium (C) (Cao et al. 2015), iron (Fe), manganese (Mn), and copper (Cu), and maintain 90% phosphorus, 60% nitrogen, and 20% total

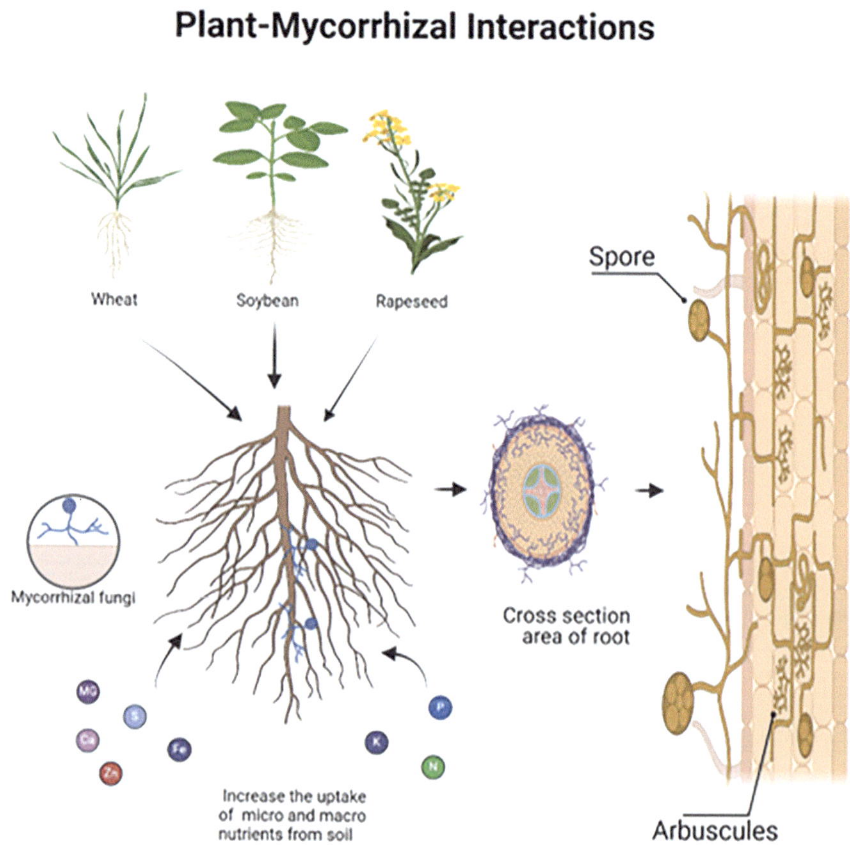

Fig. 2 Mycorrhizal interactions at the root level with cereal, pulse, and oilseed crops. (Smith and Read 2010)

carbon in plants (Khan et al. 2021b; Mei et al. 2019; Miransari 2013; Mustafa et al. 2016; Smith and Read 2008). AMF may contribute to the stimulation of many physiological and biochemical mechanisms that facilitate plant growth and stress tolerance in cereal crops. AMF facilitate enzymatic activity in cereals, which enhances chlorophyll content and photosynthesis (Pepe et al. 2018; Ryan and Angus 2003). Many AMF species, such as *Funneliformis mosseae*, *Rhizophagus irregularis,* and *Glomus etunicatum,* accelerate primary and secondary metabolite production, which is responsible for basic growth and tolerance to biotic and abiotic stressors (Ben Rejeb et al. 2014; Ganugi et al. 2019; Kulkarni and Goswami 2019).

Table 1 Key references on mycorrhizal studies over the past years highlight the significant advancement in understanding fungal plant symbiosis

Title	Journal/book	Year	Author
Genes conserved for arbuscular mycorrhizal symbiosis identified through phylogenomics	Nature Plants	2016	Bravo et al. (2016)
An N-acetylglucosamine transporter required for arbuscular mycorrhizal symbioses in rice and maize	Nature Plants	2017	Nadal et al. (2017)
Arbuscular mycorrhiza alters the nutritional requirements in *Salvia miltiorrhiza* and low nitrogen enhances the mycorrhizal efficiency	Nature Scientific Reports	2022	Pu et al. (2022)
Mycorrhizal fungi-mediated uptake of tree-derived nitrogen by maize in smallholder farms	Nature Sustainability	2022	Dierks et al. (2022)
Soil microbiome indicators can predict crop growth response to large-scale inoculation with arbuscular mycorrhizal fungi	Nature Microbiology	2023	Lutz et al. (2023)
Control of arbuscule development by a transcriptional negative feedback loop in Medicago	Nature Communications	2023	Zhang et al. (2023)
Uncovering the impact of AM fungi on wheat nutrient uptake, ion homeostasis, oxidative stress, and antioxidant defense under salinity stress	Nature Scientific Reports	2023	Huang et al. (2023)
Assessing the potential role of arbuscular mycorrhizal fungi in improving the phytochemical content and antioxidant properties in *Gomphrena globosa*	Nature Scientific Reports	2024	Dhalaria et al. (2024)
Tissue-specific accumulation of PIP aquaporins of a particular heteromeric composition is part of the maize response to mycorrhiza and drought	Nature Scientific Reports	2024	Paluch-Lubawa and Polcyn (2024)
Phytophagy impacts the quality and quantity of plant carbon resources acquired by mutualistic arbuscular mycorrhizal fungi	Nature Communications	2024	Bell et al. (2024)

3.1 Wheat (*Triticum aestivum*)

AMF enhance the access of host plants to unavailable nutrients, which promotes plant growth (Mardukhi et al. 2011). For example, the wheat plants inoculated with two AMF species, *F. mosseae* and *R. intraradices* uptakes Zn 2.5 times than non-inoculated plants. Similar results were reported by Pellegrino et al. (2015) and Schweiger and Jakobsen (1999), where AMF contributed to soil fertility and overall P uptake in winter wheat. Similar results were reported in Canadian prairie, where N and P uptake increased by 2.3% in *Glomus* species of AMF (Dai et al. 2014); however, it was hypothesized that it could produce non-beneficial effects on wheat crops, depending on the availability of P in the soil and AMF status in the soil microbiome (Graham and Abbott 2000). In different crops, such as wheat, rice, and maize, AMF have been reported to increase the photosynthetic rate by facilitating P and N uptake from soil (Beltrano and Ronco 2008), and under drought conditions, AMF-treated wheat plants showed higher chlorophyll content than uninoculated plants (Peterson et al. 2004). In another study, AMF *F. mosseae* mitigated the effects of the environment and increased the photosynthetic activity of wheat growing in salty soil (Abdel-Fattah and Asrar 2012). AMF have a potential role in the formation of essential hormones, such as abscisic acid, for example, in wheat AMF (*F. mosseae*), which increases the carbohydrate concentration and elevates the negative effect of heavy metals (Rabie 2005; Shahabivand et al. 2012). In terms of productivity, AMF effectively increased the grain yield in Brookton wheat cultivar as compared to Kirchauf cultivar, AMF was used with P in Brookton in highly calcareous soil and without the addition of P grain yield was lower in both cultivars (Sukumaran et al. 2016). Flower initiation was also increased by the increased colonization of AMF which enhances the P absorption (Li et al. 2005).

3.2 Maize *(Zea mays)*

A longer root system was observed in maize inoculated with *R. irregularis* under alkaline conditions because AMF facilitate the absorption of P and N from the soil (Merlos et al. 2016). Similar results were reported in another study where Zn and P were boosted in maize shoots after 45–75 days of inoculation with *Glomus intraradices* (Subramanian et al. 2009). Plants inoculated with AMF have been reported to have high contents of N and P in the roots and shoots of plants during the grain-filling stage of the host (Azaizeh et al. 1995). Other studies suggested that the use of bacteria with AMF, that is, *F. mosseae* and *Rhizobium* resulted in increased N uptake by AMF colonization and nodulation, suggesting the positive role of combined usage of AMF and bacteria (Meng et al. 2015). The application of *G. mosseae* and *G. intraradices* on maize protected the water splitting complex under high temperature by enhancing the (PSII) photochemistry (Amerian et al. 2001), compared with un-inoculated maize plants the photosynthetic rate and chlorophyll content increased up to 29% (Almagrabi and Abdelmoneim 2012). Similar results have been reported for rice plants (de Andrade et al. 2015). The reason for the increased photosynthetic rate was the proper supply of P to maize plants by AMF. Moreover, the AMP species *Glomus tortuosum* is used to improve physiological metabolism by increasing light energy utilization, Rubisco activity, and chlorophyll content in saline soils (Xu et al. 2018). Seed coating with AMF increases the production of organic acids, including malonic, fumaric, citric, T-aconitic, and malic acids in maize roots (Baum et al. 2015). The application of *G. intraradices* successfully enhanced the dry matter, green matter, and silage yield in both irrigated and non-irrigated conditions and increased the tolerance of plants to biotic and abiotic factors (Celebi et al. 2010). Furthermore, AMF with bacteria (*Azotobacter chroococum*) assisted maize plants in N and P uptake, resulting in increased cob weight, grain weight, cob length, and starch content (Gao et al. 2020).

3.3 Barley *(Hordeum vulgare)*

In *Hordeum vulgare,* AMF increased Zn uptake compared to non-treated plants (Watts-Williams et al. 2020). In another study, AMF was used with humic acid, resulting in better AMF performance and increased Zn absorption (Maji et al. 2017) Furthermore, AMF *G. intraradices* application on barley under saline conditions inhibits the uptake of sodium and promotes Zn, P, and Fe uptake in host plants (Mohammad et al. 2003). In other cereals such as buckwheat and millet, AMF positively enhances the absorption of N, P, K, Mg, Ca, and Zn in inoculated plants compared to non-inoculated plants (Bagayoko et al. 2000; Boglaienko et al. 2014). Increased Rubisco and PSII activities by AMF colonization in barley plants improve tolerance to drought and salt stress (Meddich et al. 2021; Shamshiri and Fattahi 2016). Multiple studies have shown increased photosynthetic rate and early plant growth in barley treated with AMF (Achatz et al. 2010; Gernns et al. 2001), and higher uptake of N was observed in sorghum, followed by increased photosynthetic rate and grain production by supplying more nutrients and water to plants (Kamali and Mehraban 2020; Ruiz-Lozano and Aroca 2010). Similar results have been obtained for other crops; for example, in quinoa, *G. mosseae* and *Glomus tortuosum* increased the efficiency of PSII and chlorophyll content (García-Parra and Plazas-Leguizamón 2019). In barley, higher colonization by AMF has been reported to increase phenolic, flavonoid, isoflavonoid, and apo-carotenoids in shoots and roots (Hajiboland et al. 2019). Improved grain yield and biochemical characteristics of barley seeds were observed in plants treated with AMF *R. irregularis* owing to increasing concentrations of Zn in plants (Al Mutairi et al. 2020). Similar results were found in another study, but instead of two, they used five species of AMF that boosted grain yield and kernel weight by 68 and 90%, respectively. In inoculated barley plants, the concentrations of Cu, K, Zn, Fe, and Ca were significantly higher than those in non-inoculated plants (Jerbi et al. 2022).

4 Mycorrhiza: A Booster of Oilseeds and Pulses

In oilseed and legume crops, mycorrhizae have been shown to have a significant influence on development and productivity. The symbiotic relationship between mycorrhiza and both crops can enhance the soil microbiota, resulting in enhanced nutrient uptake from the soil and improved water-use efficiency, ultimately leading to increased yields.

4.1 Pulses/Legumes

The rhizosphere of leguminous crops is dependent on microbial communities for nutrient uptake. AMF and N2 fixing bacteria establish symbiotic relationships with legumes which help the crops attain the highest growth and productivity (Azcón and Barea 2010), ultimately improving nodulation in pulses (Barea and Azcon-Aguilar

1983). In sustainable agriculture and natural ecosystems, AMF and nodulated legumes play a significant role in supplying nutrients, such as N and P, to plants (Shtark et al. 2011). For example, studies on chickpea inoculated with AMF positively increased shoot dry weight, phosphorus content, and root nodulation compared to uninoculated chickpea plants (Alloush et al. 2000; Basiru et al. 2024). Phosphorus is an important element for increasing the root nodulation in legumes. However, nodulation can occur in non-nodulating crops when planted with nodulating plants. For example, intercropping maize with pigeonpea can result in fixed N transfer from pigeonpea to maize crops (Lambert et al. 1979; Shibata and Yano 2003). The intercropping of crops, i.e., pigeonpea (nodulating), with cereals such as maize can enhance productivity (Egbe and Adeyemo 2006). AMF play a vital role in the availability of P to plants, which increases nodulation in the roots of legume crops, resulting in better growth and productivity (Dania et al. 2009; Fagbola et al. 2005).

4.2 Oilseed Crops

AMF establish a mutualistic relationship with 90% of agricultural plants, specifically oilseed and cereals. Some AMF species, such as *F. mosseae*, *R. irregularis*, and *G. versiforme,* are considered an inherent part of the soil microbiome (Wertheim et al. 2014). In the mutualistic relationship, AMF depends on the host (C) compound, and more than 20% of the fixed carbon contributes to survival. AMF use their special structures called arbuscules to enhance P and N uptake in host plants (Battini et al. 2017). Mutualistic relationships potentially suppress abiotic and biotic factors by modifying different physiological responses, such as reducing oxidative damage and facilitating osmotic adjustment (Cao et al. 2015; Mei et al. 2019; Mustafa et al. 2016; Smith and Read 2008). Under extreme weather conditions, AMF maintain the ion balance of host plants (Frew et al. 2018; Li et al. 2005; Miransari 2013), which results in increased biomass of inoculated plants compared to non-inoculated plants (Smith and Smith 2011). AMF help the host plant mitigate biotic stress, particularly pathogenic infections. As a result of mutualism, AMF influence root microbiota and alter root exudation, which enhances nutrient uptake, resulting in stronger induction of plant defenses (Smith and Gianinazzi-Pearson 1988; Tariq and Iqbal 2010). In oilseed crops, AMF have been shown to increase mineral uptake, enhance biomass, secondary metabolism, osmoregulation, and plant defense machinery (Achatz et al. 2010; Gernns et al. 2001; Meddich et al. 2021; Shamshiri and Fattahi 2016), and are widely observed in oilseed crops such as soybean, sesame, and linseed (Borde et al. 2011; García-Parra and Plazas-Leguizamón 2019; Kamali and Mehraban 2020; Ruiz-Lozano and Aroca 2010). For example, a study conducted on soybean plants inoculated with *R. clarus* resulted in physiological traits improvements, nutrient acquisition, and increased quantum yield of photosystem II (Bowles et al. 2016), furthermore, result of another study revealed that plant inoculated with AMF "*F. mosseae*" increased the yield of oilseed crops and nutrient content in sesame (Talaat and Shawky 2014). Saline soil is a major problem for crops, especially oilseed crops, which impacts the growth stage and delays flowering, ultimately reducing crop yield; however, studies have suggested that AMF enhances salt tolerance in oilseed, including peanuts, dragon head, safflower, and sunflower plants (Hajiboland et al. 2019; Li et al. 2013; Rabie 2005; Shahabivand et al. 2012). The mechanism of AMF salinity tolerance starts with better nutrient uptake, decreased Na^{+} levels, and enhanced phytochrome production. For instance, the results of an experiment showed a positive impact of AMF "*F. mosseae*, *C. etunicatum,* and *R. intraradices*" inoculation in soybean plants. The inoculated plants had high Indole-3-acetic-acid (IAA) biosynthesis which plays a vital role in root nodulation, indicating that AMF developed a better root system for nutrient uptake (Baum et al. 2015). In sunflower and soybean, a study was conducted to determine the response of plants inoculated with AMF in soil with high Heavy metals (HMs). The results of this study demonstrated improved physiological conditions and nutrient uptake in AMF-inoculated plants (Alarcón et al. 2008; Evelin et al. 2009; Huang et al. 2008). Furthermore, soybean plants inoculated with *R. intraradices* show improved P acquisition and high grain yield (Alarcón et al. 2008).

5 Mycorrhizal Mitigation of Biotic and Abiotic Stresses in Cereals, Oilseeds, and Pulses/Legumes

The symbiotic relationship between AMF and plants enhances resistance against a variety of pathogens, including bacteria and fungi, but resistance can vary with the species type of AMF. In wheat and maize, AMF improves the efficiency of water use, resulting in the mitigation of drought stress in both crops (Abdel-Fattah and Asrar 2012; Beltrano and Ronco 2008; Ibrahim et al. 2011; Zhu et al. 2017). Two AMF species, *R. irregularis* and *Rhizoglomus intraradices,* regulate the expression of aquaporin genes in wheat and maize, facilitating the transport of neutral molecules and water across the cell membrane and resulting in enhanced water efficiency in host plants (Mathur et al. 2019). Moreover, the extended hyphal network of AMF with the host plant root system enhanced the area of the roots to a larger soil surface and water retention. For example, wheat plants inoculated with AMF (*F. mosseae*) showed improved water use efficiency under water limitation compared with non-inoculated plants. Ghorchiani et al. 2018 reported similar results for sorghum plants inoculated with *F. mosseae*. Further studies revealed the contribution of AMF in discovering solution-filled soil pores which are inaccessible to plant roots and enhancing the P and N in wheat and maize (Amerian et al. 2001; Mathur et al. 2019), which enhances the acid phosphatase and alkaline level of soil, which helps in the availability and assimilation of P (Khan et al. 2021a; Zhu et al. 2017). Phytohormones play key roles in regulating AMF interactions during later stages. Under drought conditions, complex patterns of phytohormones have been revealed, which involve modified responses and alterations in hormone levels in mycorrhizal plants. Auxins, brassinosteroids, and abscisic acid were identified as positive regulators of AMF symbiosis in different cereal fungal systems, whereas salicylic acid

(SA) and gibberellins (Gas) were identified as negative regulators during interaction (Porcel et al. 2015; Symanczik et al. 2018; Zhang et al. 2017). Different studies have revealed that AMF in cereals such as wheat, maize, and sorghum improves the maximum quantum efficiency of photosystem I and PSII by regulating and increasing the stomatal aperture and density (Rezvani et al. 2015; Watts-Williams and Cavagnaro 2018). Additionally, increased chlorophyll content and plant biomass have been reported in AMF colonization (Jannoura et al. 2014; Zhao et al. 2014) and increase the area for solar assimilation by inducing leaf CO_2 absorption (Boglaienko et al. 2014; Tummaramatti et al. 2016). AMF play an important role in the alleviation of biotic stress in most cereal crops through diverse mechanisms including osmotic regulation, enhancement of photosynthesis, and mitigation of Heavy metals (HMs) stress. For example, numerous studies have demonstrated the resistance of wheat to pathogenic fungi, bacteria, and insects (Bhantana et al. 2021; Watts-Williams et al. 2020). For instance, in winter wheat, AMF inoculation improves the defense mechanism of host plants by competing against pathogenic fungi (Fusarium crown and root rot) for colonization sites and root space which induces anatomical changes in the root system, and helps in the detoxification of H_2O_2 to mitigate disease severity (Aliasgharzad et al. 2009). In maize, plants inoculated with *Gigaspora* sp. and *Entrophospora* sp. show improved tolerance to downy mildew (Prasetyo et al. 2021). A similar mechanism has been observed in wheat inoculated with *F. mosseae* against bacterial leaf streak disease (Çakar et al. 2024).

In oilseed crops such as soybean, linseed, and sesame, different mechanisms are enhanced by AMF which increase mineral and water uptake, photosynthesis, and defense mechanisms (Achatz et al. 2010; Borde et al. 2011; Gernns et al. 2001; Meddich et al. 2021; Ruiz-Lozano and Aroca 2010)-101). Some studies have suggested that AMF significantly enhances salinity tolerance in sunflower, peanut, safflower, and dragon heads (Shahabivand et al. 2012) by increasing phytohormone production. Heavy metals in the soil cause disturbances in plant mechanisms, which affect plant growth and yield. However, the use of AMF in soybean and sunflower helped the plants to detoxify the effect of Heavy metals (HMs) by improving the antioxidant machinery and decreasing the MDA content (Alarcón et al. 2008). AMF bind the Heavy metals (HMs) in the mycelium and act as a barrier in plant roots, preventing Heavy metals (HMs) translocation to oilseed roots and seeds. AMF exhibited a positive trend in response to biotic stress. For example, sunflower plants inoculated with the genus *Glomus* showed systemic resilience to white rot disease (Bernaola et al. 2018) and enhanced resilience against nematodes. Collectively, these findings provide a deeper understanding of mycorrhiza and their use in sustainable agriculture systems.

6 Conclusion and Future Perspective

The application of mycorrhizal fungi in agriculture holds considerable promise for enhancing nutrient uptake, crop growth, and stress tolerance across diverse crop groups, including cereals, pulses, and oilseeds. These symbiotic associations are particularly beneficial for improving phosphorus acquisition, increasing water-use efficiency, and enhancing plant resilience to both biotic and abiotic stressors. Despite these advantages, several challenges hinder the widespread adoption of mycorrhizal inoculants. Key constraints include soil fertility, pH, temperature variability, and the compatibility between host plants and specific fungal strains. Furthermore, the heterogeneity of soil environments and cropping systems complicates the standardization of inoculation practices across different agroecological zones. While the potential of AMF in major crops is well-recognized, targeted research is needed to address existing limitations. For instance, the genetic exchange among AMF strains through mechanisms such as outbreeding depression may either enhance or reduce mutualistic efficiency (Angelard and Sanders 2011), warranting continued investigation. It is crucial to understand both the direct and indirect effects of AMF on plant systems, particularly as newly introduced strains may not always adapt effectively to local environmental conditions. To support the commercialization of AMF, the development of a predictive framework is essential. Such a framework should identify the environmental and biological conditions under which inoculation is most likely to improve crop yields and contribute to sustainable agricultural practices. Importantly, this model must also integrate economic parameters to ensure feasibility and profitability at scale.

References

Abdel-Fattah GM, Asrar A-WA (2012) Arbuscular mycorrhizal fungal application to improve growth and tolerance of wheat (*Triticum aestivum* L.) plants grown in saline soil. Acta Physiol Plant 34:267–277

Achatz B, von Rüden S, Andrade D, Neumann E, Pons-Kühnemann J, Kogel K-H, Franken P, Waller F (2010) Root colonization by *Piriformospora indica* enhances grain yield in barley under diverse nutrient regimes by accelerating plant development. Plant Soil 333:59–70

Agarwal P, Sah P (2009) Ecological importance of ectomycorrhizae in world forest ecosystems. Nat Sci 7(2):107–116

Al Mutairi AA, Cavagnaro TR, Khor SF, Neumann K, Burton RA, Watts-Williams SJ (2020) The effect of zinc fertilisation and arbuscular mycorrhizal fungi on grain quality and yield of contrasting barley cultivars. Funct Plant Biol 47(2):122–133

Alarcón A, Davies FT Jr, Autenrieth RL, Zuberer DA (2008) Arbuscular mycorrhiza and petroleum-degrading microorganisms enhance phytoremediation of petroleum-contaminated soil. Int J Phytoremediation 10(4):251–263

Aliasgharzad N, Shirmohamadi E, Oustan S (2009) Siderophore production by mycorrhizal sorghum roots under micronutrient deficient condition. Soil Environ 28(2):119–123

Al-Karaki GN (2000) Growth of mycorrhizal tomato and mineral acquisition under salt stress. Mycorrhiza 10:51–54

Alloush G, Zeto S, Clark R (2000) Phosphorus source, organic matter, and arbuscular mycorrhiza effects on growth and mineral acquisition of chickpea grown in acidic soil. J Plant Nutr 23(9):1351–1369

Almagrabi O, Abdelmoneim T (2012) Using of arbuscular mycorrhizal fungi to reduce the deficiency effect of phosphorous fertilization on maize plants (*Zea mays* L.). Life Sci J 9(4):1648–1654

Amerian M, Stewart W, Griffiths H (2001) Effect of two species of arbuscular mycorrhizal fungi on growth, assimilation and leaf water relations in maize (*Zea mays*). Asp Appl Biol 63:71–76

Angelard C, Sanders IR (2011) Effect of segregation and genetic exchange on arbuscular mycorrhizal fungi in colonization of roots. New Phytol 189(3):652–657

Armada E, Azcón R, López-Castillo OM, Calvo-Polanco M, Ruiz-Lozano JM (2015) Autochthonous arbuscular mycorrhizal fungi and bacillus thuringiensis from a degraded Mediterranean area can be used to improve physiological traits and performance of a plant of agronomic interest under drought conditions. Plant Physiol Biochem 90:64–74

Azaizeh H, Marschner H, Römheld V, Wittenmayer L (1995) Effects of a vesicular-arbuscular mycorrhizal fungus and other soil microorganisms on growth, mineral nutrient acquisition and root exudation of soil-grown maize plants. Mycorrhiza 5:321–327

Azcón R, Barea J-M (2010) Mycorrhizosphere interactions for legume improvement. Springer

Bagayoko M, George E, Römheld V, Buerkert A (2000) Effects of mycorrhizae and phosphorus on growth and nutrient uptake of millet, cowpea and sorghum on a West African soil. J Agric Sci 135(4):399–407

Barea J, Azcon-Aguilar C (1983) Mycorrhizas and their significance in nodulating nitrogen-fixing plants. Adv Agron 36:1–54

Basiru S, Aitsimhand K, Elfermi R, Khatour I, Errafii K, Legeay J, Hijri M (2024) Enhancing chickpea growth via arbuscular mycorrhizal fungal inoculation: facilitating nutrient uptake and shifting potential pathogenic fungal communities. Sci Rep 14:1–12

Battini F, Grønlund M, Agnolucci M, Giovannetti M, Jakobsen I (2017) Facilitation of phosphorus uptake in maize plants by mycorrhizosphere bacteria. Sci Rep 7(1):4686

Baum C, El-Tohamy W, Gruda N (2015) Increasing the productivity and product quality of vegetable crops using arbuscular mycorrhizal fungi: a review. Sci Hortic 187:131–141

Bell C, Magkourilou E, Ault J, Urwin P, Field K (2024) Phytophagy impacts the quality and quantity of plant carbon resources acquired by mutualistic arbuscular mycorrhizal fungi. Nat Commun 15(1):801

Beltrano J, Ronco MG (2008) Improved tolerance of wheat plants (*Triticum aestivum* L.) to drought stress and rewatering by the arbuscular mycorrhizal fungus *Glomus claroideum*: effect on growth and cell membrane stability. Braz J Plant Physiol 20:29–37

Ben Rejeb I, Pastor V, Mauch-Mani B (2014) Plant responses to simultaneous biotic and abiotic stress: molecular mechanisms. Plants 3(4):458–475

Bernaola L, Cange G, Way MO, Gore J, Hardke J, Stout M (2018) Natural colonization of rice by arbuscular mycorrhizal fungi in different production areas. Rice Sci 25(3):169–174

Bhantana P, Malla R, Vista SP, Rana MS, Mohamed G, Joshi BD, Hu C (2021) Use of arbuscular mycorrhizal fungi (AMF) and zinc fertilizers in an adaptation of plant from drought and heat stress. Biomed J Sci Tech Res 38:30357–30373

Birhane E, Sterck FJ, Fetene M, Bongers F, Kuyper TW (2012) Arbuscular mycorrhizal fungi enhance photosynthesis, water use efficiency, and growth of frankincense seedlings under pulsed water availability conditions. Oecologia 169:895–904

Boglaienko D, Soti P, Shetty K, Jayachandran K (2014) Buckwheat as a cover crop in Florida: mycorrhizal status and soil analysis. Agroecol Sustain Food Syst 38(9):1033–1046

Bonfante P, Genre A (2010) Mechanisms underlying beneficial plant–fungus interactions in mycorrhizal symbiosis. Nat Commun 1(1):48

Borde M, Dudhane M, Jite P (2011) Growth photosynthetic activity and antioxidant responses of mycorrhizal and non-mycorrhizal bajra (*Pennisetum glaucum*) crop under salinity stress condition. Crop Prot 30(3):265–271

Bowles TM, Barrios-Masias FH, Carlisle EA, Cavagnaro TR, Jackson LE (2016) Effects of arbuscular mycorrhizae on tomato yield, nutrient uptake, water relations, and soil carbon dynamics under deficit irrigation in field conditions. Sci Total Environ 566:1223–1234

Bravo A, York T, Pumplin N, Mueller LA, Harrison MJ (2016) Genes conserved for arbuscular mycorrhizal symbiosis identified through phylogenomics. Nat Plants 2(2):1–6

Çakar K, Akköprü A, Uçar CP (2024) Effect of *Pseudomonas* spp. and *Funneliformis mosseae* on bacterial spot disease and growth response of pepper. J Plant Pathol 106:1733

Cao J, Wang C, Huang Y (2015) Interactive impacts of earthworms (Eisenia fetida) and arbuscular mycorrhizal fungi (*Funneliformis mosseae*) on the bioavailability of calcium phosphates. Plant Soil 396:45–57

Celebi SZ, Demir S, Celebi R, Durak ED, Yilmaz IH (2010) The effect of arbuscular mycorrhizal fungi (AMF) applications on the silage maize (*Zea mays* L.) yield in different irrigation regimes. Eur J Soil Biol 46(5):302 305

Choudhury B, Azad P (2004) Dual inoculation of native Rhizobium spp. and arbuscular mycorrhizal fungi: an impact study for enhancement of pulse production. Mycobiology 32(4):173–178

Dai M, Hamel C, Bainard LD, Arnaud MS, Grant CA, Lupwayi NZ, Malhi SS, Lemke R (2014) Negative and positive contributions of arbuscular mycorrhizal fungal taxa to wheat production and nutrient uptake efficiency in organic and conventional systems in the Canadian prairie. Soil Biol Biochem 74:156–166

Dania S, Dania M, Fagbola O (2009) Performances of maize as influenced by Cajanus cajan compost and mycorrhizal inoculation. Niger J Mycol 2(1):117–122

De Andrade SAL, Domingues AP Jr, Mazzafera P (2015) Photosynthesis is induced in rice plants that associate with arbuscular mycorrhizal fungi and are grown under arsenate and arsenite stress. Chemosphere 134:141–149

Dhalaria R, Verma R, Sharma R, Jomova K, Nepovimova E, Kumar H, Kuca K (2024) Assessing the potential role of arbuscular mycorrhizal fungi in improving the phytochemical content and antioxidant properties in Gomphrena globosa. Sci Rep 14(1):22830

Dierks J, Blaser-Hart W, Gamper H, Six J (2022) Mycorrhizal fungi-mediated uptake of tree-derived nitrogen by maize in smallholder farms. Nat Sustain 5:64–70

Douds D Jr, Lee J, Rogers L, Lohman M, Pinzon N, Ganser S (2012) Utilization of inoculum of AM fungi produced on-farm for the production of *Capsicum annuum*: a summary of seven years of field trials on a conventional vegetable farm. Biol Agric Hortic 28(2):129–145

Egbe O, Adeyemo M (2006) Estimation of the effect of intercropped pigeon pea on the yield and yield components of maize in southern Guinea savannah of Nigeria. J Sustain Dev Agric Environ 2(1):107–119

Evelin H, Kapoor R, Giri B (2009) Arbuscular mycorrhizal fungi in alleviation of salt stress: a review. Ann Bot 104(7):1263–1280

Fagbola O, Oyetunji OJ, Osonubi O, Mulongoy K (2005) Greenhouse evaluation of two woody hedgerows as affected by arbuscular mycorrhizal fungus (AMF), soil quality and moisture. Arch Agron Soil Sci 51(3):335–349

Frew A, Powell JR, Glauser G, Bennett AE, Johnson SN (2018) Mycorrhizal fungi enhance nutrient uptake but disarm defences in plant roots, promoting plant-parasitic nematode populations. Soil Biol Biochem 126:123–132

Ganugi P, Masoni A, Pietramellara G, Benedettelli S (2019) A review of studies from the last twenty years on plant–arbuscular mycorrhizal fungi associations and their uses for wheat crops. Agronomy 9(12):840

Gao C, El-Sawah AM, Ali DFI, Alhaj Hamoud Y, Shaghaleh H, Sheteiwy MS (2020) The integration of bio and organic fertilizers improve plant growth, grain yield, quality and metabolism of hybrid maize (*Zea mays* L.). Agronomy 10(3):319

García-Parra MÁ, Plazas-Leguizamón NZ (2019) Análisis del ciclo de vida de las publicaciones sobre la producción de quinua (Chenopodium quinoa Willd), a través de curvas en S. Rev Investig Desarro Innov 9(2):379–391

Gernns H, Alten H, Poehling HM (2001) Arbuscular mycorrhiza increased the activity of a biotrophic leaf pathogen–is a compensation possible? Mycorrhiza 11:237–243

Ghorchiani M, Etesami H, Alikhani HA (2018) Improvement of growth and yield of maize under water stress by co-inoculating an arbuscular mycorrhizal fungus and a plant growth promoting rhizobacterium together with phosphate fertilizers. Agric Ecosyst Environ 258:59–70

Gianinazzi S, Schüepp H (1994) Impact of arbuscular mycorrhizas on sustainable agriculture and natural ecosystems. Springer

Graham J, Abbott L (2000) Wheat responses to aggressive and non-aggressive arbuscular mycorrhizal fungi. Plant Soil 220(1):207–218

Hajiboland R, Joudmand A, Aliasgharzad N, Tolrá R, Poschenrieder C (2019) Arbuscular mycorrhizal fungi alleviate low-temperature stress and increase freezing resistance as a substitute for acclimation treatment in barley. Crop Pasture Sci 70(3):218–233

Huang L, Yang C, Zhao Y, Xu X, Xu Q, Li G, Cao J, Herbert S, Hao L (2008) Antioxidant defenses of mycorrhizal fungus infection against SO_2-induced oxidative stress in *Avena nuda* seedlings. Bull Environ Contam Toxicol 81:440–444

Huang S, Gill S, Ramzan M, Ahmad MZ, Danish S, Huang P, Al Obaid S, Alharbi SA (2023) Uncovering the impact of AM fungi on wheat nutrient uptake, ion homeostasis, oxidative stress, and antioxidant defense under salinity stress. Sci Rep 13(1):8249

Ibrahim A, Abdel-Fattah G, Eman F, Abd El-Aziz M, Shohr A (2011) Arbuscular mycorrhizal fungi and spermine alleviate the adverse effects of salinity stress on electrolyte leakage and productivity of wheat plants. Phyton 51(2):261–276

Jannoura R, Joergensen RG, Bruns C (2014) Organic fertilizer effects on growth, crop yield, and soil microbial biomass indices in sole and intercropped peas and oats under organic farming conditions. Eur J Agron 52:259–270

Jansa J, Mozafar A, Frossard E (2003) Long-distance transport of P and Zn through the hyphae of an arbuscular mycorrhizal fungus in symbiosis with maize. Agronomie 23(5–6):481–488

Jerbi M, Labidi S, Laruelle F, Tisserant B, Jeddi FB, Sahraoui AL-H (2022) Mycorrhizal biofertilization improves grain yield and quality of hulless barley (*Hordeum vulgare* ssp. *nudum* L.) under water stress conditions. J Cereal Sci 104:103436

Kamali S, Mehraban A (2020) Effects of Nitroxin and arbuscular mycorrhizal fungi on the agro-physiological traits and grain yield of sorghum (Sorghum bicolor L.) under drought stress conditions. PLoS One 15(12):e0243824

Kayama M, Yamanaka T (2014) Growth characteristics of ectomycorrhizal seedlings of *Quercus glauca*, *Quercus salicina*, and *Castanopsis cuspidata* planted on acidic soil. Trees 28:569–583

Khan Y, Sohail A, Yaseen T, Rehman KU, Noor M, Akbar K (2021a) Arbuscular mycorrhizal fungi improved the growth and yield productivity of *Lens esculenta* under the influence of poultry litter. Pak J Phytopathol 33(1):145–151

Khan Y, Yang X, Zhang X, Yaseen T, Shi L, Zhang T (2021b) Arbuscular mycorrhizal fungi promote plant growth of *Leymus chinensis* (Trin.) Tzvelev by increasing the metabolomics activity under nitrogen addition. Grassl Sci 67(2):128–138

Kulkarni S, Goswami A (2019) Effect of excess fertilizers and nutrients: a review on impact on plants and human population. In: Proceedings of International Conference on Sustainable computing in science, technology and management (SUSCOM), Amity University Rajasthan, Jaipur, India

Laad P, Maheshwari HS, Sharma MP (2019) Utilization of mycorrhizal fungi in the mobilization of macro- and micronutrients to important pulses and oil-seed crops. In: Biofertilizers and biopesticides in sustainable agriculture. Apple Academic Press, pp 299–318

Lal R (2015) Restoring soil quality to mitigate soil degradation. Sustainability 7(5):5875–5895

Lambert D, Baker DE, Cole H Jr (1979) The role of mycorrhizae in the interactions of phosphorus with zinc, copper, and other elements. Soil Sci Soc Am J 43(5):976–980

Li H, Zhu Y, Marschner P, Smith F, Smith S (2005) Wheat responses to arbuscular mycorrhizal fungi in a highly calcareous soil differ from those of clover, and change with plant development and P supply. Plant Soil 277:221–232

Li T, Hu YJ, Hao ZP, Li H, Wang YS, Chen BD (2013) First cloning and characterization of two functional aquaporin genes from an arbuscular mycorrhizal fungus *Glomus intraradices*. New Phytol 197(2):617–630

Liu T, Sheng M, Wang C, Chen H, Li Z, Tang M (2015) Impact of arbuscular mycorrhizal fungi on the growth, water status, and photosynthesis of hybrid poplar under drought stress and recovery. Photosynthetica 53(2):250–258

Lutz S, Bodenhausen N, Hess J, Valzano-Held A, Waelchli J, Deslandes-Hérold G, Schlaeppi K, van der Heijden MG (2023) Soil microbiome indicators can predict crop growth response to large-scale inoculation with arbuscular mycorrhizal fungi. Nat Microbiol 8(12):2277–2289

Maji D, Misra P, Singh S, Kalra A (2017) Humic acid rich vermicompost promotes plant growth by improving microbial community structure of soil as well as root nodulation and mycorrhizal colonization in the roots of *Pisum sativum*. Appl Soil Ecol 110:97–108

Mardukhi B, Rejali F, Daei G, Ardakani MR, Malakouti MJ, Miransari M (2011) Arbuscular mycorrhizas enhance nutrient uptake in different wheat genotypes at high salinity levels under field and greenhouse conditions. C R Biol 334(7):564–571

Marschner H (2011) Marschner's mineral nutrition of higher plants. Academic

Mathur S, Tomar RS, Jajoo A (2019) Arbuscular mycorrhizal fungi (AMF) protects photosynthetic apparatus of wheat under drought stress. Photosynth Res 139:227–238

Meddich A, Ouhaddou R, Anli M, Boutasknit A (2021) Role of phosphorus and arbuscular mycorrhizal fungi in the growth performances and tolerance of barley to water stress. Plant Cell Biotechnol Mol Biol 22:45–67

Mei L, Yang X, Cao H, Zhang T, Guo J (2019) Arbuscular mycorrhizal fungi alter plant and soil C:N:P stoichiometries under warming and nitrogen input in a semiarid meadow of China. Int J Environ Res Public Health 16(3):397

Meng L, Zhang A, Wang F, Han X, Wang D, Li S (2015) Arbuscular mycorrhizal fungi and rhizobium facilitate nitrogen uptake and transfer in soybean/maize intercropping system. Front Plant Sci 6:339

Merlos MA, Zitka O, Vojtech A, Azcón-Aguilar C, Ferrol N (2016) The arbuscular mycorrhizal fungus *Rhizophagus irregularis* differentially regulates the copper response of two maize cultivars differing in copper tolerance. Plant Sci 253:68–76

Miransari M (2013) Soil microbes and the availability of soil nutrients. Acta Physiol Plant 35(11):3075–3084

Mohammad MJ, Malkawi HI, Shibli R (2003) Effects of arbuscular mycorrhizal fungi and phosphorus fertilization on growth and nutrient uptake of barley grown on soils with different levels of salts. J Plant Nutr 26(1):125–137

Munkvold L, Kjøller R, Vestberg M, Rosendahl S, Jakobsen I (2004) High functional diversity within species of arbuscular mycorrhizal fungi. New Phytol 164(2):357–364

Mustafa G, Randoux B, Tisserant B, Fontaine J, Magnin-Robert M, Lounès-Hadj Sahraoui A, Reignault P (2016) Phosphorus supply, arbuscular mycorrhizal fungal species, and plant genotype impact on the protective efficacy of mycorrhizal inoculation against wheat powdery mildew. Mycorrhiza 26:685–697

Nadal M, Sawers R, Naseem S, Bassin B, Kulicke C, Sharman A, An G, An K, Ahern KR, Romag A (2017) An N-acetylglucosamine transporter required for arbuscular mycorrhizal symbioses in rice and maize. Nat Plants 3(6):1–7

Olsson P, Thingstrup I, Jakobsen I, Bååth E (1999) Estimation of the biomass of arbuscular mycorrhizal fungi in a linseed field. Soil Biol Biochem 31(13):1879–1887

Ortas I, Sari N (2003) Enhanced yield and nutrient content of sweet corn with mycorrhizal inoculation under field conditions

Ortas I, Kaya Z, Cakmak I (2001) Influence of arbuscular mycorrhizae inoculation on growth of maize and green pepper plants in phosphorus- and zinc-deficient soil. In: Plant nutrition-food security and sustainability of agro ecosystems, pp 632–633

Paluch-Lubawa E, Polcyn W (2024) Tissue-specific accumulation of PIP aquaporins of a particular heteromeric composition is part of the maize response to mycorrhiza and drought. Sci Rep 14(1):21712

Parniske M (2008) Arbuscular mycorrhiza: the mother of plant root endosymbioses. Nat Rev Microbiol 6(10):763–775

Pellegrino E, Öpik M, Bonari E, Ercoli L (2015) Responses of wheat to arbuscular mycorrhizal fungi: a meta-analysis of field studies from 1975 to 2013. Soil Biol Biochem 84:210–217

Pepe A, Giovannetti M, Sbrana C (2018) Lifespan and functionality of mycorrhizal fungal mycelium are uncoupled from host plant lifespan. Sci Rep 8(1):10235

Peterson RL, Massicotte HB, Melville LH (2004) Mycorrhizas: anatomy and cell biology. NRC Research Press

Porcel R, Redondo-Gómez S, Mateos-Naranjo E, Aroca R, Garcia R, Ruiz-Lozano JM (2015) Arbuscular mycorrhizal symbiosis ameliorates the optimum quantum yield of photosystem II and reduces non-photochemical quenching in rice plants subjected to salt stress. J Plant Physiol 185:75–83

Prasetyo J, Ginting C, Akin H M, Suharjo R, Niswati A, Afandi A, Adiwijaya R, Sudiono S, Nurdin M (2021) The effect of biological agent and botanical fungicides on maize downy mildew. Biodiversitas J Biol Divers 22(4)1652-1657

Pu C, Yang G, Li P, Ge Y, Garran TA, Zhou X, Shen Y, Zheng H, Chen M, Huang L (2022) Arbuscular mycorrhiza alters the nutritional requirements in Salvia miltiorrhiza and low nitrogen enhances the mycorrhizal efficiency. Sci Rep 12(1):19633

Rabie GH (2005) Contribution of arbuscular mycorrhizal fungus to red kidney and wheat plants tolerance grown in heavy metal-polluted soil. Afr J Biotechnol 4(4):332–345

Redecker D, Kodner R, Graham LE (2000) Glomalean fungi from the Ordovician. Science 289(5486):1920–1921

Rezvani M, Ardakani MR, Rejali F, Zaefarian F, Teimouri S, Noormohammadi G, Miransari M (2015) Uptake of heavy metals by mycorrhizal barley (Hordeum vulgare L.). J Plant Nutr 38(6):904–919

Ruiz-Lozano JM, Aroca R (2010) Host response to osmotic stresses: stomatal behaviour and water use efficiency of arbuscular mycorrhizal plants. In: Arbuscular mycorrhizas: physiology and function, pp 239–256

Ryan M, Angus J (2003) Arbuscular mycorrhizae in wheat and field pea crops on a low P soil: increased Zn-uptake but no increase in P-uptake or yield. Plant Soil 250:225–239

Schweiger P, Jakobsen I (1999) Direct measurement of arbuscular mycorrhizal phosphorus uptake into field-grown winter wheat. Agron J 91(6):998–1002

Shahabivand S, Maivan HZ, Goltapeh EM, Sharifi M, Aliloo AA (2012) The effects of root endophyte and arbuscular mycorrhizal fungi on growth and cadmium accumulation in wheat under cadmium toxicity. Plant Physiol Biochem 60:53–58

Shamshiri M, Fattahi M (2016) Effects of arbuscular mycorrhizal fungi on photosystem II activity of three pistachio rootstocks under salt stress as probed by the OJIP-test. Russ J Plant Physiol 63:101–110

Shibata R, Yano K (2003) Phosphorus acquisition from non-labile sources in peanut and pigeonpea with mycorrhizal interaction. Appl Soil Ecol 24(2):133–141

Shrivastava P, Kumar R (2015) Soil salinity: a serious environmental issue and plant growth promoting bacteria as one of the tools for its alleviation. Saudi J Biol Sci 22(2):123–131

Shtark O, Provorov N, Mikić A, Borisov A, Ćupina B, Tikhonovich I (2011) Legume root symbioses: natural history and prospects for improvement. Field Veg Crop Res 48(2):291

Smith S, Gianinazzi-Pearson V (1988) Physiological interactions between symbionts in vesicular-arbuscular mycorrhizal plants. Annu Rev Plant Physiol Plant Mol Biol 39(1):221–244

Smith SE, Read D (2008) Mineral nutrition, toxic element accumulation and water relations of arbuscular mycorrhizal plants. In: Mycorrhizal Symbiosis, p 145–VI

Smith SE, Read DJ (2010) Mycorrhizal symbiosis. Academic

Smith SE, Smith FA (2011) Roles of arbuscular mycorrhizas in plant nutrition and growth: new paradigms from cellular to ecosystem scales. Annu Rev Plant Biol 62(1):227–250

Subramanian K, Tenshia V, Jayalakshmi K, Ramachandran V (2009) Role of arbuscular mycorrhizal fungus (*Glomus intraradices*) in zinc nutrition of maize. J Agric Biotechnol Sustain Dev 1(1):29

Sukumaran S, Li X, Li X, Zhu C, Bai G, Perumal R, Tuinstra MR, Prasad PV, Mitchell SE, Tesso TT (2016) QTL mapping for grain yield, flowering time, and stay-green traits in sorghum with genotyping-by-sequencing markers. Crop Sci 56(4):1429–1442

Symanczik S, Lehmann MF, Wiemken A, Boller T, Courty P-E (2018) Effects of two contrasted arbuscular mycorrhizal fungal isolates on nutrient uptake by *Sorghum bicolor* under drought. Mycorrhiza 28(8):779–785

Talaat NB, Shawky BT (2014) Protective effects of arbuscular mycorrhizal fungi on wheat (*Triticum aestivum* L.) plants exposed to salinity. Environ Exp Bot 98:20–31

Tariq M, Iqbal H (2010) Maize in Pakistan–an overview. Agric Nat Resour 44(5):757–763

Tummaramatti S, Hegde L, Vijaykumar N (2016) Pot experiment: studies on effect of bio-fertilizers on buckwheat (*Fagopyrum esculentum* Moench). Ecol Environ Conserv 22:S55–S58

Watts-Williams SJ, Cavagnaro TR (2018) Arbuscular mycorrhizal fungi increase grain zinc concentration and modify the expression of root ZIP transporter genes in a modern barley (*Hordeum vulgare*) cultivar. Plant Sci 274:163–170

Watts-Williams SJ, Nguyen TD, Kabiri S, Losic D, McLaughlin MJ (2020) Potential of zinc-loaded graphene oxide and arbuscular mycorrhizal fungi to improve the growth and zinc nutrition of Hordeum vulgare and *Medicago truncatula*. Appl Soil Ecol 150:103464

Wertheim F, Douds D, Handley D, Hutton M (2014) Evaluating the potential of arbuscular mycorrhizal fungi to boost yields in field-grown leeks. J NACAA 7(1)

Xu H, Lu Y, Tong S (2018) Effects of arbuscular mycorrhizal fungi on photosynthesis and chlorophyll fluorescence of maize seedlings under salt stress. Emirates J Food Agric 30(3):199–204

Zhang X, Cai X (2011) Climate change impacts on global agricultural land availability. Environ Res Lett 6(1):014014

Zhang X, Wang L, Ma F, Yang J, Su M (2017) Effects of arbuscular mycorrhizal fungi inoculation on carbon and nitrogen distribution and grain yield and nutritional quality in rice (*Oryza sativa* L.). J Sci Food Agric 97(9):2919–2925

Zhang Q, Wang S, Xie Q, Xia Y, Lu L, Wang M, Wang G, Long S, Cai Y, Xu L, Wang E (2023) Control of arbuscule development by a transcriptional negative feedback loop in Medicago. Nat Commun 14(1):5743

Zhang J, Zhao R, Xia LI, Zhang J (2024) Potential of arbuscular mycorrhizal fungi for soil health: a review. Pedosphere 34(2):279–288

Zhao J, Zhong L, Zou L, Zhang C, Peng L, Xiao W, Zhao G (2014) Efficient promotion of the sprout growth and rutin production of tartary buckwheat by associated fungal endophytes. Cereal Res Commun 42:401–412

Zhu X, Song F, Liu F (2017) Arbuscular mycorrhizal fungi and tolerance of temperature stress in plants. In: Arbuscular mycorrhizas and stress tolerance of plants, pp 163–194

Cultivating Excellence: Mycorrhizal Fungi Quality Management, Production, and Application

Prasann Kumar, Shipa Rani Dey, and Joginder Singh

Abstract

The chapter "Cultivating Excellence: Mycorrhizal Fungi Quality Management, Production, and Application" delves into the crucial significance of mycorrhizal fungi in augmenting soil health, advancing sustainable agriculture, and enhancing plant productivity. This study examined the complete life cycle of mycorrhizal fungi, encompassing production and application, emphasising the need to uphold quality standards throughout the process. This chapter outlines innovative methods for large-scale production, standardised quality assurance protocols to ensure consistent performance, and practical approaches to integration into modern agricultural systems. It offers a strategic framework to maximise the benefits of mycorrhizal fungi across diverse agricultural settings, thereby enhancing crop yields and contributing to sustainable farming practices by implementing effective management and application techniques.

Keywords

Agriculture · Environment · Mycorrhiza · Productivity · Quality · Soil health

P. Kumar (✉)
Department of Agronomy, School of Agriculture, Lovely Professional University, Phagwara, Punjab, India

S. R. Dey
Department of Agronomy, Faculty of Agricultural Sciences, GLA University, Chaumuhan, Uttar Pradesh, India

J. Singh
Department of Botany, Nagaland University, Lumami, Nagaland, India

Y. Siddiqui et al. (eds.), *Sustainable Mycorrhizal Cultivation*,
https://doi.org/10.1007/978-981-95-1778-7_11

1 Introduction

Mycorrhizal fungi have broader implications than their direct interactions with plants. They are significant players in sustainable agriculture and environmental health. Modern agricultural practices face challenges such as soil degradation, nutrient depletion, loss of biodiversity, and excessive use of chemical fertilisers and pesticides (Dey et al. 2023a; Hidangmayum et al. 2023). Mycorrhizal fungi present nature-based solutions as bio-fertilisers, biocontrol agents, and soil enhancers. The most widely documented role of mycorrhizal fungi is to enhance nutrient acquisition, which primarily influences phosphorus acquisition. Phosphorus is abundant in soils but occurs in immobile forms that are scarcely available to plants. The extensive network of mycorrhizal hyphae considerably extends root length, facilitating nutrient acquisition across larger soil areas other than the rhizosphere. The role of mycorrhizal fungi in nutrient acquisition reduces dependence on synthetic fertilisers that negatively affect the environment (Devi et al. 2023b; Sharma et al. 2023b). Additionally, they facilitate better nitrogen uptake by directly facilitating nitrogen uptake and competition by dissolving the nitrogen-fixing bacteria. Several biotic and abiotic stress factors were present or alleviated by the presence of mycorrhizal fungi. For instance, fungal interactions with plants enhance plant water uptake and status, thereby enhancing drought tolerance. Extensive mycorrhizal networks enhance the soil structure and tillage, making the soil more porous and infiltrative. They also play a role in salt and heavy metal stress, and in alleviating soil-borne diseases. However, this role is vital in mitigating the effects of climate change and desertification and maintaining high productivity. Carbon sequestration and soil health: mycorrhizal fungi are critical in soil organic carbon and carbon sequestration. The arbuscular mycorrhizal fungi (AMF) produce glomalin, a glycoprotein essential for creating stable soil aggregates and serving in soil carbon storage (Kumar and Pathak 2019; Sharma et al. 2023c). The role of mycorrhizal fungi in soil carbonisation and aggregation prevents soil erosion, which is a vital factor for maintaining environmental health and preventing climate change.

Mycorrhizal fungi promote plant species biodiversity at multiple trophic levels. The interaction networks promote plant–plant interactions and competitiveness, influencing the plant community structure. The role of mycorrhizal fungi in biodiversity is critical in natural systems that are exposed to various forms of stress. However, mycorrhizal fungi promote crop rotation and intercrop development and support diverse agricultural land use practices. In addition, mycorrhizal fungi have the potential to mitigate environmental pollution. Mycorrhizal fungi's ability to immobilise heavy metals and degrade organic pollutants may be benefited from high soil pollution levels. Mycorrhizal fungi create a relationship that enhances the density of heavy metal accumulation in the soil or plants and helps in the biodegradation of organic pollutants. Such interactions are essential to ensure food security and mitigate pollution (Kumar et al. 2019; Kumar and Naik 2020).

This study provides key insights into mycorrhizal fungi's ecological and agricultural significance, particularly their role in promoting plant growth, enhancing soil health, and supporting ecosystem sustainability. It bridges fundamental biological research with practical agriculture and environmental management applications, offering farmers, policymakers, and researchers actionable knowledge.

This work aligns with global sustainability priorities, notably the United Nations Sustainable Development Goals (SDGs): Goal 2—Zero Hunger, Goal 13—Climate Action, and Goal 15—Life on Land. It also addresses the challenges in applying mycorrhizal fungi across various agroecological zones. According to the UN Convention to Combat Desertification, an estimated 10–20 million square kilometres of productive land in arid and semi-arid regions faces resource depletion (Table 1).

This study emphasises integrating mycorrhizal technologies into conventional and organic farming systems through capacity-building and public awareness initiatives. As a keystone species, mycorrhizal fungi contribute to nutrient uptake, soil structure, and ecosystem stability, making them essential for sustainable agriculture and environmental stewardship. Ultimately, this work highlights the transformative potential of mycorrhizal fungi in addressing global challenges such as food security and land degradation, advancing scientific research, and improving quality of life (Pathak et al. 2017; Kumar et al. 2018e).

The main goal of this study was to yield insights into mycorrhizal fungi as a whole and their importance for plant growth, soil health, and sustainability within ecosystems. This work is a good example of bridging the gap between too basic questions, like in this case, about mycorrhizal biology, and more applied outcomes for agriculture or environmental management. This study section is vast, starting with basic research and moving on to applied sciences and policy implications. The study aims to derive actionable insights for farmers, policymakers, and researchers by building on this knowledge base. The emphasis on sustained agriculture also accords with a critical area of global development: the UN SDG targets, particularly Goal 2—Zero Hunger, Goal 13—Climate Action, and Goal 15—Life on Land. Regional issues encountered when mycorrhizal fungi are applied in different agroecological zones must also be addressed. The UN Convention to Combat Desertification estimates that some 10–20 million square kilometres of the world's biological productivity, which humans use, lands in arid and semi-arid zones, are affected by resource depletion (Table 1). The study also underlines the imperative to mainstream mycorrhizal technologies into conventional and organic farming systems through programmes to enhance capacity and create public awareness. Mycorrhizal fungi are a keystone species in sustainable agriculture and environmental stewardship. Their multiple roles in plant nutrition, soil health, and ecological stability make them essential ingredients to meet the global challenges of food security and environmental degradation. The work also empowers mycorrhizal fungi to be potential game changers for improving the quality of life on Earth and advancing research (Pathak et al. 2017; Kumar et al. 2018e).

Table 1 Cultivating excellent mycorrhizal fungi

Category	Aspect	Best practices	Key considerations	Expected outcomes
Quality management	Strain selection	Use well-documented, high-performing strains suitable for target crops and environments. Ensure strains are sourced from reputable suppliers or research institutions	Match fungal strains with crop species and soil conditions. Consider local adaptability and symbiotic compatibility	Optimal colonisation, nutrient exchange, and crop growth. Enhanced resilience to environmental stress
	Viability testing	Regularly test spore viability before application. Employ quantitative PCR or microscopy to measure spore density and viability	Ensure fresh or well-stored inoculants. Maintain proper storage conditions in a calm, dry environment	Consistent efficacy and reliable inoculation results
	Contamination control	Use sterile production and packaging methods—test batches for contamination by other microbes or pathogens	Regular monitoring during production. Implement quality assurance protocols	Reduced risk of introducing unwanted pathogens. Improved product consistency
Production	Cultivation methods	Employ in vitro or substrate-based cultivation in a controlled environment or host plant system	Select cost-effective and scalable production methods. Optimise conditions for spore and hyphal development	High spore yield and robust fungal growth. Economic production processes
	Nutrient media	Use nutrient-rich media specific to mycorrhizal fungi, such as carbon-rich substrates	Balance nutrient levels to avoid overgrowth of non-target microbes. Use bio-safe additives to enhance fungal productivity	Improved growth and sporulation. Higher-quality inoculants

(continued)

Table 1 (continued)

Category	Aspect	Best practices	Key considerations	Expected outcomes
	Harvesting and formulation	Harvest spores carefully to maintain viability. Formulate inoculants as powders, liquids, or granules for easy application	Minimise spore damage during harvesting. Use stabilisers or carriers to prolong shelf life	Versatile products suitable for diverse agricultural systems. Increased product shelf life
Application	Delivery methods	Apply via seed coating, soil drenching, root dipping, or direct soil inoculation. Use precision tools for even distribution	Match the application method with the crop type and growth stage. Consider timing to coincide with critical growth phases such as transplanting or early root development.	Effective fungal colonisation and enhanced plant–fungal symbiosis.
	Integration with cropping systems	Combine fungi with compatible practices like reduced tillage, crop rotation, and organic matter management.	Avoid practices that disrupt fungal networks, such as deep ploughing or excessive fertiliser use. Integrate with other microbial inoculants or biostimulants	Sustainable soil health improvements. Increased crop yields and reduced input dependency
	Environmental conditions	Ensure favourable soil pH between 6 and 7 and adequate moisture for fungal colonisation. Apply during moderate temperatures to avoid stress on spores	Test the soil for chemical residues or conditions such as high phosphorus that inhibit fungal activity. Tailor strategies to specific climate and seasonal conditions	Higher colonisation success rates. Improved plant nutrient uptake and growth
Research and development	Genetic enhancement	Investigate genetic modification or selection to improve fungal resilience to extreme conditions such as drought or salinity	Ensure compliance with GMO regulations and public acceptance. Focus on ecological safety and ethical considerations	Advanced fungal strains with greater adaptability and performance

(continued)

Table 1 (continued)

Category	Aspect	Best practices	Key considerations	Expected outcomes
	Synergistic applications	Explore combinations with nitrogen-fixing bacteria, biocontrol agents, or biofertilisers	Study interactions between fungi and other soil microbes. Identify synergistic benefits for specific crops	Comprehensive soil and plant health solutions. Enhanced productivity and ecological benefits
Economic considerations	Cost-effectiveness	Use scalable production methods to reduce costs. Provide subsidies or incentives to encourage adoption by small-scale farmers	Assess the cost–benefit ratio in comparison to synthetic inputs. Develop region-specific pricing strategies	Broader adoption among farmers. Improved access to sustainable farming inputs
	Market accessibility	Develop user-friendly products with clear application guidelines. Expand distribution networks to rural and underserved areas	Leverage e-commerce and local agri-dealers for distribution. Offer training programmes and demonstrations to build farmer confidence	Increased adoption rates and market penetration
Environmental impact	Sustainability goals	Promote use as part of regenerative agriculture practices to restore soil health and reduce chemical input reliance	Align with global sustainability initiatives like carbon sequestration and biodiversity enhancement. Monitor long-term ecological impacts of fungal applications	Contribution to climate resilience and sustainable farming. Enhanced ecosystem services
	Monitoring and feedback	Implement feedback loops where farmers can report results and challenges. Conduct field trials to assess efficacy under diverse conditions	Build partnerships between farmers, researchers, and product developers. Use digital platforms for data collection and knowledge sharing	Continuous improvement in product design and application techniques

Source: Pathak et al. (2017) and Kumar et al. (2018e)

2 The Role of Mycorrhizal Fungi in Soil Health

2.1 How Mycorrhizal Fungi Improve Nutrient Uptake and Enhance Soil Structure

These organisms are essential components of soil for enhancing nutrient uptake, ensuring a good soil structure, and making them vital to sustainable agriculture and ecosystem health. These symbiotic fungi drastically increase the efficiency of nutrient acquisition by extending the root system across the soil space using a network of fungal hyphae. These hyphae grow deeper into the soil matrix beyond the plant root depletion zones, tapping nutrient pools otherwise inaccessible to plants. Moving immobile nutrients, such as phosphorus, is crucial for plant growth. They are often insoluble in many soils, so plants cannot use them. Including mycorrhizal fungi in the growing medium ensures that this life-essential element remains readily available in the more active orthophosphate form. For example, forming arbuscules by AMF within the root cortical cells is a key adaptation that effectively transfers phosphorus and other nutrients (nitrogen, potassium, zinc, and copper) from the fungus to the host plant. This transfer of nutrients is bidirectional, with plants providing fungal carbon from photosynthesis, resulting in mutualistic associations that are overarching for plant nutrition and growth (Pankaj et al. 2012a; Kumar et al. 2018b).

In addition to nutrient uptake, mycorrhizal fungi improve the soil structure, which supports reuniting all these nutrients and fixing them in a mineralised form attached to the roots. The hyphal networks of these fungi physically hold soil particles, promoting the formation of soil aggregates and leading to increased porosity, water infiltration, and aeration in soils. AMF further enhances antigenicity by releasing glycoprotein glomalin, a sticky glue exudate that adheres to soil particles to protect them from erosion. Improved soil aggregation not only leads to improved physical properties of the soil but also creates micro-environments to support the diversity and activity of microbes. The association with soil organic matter aids in the decomposition of organic residues, allowing nutrients to be released into the soil; thus, nutrient cycling is further enhanced (Kumar et al. 2013).

In addition, mycorrhizal fungi improve soil fertility by altering the chemical composition of the rhizosphere of the plant roots (root zone). This process of fungi's release of organic acids and enzymes helps to solubilise bound nutrients and makes them available to plants. These are vital biochemical processes in soils that lack sufficient nutrients, a common feature in many regions. By enhancing nutrient uptake and reducing synthetic fertiliser requirements, mycorrhizal fungi promote sustainable agriculture, thereby reducing environmental pressure on chemical inputs. In addition, beneficial effects on soil structure and fertility promote increased water retention and resistance to compaction, underpinning sustainable land use in areas with changing climates.

Altogether, mycorrhizae act as natural collaborators for improving nutrient absorption and soil formation, providing a high ecological sustainability level for today's intensive farming practices. These symbioses help to cycle nutrients, stabilise soils, and reduce dependency on chemical inputs, thus supporting soil health, productivity, and long-term ecosystem functioning. Through their multi-functional roles, they could serve as pivotal tools in agroecological practice; therefore, it is indispensable to incorporate them within the necessary agricultural systems for food security and environmental homeostasis (Pankaj et al. 2012b).

2.2 Impact on Water Retention, Root Development, and Plant Resilience to Stress

Mycorrhizal fungi play a crucial role in the capacity of plants to conserve water, develop roots, and resist stress, thus rendering them exceedingly valuable allies in sustainable agricultural practices and ecosystem management. Mycorrhizal fungi create an extensive network of hyphae that span thousands of particles away from roots and effectively enhance the absorptive surface area of plants. The extended network improves the capacity of plants to absorb water from the soil, particularly under drought conditions. Mycorrhizal hyphae enter micropores in the soil, which are too small for roots to access, enabling plants to extract moisture from otherwise unavailable resources. This increased water uptake not only helps plants in drought conditions stay hydrated but also benefits overall plant performance and might play an essential role in maintaining the health of plants and productivity under limited water availability (Kumar and Mistri 2020; Kumar et al. 2021c). Additionally, mycorrhizal fungi help soil aeration and hold more water. The fungi help form soil aggregates that improve soil porosity and water-holding capacity. The granularity of grouped soils allows water to remain while maintaining adequate seepage and air circulation to nourish the plant ecosystem.

Mycorrhizal fungi promote water retention and influence root structure development by producing greater and quicker root systems. The symbiotic connection encourages root branches and growth, increasing the soil area within reach for nutrient and water uptake. This is extremely important in poor soils, where plants depend more heavily on fungi growing in the soil matrix. Mycorrhizal fungi also help absorb vital nutrients such as phosphorus and nitrogen for root structure and function (Das et al. 2022; Kumari et al. 2022). Better plant nutrition allows plants to create more profound and extensive root systems to explore greater soil volumes for water and nutrients. Increased root development is significant for crops grown in marginal soils or under stress conditions, as it allows them to access water and nutrients more economically when they fluctuate.

Another prominent feature of the ecological significance of mycorrhizae is that mycorrhizal fungi enhance the resistance to stress in plants. Plants colonised by mycorrhizal fungi exhibit improved tolerance to a wide range of biotic and abiotic stresses such as drought, salinity, heavy metal toxicity, and pathogen infection. In drought situations, mycorrhizal fungi exert control over water uptake and its movement in the plant by aiding root hydraulic properties related to cellular hydration status and transport, and calibrating solute content in stomata, lowering transpiration. These physiological adaptations help reduce drought-induced stress on the crop, allowing it to photosynthesise better and maintain its growth and yield (Goud et al. 2022; Upadhyay et al. 2023). The key functions of mycorrhizal fungi in salt stress alleviation in saline soils are improved supply and selective uptake of water and mineral nutrients, while excluding the excessive entry of sodium ions. Selective ion transport is essential for preventing ionic imbalance and the harmful effects of salinity on plant cells. Moreover, mycorrhiza can improve the antioxidant defence mechanisms of plants by decreasing oxidative damage due to environmental stresses (Chakraborty et al. 2021; Aley et al. 2022).

Mycorrhizal fungi contribute to crop health in at least as many ways as they help in drought resistance, and nine proven mechanisms can be identified with the combined effect of improving general plant growth performance even under optimal watering conditions and when not having to cope with soil drying, including protection from several forms of both abiotic and biotic stressors. Root colonisation by mycorrhiza occurring after infection induces systemic resistance, thus protecting the plant from disease severity by inducing its defence pathways. This protective effect is explained by producing bioactive compounds and signalling molecules that promote plant immunity. Additionally, the increase in nutrient acquisition due to mycorrhizal fungi results in the overall enhancement of plant vigour, providing increased capabilities for plants to recover from and withstand stressful events. Together, these features and the capacity of mycorrhizal fungi to increase water-holding potential and stimulate root growth, in addition to their resistance to stress, highlight the highly interactive role these organisms play in agricultural sustainability and ecosystem stability. Including them in farm practices provides a sustainable and natural way to meet the challenges of climate change, resource depletion, and environmental deterioration, guaranteeing the future productivity and health of crops and ecosystems (Kumar et al. 2021b; Reddy et al. 2022).

2.3 Benefits to Various Crops and Ecosystems

Mycorrhizal fungi offer many advantages to diverse crops and ecosystems, making them indispensable components for sustainable agricultural systems and natural ecosystems. Plant root colonisation, reciprocal nutrient and water uptake, stress tolerance, and plant health associated with acid-bacteria-promoted beneficial effects are key contributors to crop productivity and ecological balance. AMF have been shown to increase nutrient efficiency, especially in phosphorus-poor soils of staple crops such as wheat, rice, and maize. Mycorrhizal fungi can expand the functional root zone of their host plant by accessing nutrient pools that would otherwise be beyond the reach of plant roots through their extraradical hyphal networks. Incorporating mycorrhizal fungi for nutrient enhancement results in increased plant yield and better grain quality, which are essential for food security worldwide (Kumar et al. 2018d, 2020). Horticultural Crops: Increase in fruit quality, flavour, and nutritional content in tomato, cucumber, and strawberry owing to mycorrhizal fungi. Boosted nutritional uptake of phosphorus support and micronutrients through fungal symbiosis promotes metabolic processes required for successful fruit development, resulting in a higher content of critical vitamins and antioxidants. In addition, thanks to mycorrhizal fungi, flowering and fruit set occur throughout the plant, which, as we know, is crucial for obtaining commercial yields in horticulture. Mycorrhizal fungi can help increase nutrient and water absorption and build capacity for stress from soil-borne pathogens and other environmental stresses in cash crops such as coffee, tea, and cacao, ultimately guaranteeing crop productivity under variable climatic conditions. In particular, resilience is essential in tropical and subtropical regions, which are commonly affected by abiotic stresses, such as drought and salinity.

Right Wingers: Besides all its benefits to individual crops, mycorrhizal fungi also help maintain ecosystem functioning. They recycle nutrients and break down organic matter, basic nutrient balance processes in natural and agricultural ecosystems. Mycorrhizal networks provide channels through which nutrients and carbon are exchanged among plants; network signalling can enhance communication and cooperation among neighbouring plants in plant communities. This results in a high connectivity level, promoting plant diversity and productivity, especially in grasslands and forests. For example, ectomycorrhizal fungi in forests aid trees' nitrogen and phosphorus uptake, promoting long-term forest health and vigour. In addition, fungi contribute to carbon sequestration by encapsulating soil organic matter, decreasing greenhouse gas emissions, and diminishing the impacts of climate change (Karam et al. 2023; Devi et al. 2024).

Mycorrhizal fungi are essential for restoration and rehabilitation initiatives in degraded ecosystems. Their ability to ease soil structure and fertility and facilitate water retention helps to establish plant cover in disturbed or marginal soils. In semi-arid and arid regions, mycorrhizal inoculation promotes drought-tolerant species, which are key to desertification control and land restoration. Similarly, mycorrhizal symbionts can reduce plant toxicity by sequestering heavy metals in hyphae or modifying rhizosphere chemistry, which may immobilise contaminants, especially in mine tailings and heavy metal-contaminated soils. These activities render mycorrhizal fungi valuable for phytoremediation and restoration.

Beyond agricultural productivity and ecosystem restoration, mycorrhizal fungi play a critical role in conserving biodiversity in response to environmental disturbances. By helping plants use resources efficiently, mycorrhizae enable plants to grow in nutrient-poor or stressful situations, thus maintaining vegetation cover and the functions it provides to ecosystems. This array of functional capacities highlights the key role that mycorrhizal fungi play in ecosystems, influencing plant community dynamics and contributing to terrestrial stability and sustainable functioning. These multiple benefits of AM fungi indicate that crops and natural ecosystems will require mycorrhizal fungi in intensive agriculture and environmental management strategies, considering global challenges related to food security, climate change, and biodiversity conservation (Kumar and Choudhury 2024a; Raj et al. 2024).

3 Mycorrhizal Fungi Production Techniques

3.1 Methods for Culturing and Producing Mycorrhizal Fungi at Scale

The scale of culturing and producing beneficial mycorrhizal fungi has significant implications for research and development, sustainable agriculture, environmental restoration, commercial applications, etc. Manufacturing mycorrhizal fungi requires productive, scalable technologies to retain fungal functionality and viability for application across various ecological arenas. More than 60% of the approaches are available in the literature for culturing AMF in vitro or with knowledge of only receptive hosts. One of the most common methods is in vivo cultivation, which is performed by growing AMF associated with a host plant under laboratory-controlled conditions (Kumar 2023; Kumar and Kumar 2023; Jagadesh et al. 2024). Usually, this technique involves a sterile or soilless substrate inoculated with mycorrhizal spores and compatible host plants such as maize or sorghum. The symbiosis eventually results in the growth of fungal structures (hyphae, arbuscules, vesicles, and spores), which can be harvested over time. This method is widely employed owing to its simplicity and ability to produce high-quality inocula consisting of all fungal propagules and rhizosphere microorganisms. However, the slow growth rate of AMF and labour-intensive processes impede this.

In vitro or axenic culture methods have been further developed to overcome these limitations and allow the fungus to grow within a controlled environment independent of any plant host. In vitro methods, such as root organ culture (ROC), one of the most well-established models, utilise excised plant roots, usually transformed with Agrobacterium rhizogenes, grown in a nutrient medium where fungal spores or hyphae can be established. This system enables the continuous propagation of AMF without contamination. In vitro methods allow for preparing high-quality and pure inoculations needed for research or commercial applications. However, these methods have high throughput and require specialised expertise and equipment, which are generally unavailable for large-scale production in resource-poor settings.

Another interesting approach for bulk production is substrate cultivation, in which mycorrhizal fungi are cultivated on organic or artificial substrates. These methods use substrates, such as vermiculite, perlite, or biochar, which are colonised by fungal spores and plant roots (host). The fungi in the colonised substrate can be harvested and used as an inoculant on a carrier. Substrate-based cultivation is suited for commercial applications because of its simplicity and scalability. The system has progressed by creating customised nutrient-rich substrates or adding growth promoters to help grow healthier mycelia (Dey et al. 2023b; Sharma and Kumar 2023).

The production upscaling of mycorrhizal fungi is essential for developing cost-effective and sustainable biofertilizers to enhance crop productivity. Furthermore, another biotechnological innovation around bioreactor systems requires further optimisation. These systems enable the mass production of fungal propagules with the help of controlled plant growth environments such as temperature, humidity, and aeration. Both liquid and solid-state fermentation systems have been used in oyster mushrooms and other ectomycorrhizal fungi (praised because they are easier to culture in axenic conditions than AMF). Bioreactors can deliver high productivity and consistency. Therefore, they are ideally suited for large-scale industrial applications but have a high upfront cost and require technical expertise.

Standardisation and formulation are mandatory for mycorrhizal inoculation to maintain viability and efficacy. Fungal propagules recovered from crop residues or culture filtrates are formulated into spore suspensions, granular, or powder-based products designed for specific agricultural and ecological settings. For these formulations to be effective, they must exhibit spore viability, purity, and infectivity. Creating shelf-stable inoculation with an enhanced shelf life is a significant area of research.

While much can be done to improve the cost-effectiveness and optimisation of cultivation methods for individual mycorrhizal fungal species in specific application contexts, challenges remain. New technologies, such as molecular tools for tracking fungal colonisation and genetic engineering to enhance specific fungal traits, are expected to address some of these issues. There was also a robust positive relationship between the increased mycorrhizal inoculum production and sustainability. When coupled with sustainable farming practices, our ability to move away from chemical fertilisers can be honest, and innovation such as this has an important part to play (Kumar et al. 2011a, 2021a; Srivastav et al. 2024).

3.2 Challenges in Production and Maintenance of Fungal Viability

There are significant obstacles to scaling the production and persistence of mycorrhiza to a scale suitable for broader application, limiting their use in agriculture, environmental restoration, and commercial ventures. The primary factors responsible for these challenges are the complexity of their biology, obligate symbiotic ways of life, and technical difficulties associated with large-scale cultivation. One of the major hurdles is that AMF are obligate biotrophs that require living hosts for growth and reproduction. AMF cannot be grown in vitro as saprotrophic fungi; the inner cells are obligate biotrophs, meaning they are non-axenic cultures and must be seen within an in vivo or root organ culture system. This dependence leads to increased production complexity, cost, and time-consuming activities, as maintaining a commercially sustainable host plant or a viable root system under controlled conditions entails considerable labour, expertise, and resources (Kumar et al. 2017; Kumar and Dwivedi 2020).

Another challenge is to keep the mycorrhizal fungi viable and infective during all stages of production. Storage conditions such as temperature, humidity, and storage duration can significantly affect fungal propagules, including spores, hyphae, and colonised root fragments. Poor handling or inappropriate storage can cause fungal propagules to lose viability quickly, and an insufficient inoculum can prevent the pathogen's spread. Some examples of the general types of stress that weaken conidia resistance include desiccation, UV light exposure, or temperature shock during storage and shipping, which can negatively impact spore germination rates and hyphal growth potential. In addition, fungal propagules are susceptible to contamination by saprophytic fungi and opportunistic pathogens during their production stages, which can outcompete the mycorrhizal species they contain, thus negatively affecting the quality of the inoculum.

The diversity of fungal species and their interactions with specific hosts increases the complexity of production. There is a need to have a variety of mycorrhizal species or strains for different crops and, in some cases, specific ecosystems within them to ensure the best result. However, scaling species such as this is difficult because they grow slowly, and it is difficult to determine whether different strains work better than others. It is also important to note that the compatibility of mycorrhizal symbiosis with agriculture, including whether agrochemicals and soil amendments are used, should be appropriately managed, because some chemicals and practices can result in fungal colonisation or survival (Dwivedi and Kumar 2011; Saini et al. 2023).

Financial reasons also inhibit the general application of mycorrhizal inoculations. The prices make these products less available for smallholder farmers and resource-limited regions because of the labour-intensiveness of production methodologies and strict quality control needs. To overcome these survival barriers, it is necessary to develop scalable and cost-effective production strategies that preserve the viability and activities of fungi. In addition to the difficulty in screening fungi as they evolve, the absence of standardised protocols can cause inconsistency or deterioration of products when produced for commercialisation.

Progress in biotechnology and molecular biology could provide solutions to some of these problems. Several tools, such as genetic engineering and molecular markers, can aid in identifying high-performing strains with better viability and infectivity, which enhance the propagating traits of fungal strains. Current bioreactor systems are not well suited to AM fungal production; however, improvements in liquid culture approaches, shared by ectomycorrhizal fungi and adapted for AMF cultivation, may offer future solutions. Moreover, preparing shelf-stable formulations, such as spore-embedded beads or granulated products, might provide better handling and distribution on the one hand and preservation of fungal viability on the other hand (Kumar et al. 2022b; Devi et al. 2023a).

However, our knowledge of mycorrhizal fungi's ecology and life cycle is limited. Interactions are similar between fungi, their host plants, and compounds characterised by soil microbiota, with less extensive understanding, making production systems complex to adapt to contrasting environmental and agricultural conditions. A comprehensive solution to these challenges requires an interdisciplinary approach incorporating breakthroughs in microbiology, agronomy, and industrial engineering. Further studies and funding may help optimise the production and conservation of stable mycorrhizal fungi, making this an even more extensive strategy for improving agricultural sustainability, ecosystem recovery, and environmental health.

3.3 Innovations in Spore Production and Carrier Materials for Distribution

The current understanding of mass sporulation processes and carrier-based production technologies is advancing the development and large-scale application of mycorrhizal fungi-based geobiont inoculants. The structure of the mycorrhizal inoculum is spore production, and the advances are due to the availability of spores as the main propagules for colonising host plants. Recent progress has included improving in vivo and in vitro production systems to increase the spore yield and quality. This method is innovative because it utilises in vivo systems employing specific host plants highly compatible with individual fungal species and can sustain prolonged colonisation. These hosts are often cultivated under controlled conditions with sterilised substrates, ensuring uniform spore production and reducing contamination risk. In addition, automated monitoring and precision agriculture tools within these systems provide insights into plant conditions, soil quality, and fungal propagation, enabling increased yields with minimal resource inputs (Kumar et al. 2023b; Kumar and Choudhury 2024b).

Root organ culture (ROC) has been well defined using various in vitro systems, including transformed roots that grow continuously on media with sterilised components. Subsequent improvements in culture media formulations have optimised the supply of nutrients and signalling compounds essential for fungal growth and sporulation, contributing to further increases in spore production. The natural rhizosphere has been mimicked with beneficial microbial consortia co-culture systems that induce spore production and ameliorate inoculum activity. There are also bioreactor technologies for ectomycorrhizal fungi and, more recently, AMF, which allow larger fermentations in the liquid or solid state. The advantages of reproducible environmental conditions, scalability, and potentially continuous production cycles are the main reasons for their potential industrial applications in bioreactors.

The selection of carrier materials that enable efficient distribution without compromising fungal viability is key to the success of mycorrhizal fungi in agricultural and ecological applications. While traditional carriers (such as soil or sand) are simple, they can be bulky and lack shelf stability owing to excessive particle-to-particle interlinking. Nowadays, innovations have leaned more heavily towards using lightweight, inert, and biodegradable materials that form a protective matrix for fungal propagules. Biochar, vermiculite, and perlite are popular because of their ability to retain moisture, which is essential for spore survival. Biochar, in particular, provides dual benefits as it traps carbon, enhances soil health, and serves as a carrier material (Devi et al. 2023c; Sharma et al. 2023a; Sinam et al. 2023).

This adequately explains why encapsulation technologies are a significant breakthrough in carrier development. Spores can be encapsulated in alginate beads or polymer-based microspheres to improve spore output during spore preparation and stability during storage and application. This serves as a controlled-release mechanism, minimising the harmful effects of desiccation and other environmental stresses. The capsules can include growth stimulants or nutrient sources, which trigger rapid colonisation of fungi upon application. An alternative innovative method is that hydrogels can carry propagules, such as protecting the inoculum from biotic stresses, increasing water reservoirs in soils, and improving fungal/host plant water regimes.

Freeze-drying and vacuum packaging methods have been used to improve shelf life and transportability, maintaining the viability of fungal propagules for many years. Antifungal agents that act only against contaminant fungi, leaving untreated mycorrhizal fungal species, have been assessed to maintain inoculum purity. Furthermore, developments in nanotechnology have allowed for the development of nanocarriers that precisely deliver spores to the root zone and provide protection against abiotic stress (Table 2). It is important to note that spore production and carrier-material innovations are the most significant gap between laboratory research and field application, ensuring the effective percolation of mycorrhizal fungi into agricultural and ecological management approaches. We further expect that as these technologies advance, sustainable agriculture will be transformed worldwide by reducing reliance on chemical inputs and promoting crop productivity and environmental restoration (Kumar et al. 2022c; Kumar and Dey 2024).

4 Quality Management in Mycorrhizal Fungi Production

4.1 Importance of Quality Control Throughout the Production Process

The production of mycorrhizal fungi follows many quality control steps to ensure the viability, efficacy, and safety of these inocula for various agricultural, ecological, and commercial purposes. Such strict control is essential because mycorrhizal fungi are incredibly sensitive organisms, and because of the importance of these species in establishing symbiotic associations with plants. The first step in spore cultivation quality control is the establishment of qualitatively distinct fungal propagules. This procedure must be maintained throughout the process, from the initial stages of spore isolation to the final inoculation distribution (Kumar and Choudhury 2024c; Kumar and Singh 2024). Quality control is required to measure the viability of fungal spores, hyphae, and other propagules. This entails examining the physical health of fungi, primarily during their life cycle and post-harvest treatment maintenance. Standard viability tests, including spore germination assays and assessments of their capacity to colonise roots, prove that fungi can form effective symbioses in specific fields. These tests are necessary because the production or storage of the inoculum under suboptimal conditions may cause degradation of fungal propagules, making them ineffective when applied. Then, very little or no colonisation can be achieved.

Another essential factor of the respective control is protection against non-target organisms, including saprophytic fungi, bacteria, and pathogens. In some cases, the efficacy of mycorrhizal inocula may be compromised indirectly by contaminants that outcompete or inhibit target fungi (which could result in a reduction in plant benefits and undesirable effects). Sterilisation of growth media, sterile injection

Table 2 Techniques for mycorrhiza cultivation

Technique	Description	Advantages	Challenges	Applications
In vivo cultivation	Culturing fungi on host plants in controlled environments such as greenhouses	Produces high-quality spores and symbiotic structures	It is time-consuming and requires extensive space	Ideal for research and premium inoculant production
In vitro cultivation	Growing fungi in axenic cultures using nutrient media in sterile conditions	Enables precise control of growth conditions and reduces contamination risks	Expensive and technically demanding	Laboratory-scale production and research purposes
Substrate-based cultivation	Organic substrates like vermiculite, peat moss, or perlite are used for fungal growth	Cost-effective and scalable for mass production	Substrate contamination and sterilisation challenges	Commercial inoculant production
Bioreactor systems	Employing bioreactors to culture fungi in liquid- or solid-state systems	High spore yield and controlled production conditions	Requires technical expertise and infrastructure investment	Large-scale industrial production
Root organ culture (ROC)	Growing fungi on excised roots in sterile media	Ensures consistent fungal quality and reduces variability	Limited scalability and high production costs	Research and specialised inoculants
Spore extraction from soil	Collecting fungal spores from naturally infected soils	Access to diverse and locally adapted fungal strains	Labour-intensive and potential contamination with non-target microbes	Initial strain isolation and biodiversity studies
Spore bank maintenance	Storing fungal spores in sterile media for future use	Long-term preservation of genetic diversity	Requires strict storage conditions to maintain viability	Germplasm conservation and inoculant development
Co-cultivation with plants	Cultivating fungi in association with live host plants in nursery systems	Promotes natural symbiotic development	Slow process and limited production scale	Nursery inoculants and plant propagation
Liquid media culture	Growing fungi in liquid media with specialised nutrients	Enables high-density fungal production	Risk of fungal aggregation and reduced viability in liquid media	Laboratory-scale and bioreactor systems

(continued)

Table 2 (continued)

Technique	Description	Advantages	Challenges	Applications
Solid-state fermentation (SSF)	Solid substrates like agricultural waste are used for fungal growth	Sustainable and suitable for large-scale production	Substrate selection and contamination risks	Commercial production and waste recycling
Carrier-based formulation	Embedding fungal spores in carriers like clay, granules, or gel	Facilitates application and prolongs shelf life	Requires compatibility testing between carriers and fungi	Field applications and seed coating
Native fungal isolation	Extracting fungi from indigenous soils and plants	Access to locally adapted strains with high ecological compatibility	Requires expertise in fungal identification and isolation	Region-specific inoculant production
Freeze-drying (lyophilisation)	Preserving fungal spores by freezing and drying under vacuum	Extends storage life and ensures spore viability	High setup costs and requires sophisticated equipment	Long-term storage and transportation
Spore encapsulation	Encapsulating spores in protective coatings to enhance viability and targeted delivery	Improved shelf life and ease of handling	Requires precision in encapsulation techniques and material selection	Controlled release formulations
Integrated multi-stage systems	Combining methods such as in vitro, bioreactors, and substrate-based cultivation for optimised yield	Maximises production efficiency and adaptability for various scales	Complex process design and high initial investment	Scalable production systems

Source: Kumar et al. (2022c) and Kumar and Dey (2024)

techniques, and periodic microbiological testing help avoid contamination. Molecular diagnostic tools (e.g. PCR) and next-generation sequencing (NGS) often confirm fungal cultures' purity and the purity of unwanted microorganisms. This precision and reliability allow the preservation of the integrity of the products throughout every stage of production to guarantee superior quality.

Standardisation of fungal propagule counts is another critical quality control area because the final procedure must have an efficient concentration of viable propagules to be effective as an inoculum. The final product was tested using accurate quantification methods (spore counting using a microscope and flow cytometry) to determine the density of fungal propagules. This guarantees continued quality in the production of each inoculum and is expected to provide consistently high efficiency for every application. Quality control also requires confirming the infectivity and functionality of the fungi, as the presence of propagules does not predict that successful rain entering or colonisation will occur, regardless of all symbiotic prospects. Under controlled conditions, bioassays with model plants are frequently used to understand better mycorrhizal fungi's colonisation efficiency and nutrient uptake capacity (Kumar et al. 2011b, 2018a).

Specific important parameters of storage and packaging influence the quality of mycorrhizal inocula. Quality control measures should also confirm that the resulting storage of fungi does not affect their viability at selected storage temperatures, relative humidity, or light exposure over a specified shelf-life period. One of the main developments has been using durable packaging materials such as vacuum and moisture-resistant containers. Standardised procedures for checking the viability and infectivity of stored inocula at regular intervals are obligatory to maintain their effectiveness until the final application.

In addition, compliance with regulatory standards and certifications is part of quality control. Many countries have regulations regarding the preparation and marketing of mycorrhizal products, such as safety, efficacy, and labelling. Producers should prove that their inoculum is free from harmful agents, claim propagule counts, and deliver expected agroecological outcomes. Certification of mycorrhizal products by accredited organisations makes farmers, researchers, and consumers more credible in marketing their products.

In addition to maintaining product quality, effective quality control systems help make mycorrhizal fungal production systems sustainable and scalable. These cost-saving measures decrease food production costs and their negative impact on the environment by reducing waste while improving efficiency. Quality control also aids in research and development, as consistency and reliability of inocula are essential for experimental studies, promoting new directions to develop the field of mycorrhizal technology (Kumar et al. 2023a, 2024).

4.2 Standardisation Techniques for Ensuring Fungal Potency and Effectiveness

Therefore, it is essential to have standardised methods to promote the viability and performance of AMF to guarantee their functional stability in agricultural, ecological, and commercial situations. Mycorrhizal fungi are crucial for plant nutrition and sustainability, as they contribute to essential functions that support plant health, nutrient availability, and stress tolerance. However, the success of their application was highly dependent on inoculum quality and performance requirements. The standardisation process includes a battery of scientific, technical, and procedural validations to verify the fungi's viability, infectivity, and efficiency. These methods are essential for earning end-user trust, controlling utility for mass adoption, and being a legally compliant organisation. Quantification, viability determination, and colonisation potential are the most important aspects to be standardised in studying fungal propagules (Devi and Kumar 2023a; Kumar and Choudhury 2024d).

The core essential calibration exercise counts fungal propagules, such as spores, hyphal fragments, and colonised root pieces. Correct counting led to the highest inoculum products with the required propagules for reaching the desired colonisation levels under field conditions. Spore counts are usually carried out by microscopic means, which enable direct counting of viable spores in a known volume of inoculum. Flow cytometry is a state-of-the-art tool for quantifying flow cytometric fungal propagules and offers better throughput and accuracy than plate-based counting. The technology can distinguish viable from non-viable propagules using fluorescence sorting, allowing them to deliver consistent and effective inoculation (Kumar et al. 2018c; Sharma et al. 2023d).

Cell viability assays also provide an essential endpoint for standardisation. This was assessed through assays that evaluated the propagules' physiological competence and germination capability. Two typical test methods are spore germination tests (which usually involve incubation of spores on nutrient media or in a germinator environment simulating the rhizosphere and quantification of total counts as well as viable counts. For example, we showed that FDA staining can quickly identify metabolically active spores, providing a simple and practical basis for viability. High viability rates are critical for the efficacy of mycorrhizal products; non-viable propagules cannot establish symbiosis with plants.

Prototype trials with model plants have evaluated the actual capacity of a fungus to attack and colonise plant tissues. A typical example of these tests involves cultivating plants in sterilised substrates inoculated with mycorrhizal fungi: root samples are taken after a pre-set time to measure the degree of fungal colonisation. The structure of the roots, where fine fibre-like arbuscules, vesicles, and hyphae (all parts of the fungus), reveals how this somewhat invisible ally interacts with its host. For standardising the infectivity of different inoculum batches, it is essential to use quantitative measures like % of root length colonised.

Molecular techniques have revolutionised the ability to standardise mycorrhizae based on accurate species and strain identification and characterisation. Molecular techniques such as (q) PCR provide potent methods for detecting specific fungal DNA sequences with high sensitivity and specificity, demonstrating that the inoculum is pure and correctly identified. High-throughput sequencing approaches, such as next-generation sequencing (NGS), enable more detailed exploration of the composition of fungal communities and confirm that particular beneficial species are present while excluding contaminants. However, they are now becoming an integral and compulsory part of quality control and industrial standardisation at large production sites (Kumar et al. 2022a; Kumari et al. 2023).

Standardisation also includes an examination of the effectiveness of mycorrhizal fungi when used under controlled and field conditions. Controlled environmental experiments measure the ability of fungi to improve plant growth, nutrient uptake, and stress tolerance under defined soil and climate conditions. These broad-scale field trials in various agricultural and ecological settings sustain and expand the findings by having more variability in terms of environmental factors. Data from these trials were used to produce performance benchmarks to ensure that the inoculum consistently provided benefits between applications.

The physicochemical properties of the inoculum, including pH, moisture content, and particle size, were standardised in addition to biological and functional parameters. From a formula standpoint, these measurements determine how long something lasts on the shelf and how well it can and should be applied to the skin for performance. For example, the proper moisture content will avoid desiccation, and particle size standardisation ensures even distribution during application (Devi and Kumar 2023b; Dey et al. 2023c).

Compliance with regulations is another reason for standardisation efforts. Many countries have defined guidelines for manufacturing and labelling mycorrhizal fungal inoculants, followed by new operational procedures for producers that require measuring propagule counts, viability, and multiple performance metrics. This regulation ensures product quality and encourages everyone to engage in international trade and market access. Standardisation procedures also include packaging and storage instructions. Although formulation technology has been refined to prevent premature loss (regarding encapsulation and carrier materials), it is now better regarded as a fuel for retaining high fungal value over longer durations. Testing the viability and functionality of stored inoculations regularly guarantees that products retain efficacy over their shelf life. Standardisation processes are key for securing mycorrhizal fungi inoculum that fulfils the demands of farmers, researchers, and environmental managers regarding biological efficiency. These technologies serve as strong counter-platforms for preserving the potency and efficacy of mycorrhizal fungi, thus promoting their operational success in sustainable agricultural scenarios or ecosystem restoration (Kumar and Dwivedi 2020; Kumar and Choudhury 2024b; Srivastav et al. 2024).

4.3 Testing Methods for Identifying Viable Spores and Contaminants

Methods for testing viable spores and contaminants require a primary emphasis on quality, effectiveness, and safety policy to ensure the production of mycorrhizal inoculum. Since mycorrhizal fungi are crucial to plant health and purity, mycorrhizal inoculation must be able to promote growth through bacteria-free media. Many methods have been designed to ensure fungal spore viability and the absence of contaminants. Regarded as the headquarters for your misery in life, the rhizosphere is a complex, interdependent world of plants and microorganisms. Various techniques are used to assess the viability of fungal spores and determine whether potential sources of shame, such as other fungi, bacteria, or pathogens, might defeat the purpose of mycorrhizal inoculation. The previously described techniques were small-scale (Kumar et al. 2018d; Kumar et al. 2021a; Kumar 2023). There are various methods to determine the viability of fungal spores and the presence of potential contaminants. These include other fungi, bacteria, or agents of disease in plants, which could become widespread with mycorrhizal inoculation. The spore germination test is one of the most commonly employed methods for evaluating spore viability. This test involves spraying or adding to a suitable substrate a small vial of fungal spores prepared using means akin to enacting what such organisms think nature is in the divine wisdom Mother Nature provided.

As an effective method to assess the viability of spores, FAD is a non-toxic compound taken up by living cells and corresponding metabolites in extracts. This fluorescent signal can be observed under UV light. Fluorescent light corresponds to the concentration of spores, which can then be used to measure metabolic activity. These properties allow rapid viability determination without the need for germination. This method is beneficial because it applies equally well to large and small sample sizes, providing results in seconds with high accuracy. Similarly, FDA staining cannot distinguish between cases potentially involved in creating mycorrhizal or arbuscular symbioses and those simply showing metabolic activity or stress. In addition to testing for viability, checking for contamination is essential in mycorrhizal inoculum production. Such contamination can originate from various sources, such as other fungi, bacteria, and viral pathogens. It may outcompete or suppress the target mycorrhizal fungi, thereby reducing the effectiveness of the inoculum in transmitting benefits. Producers often use selective media to detect contaminants that allow mycorrhizal fungi to grow but inhibit other microorganisms (Kumar et al. 2011b, 2018d, 2021a; Kumar 2023; Sinam et al. 2023; Kumar and Dey 2024). For example, media containing antifungal or antibacterial agents can selectively isolate mycorrhizal fungi and minimise the growth of unwanted organisms. Once fungal growth is detected, morphological identification can help differentiate between the target species and any potential contaminants.

Molecular techniques have become increasingly crucial for contaminant detection. For instance, PCR can amplify specific DNA sequences, even at low concentrations, and identify the fungal species on these targets. For example, PCR assays with unique genetic markers, such as ribosomal RNA genes or specific fungal ITS regions, can show that a species is correctly identified and no non-target organisms are present. In addition, quantitative PCR can be carried out to estimate the number of fungal spores in the sample and determine the spore concentration more accurately. This method has high sensitivity and specificity; therefore, such systems have become indispensable tools for modern-day quality assurance during mushroom growth.

Another cutting-edge molecular technique for evaluating contamination and fungal diversity is next-generation sequencing (NGS). This method is especially effective at detecting low-level contaminants that would otherwise escape detection using standard culturing techniques. By sequencing the DNA from the inoculum, producers can obtain a detailed and objective picture of the microbial and fungal communities, enabling them to identify and control contamination risks effectively.

Tests for identifying separate viable spores from inorganic impurities represent a critical step in maintaining a mycorrhizal inoculum's analytical accuracy and effectiveness. Microscopic techniques are still essential for defining fungal spores, contaminants, or residues. Through microscopy, one can observe the visible characteristics of fungal spores, cysts, septa, and, if contamination is possible, even obscure microorganisms. In addition to identifying the form and structure of fungal spores, microscopy can also observe regular or irregular contact between mycorrhizal fungi and plant roots. This gives us an insight into whether an inoculum will likely colonise a plant. Microscopy is an ancient and labour-intensive testing method. However, it is still one of the best methods for detecting the presence of pathogens and determining the physical characteristics of contaminated spores.

In addition, culturing techniques that require mycelial or spore inoculum on a growth medium and incubating the sample remain essential for contaminant identification. They can separate bacteria, fungi, and other pathogenic microorganisms from the inoculum. Producers can identify unwanted organisms by their growth mode and colony morphology, and take necessary measures to eliminate them.

Finally, tests for identifying viable spores and contaminants are indispensable for maintaining the quality and effectiveness of mycorrhizal inocula. Germination tests, fluorescence staining, PCR, next-generation sequencing (NGS), and minimal (traditional) microscopy offer a universal approach to determine whether they are dead or contaminated fungal propagules. By using these methods in equal parts, reproducers are assured that their product includes no harmful microorganisms on one hand but is full of live spores to establish symbiotic relationships with plants. These testing techniques are indispensable for the successful real-world application of mycorrhizal inoculation in agriculture, forestry, and environmental restoration. They also make sure that these fungi lose all their abilities to support plant growth and maintain ecological balance.

5 Application Methods for Mycorrhizal Fungi in Agriculture

5.1 Techniques for Applying Mycorrhizal Fungi to Various Crops

Mycorrhizal fungi are symbiotic with plant roots and provide several methods for increasing crop growth, nutrient extraction, and disease management. Using mycorrhizal fungi in many crops helps to improve soil health and yield, especially in nutrient-poor soils. Applying mycorrhizal fungi to seeds is one of the most conventional strategies, and it can be used to treat seeds before sowing with fungal spores or mycelia. Thus, delivering the fungi at this stage of life ensures that the plants create a robust mycorrhizal network and helps cultivate their young. The spores or fungal culture are mixed with a carrier solution in liquid or powder form for seed inoculation and applied to coat seeds (Kumar et al. 2011a; Saini et al. 2023). This is wildly successful with crops such as cereals and legumes, for which early fungal establishment can extend the benefits of nutrient uptake (in particular phosphorus, which is often scarce in soil) over a longer term. A typical application is soil inoculation, for example, adding mycorrhizal fungi directly to the soil before (in a controlled manner) or after planting. This may include broadcasting a suspension of fungal spores, pre-inoculated organic carriers (compost) over the soil surface, or incorporation into the topsoil. Soil inoculation is used for crops grown in fields or larger agricultural systems because it encourages the fungal network to spread throughout the root zone of many plants. It is a common farming practice, especially for crops such as corn, wheat, and fruit trees, because it can increase their ability to absorb nutrients more efficiently throughout vast areas. For instance, foliar application has become a less traditional method for delivering mycorrhizal fungi, but is growing in popularity. Therefore, fungal spores are mixed with water and applied directly to the plant's foliage. Although not as effective at creating sustainable mycorrhizal connections in plants as seed or soil inoculation, foliar application can be a helpful way to quickly initiate mycorrhizae, giving the plant some resistance against stressors such as drought. It has particular utility for crops grown at high density or those that do not naturally develop strong root–mycorrhizal associations. Hydroponic inoculation represents a different approach for applying mycorrhizal fungi; hydroponic cultivation is widely used without soil. Mycorrhizal fungi can also be added to the nutrient solution in hydroponics, and the fungi can colonise the root systems of plants without any actual need for soil. This is particularly effective for crops such as tomatoes, lettuce, and peppers, which have their root environment precisely controlled and grow in a fungi-dominant state without competing with other soil microorganisms. Finally, mycorrhiza can be applied with different soil management practices, such as organic amendments, mulch, and crop rotation, to maximise their effectiveness. For example, mycorrhizal inoculation can be used with organic matter, such as compost or biochar, to augment the establishment and activity of fungi by providing additional nutrients and improving soil structure. Furthermore, intercropping with species that support mycorrhizal growth can also enhance the ability to help other crops. Growers can use mycorrhizal fungi to increase crop productivity, reduce synthetic fertiliser inputs, improve soil health, and make agriculture more sustainable by choosing an appropriate technique for their particular crop composition, soil conditions, and farming practices (Kumar et al. 2024; Kumar and Singh 2024).

5.2 Best Practices for Inoculating Seeds, Seedlings, and Mature Plants

Inoculation of seeds, seedlings, and mature plants with beneficial microorganisms, such as mycorrhizal fungi or nitrogen-fixing bacteria, may be helpful from the perspective of plant growth, nutrient uptake, and disease resistance. Immunisation works best when there is a way and method by which the vaccine is administered. The most effective methodology should be applied to each growth level. The type of inoculant and the uniformity of its application are crucial for effective seed inoculation. Beneficial microorganisms can be formulated into liquid or powdered forms that can be applied to seeds. They often include excipients such as talc or clay for easy application and seed coverage. Placement should be handled carefully not to disturb the delicate seed exterior, so the inoculant will easily adhere without being flushed out during planting or sprouting. A popular approach is to coat the seed surfaces with microorganisms by adding a wetting agent to mix the inoculant with the slurry. The seeds were dried in a shaded area after coating the plant. It is particularly advantageous for legumes, cereals, and vegetables, where colonisation of the root rapidly and efficiently can make a huge difference in stress resistance and nutrient acquisition by the plant. The key to this is root inoculation when treating the seedlings. This method involves drenching or dipping seedling roots into a microbial solution before they reach the soil for planting or transplantation. Root dips are the most concentrated, as the inoculant is suspended in water, and the entire root system should be coated (Kumar et al. 2018c, 2023a). This was best done before the seedlings were sown in open soil to enable their strong symbiotic association with the roots. This is important, as in the absence of a host plant, these organisms grow better than what they will be allowed to do inside or on plants, leading to competition for sites once inoculated into water or onto seeds, and the individual handling of fragile seedlings may cause damage that can restrict root colonisation. It is optimally taken up by planting the inoculant in the morning or evening when seedlings are not heat- or light-stressed. The inoculation of mature plants requires a different method to be effective, as the plant has an established root system. Mature plants were best treated with soil injections. This technique involves the application of a microbial inoculant at the plant root zone by soil application. Inoculants can be used through broadcasting, with the soil before planting, or as a top dressing. A critical aspect is ensuring that the inoculant enters the root zone so that beneficial microorganisms begin building a symbiotic relationship with plant roots. For more significant applications, liquid inoculants can be applied through irrigation systems or as foliar sprays to the plant root zone. It is also helpful in conditions where there may be a need to quickly inoculate the soil matrix, especially in soil disturbance or high stress (Dwivedi and Kumar 2011; Sharma et al. 2023a; Kumar and Choudhury 2024b). A favourable soil environment leads to conducive conditions for microbial growth, which is ultimately required for effective seedling or adult plant inoculation. Helpful practices, such as proper irrigation, reduced chemical fertiliser input, and sufficient soil aeration, will increase the effectiveness of inoculants. Soil inoculation is also recommended to prevent the use of high doses of chemical pesticides or fungicides for the survival and activity of microorganisms. When inoculating crops with beneficial organisms, it is essential to do so at the right time and most effectively, given the operational constraints or environmental conditions for optimal growth. Inoculation in the seed (seed inoculum) or seedling phase should be performed when the plant is under non-stress conditions, and a favourable environment must be maintained for the survival of microorganisms associated with the rhizosphere, which may establish a beneficial relationship with the plant. Using these best practices will allow growers to take full advantage of the many benefits of inoculants, including enhanced plant growth, increased nutrient availability, improved disease resistance, and lower reliance on traditional chemical inputs.

5.3 Timing and Environmental Considerations for Maximising Efficacy

When applied to seeds, seedlings, or mature plants, the timing and environmental conditions during the inoculation process are critical factors for maximising the effectiveness of beneficial microorganisms, such as mycorrhizal fungi or nitrogen-fixing bacteria. These factors directly influence the establishment and activity of organisms and their ability to form successful symbiotic relationships with plants. Understanding the optimal conditions for inoculation can significantly enhance the benefits of microbial application, leading to better crop growth, nutrient uptake, and disease resistance.

Timing is key in ensuring beneficial microorganisms can establish themselves effectively. The best time to apply the inoculant for seed inoculation was immediately before planting, to ensure that the organisms were present immediately after germination. The inoculant should be applied when the seed is in the ideal moisture range for germination, as excessive moisture can cause the inoculant to wash off. At the same time, too little humidity can prevent the seeds from properly absorbing the microbial inoculant. Seed coating should be performed when temperatures are mild, typically during the early morning or late afternoon, to avoid high temperatures that could damage microorganisms. Early application allows fungi or bacteria to become well established for crops with a longer growing season when the plant enters its critical growth phases (Reddy et al. 2022; Kumar and Choudhury 2024a).

For seedling injection, the timing should coincide with transplantation or the early stages of root development. It is essential to inoculate seedlings before they are placed in the soil to ensure that beneficial microorganisms can colonise the root system as soon as possible. This is especially important for mycorrhizal fungi, which form a symbiotic relationship with plant roots and improve nutrient uptake, particularly phosphorus. Seedlings should be inoculated when healthy and not under environmental stress, as stressed plants may have a reduced capacity to support microbial colonisation. Inoculation should be performed when the soil moisture is adequate, but not excessive, because overly wet or compacted soils may limit microbial activity and root penetration.

In mature plants, inoculation timing is primarily influenced by the growth cycle. The best time for applying inoculants to established plants is during early spring or autumn when plants are actively growing, and the root system is the most receptive. Soil inoculation can be performed before planting, but it is also effective for established crops in rapid root growth or during soil preparation. Applying inoculants during plant dormancy or high-stress periods, such as drought or extreme heat, is less practical. Plants are less likely to benefit from microbial inoculation if stressed because the physiological responses of plants may limit their ability to engage with microorganisms.

Environmental considerations are as crucial as timing. Soil temperature, moisture, and pH play significant roles in the success of microbial inoculation. Soil temperature should be within a range conducive to microbial activity, typically between 15 °C and 30 °C, for most beneficial fungi and bacteria. Temperatures that are too high or too low can either inhibit microbial growth or reduce the effectiveness of inoculants. Soil moisture is another critical factor, as the inoculant must remain moist long enough to establish and colonise plant roots. However, saturated soils can cause anaerobic conditions that harm beneficial microbes; therefore, maintaining proper drainage and avoiding over-irrigation is essential (Pankaj et al. 2012a; Kumar and Naik 2020; Das et al. 2022).

Soil pH is another consideration because many beneficial microorganisms have specific pH preferences. Mycorrhizal fungi, for example, tend to thrive in slightly acidic to neutral soils (pH 5.5 to 7.0), whereas nitrogen-fixing bacteria often prefer slightly alkaline conditions. Adjusting soil pH to suit the needs of microorganisms may enhance their effectiveness. However, some commercial inoculants are formulated to withstand a wider pH range; additionally, organic matter content and soil structure impact the success of microbial inoculation. Well-aerated, loamy soils with good organic matter content provide the best environment for microbial activity because they allow for proper water retention and oxygen flow to the roots. It is vital to consider both timing and environmental conditions to maximise the efficacy of inoculating seeds, seedlings, or mature plants. Vaccination should be performed during periods of active plant growth when conditions such as temperature, moisture, and pH are favourable for microbial survival and colonisation. By aligning the inoculation practices with these considerations, growers can achieve excellent plant health, enhanced nutrient uptake, and sustainable farming practices (Table 3).

6 Mycorrhizal Fungi in Commercial Agriculture

Effective mycorrhizal fungi applications during the last 60 years have increased crop yields and improved soil health, transforming this method into the foundation of sustainable farming. They form symbiotic relationships with plant roots and create vast networks of hyphal strands far beyond the rooting zone to improve nutrient and water uptake by the plant. Along with plant growth-promoting biofertilisation traits, this relationship also enhances tolerance to abiotic stresses, such as drought, salinity, and nutrient deficiency in plants (Kumar et al. 2018b; Kumar and Naik 2020; Sharma et al. 2023b). Mycorrhizal fungi play key roles in acquiring plant nutrients, such as phosphorus (P), an essential nutrient for plant growth at low concentrations in many soils. They also enhance nitrogen assimilation and stimulate the release of micronutrients such as zinc and copper. For example, wheat, corn, and rice have shown significant yield increases owing to inoculation with mycorrhizal fungi. These staple crops, which feed billions, can benefit from improved root development and nutrient acquisition, thereby decreasing dependence on chemical fertilisers. About agri-economy, various studies have proven the potential of mycorrhizae in improving crop response to stresses and extending opportunities for expanding the area under cultivation, both through crop intensification and diversification, as mycorrhizal inoculation in wheat has been reported to increase competitive advantage over uninfected plants by 30% or more. Similarly, rice culture under field conditions demonstrated higher water efficiency if cropped under mycorrhizal influence, which is vital in hazardous dryland agriculture. Similarly, legumes such as soybeans and lentils often enjoy a mutually beneficial relationship with mycorrhizal fungi and nitrogen-fixing bacteria, significantly enhancing nutrient cycling for plant growth (Kumar et al. 2021b; Karam et al. 2023; Kumar 2023). This is particularly important for farmers in a system with crop rotation, so residual soil fertility from the previous season will help subsequent crops. Some vegetables, such

Table 3 Application methods for mycorrhizal fungi in agriculture

Method	Description	Advantages	Challenges	Applications
Seed coating	Applying fungi to seeds using adhesives or binders before planting	Ensures fungi are in direct contact with germinating roots	Requires uniform coating and compatibility with seed treatments	Row crops, grains, and legumes
Root dipping	Immersing plant roots in a fungal suspension before transplanting	Promotes immediate colonisation of roots	Labour-intensive and less suitable for large-scale operations	Vegetable seedlings, trees, and ornamentals
Soil incorporation	Mixing fungi into soil or planting media before seeding or transplanting	Ensures even distribution in the root zone	Potential loss of viability during mixing or prolonged storage	Nurseries, greenhouses, and raised beds
Soil drenching	Applying a liquid fungal suspension directly to the soil around plants	Effective for targeted application and ease of use	Requires water and precise application methods	High-value crops and container-grown plants
Furrow application	Placing fungal inoculants in furrows or planting rows during seeding	Simple and integrates well with planting operations	May require specialised equipment for even distribution	Cereal crops, vegetables, and legumes
Broadcast application	Spreading fungal inoculants evenly over the field before planting	Covers large areas efficiently	Risk of uneven distribution and exposure to environmental factors	Pastures and cover crops
Seed pelletising	Embedding fungi within nutrient-rich seed coatings for enhanced colonisation	Combines fungal inoculants with protective nutrients	Requires specialised equipment and testing for compatibility	Precision planting of small-seeded crops
Foliar spray	Fungal spores or mycelium can be applied as a liquid spray to plant foliage	Easy to apply and suitable for specific fungal formulations	Limited efficacy due to the need for root colonisation	Supplementary use in high-value crops
Injection into irrigation	Adding fungi to drip or sprinkler irrigation systems	Provides even distribution across large fields	Requires compatibility with irrigation systems and water quality	Row crops, orchards, and vineyards
Pelleted formulations	Incorporating fungi into granular carriers for easy application	Provides slow release and ease of handling	It may require adjustments for specific soil types or conditions	Dryland farming and sandy soils

(continued)

Table 3 (continued)

Method	Description	Advantages	Challenges	Applications
Hydroponic integration	Adding fungi to nutrient solutions in hydroponic or soilless systems	Enhances nutrient uptake in controlled environments	It requires careful monitoring of fungal viability in the solution	Hydroponic vegetable and herb production
Tree pit inoculation	Adding fungi directly to planting holes for trees or shrubs	Direct inoculation ensures root contact	It is time-consuming and suitable only for perennial crops	Orchards, forestry, and landscape planting
Band application	Placing fungi in concentrated bands near seed rows or root zones	Targets root development zones for maximum impact	Requires precise placement equipment	Row crops and high-value vegetables
Slurry application	Creating a thick suspension of fungi and applying it during transplanting	Provides a concentrated dose of fungi directly to roots	Requires preparation time and care during application	Ornamental plants and greenhouse crops
Biochar-based inoculation	Embedding fungi in biochar for soil amendment and inoculation	It enhances fungal survival and provides additional soil benefits	Requires production or sourcing of biochar	Sustainable farming and degraded soils

Source: Reddy et al. (2022) and Kumar and Choudhury (2024a)

as tomatoes, peppers, and carrots, benefit from enhanced growth by inoculating plants with mycorrhizal fungi. For example, tomatoes build stronger root systems and grow more fruits because they get better phosphorus and potassium uptake. By seeking deeper nutrients in the soil horizons, fungi play a vital role in increasing yield, uniformity, and quality of root crops, such as carrot (*Daucus carota*) and potato (*Solanum tuberosum*). Tree farms, including fruit trees such as apples, mangoes, and citrus, also exhibit significant growth; fruit quality and output have increased. One of the most important functions of mycorrhizal fungi in these perennial systems is to reduce transplant shock and improve root establishment and plant resistance to soil pathogens. These fungi have even reduced the effects of drought on grape quality and water stress in grapevines, which are important for crops that have adapted to stabilise the availability of water and nutrients. In addition to these direct benefits to crops, mycorrhizal fungi play a significant role in soil health. Their hyphal networks also stabilise the soil by creating aggregates, which increase aeration and water infiltration. This process also sequesters carbon, meaning that organic carbon is stabilised within soil aggregates, and climate change is exacerbated. Mycorrhizal fungi also reduce the number of soil-borne pathogens by displacing organic-matter-decomposing organisms and inducing systemic resistance in plants. This contributes to healthier, raised soils and more resilience, allowing us to farm the land for decades. Mycorrhizal fungi have significant economic and environmental implications for large-scale farming operations. Farmers avoid using synthetic fertilisers and pesticides by cutting their production costs, which do not lead to environmental pollution. The abuse of fertilisers has a long record of nutrient run-off, leading to water contamination and greatly affecting aquatic life. Mycorrhizal fungi improve nutrient uptake efficiency, reduce the risk of eutrophication in soils associated with the high application of fertilisers, and improve water quality. Moreover, the improved drought tolerance afforded by mycorrhizal fungi also mitigates some economic risks, with water shortages increasing due to climate change. This helps support farmers further because they can achieve high yields without using much water, making agriculture more profitable and sustainable. In addition, the wider use of mycorrhizal fungi on a mass scale contributes to lowering greenhouse gas emissions from agriculture. Fertiliser: A major producer of carbon dioxide and nitrous oxide emissions is fertiliser manufacturing and utilisation. Mycorrhizal fungi help reduce these emissions by diminishing the necessity of fertilisers and, hence, indirectly help as a part of the global balance to eliminate issues related to climate change. Furthermore, they play a significant role in carbon sequestration, which is even better for the environment. Although more benefits will be gained, implementing mycorrhizal fungi on an industrial-scale farming basis requires fixing a few bottlenecks. The success of vaccination can be patchy and depends on soil type, crop species, and management practices (Saini et al. 2023; Kumar and Choudhury 2024b; Kumar and Dey 2024). Some soils that are high in available phosphorus may suppress mycorrhizal colonisation, reducing the effectiveness of the mycorrhizal inoculum. Similar to reduced-tillage or no-till farming, intensive tillage can devastate fungal networks. However, this has been addressed by developments in biotechnology and microbial formulation. Scientists have been working on creating multi-stress and species inoculants, that is, quality and broad-spectrum mycorrhiza, which are suitable for use under various conditions. Targeted applications for optimal benefit, customised formulations are also becoming more common and allow highly targeted applications depending on the crop or soil type. Economically, the high up-front cost of mycorrhizal inoculants may be a drawback for many farmers, especially smallholders. The initial investments may appear massive, but the long-term benefits of reduced input costs, increased yield, and better soil health are far greater. The government and agro-industry play a huge role in offering some form of willingness to pay or subsidies to get low-income farmers to use mycorrhizal fungi, especially when conventional agriculture has reached its limit, such as resource-poor farmers under environmental constraints. The mycorrhizal fungi used in farming also have the potential for large-scale agricultural operations. The infamous monoculture systems, blamed for destroying soil health, might benefit from mycorrhizal application. Such measures help address climate change, improving soil health and increasing farm productivity. Mycorrhizal fungi have been proven to provide a transformative solution that significantly increases yield and regenerates the health of soils for many plant species. Examples of cereals, legumes, vegetables, and fruit trees show the advantages of fungi in these crops for better agriculture. Among the financial and ecological damages wrought by industrial agriculture, the same implications would be true for other massive agricultural enterprises, all potentially cutting costs, increasing sustainability, and fighting against global issues such as food security and climate change. Given the ongoing progress in research and technology, there is enormous potential to integrate mycorrhizal fungi into conventional agriculture to facilitate sustainable and resilient farming systems.

7 Challenges and Future Directions in Mycorrhizal Fungi Use

Quiescent but not despaired agricultural applications of mycorrhizal fungi remain limited owing to various logistical challenges for their broad use. A significant challenge is the ignorance and unawareness of farmers regarding the benefits of mycorrhizal fungi. Some growers do not know how these fungi work, how they can help their yields, or how they can be used to manage soils sustainably. This knowledge gap is often further compounded by the widespread use of conventional agricultural inputs, such as synthetic fertilisers and pesticides, which are marketed aggressively and understood as panaceas for farming problems. Moreover, farmers who use these traditional practices might disagree with the use of microbial inoculants, as they consider them ineffective, mainly in terms of soil and environmental variability. Soil conditions such as elevated phosphorus levels, which can arise from excess fertiliser use, can discourage fungal colonisation and leave farmers feeling disenchanted by what appears to be the lack of results. The cost of these inoculants, albeit modest relative to their potential long-term benefits, is another key barrier, especially for smaller and resource-limited farmers. These investments may discourage adoption without subsidies or other financial incentives (Kumar et al. 2023a; Kumar and Choudhury 2024d).

Another significant challenge is the variable quality and performance of commercial mycorrhizal inoculants. The effectiveness of these products depends on the strain, formulation, and application method, among other factors. The results in the field are mixed. The second challenge arises from a lack of standardised regulatory frameworks and certifications for these mycorrhizal products, which may lead sceptical farmers to doubt the quality of the available options. Logistical challenges such as storage, handling, and the short shelf life of live microbial products can also inhibit widespread distribution and implementation. In addition, deep ploughing and monocropping, among other (modern) agricultural practices, can destroy mycorrhizal networks in the soil, thus compromising the benefits of vaccination. They emphasise moving to farm systems that favour mycorrhizae, which include reducing tillage, promoting crop diversity, and managing organic matter (Devi and Kumar 2023a; Kumari et al. 2023; Kumar and Singh 2024).

Mycorrhizal performance is limited by several cell biological barriers that can be overcome by genetic modifications and bioengineering strategies that drive further innovation. Biotechnology is crucial for developing fungal strains that are more robust to harsh environmental conditions such as temperature extremes, soil salinity, and drought. For instance, the mycorrhizal genetic modification of fungal symbionts could be designed to forge symbioses with new crops not currently adapted to these symbiotic relationships. Bioengineered fungi can be tailored to produce enzymes or compounds that increase nutrient availability, pest resistance, and plant growth. These fungi could further benefit crops by transferring genes that enhance their chemical appetite, such as those involved in phosphorus solubilisation or nitrogen fixation. The use of genetically modified organisms (GMOs) in agriculture is still contentious, and GMOs face regulatory and public acceptance hurdles. As with any new technology, addressing these concerns through transparent research, rigorous testing, and clear communication about the safety and benefits of bioengineered mycorrhizal fungi is necessary for adoption.

Mycorrhizal fungi will probably play a role in sustainable agriculture in the future, with priorities in the rest of the world moving towards sustainable farming methods. Rising concerns regarding climate change, soil degradation, and the environmental costs of conventional agricultural practices have fuelled interest in alternative solutions that can sustain productivity over time without damaging the surrounding ecosystems. Mycorrhizal fungi fit perfectly with these goals, providing a natural and cost-effective way to maintain soil health, decrease chemical inputs, and increase crop resilience to stress. Read new species discoveries and emerging beneficial organisms that aid agriculture. Fungi are known to interact with a range of beneficial microbes, including nitrogen-fixing bacteria and biocontrol agents, and research on their synergistic effects is likely to gain traction. These advances have paved the way for integrated microbial consortia targeting specific crops and geographies. Advances in delivery systems, including seed coatings, granules, or liquid formulations, may also be developed to simplify mycorrhizal inoculant application, increasing farmers' convenience and ease of use (Kumar and Kumar 2023; Devi et al. 2024; Jagadesh et al. 2024).

General advancements in precision agriculture and digital farming technologies also contribute to the optimal use of mycorrhizal fungi. Soil sensors, remote sensing, and data analytics can be used to determine the optimal conditions for fungal inoculation and enact precision applications for improved efficacy. Some of these applications, such as the real-time measurement of soil health, could make decisions about whether and when to apply mycorrhizal fungi much more straightforward, ensuring they are being used to their fullest potential. The target sites for these fungi should be chosen with educational extension efforts that can inform farmers of the soil management principles that can help them in the long term. This knowledge gap can be significantly narrowed through outreach initiatives, including extension programmes, demonstration farms, and collaboration between researchers, policymakers, and industry stakeholders. With more studies on this subject, the clash between fungi and plants in soils and their other ecological contributions, not limited to agriculture, can be explored further. When solving a substantial part of the human impact on Mother Earth, there are no better allies than trees. Their capacity to hold carbon, prevent soil erosion, and provide habitats for biodiversity makes them ideal tools for combating global environmental issues. Governments and international organisations can support these efforts by integrating mycorrhizal fungi into policies and programmes to achieve climate and sustainability objectives, including carbon farming programmes or regenerative agriculture incentives.

Although there are significant hurdles to the widespread use of mycorrhizal fungi in agricultural systems (knowledge gaps, differences among commercial products, or logistical challenges), the fields of bioengineering, research, and education provide significant opportunities to advance our understanding and the effective use of mycorrhizal fungi in commercial cropping systems. Genetic enhancement of fungal performance and its application in precision agriculture systems represent a new frontier in maximising their benefits. With sustainable agriculture rising to the forefront of global concerns, mycorrhizal fungi are expected to become a crucial species for increasing crop yields, improving soil quality, and solving environmental problems. Breeding, continued investment in research, innovation, and farmer outreach are all crucial to unlocking their potential and shaping the future of agriculture.

8 Conclusion

In conclusion, mycorrhizal fungi offer hope for the future of sustainable farming: they can improve yield and quality without resorting to artificial fertilisers, which can damage the environment. They have a unique, mutually beneficial relationship with plants that break down nutrients and efficiently absorb water, enabling resistance to environmental adversity and persistent soil fertility. These are evident in various crops, from staple foods in the north to delicious tropical fruits and vegetables. This demonstrates how widely adaptable mycorrhizal fungi are when dealing with global human needs. Their achievements in carbon sequestration, improved soil structure, and environmental support for biodiversity contribute to the international goals of sustainable development and ecological restoration. However, mycorrhizal fungi's potential is minimal given their relative lack of public awareness, variability in product efficacy, and transportation difficulties. However, advances in biotechnological techniques, such as genetically enhancing fungi and using novel means of distribution, offer hope in this regard. Mycorrhizal fungi tailored to individual crops and environmental conditions can optimise their effectiveness when used correctly. Precision agriculture technologies help integrate them into farming systems.

References

Aley P, Singh J, Kumar P (2022) Chapter 23 – Adapting the changing environment: microbial way of life. In: Kumar A, Singh J, Ferreira LFR (eds) Microbiome under changing climate. Woodhead Publishing, pp 507–525

Chakraborty S, Kumar P, Sanyal R, Mane AB, Arvind Prasanth D, Patil M, Dey A (2021) Unravelling the regulatory role of miRNAs in secondary metabolite production in medicinal crops. Plant Gene 27:100303. https://doi.org/10.1016/j.plgene.2021.100303

Das T, Saha SC, Sunita K, Majumder M, Ghorai M, Mane AB, Prasanth DA, Kumar P, Pandey DK, Al-Tawaha AR, Batiha GE-S, Shekhawat MS, Ghosh A, Sharifi-Rad J, Dey A (2022) Promising botanical-derived monoamine oxidase (MAO) inhibitors: pharmacological aspects and structure-activity studies. S Afr J Bot 146:127–145. https://doi.org/10.1016/j.sajb.2021.09.019

Devi P, Kumar P (2023a) Effect of microbial oxidative enzymes on soil health. In: Microbial oxidative enzymes: biotechnological applications. De Gruyter, Department of Agronomy, School of Agriculture, Lovely Professional University, Phagwara, Punjab, pp 37–61

Devi P, Kumar P (2023b) Role of oxidative enzymes in the alleviation of oxidative stress in plant system. In: Microbial oxidative enzymes: biotechnological applications. De Gruyter, Department of Agronomy, School of Agriculture, Lovely Professional University, Phagwara, Punjab, pp 109–128

Devi P, Dey SR, Kumar P (2023a) Integration of water resources management in rural areas. In: Water resources management for rural development: challenges and mitigation. Elsevier, Department of Agronomy, School of Agriculture, Lovely Professional University, Punjab, Phagwara, pp 139–152

Devi P, Dey SR, Saini L, Kumar P, Panigrahi S, Dwivedi P (2023b) Toward sustainable agriculture: strategies involving phytoprotectants against reactive oxygen species. In: Reactive oxygen species: prospects in plant metabolism. Springer, Department of Agronomy, School of Agriculture, Lovely Professional University, Phagwara, pp 229–247

Devi P, Kumar P, Sharma K, Singh J (2023c) Is there a relationship between myconanotechnology and sustainable development? In: Myconanotechnology and application of nanoparticles in biology: fundamental concepts, mechanism and industrial applications. Elsevier, Department of Agronomy, School of Agriculture, Lovely Professional University, Punjab, Phagwara, pp 175–207

Devi P, Dogra P, Kumar P, Singh J (2024) Global status of biomass energy programmes-challenges and roadmap. In: Biomass energy for sustainable development. CRC Press/Lovely Professional University, Phagwara, pp 394–414

Dey SR, Devi P, Kumar P (2023a) Organic contaminants in aquatic environments: sources and impact assessment. In: Metal organic frameworks for wastewater contaminant removal. Wiley, Department of Agronomy, School of Agriculture. Lovely Professional University, Punjab, pp 299–317

Dey SR, Kumar P, Singh J (2023b) Growth of mustard under open field conditions with salicylic acid and antimicrobial Thiourea. Biopestic Int 19(2):183–192. https://doi.org/10.59467/BI.2023.19.183

Dey SR, Sharma M, Kumar P (2023c) Effects and responses of chromium on plants. In: Environmental science and engineering. Springer/Department of Agronomy, School of Agriculture, Lovely Professional University, Punjab, Phagwara, pp 385–427

Dwivedi P, Kumar P (2011) Anti-diabetic medicinal plants and their conservation: waging green war on diabetes. Med Plants 3(3):181–189. https://doi.org/10.5958/j.0975-4261.3.3.031

Goud EL, Singh J, Kumar P (2022) Chapter 19 – Climate change and their impact on global food production. In: Kumar A, Singh J, Ferreira LFR (eds) Microbiome under changing climate. Woodhead Publishing, pp 415–436

Hidangmayum A, Dwivedi P, Kumar P, Upadhyay SK (2023) Seed priming and foliar application of chitosan ameliorate drought stress responses in mungbean genotypes through modulation of morpho-physiological attributes and increased antioxidative defense mechanism. J Plant Growth Regul 42(10):6137–6154. https://doi.org/10.1007/s00344-022-10792-1

Jagadesh M, Dash M, Kumari A, Singh SK, Verma KK, Kumar P, Bhatt R, Sharma SK (2024) Revealing the hidden world of soil microbes: metagenomic insights into plant, bacteria, and fungi interactions for sustainable agriculture and ecosystem restoration. Microbiol Res 285. https://doi.org/10.1016/j.micres.2024.127764

Karam N, Kumar P, Anand YR, Choudhury D (2023) Management of rice blast-a review of the strategies available. Biopestic Int 19(2):107–116. https://doi.org/10.59467/BI.2023.19.107

Kumar P (2023) Degradation of xenobiotics by bacteria and fungi: an overview. Biopestic Int 19(2):117–132. https://doi.org/10.59467/BI.2023.19.117

Kumar P, Choudhury D (2024a) Toxicity of rhizospheric cadmium contaminated soil and its phytoremediation. In: Springer water. Springer Nature, Department of Agronomy, School of Agriculture, Lovely Professional University, Punjab, Phagwara, pp 227–264

Kumar P, Choudhury D (2024b) Microbial native soil bacteria against cadmium toxicity. In: Springer water. Springer, Department of Agronomy, School of Agriculture, Lovely Professional University, Punjab, Phagwara, pp 193–226

Kumar P, Choudhury D (2024c) Hydroponic removal of organic contaminants from water. In: Springer water. Springer, Department of Agronomy, School of Agriculture, Lovely Professional University, Punjab, Phagwara, pp 143–164

Kumar P, Choudhury D (2024d) Bioremediation of wastewater using hydroponics. In: Springer water. Springer, Department of Agronomy, School of Agriculture, Lovely Professional University, Punjab, Phagwara, pp 27–64

Kumar P, Dey SR (2024) Hydroponics phytoremediation: an overview. In: Springer water. Springer, Department of Agronomy, School of Agriculture, Lovely Professional University, Punjab, Phagwara, pp 361–396

Kumar P, Dwivedi P (2020) Lignin estimation in sorghum leaves grown under hazardous waste site. Plant Arch 20:2558–2561

Kumar VA, Kumar P (2023) Revolutionizing agriculture with satellite technology for farmers: a review. Biopestic Int 19(2):97–106. https://doi.org/10.59467/BI.2023.19.97

Kumar P, Mistri TK (2020) Transcription factors in SOX family: potent regulators for cancer initiation and development in the human body. Semin Cancer Biol 67:105–113. https://doi.org/10.1016/j.semcancer.2019.06.016

Kumar P, Naik M (2020) Biotic symbiosis and plant growth regulators as a strategy against cadmium and lead stress in chickpea. Plant Arch 20:2495–2500

Kumar P, Pathak S (2019) Responsiveness index of sorghum (Sorghum bicolor (L.) Moench) grown under cadmium contaminated soil treated with putrescine and mycorrhiza. Bangladesh J Bot 48(1):139–143

Kumar P, Singh J (2024) Hydroponic: an eco-friendly future. In: Springer water. Springer, Department of Agronomy, School of Agriculture, Lovely Professional University, Punjab, Phagwara, pp 223–267

Kumar P, Mandal B, Dwivedi P (2011a) Heavy metals scavenging of soils and sludges by ornamental plants. J Appl Hortic 13(2):144–146

Kumar P, Mandal B, Dwivedi P (2011b) Heavy metal scavenging capacity of Mentha spicata and Allium cepa. Med Plants 3(4):315–318. https://doi.org/10.5958/j.0975-4261.3.4.053

Kumar P, Mandal B, Dwivedi P (2013) Combating heavy metal toxicity from hazardous waste sites by harnessing scavenging activity of some vegetable plants. Vegetos 26(2):416–425. https://doi.org/10.5958/j.2229-4473.26.2.106

Kumar P, Singh BN, Dwivedi P (2017) Plant growth regulators, plant adaptability and plant productivity: a review on abscisic acid (ABA) signaling in plants under emerging environmental stresses. In: Sustaining future food security in changing environments. Nova Science Publishers, Inc., Department of Plant Physiology, Institute of Agricultural Sciences, Banaras Hindu University, Varanasi, pp 81–97

Kumar P, Harshavardhan M, Kumar PS, Yumnam J, Jyoti N, Naik M, Misao L, Purnima KS (2018a) Effect on chlorophyll a/b ratio in cadmium contaminated maize leaves treated with putrescine and mycorrhiza. Ann Biol 34(3):281–283

Kumar P, Kumar S, Harshavardhan M, Naik M, Yumnam J, Kumar PS, Jyoti N, Misao L, Purnima (2018b) Evaluation of plant height and leaf length of sorghum grown under different sources of nutrition. Ann Biol 34(3):284–286

Kumar P, Pathak S, Amarnath KS, Veerendra Brahma Teja P, Dileep B, Kumar K, Singh M, Siddique A (2018c) Effect of growth regulator on morpho-physiological attributes of chilli: a case study. Plant Arch 18(2):1771–1776

Kumar P, Pathak S, Kumar M, Dwivedi P (2018d) Role of secondary metabolites for the mitigation of cadmium toxicity in sorghum grown under mycorrhizal inoculated hazardous waste site. In: Biotechnological approaches for medicinal and aromatic plants, pp 199–212

Kumar P, Yumnam J, Kumar PS, Misao L, Jyoti N, Naik M, Purnima KS, Harshavardhan M (2018e) Cadmium induced changes in germination of maize seed treated with mycorrhiza. Ann Agric Bio Res 23(2):169–170

Kumar D, Rameshwar SD, Kumar P (2019) Effect of intergated application of inorganic and organic fertilizers on the roots of chickpea. Plant Arch 19(1):857–860

Kumar P, Kumar T, Singh S, Tuteja N, Prasad R, Singh J (2020) Potassium: a key modulator for cell homeostasis. J Biotechnol 324:198–210

Kumar P, Devi P, Dey SR (2021a) Fungal volatile compounds: a source of novel in plant protection agents. In: Volatiles and metabolites of microbes. Elsevier, pp 83–104

Kumar P, Sharma K, Saini L, Dey SR (2021b) Chapter 8 – Role and behavior of microbial volatile organic compounds in mitigating stress. In: Kumar A, Singh J, Samuel J (eds) Volatiles and metabolites of microbes. Academic, pp 143–161

Kumar V, Dwivedi P, Kumar P, Singh BN, Pandey DK, Kumar V, Bose B (2021c) Mitigation of heat stress responses in crops using nitrate primed seeds. S Afr J Bot 140:25–36. https://doi.org/10.1016/j.sajb.2021.03.024

Kumar P, Goud EL, Devi P, Dey SR, Dwivedi P (2022a) Heavy metals: transport in plants and their physiological and toxicological effects. In: Plant metal and metalloid transporters. Springer/Department of Agronomy, School of Agriculture, Lovely Professional University, Punjab, Jalandhar, pp 23–54

Kumar P, Goud EL, Devi P, Koul B (2022b) Metal pollutants in the environment. In: Environmental microbiology: emerging technologies. De Gruyter, Department of Agronomy, School of Agriculture, Lovely Professional University, Jalandhar, Punjab, pp 291–323

Kumar P, Koul B, Sharma M (2022c) Phytoremediation of heavy metals. In: Heavy metals in plants: physiological to molecular approach. CRC Press, Department of Agronomy, School of Agriculture, Lovely Professional University, Punjab, Jalandhar, pp 369–388

Kumar P, Latha DII, Singh J (2023a) Rhizoremediation of organic emerging soil contaminants: green technology. In: Clean technologies toward a sustainable future: physicochemical, biochemical and biotechnological approaches. IWA Publishing, Department of Agronomy, School of Agriculture, Lovely Professional University, Phagwara, Punjab, pp 211–229

Kumar P, Saini L, Sharma M (2023b) Assessment of the phytochemical constituents and metabolites in the medicinal plants and herbal medicine used in the treatment and management of respiratory diseases. In: Reference series in phytochemistry. Springer/Department of Agronomy, School of Agronomy, Lovely Professional University, Phagwara

Kumar P, Dey SR, Choudhury D (2024) Effectiveness of cadmium on biochemical shift of pea plant treated with mycorrhiza and putrescine. Nat Environ Pollut Technol 23(1):275–285. https://doi.org/10.46488/NEPT.2024.v23i01.023

Kumari P, Singh J, Kumar P (2022) Chapter 21—Impact of bioenergy for the diminution of an ascending global variability and change in the climate. In: Kumar A, Singh J, Ferreira LFR (eds) Microbiome under changing climate. Woodhead Publishing, pp 469–487

Kumari P, Kumar P, Singh J (2023) Toxic contamination of aquatic habitat and its remediation. In: Spatial modeling of environmental pollution and ecological risk. Elsevier/Department of Agronomy, School of Agriculture, Lovely Professional University, Punjab, Phagwara, pp 209–220

Pankaj PK, Kumar P, Nigam RC, Mishra PK (2012a) To studies organochlorine insecticides in different brands of cold drinks. J Chem Pharm Res 4(8):3934–3938

Pankaj PK, Kumar P, Nigam RC, Mishra PK (2012b) Monitoring and surveillance of synthetic pyrethroids and organophosphate in different brands of soft drinks. J Chem Pharm Res 4(8):3939–3943

Pathak S, Kumar P, Mishra PK, Kumar M (2017) Mycorrhiza assisted approach for bioremediation with special reference to biosorption. Pollut Res 36(2):329–332

Raj A, Goutam E, Kumar P, Singh J (2024) Polyphenol oxidase: an enzyme of plants. In: Polyphenol oxidases: function, wastewater remediation, and biosensors. De Gruyter, Department of Agronomy, School of Agriculture, Lovely Professional University, Phagwara, Punjab, pp 105–128

Reddy SS, Kumar P, Dwivedi P (2022) Heavy metal transporters, phytoremediation potential, and biofortification. In: Plant metal and metalloid transporters. Springer, pp 387–405

Saini L, Sharma K, Kumar P (2023) Water efficiency in agriculture as a vital approach toward water management in rural areas. In: Water resources management for rural development: challenges and mitigation. Elsevier/Department of Agronomy, School of Agriculture, Lovely Professional University, Punjab, Phagwara, pp 127–138

Sharma K, Kumar P (2023) Microbe-mediated impact on cell membrane stability and leaf pigmentation in pearl millet under toxic mercury stress in soil. Biopestic Int 19(2):217–228. https://doi.org/10.59467/BI.2023.19.217

Sharma K, Devi P, Dey SR, Kumar P (2023a) Mercury phytovolatilization: an overview of the mechanism and mitigation. In: Role of green chemistry in ecosystem restoration to achieve environmental sustainability. Elsevier/Department of Agronomy, School of Agriculture, Lovely Professional University, Punjab, Phagwara, pp 325–331

Sharma K, Devi P, Kumar P (2023b) Metal-organic frameworks for wastewater contaminants removal. In: Metal organic frameworks for wastewater contaminant removal. Wiley, Department of Agronomy, School of Agriculture, Lovely Professional University, Punjab, Phagwara, pp 95–117

Sharma K, Devi P, Kumar P, Dey A, Dwivedi P (2023c) Hazardous phytotoxic nature of reactive oxygen species in agriculture. In: Reactive oxygen species: prospects in plant metabolism. Springer, Department of Agronomy, School of Agriculture, Lovely Professional University, Phagwara, pp 135–146

Sharma M, Dey SR, Kumar P (2023d) Antioxidant defence: a key mechanism of chromium tolerance. In: Environmental science and engineering. Springer, Department of Agronomy, School of Agriculture, Lovely Professional University, Punjab, Phagwara, pp 91–116

Sinam V, Kumar P, Singh J (2023) Bioremediation and ecorestoration strategies of aquatic environment. In: Spatial modeling of environmental pollution and ecological risk. Elsevier, Department of Agronomy, School of Agriculture, Lovely Professional University, Punjab, Phagwara, pp 483–499

Srivastav AL, Patel N, Rani L, Kumar P, Dutt I, Maddodi BS, Chaudhary VK (2024) Sustainable options for fertilizer management in agriculture to prevent water contamination: a review. Environ Dev Sustain 26(4):8303–8327. https://doi.org/10.1007/s10668-023-03117-z

Upadhyay SK, Devi P, Kumar V, Pathak HK, Kumar P, Rajput VD, Dwivedi P (2023) Efficient removal of total arsenic (As3+/5+) from contaminated water by novel strategies mediated iron and plant extract activated waste flowers of marigold. Chemosphere 313:137551. https://doi.org/10.1016/j.chemosphere.2022.137551

Mechanisms of Mycorrhizal Symbiosis and Its Impact on Soil Fertility and Crop Nutrition

Krishnagowdu Saravanan, Pandian Rajendran, Smrutirekha Sahu, Subramanian Kaliyannan, Muthukumar Elankumaran, Sundarasamy Dhanapal, Nagarajan Kiruthiga, and Chinnappan Deepa Joan of Arc

Abstract

Mycorrhizal symbiosis, or a beneficial partnership between plant roots and fungus, is critical for boosting crop nutrition and soil fertility. Mycorrhizal fungi, a varied collection of organisms from the phylum Glomeromycotina, develop extensive hyphal networks in the soil matrix. These networks considerably boost the plant's ability to absorb important nutrients such as micronutrients, phosphate, and nitrogen. Different mycorrhizal fungus species, including ectomycorrhizae, ericoid mycorrhizae, and arbuscular mycorrhizae, provide distinctive advantages to various species of plants and habitats. In exchange, the plant provides the fungus with carbohydrates. In addition to improving the physiological and chemical features of the soil, this mutualistic relationship promotes microbial diversity, avoids soil erosion, speeds up the decomposition of organic materials, and stimulates the manufacture of plant growth hormones. Because of all these factors, soil fertility has grown. Mycorrhizae also bolster crops' defenses against soil-borne pathogens, improve nutrient absorption, and increase their resistance to abiotic stresses. Increased yields and better nutritional value are thus advantageous to crops. Taking everything into account, these mycorrhizal activities are crucial for boosting soil fertility and are a practical strategy in modern agricultural systems and crop yields. The remainder of this chapter gives a summary of the molecular principles behind the symbiotic relationship between

K. Saravanan (✉) · S. Dhanapal · N. Kiruthiga · C. Deepa Joan of Arc
Post Graduate and Research Centre in Biotechnology, School of Life Sciences, Arignar Anna College (Arts and Science), Krishnagiri, Tamil Nadu, India
e-mail: saravanankrishnagowdu95@gmail.com

P. Rajendran · S. Sahu · S. Kaliyannan
ICAR – National Institute of Biotic Stress Management, Raipur, Chhattisgarh, India

M. Elankumaran
Department of Botany, Bharathiar University, Coimbatore, Tamil Nadu, India

Y. Siddiqui et al. (eds.), *Sustainable Mycorrhizal Cultivation*,
https://doi.org/10.1007/978-981-95-1778-7_12

mycorrhizal fungi and discusses strategies to enhance crop nutrition and soil fertility for sustainable agricultural operations.

Keywords

Mycorrhizal fungi · Soil fertility · Nutritional enhancement · Symbiosis

1 Introduction

Mycorrhizal fungi (a diverse group of phyla Glomeromycotina) create vast networks of hyphae within the soil matrix. The mycorrhizal symbiosis, which is the association between plant roots and fungi, plays a pivotal role in the uptake of nutrients and the maintenance of soil health. Two principal mycorrhizal groups were distinguished on the basis of their anatomical characteristics and the nature of their colonization of the roots, namely ectomycorrhizae and endomycorrhizae. These networks significantly enhance the plant's capacity to absorb essential nutrients, including phosphorus (P), nitrogen (N), and micronutrients (Dong et al. 2018). In ectomycorrhizae, the fungus forms a sheath around the root. Subsequently, the sheath produces hyphae that penetrate the root and spread into the soil. In contrast to the formation of a sheath by ectomycorrhizae, hyphae of endomycorrhizae grow within root cells and extend into the soil. These are considerably more prevalent than ectomycorrhizae. In hilly regions, trees belonging to the families Pinaceae, Fagaceae, Dipterocarpaceae, and Caesalpinoidaceae interact with hundreds of species of Basidiomycetes and Ascomycetes. These fungi have played a significant role in shaping the present forests over the last 180 million years (Bonfante and Genre 2010; Horan et al. 1988). The symbiotic relationship between ectomycorrhizal (ECM) fungi and tree roots results in alterations to the root structure, facilitating the exploration of the soil by extrametrical mycelium (EMM) systems and enhancing P acquisition beyond the root depletion zone. The mycelial foraging strategies of different ECM fungal taxa vary considerably. Some taxa produce only a small amount of EMM, while others produce extensive explorative mycelial systems that can spread over considerable distances into the soil. The mycelial networks of these fungi enhance soil aggregation, improve soil structure, and stabilize microbial communities. Furthermore, the hyphal density at the growing mycelial front varies between different ECM fungal taxa (Cope et al. 2019). It is of great importance to gain an understanding of the symbiotic mechanisms of the ECM fungus, as forests play a significant role in the buffering of rising CO_2 levels. Furthermore, recent years have seen significant discoveries in this field with regard to molecular pathways that drive symbiosis growth, stress-tolerant mechanisms, and nutrient exchange processes. The following review will illustrate the underlying mechanisms of mycorrhizal symbiosis and its contribution to crop nutrition and soil fertility.

2 Mechanism of Plant-Mycorrhizal Interaction/Symbiosis

2.1 Formation of Hyphal Network and Its Development

2.1.1 Ectomycorrhizal Symbiosis

Horan et al. (1988) studied the mycorrhizal infection process in *Eucalyptus* using light microscopy and the papersandwich inoculation technique, allowing sequential events to be time-tracked. One day after inoculation, chemical interactions between the fungus and root were evident in the cap cells, as indicated by the presence of brown flecks in root apices of inoculated plants, but absent in uninfected roots. At this stage, however, no physical attachment of the mycorrhizal fungus was observed. These chemical deposits were later confirmed to be phenolic compounds, as demonstrated by Ling Lee et al. (1977), who showed their reaction with toluidine blue resulting in a green coloration. Two days after inoculation, the hyphae attached to the root apex by growth between and into cap cells, and there were the first indications of swelling of root tissues. This was associated with a dense buildup of hyphae around the cap region, which is referred to as the extramatrical mycelium, the mantle. The mantle is the living tissue barrier between rootlet and soil, suppressing root hair development and forming a sock-like sheath. From that sheath, hyphae grow as root hairs. By 4 days of inoculation, all the basic features of a typical ectomycorrhiza were present, including a blunt swollen apex surrounded by dense fungal sheath and blocky epidermal cells with some development of the Hartig net between them. Hartig net is a characteristic of ectomycorrhizal symbiosis. Hyphae penetrate the apoplastic space outside the host cell plasma membrane, and hyphae remain outside the root endodermis. In angiosperms, hyphal penetration only occurs in the epidermal layer, but in gymnosperms, hyphae also penetrate between the outer cortical cells. The shortest time required for the formation of an ectomycorrhiza after contact between fungus and root was confirmed as 4 days by the experiment of Horan et al. in 1988. Pigments in hyphae represent each morphotype by its own characteristic colors, such as purple, black, white, and shades between dark brown and pale yellow. Massicotte et al. (1989) experimentally proved that root exudates trigger pink color pigment synthesis in colorless fungal inoculum plugs, which shows that root exudates act as the carbon source for the activation of fungal colonization in *B. alleghaniensis* and *P. resinosa* roots by the strain of *L. bicolor*. The Hartig net acts as the interface between fungal hyphae and plant root cells, where the development and physiology of partners become integrated into a functional unit. A fungus is chemotropically attracted to grow toward a nearby root of its host, as mentioned earlier. Root exudates released by the host under nutrient-deficient soil conditions include important signaling molecules such as flavonoids and strigolactones.

2.1.2 Molecular Study of ECM Symbiosis

The work of Cope et al. (2019) on *L. bicolor* as a model ECM with *Populus* as the host demonstrated that *L. bicolor* produces both non-sulfated lipochito oligosaccharides (LCO) and sulfated LCOs, which act as a signal molecule from *L. bicolor* and trigger root hair branching toward it. The activity of different LCOs varies at different stages of the symbiosis development process. Additionally, LCOs have been observed to induce calcium spiking in *Populus* in a manner dependent on the CASTOR/POLLUX protein. Non-sulfated LCOs enhanced lateral root development in *Populus* in a manner dependent on CCaMK, while sulfated LCOs improved the colonization of *L. bicolor* in *Populus*. RNA-based gene silencing using the RNAi technique was employed to silence two homologs of CASTOR, POLLUX, and CCaMK in the *Populus* genome, which resulted in a reduction in the colonization of *L. bicolor*. In the region of Hartig net development, contact between host and mycorrhizal hyphae results in the release of a small secreted protein, designated MiSSP7, by the fungus. MiSSP7 enters the plant cell via active phosphatidylinositol 3-phosphate-mediated endocytosis and subsequently reaches the nucleus, where it acts as a transcription factor affecting the expression of genes involved in defense, cell wall remodeling, and signaling. These effector proteins are conserved effectors of plant pathogenic fungi, nematodes, and bacteria. Additionally, the plant's resistance is similarly suppressed, and interaction is favored. Transformants of *L. bicolor* with reduced expression of MiSSP7 were unable to establish a symbiosis with poplar roots (Plett et al. 2011). Following colonization, the plant's auxin responses undergo a transformation, resulting in the formation of lateral roots and the emergence of the typical ECM morphotype (Watkinson 2016). In their study of the relative abundance of transcripts in ectomycorrhizal symbiosis, Voiblet et al. (2001) employed systematic sequencing of cDNA clones, SSH, and cDNA mini-arrays to compare the free-living partners and a 4-day-old *Eucalyptus globulus-Pisolithus tinctorius* ectomycorrhiza. This evidence supports the hypothesis that *P. tinctorius* hyphae preferentially express specific sets of structural proteins in the wall during the formation of symbiosis. Additionally, ectomycorrhizal tissues exhibit an upregulation of signal transduction and elicitor-inducible defense-related genes. Therefore, the plant cell wall and its surface play a crucial role in symbiosis (Voiblet et al. 2001).

2.1.3 Endomycorrhizal Symbiosis

Arbuscular mycorrhizal (AM), one of the endomycorrhizal symbionts, is a member of the phylum Mucoromycota, subphylum Glomeromycotina. From an evolutionary standpoint, one of the pivotal factors that facilitated the transition from aquatic to terrestrial life for plants was the early interaction with soil microorganisms. Fossil evidence indicates that plants engaged in symbiotic associations with AM fungi as early as 400 million years ago (Escudero-Martinez and Bulgarelli 2019). AM have frequently been overlooked by foresters due to their association with species-rich ecosystems. However, these characteristics are also present in valuable trees such as Acer, Araucaria, Podocarpus, and Agathis, as well as all Cupressaceae, Taxodiaceae, Taxaceae, Cephalotaxaceae, and the majority of tropical hardwoods. AM symbiosis can be categorized as a common symbiosis pathway (CSP), comprising three stages: the pre-symbiotic stage, the pre-contact stage, and the colonization stage.

2.1.3.1 Pre-symbiotic Stage

A deficiency in nutrients in the soil prompts the plant to release specific metabolites into the rhizosphere, which then act as signaling molecules. In the context of AM symbiosis, root exudates play a pivotal role in facilitating distinctive hyphal branching patterns and in identifying host plants (Besserer et al. 2006). Carotenoid-based phytohormones, strigolactones (SLs), serve as the primary pre-colonization signals. Additionally, *N*-acetylglucosamine (GlcNAc), flavonoids, and 2-OH fatty acids have been identified as factors involved in the reprogramming of fungal transcription, which is conducive to the formation of a symbiotic relationship (Banasiak et al. 2021). Kretzschmar et al. (2012) reported that SLs are released into the rhizosphere through suberin-free hypodermal passage cells (HPCs) by the Pleiotropic Drug Resistance 1 (PDR1) exporter. This hypothesis has been confirmed by the results obtained from Pdr1 mutants, which secrete less SL and consequently exhibit reduced mycorrhizal colonization. N-acetylglucosamine (GlcNAc)-like compounds have been identified as symbiotic signals following the treatment of *Rhizophagus irregularis* with the rice root exudates of the nope1 mutant transporter of GlcNAc, which is unable to form the AM symbiosis (Nadal et al. 2017).

2.1.3.2 Pre-contact Stage

A diverse range of substances are released by microbes into their surrounding environment, which are used for intra- and interspecies communication. Tsuzuki et al. (2016) reported the initial discovery of a proteinaceous AM fungal rhizosphere signal, which is essential for root colonization. The application of SL (strigolactone) to *R. irregularis,* one of the AM fungi, results in the secretion of a protein designated SL-induced putative secreted protein 1 (SIS1) into the rhizosphere. The results of host-induced gene silencing (HIGS) on SIS1 transcripts demonstrated a notable reduction in root colonization, which substantiates the assertion that SIS1 plays a pivotal role in facilitating fungal root colonization. AM fungi secrete two distinct classes of oligosaccharides: mycorrhizal-lipochitooligosaccharides (Myc-LCOs) and short-chain chitooligosaccharides. The former includes chitotetraose and the latter, chitopentaose (CO5). These oligosaccharides possess a shared β-1–4-linked glucosamine (GlcNAc) backbone. LCOs undergo further modification with oleic or palmitic acid at the N-terminus and also exhibit O-sulfation or non-sulfation at the C-terminus, as observed in the aforementioned ECM signals (Maillet et al. 2011). What differentiates GlcNAc-based microbial signal molecules to enable signaling specificity for interactions with AMF, as opposed to rhizobia or pathogenic fungi? To address this question, it is essential to gain a fundamental understanding of how plants recognize AMF.

Receptor-like kinases (RLKs) on the surface of plant cells are responsible for receiving the microbial signals in the form of GlcNAc oligosaccharides. Receptor-like kinases (RLKs) possess lysin motifs (LysMs) within their extracellular domain, which facilitate GlcNAc binding. Research in *Lotus japonicus* and *Medicago truncatula* has led to the identification of NFR1 and NFR5, which have been shown to bind to Nod-LCOs and activate the common symbiosis signaling pathway (CSP) (Zipfel and Oldroyd 2017). In a mutational study, Rasmussen et al. (2016)

confirmed the Nod-factor receptors *LYS11* of *Lotus japonicus* and NFR1 as receptors involved in AMF perception. However, mutants of NFR5 demonstrate wild-type AMF colonization. The disruption and RNAi-mediated downregulation of the rice homolog of NFR1, *chitin elicitor receptor kinase 1* (*CERK1*), have been observed to result in a reduction in fungal colonization and the formation of abnormal AMF infections (Zhang et al. 2015). As observed in legumes, mutation of the rice NFR5 homolog had no impact on AMF colonization (Miyata et al. 2016). The application of RNAi or virus-induced gene silencing (VIGS) to *Parasponia andersonii* and *Solanum lycopersicum* resulted in the strong AM colonization phenotypes observed in other species, indicating that the recognition of AMF may involve different LysM-RLKs in different plant species (Choi et al. 2018).

2.1.3.3 Colonization Stage

Wang et al. (2012) reported that aliphatic compounds linked to the plant polyester cutin have been suggested as another class of signaling molecules that are important for promoting mycorrhizal colonization. RAM2/GPAT protein (Required for arbuscular Mycorrhization/Glycerol-3-Phosphate Acyl-Transferase), which is involved in cutin monomer biosynthesis, is required for both appropriate hyphopodium formation on the root surface and arbuscule development. Once fungal hyphae cross the epidermis, they enter into the outer cortical cell layer; this layer is suberized and is referred to as the exodermis in most of the angiosperms. Suberized cells are impenetrable to AM fungi. Suberin-free hypodermal passage cells (HPCs) are access points for further fungal colonization of the outer cortex. HPC distribution may influence root penetration by AM fungi and is controlled by hormonal signaling and environmental conditions (Banasiak et al. 2021). AM fungi produce unique structures known as hyphopodia, which originate from mature hyphae rather than germination tubes. Due to a series of chemical and mechanical stimuli, plant cells generate a pre-penetration apparatus (PPA). Afterward, the fungal hypha emerging from the hyphopodium enters the PPA, leading the fungus through root cells toward the cortex. In this case, the fungus exits the plant cell and goes into the apoplast, where it spreads and expands horizontally alongside the root axis. These hyphae stimulate the formation of PPA-like structures in inner cortical cells, then invade these cells, and divide to create arbuscules (Parniske 2008).

2.2 Nutrient Exchange

The establishment of mycorrhizal symbiosis in plants is a major benefit, which improves the acquisition of mineral nutrients. Exchange of nutrients occurs in both directions. The characteristic of the AM symbiosis is the mutual sharing of nutrients. AM fungi give soil-derived phosphorus and N in return for organic carbon. Fungi transport polyphosphates to arbuscules, where they are converted to the inorganic form and released into the periarbuscular interface (PAI). Symbiosis-specific phosphate transporters of the *MtPT4*/*OsPT11* type obtain phosphate through the periarbuscular membrane (PAM) from this location. In the same way, N obtained from soil enters the arbuscule as amino acids (glutamate and arginine), which are then converted to urea and ammonia before the release of the latter into the PAI. Ammonium is absorbed through the PAM by plant ammonium transporters, which may include the *MtAMT2;4 type*. In AM symbiosis, roots increase their capacity to supply sugars and lipids to AMF by gaining sink strength. Within the PAI, sucrose is converted into hexose, where fungi mainly take up glucose and transform it into storage forms like trehalose and glycogen. AMF do not possess the genes needed for producing long-chain fatty acids (FAs) and rely on the host to provide them. New synthesis of fatty acids takes place in arbusculated cells through the activation of the fatty acid biosynthetic genes *WRI5b/EFR1* (transcription factor), *DIS* (*fatty acid synthase*), *FatM* (*acyl transferase*), *RAM2* (*GPAT6*), and lipid transporter candidates STR/STR2 located in the peroxisome-associated membrane. The long-chain fatty acids given are either turned into triacylglycerol (TAG) for storage or they are desaturated for growth and development purposes. The key transcription factor RAM1 triggers the activation of genes related to nutrient exchanges (Choi et al. 2018).

2.3 Expression of Plant Growth Hormones

2.3.1 Signaling of Mycorrhizae

Rhizospherical space creates a complex interplay of biochemical exchanges between numerous organisms residing in the underground arena and the root system of plants. Out of the four classes of mycorrhizal associations (i.e., Glomeromycota, Ectomycorrhizae, ericoid, and orchid), the signaling for ericoid and orchid associations is poorly deciphered. The description of the other two, i.e., AM and ECM, is well defined and have been stated as follows:

Strigolactones are key players in serving as signaling molecules and are triggered by the deficiency of P content in the plants corresponding to the low P levels of soil (Umehara et al. 2008). Upon sensing the Strigolactones signals via ABC transporter *PDR1* gene (Kretzschmar et al. 2012), the AM fungi start germination of spores, hyphal branching, and production of lipochitooligosaccharides (Maillet et al. 2011) and short-chain (four- to five-chain) chitin oligomers (Genre et al. 2013). AM fungi utilize these complex Myc factors to carry out a dialogue with the host plant. Host plants recognize these compounds on their plasma membrane through lysine-motif receptor-like kinases acting in tandem with a coreceptor called NORK/DMI/SymRK which is a leucine-rich repeat receptor-like kinase (Stracke et al. 2002; Op den Camp et al. 2011; Miyata et al. 2014; Zhang et al. 2015). A cascade of reactions starts from this point onward involving various players. Interaction of NORK with HMGR1 leads to mevalonate production, and through an unknown chain of events, mevalonate leads to the activation of nuclear ion channels as well as cyclic-nucleotide-gated calcium channels. Thus, it leads to calcium spiking with repeated oscillations of the release of Ca^{2+} ions into and out of perinuclear space. This further sets in motion the expression of downstream targets like DMI3/CCaMK (calcium- and calmodulin-dependent protein kinase) which results in the phosphorylation of the transcription factor CYCLOPS, regulating the expression of several transcription factors like a GRAS transcription factor (RAM1) and a glycerol-3-phosphate acyl transferase (RAM2) for the development of AM symbiosis (Luginbuehl and Oldroyd 2017; Venkateshwaran et al. 2013). ECM is considered the lifeline for forest ecosystems that significantly aid the growth and performance of the host plants by augmenting the uptake of nutrients. The establishment of association between ECM and its host plants begins with the recognition of signal molecules which was first deciphered in the 1950s (Melin 1954). Reports suggest the presence of flavonoid compounds like rutin and quercetin to enhance hyphal growth and influence further colonization of ECM fungi (Cope et al. 2019). ECM fungi release chitin-derived molecules constitutively into the environment that could allow for cell wall remodeling. MiSSP7 is a well-studied fungal effector protein, and its expression is associated with studies of ECM symbiosis (Plett and Martin 2012). It is a 7 kDa protein that gets accumulated in the hyphae and gets secreted into the extracellular environment after perceiving the diffusible plant signals via phosphatidylinositol-3-phosphate-mediated endocytosis and is thereby translocated to the root cell nucleus within a few minutes. ECM genes show polyphyletic evolution, requiring specificity to be built for every pair of fungus-host association; for instance, *L. bicolor* and poplar root have genes that contribute to a common symbiosis signaling pathway as described above for AM symbiosis (Cope et al. 2019). However, this common symbiotic signaling does not hold true in the case of Pinaceae trees as they have lost significant genes particular to this pathway which are responsible for their association with ECM fungi (Garcia et al. 2015). The signaling pathway for ericoid and orchid mycorrhiza still remains largely unknown and opens a frontier area of research.

2.3.2 Mycorrhizae and Hormone Crosstalk

MiSSP7, produced by *L. bicolor*, interacts with JAZ6, a transcriptional repressor in the host nucleus, to prevent proteasomal degradation of JAZ6. This interaction prevents the accumulation of jasmonate in the host cell nuclei, which would otherwise activate transcriptional repression. This interaction also aids in preventing the degradation of JAZ6. In response to a wounding stimulus such as hyphal or bacterial cell penetration, the release of jasmonate is in its isoleucine-conjugated form. This is detected by the Fbox protein *CORONATINE INSENSITIVE 1* (COI1). By ubiquitinating *JAZ6*, COI1 contributes to the E3 ubiquitin ligase complex SCFCOI1, which targets JAZ6 for proteasomal degradation and relieves transcriptional repression. In the nucleus of host cells like *L. bicolor*, MiSSP7 physically attaches to JAZ6, resembling the associations of transcriptional activators and COI1. As a result, the proteasomal degradation of JAZ6 that would otherwise be caused by the buildup of jasmonate brought on by the fungal infection is prevented by the interaction between MiSSP7 and JAZ6. Auxin's role serves as that of a signaling molecule and modification of root architecture by stimulating lateral root architecture. However, Hartig net development was limited due to differential expression influenced by the integration of ethylene and jasmonate signaling (Martin et al. 2016).

It has been shown that DELLA proteins which act as repressors of gibberellin signaling, function upstream of RAM1 to restore arbuscule production in roots treated with gibberellic acids. DELLA proteins are crucial for arbuscular growth. Together with CCaMK and CYCLOPS, they form a complex that leads to the transcriptional induction of RAM1. Although its precise function in mycorrhization is yet unknown, the possible involvement of gibberellic acid in mycorrhization may provide an exciting avenue with regard to its further coordination of mycorrhization with plant development. DELLA proteins are also shown to be associated with proteins like RAD1 and MIG1 for the formation of arbuscules in colonized roots and expansion of radial cells in order to accommodate fungal infection structures at the time of AM symbiosis, respectively. Fungal colonization in plants involves morphological and gene expression changes, altering signaling cascades. AM fungi produce compounds similar to plant hormones, such as auxins and cytokinins. ABA plays a role in mycelium production and growth which has been substantiated with studies on ABA-deficient mutants. It has been shown that ABA deficiency hampers mycorrhizal intensity, reduces arbuscular formation, and increases ethylene levels which could be likely due to increased expression of defense-related genes and cell wall modifications (Basu et al. 2018). Recent research has demonstrated similarities between mycorrhizal and N symbiosis, such as the induction of ENOD11 (*Medicago truncatula* Early Nodulin) by both *G. intraradices* and *Sinorhizobium meliloti* (Mukherjee and Ané 2011). Jasmonic acid (JA) has been studied for its role in plant responses to pathogens, but its role in AM colonization needs further investigation. Exogenous application of JA has shown both positive and inhibitory effects on fungal colonization. Studies have shown that downregulation of the allene oxide cyclase gene in transgenic plants results in delayed AM colonization. Salicylic acid, like JA, protects plants against biotic stresses and is believed to be induced during AM colonization.

3 Effect of Mycorrhizal Symbiosis in Soil Fertility

3.1 Impact on Mycorrhizal Symbiosis on Soil Chemical Properties and Improved Soil Structure

Phosphorus is one of the major elements that are essential for plant growth and responsible for producing nucleotides, ATP molecules, phospholipids, and enzymes. It also helps in the transportation of energy, development of root structure, and overall plant growth (Agledal et al. 2010). Mycorrhizal fungi can enhance and control soil particle aggregation to different extents, such as plant communities, individual host plant roots, and fungal mycelium (Rillig and Mummey 2006), which is triggered by the release of glomalin. The glomalin is a binding agent that acts as a glue substance, increasing the soil porosity, water retention, and aeration in the soil particles. This helps to avoid soil erosion and compaction. Mycorrhizal symbiosis can enhance soil stability and reduce compaction, finally increasing the ability to sustain plant growth (Xie et al. 2015). It increases the soil aggregate stability and hydraulic conductivity between roots and the soil, promoting root penetration and water absorption in the soil ecosystem (Leifheit et al. 2014). Arbuscular mycorrhizae can form macro aggregates to maintain better soil structure and improve root growth metabolism (Six et al. 2004).

3.1.1 Soil Nitrogen and Phosphorus

In plants, mycorrhizal fungi play an important role in nutrient absorption, particularly in absorbing N and P elements. They are involved in the enhancement of these important nutrients by enlarging the root surface area and retrieving the nutrient components that the plants cannot reach independently. Along these networks, mycorrhizal fungi can improve P absorption by obtaining the phosphate ions bound to soil particles. Hence, plant productivity has been improved (Smith and Read 2008). These fungi support N uptake in organic soils by associating with N-fixing bacteria and offering the decomposition of organic substances. These interactions help in the absorption of N, improve nutrient acquisition, and stimulate plant growth. This fungal colonization triggers plant N content by the ability of roots to access N-rich areas in the soil, especially in N-deficient environments (Hodge and Storer 2015).

3.1.2 Improved Nitrogen Uptake

Nitrogen is a major element that is involved in plant growth, and mycorrhizal fungi enhance its availability in soils. AM fungi improve plants' access to ammonium and nitrate, which increases N absorption even though they do not directly fix ambient N. Their vast hyphal networks reach deep into the soil and draw from N-rich regions, which is particularly advantageous in soils that are low in N. Additionally, they work with bacteria that fix N, increasing the amount of N available to host plants and improving root health and nutrient absorption. Ammonium is preferentially absorbed by AM fungus over nitrate, which might lessen N loss and nitrate leaching (Hodge and Storer 2015).

3.1.3 Enhanced Phosphorus Availability

Phosphorus is very crucial for energy transfer in plants, but it is slightly difficult to access in many soils. Mycorrhizal fungi can improve P uptake, particularly in P-lacking environmental conditions. Organic acids are produced by these fungi that can be solubilized in organic P and made available to plants. They help mobilize P from calcium-phosphate complexes in calcareous soils (Smith and Read 2008). Mycorrhizal fungi extend their hyphae beyond the root zone and significantly increase the surface area for absorption of P, allowing plants to access P from previously unreachable areas. The presence of AM fungi can suppress the need for synthetic P fertilizers by improving natural P acquisition (Koide and Mosse 2004).

3.1.4 Trace Element Mobilization: Zinc, Copper, and Iron

Mycorrhizal fungi are the major players in the absorption and transportation of essential micronutrients copper (Cu), iron (Fe), and zinc (Zn), which are commonly present in the soil, but the plants cannot directly absorb them from the soil. These fungi increase their contact with the soil to improve the breakdown of mineral components, increase the availability of these trace elements, and become more supportive in critical situations.

3.1.5 Copper Mobilization

The absorption of Cu is one of the important roles in mycorrhizae, which is involved in the enzymatic process, and it can promote the photosynthesis process. Mycorrhizae can grow toward the roots with the help of hyphae, and they increase their surface area for the absorption of the Cu molecules in the soil. It will lead to the enhancement of Cu availability, improving their enzymatic activity and plant growth metabolism (Rillig 2004). High concentrations of Cu (20–100 mg kg^{-1}) can cause toxic effects on soil microorganisms, which are essential for plant growth and inhibit the essential minerals in the soil (Azeez et al. 2015).

3.1.6 Iron Mobilization

The availability of Fe in plants is limited since it is frequently contained in insoluble forms such as Fe oxides. To assist, mycorrhizal fungus release siderophores, or organic acids, which chelate Fe and change it into a form that plants can absorb. Given that Fe availability impacts plant development and chlorophyll production in soils lacking Fe, this process is essential. For plants to get access to this vital mineral, AM fungus helps mobilize Fe from these insoluble forms. In addition to employing phytosiderophores to maximize Fe absorption efficiency, AM fungi are crucial for the uptake or transfer of Fe chelated with siderophores made by other soil microorganisms such as bacteria, fungi, and AM fungi (Rajapitamahuni et al. 2023).

3.1.7 Zinc Mobilization

Although Zn is frequently absent in its native soil form, it is essential for plant processes, including protein synthesis and enzyme activation. By solubilizing Zn-containing minerals through acidity and enzymatic activity, AM fungus enhances Zn absorption. This increases the availability of Zn, especially in soils that are lacking it, which improves plant development and output. These fungi release trace elements such as Zn and Fe from the soil through plant root systems, which the plants cannot uptake in nature (Clark and Zeto 2000), and their presence enhances Zn levels in deficient soils, improving plant growth performance (Bhantana et al. 2021).

3.2 Role of Mycorrhizal Symbiosis in Carbon Sequestration

3.2.1 Increased Carbon Input Through Plant Growth

By facilitating better nutrient uptake, particularly of N and P, mycorrhizal fungi increase photosynthetic activity and plant productivity. Carbon fixation and carbon deposition in plant biomass increase as a result. Mycorrhizal fungi increase photosynthetic rates by enhancing plant nutrition. This increases the amount of carbon assimilated into plant tissues and adds to soil carbon through plant wastes and root exudates (Smith and Read 2008). By transmitting sugars and organic molecules through their root systems, plants provide mycorrhizal fungi with a portion of their carbon. Rhizodeposition is a mechanism that helps create stable carbon pools by increasing the amount of carbon in the soil and encouraging microbial activity (Leake et al. 2006).

3.2.2 Hyphal Networks and Soil Organic Carbon Stabilization

Soil organic carbon has been stabilized by the enhancement of nutrient uptake through mycorrhizal fungal hyphae; these fungal hyphae will be involved in the slow decomposition due to the presence of carbon in the hyphae and lead to long-term SOC pools. So this ensures that the hyphae act as carbon storage in the soil (Rillig 2004). In general, AM fungi can produce a resilient glycoprotein (glomalin) that helps in carbon storage activities. It binds toward the soil particles and the organic substances in the soil to form a protective shield of carbon, which safeguards the carbon from the decomposition by microbes; hence, the carbon stability is maintained in the soil (Rillig et al. 2001).

3.2.3 Soil Aggregation and Physical Protection of Organic Matter

In carbon sequestration, soil aggregation plays a major role in the protection of organic substances that have been decomposed by the microbes. Mycorrhizal fungi can physically attach the soil particles with the help of hyphae, which produce glomalin. The main purpose is to protect the carbon, which is encapsulated in the soil aggregates and avoid the microbial breakdown in the soil ecosystem, leading to the long-term storage of carbon, especially in the huge mycorrhizal ecosystems (Wilson et al. 2009).

3.2.4 Increased Soil Carbon Through Fungal-Plant Interactions

These fungi can mainly influence the carbon regulation ecosystems along with symbiotic association with plants, where plants can transfer the carbon to the soil. It acts as an efficient carbon sink and can fix 20% of carbon through photosynthesis (Hobbie 2006). It contributes to fungal growth and enhances metabolic activities. It is functionally merged into the carbon pools of the soil as a fungal biomass decomposer. It helps in the regulation of carbon input and output ecosystems by taking part with decomposers for nutrient intake. This process reduces the organic matter decomposition and enhances the carbon sequestration in the soil (Averill et al. 2014).

3.2.5 Role of Mycorrhizal Fungi in Climate Change Mitigation

Mycorrhizal fungi are also responsible for climatic change by capturing the carbon in the soil and roots of the plants, which optimize the CO_2 levels in the atmosphere. They play an important role in carbon sequestration across various ecosystems (Phillips et al. 2012). Mycorrhizal fungi support the soil ecosystem's resilience to climate change by promoting plant tolerance to stress factors such as drought and nutrient deficiency. This acts as a carbon sink at different environmental adaptations (van der Heijden et al. 2015).

3.3 Role of Mycorrhizal Symbiosis in Soil Erosion and Nutrient Leaching Prevention

3.3.1 Reduction in Soil Erosion

Soil erosion is controlled by a particular binding agent, which is secreted by the mycorrhizal fungus, and interacts with soil to reduce erosion, named glomalin, which is a glycoprotein (Rillig 2004). Stronger soil particles are less vulnerable to soil erosion. These fungal hyphal networks can bind the soil particles to produce a more stable soil structure in between the roots (Wilson et al. 2009). When roots can be paired with hyphae to increase root growth, they tighten the soil and reduce soil erosion and cohesion, mainly in sloped areas (Leifheit et al. 2014).

3.3.2 Prevention in Nutrient Leaching

Through their hyphal networks, mycorrhizal fungi expand the surface area available for nutrient absorption, so absorbing vital nutrients such as N and P before they can escape (Smith and Read 2008). Mycorrhizal fungi enhance the solubilization of P and are involved in the reduction of P leaching (Koide and Kabir 2000). They enhance N uptake through ammonia and also stop the nitrate leaching in the soil when it is in high form, which will affect the water bodies (Hodge et al. 2001). These associations will be more than enough for plant growth; using these methods will avoid artificial fertilizers. Thus, nutrient leaching is reduced due to the application in soil (Bender et al. 2016).

3.3.3 Water Retention and Reduced Surface Runoff

Water infiltration and retention are controlled by the mycorrhizal fungi, which leads to a reduction in the formation of surface runoff and leaching. The soil particles aggregate to improve root development and decrease the impact of heavy rainfall, shortening the runoff and reducing soil erosion, especially in slope areas, and nutrient loss is also reduced (Augé 2001).

3.3.4 Role in Organic Matter Decomposition and Nutrient Recycling

In the soil ecosystem, organic matter is decomposed with the help of mycorrhizal fungi, which interact with saprotrophic living things and bring back the P and N to the soil (Hodge et al. 2001). These fungi tightly bind with the soil and roots to reduce the mobility of essential nutrients in the forest ecosystems (van der Heijden et al. 2015) (Table 1).

Table 1 Effect of mycorrhiza on soil improvement, carbon sequestration, and nutrient leaching reduction

Aspect	Effect of mycorrhizal fungi	References
Soil chemical properties	Enhances nutrient availability, particularly nitrogen (N), phosphorus (P), and trace elements (e.g., Zn, Cu, Fe)	Bhantana et al. (2021)
Soil structure improvement	Promotes soil aggregation by binding soil particles, improving porosity and water retention	Pauwels et al. (2023)
Carbon sequestration	Increases carbon storage by enhancing plant biomass and contributing to soil organic matter	Wu et al. (2022)
Soil erosion control	Reduces soil erosion by stabilizing soil through hyphal networks that anchor soil particles	Fall et al. (2022)
Nutrient leaching reduction	Minimizes nutrient leaching by improving nutrient uptake efficiency and retaining nutrients in the root zone	Zhang et al. (2021)

4 Impact on Crop Nutrition

4.1 Mechanism of Macro (N, P, K) and Micro (Zn, Cu) Element Solubilization and Their Uptake

Excessive land use can significantly affect biodiversity and ecosystem function. The incorporation of organic carbon into plants and animals through symbiotic relationships can facilitate the uptake of nutrients. In order to increase photosynthetic output and biomass accumulation, AMF colonization can raise macro- and micronutrient concentrations which additionally help improve nutrient-deficient soils by increasing the uptake of inorganic nutrients, especially phosphate. Mycorrhizal association enhances P supply to infected roots in P-limited conditions, leading to improved Pi uptake rate in maize plants. This is linked to increased photosynthetic activities and leaf functions, which are influenced by the uptake of N, P, and carbon. AM fungus uses Pi transporters present on the plasma membrane surface of extraradical hyphae to absorb Pi from the soil in mycorrhizal plants. The Pi absorbed gets quickly transformed into polyphosphate (poly-P) and separated in tubular vesicles, where endo-polyphosphatase present in the vacuole breaks the IRM long-chain polyphosphate into shorter chains. Following this, Pi is released from the terminal residues of short-chain polyphosphate by polyphosphatase, and this permits Pi to be transmitted to the host plant via the mycorrhizal interface. It has been determined that *Rhizophagus irregularis* (*GiPT*), *Funneliformis mosseae* (*GmosPT*), *Rhizophagus clarus* (*GvPT*), and *Gigaspora margarita* (*GigmPT*) are the four Pi transporters of AM fungus (Garcia et al. 2016; Harrison and Buuren 1995; Xie et al. 2016). It is currently understood that N is taken up by AM fungi in the extraradical mycelium (ERM) and assimilated through the GS/GOGAT cycle, primarily in the form of ammonium, and metabolized to arginine. This is the primary form of N that is transported from the ERM to the intraradical mycelium (IRM), where it is broken down into urea and ornithine. After urea is hydrolyzed, ammonium is produced. This is then released into the symbiotic interface, where plants absorb it. The first plant

AMT gene identified as being involved in N uptake during AM symbiosis is *LjAMT2.2* and gets highly unregulated upon colonization of *Lotus japonica* roots with *Gigaspora margarita*. When AM fungi colonized the roots, *SlNRT2.3* was expressed in the inner cortical cells, and the *SlNRT2.3* transcript levels in the colonized roots were higher than in the non-colonized controls. This could potentially mediate the beneficial effects of AM fungi on the uptake of NO_3^- from soil and its distribution (Hildebrandt et al. 2002).

ECM fungi have the ability to be considered as facultative saprotrophs. They also contain several hydrolytic enzymes to help them degrade the litter polymers and utilize the organic nutrient sources. But they have lost the key ability to degrade plant cell walls which restricts their penetration into the intercellular spaces of the root. Numerous tracer studies have found key evidence regarding enhanced host plant nutrition in the event of formation of ECM symbioses. Species variation in response to nutrient acquisition is seen due to the polyphyletic nature of ECM symbiosis, and thus, fungal diversity is essential to take into account while considering any aspect of ECM fungal lifestyle. In model species of ECM fungi, studies are replete with encoded genes for nitrate and ammonium transporters which help them acquire nutrients from the soil along with necessary enzymes for their utilization. Ammonium is mostly preferred, and functional characterization has revealed high-affinity ammonium importers AMT1 and AMT2 in ECM fungal species like *Hebeloma cylindrosporum, Tuber borchii,* and *Amanita muscaria,* and their expression is found to be upregulated in low-ammonium conditions. While ECM has a wide capability to grow on nitrate as the sole N source, it is not favored due to high assimilation costs. The supply of P is a major limiting factor for many terrestrial plants. High-affinity Pi transporters have been found in the basidiomycete *Hebeloma cylindrosporum* such as HcPT1.1 and HcPT2 (Garcia et al. 2013) which have been characterized as proton-dependent Pi importers. Orchid mycorrhizal fungi like *Serendipita vermifera* and *Tulasnella calospora* have a diverse range of polysaccharide-dependent enzymes (PCWDEs), including secreted endo cellulases, lytic polysaccharide monooxygenases, and gene families which are related to hemicellulose degradation, for instance, CE1 and CE16. Thus, these fungi indirectly supply nutrients to seeds with insufficient carbohydrate storage for development. *Oidiodendron maius*, an ericoid mycorrhizal fungus belonging to the class Leotiomycetes of the Myxotrichaceae family, possesses multiple copies of genes for secreted GH6 and GH7 cellulases as well as LPMOs. These genes give the fungus with a variety of PCWDEs that aid in the decomposition of peat moss. It is interesting to note that the majority of these PCWDEs are induced in cultures containing organic matter and not expressed during the biotrophic phase of fungal growth but in the case of *S. indica*, they are expressed during the necrotrophic phase of growth that follows plant cell death on specific hosts (Martin et al. 2016).

In order to provide mineral nutrients to AM fungi, plants transport photosynthetically fixed carbon to them. The fungus can absorb up to 20% of the photosynthates from plants, which strengthens the carbon sink and causes sugars and lipids to build up in mycorrhized roots. Studies using radio-labeling reveal that whereas fungal spores and extraradical hyphae cannot take up hexoses, especially glucose, intraradical hyphae can. The efficient transport of monosaccharides such as glucose, mannose, and fructose is made possible by the identification and characterization of mycorrhizal-induced high-affinity monosaccharide transporter MST2 in *Rhizophagus irregularis*. Some studies suggest that plants might also supply fatty acids as carbon sources with the help of transporters like ABCG transporters, STR and STR2. The mechanism of lipid delivery to the fungus remains to be understood, as a significant amount of nutritional needs of the fungus is thought to be derived from fatty acid molecules (Luginbuehl and Oldroyd 2017).

4.2 Tolerance Against Stress (Abiotic and Biotic)

Mycorrhizal fungi increase tolerance to biotic and abiotic stresses in plants. Fungal symbionts have a mutual association with the plant's roots, so this confers many advantages on the host plants (Kumar et al. 2017; Zhu et al. 2022). Mycorrhizae enhance plant growth, development, and immunity by improving uptake of nutrients like P, Zn, and Cu, which are immobile elements. In abiotic stress tolerance, fungi enable the plant to tolerate other environmental stresses, such as drought stress, salinity, high temperature, and nutrient availability stress (Kumar et al. 2017; Mitra et al. 2021). In the case of these stresses, the microorganism enhances plant defense mechanisms through an increase in the uptake of water and nutrients, regulation of the osmotic balance, and an increase in the activity of antioxidant compounds (Kumar et al. 2017; Mitra et al. 2021). For example, mycorrhizal inoculation enhanced photosynthesis, gas exchange, and water use efficiency under stress conditions (Mitra et al. 2021). Additionally, mycorrhizae enhance the tolerance of biotic stress by promoting systemic resistance to pathogens and herbivores (Bueno et al. 2022; Kumar and Verma 2017). In this regard, ISR also allows plants to fight different biotic aggressions (Kumar and Verma 2017). This mycorrhizal fungus may thus act as a biological control agent also by inducing crop plants for greater synthesis of defense-related compounds. Shortly, mycorrhizal symbiosis becomes a friendly habitat to raise stress tolerance in plants. Related to nutrient uptake, water relations, and enhancement of plant defense, mycorrhizae help increase the resistance of plants against biotic as well as abiotic stresses (Lone et al. 2017). This symbiotic relationship, in recent times, has emerged as one of the cornerstone principles of sustainable agriculture, offering a possible alternative to the use of chemical fertilizers and pesticides, contributing to improving plant productivity and health in the ecosystem.

5 Summary and Conclusion

The mycorrhizal fungi improve soil fertility and nutrition in plants through mycorrhizal symbiosis. These mutualistic associations between the hosts and fungi enhance the nutrients, particularly P absorption, while the growth and development of plants are improved alongside their health (Jeffries et al. 2003; Ramasamy et al.

2011). The symbiotic association is also involved with the formation and stabilization of soil aggregates through hyphal networks (Duponnois et al. 2008). It is indeed interesting that although the overall benefit of mycorrhizal symbiosis can be gained, its efficacy still differs from one condition of the soil to another, not to mention different plant genotypes. A number of studies reported no correlation of plant mycorrhizal status with nutrient availability in some European herbaceous habitats (Bitomský et al. 2021). However, modern cultivars still tend to have unintended effects with respect to mycorrhizal symbiosis practices: some modern cultivars tend to show increased development of mycorrhiza, but the responses still seem to be inconsistent when related to the symbiosis (Ellouze et al. 2015). In conclusion, mycorrhizal symbiosis also holds multiple benefits for sustainable agriculture; this includes improved nutrition, increased tolerance to adverse conditions, and hence may offer biocontrol protection against pathogens (Alves Cardoso Filho 2023; Slimani et al. 2024). The complex interactions between AM fungi, plants, and other soil microorganisms in the mycorrhizosphere contribute to overall soil health and plant community structures (Duponnois et al. 2008). Further research using molecular and systems biology approaches may lead to better harnessing of mycorrhizal symbiosis for improved crop yields and sustainable agricultural practices (Slimani et al. 2024).

6 Future Prospects

Numerous transcription factors are involved in arbuscular development, but only a few target genes are known, and the description of molecular crosstalk needs to be elucidated, especially for ECM, cricoid, and orchid mycorrhizal associations. Interaction studies show that transcription factors can physically interact, suggesting they act in large transcription-factor complexes to regulate gene expression during arbuscular establishment. Genome and transcriptome sequences for AM fungi reveal their consumption of lipids. This could explain their obligate biotrophy, requiring them to rely on plants for their life cycle. New tools like HIGS and genome and transcriptome sequences could help investigate fungal processes for plant root colonization, enhancing research on plant genetic components. It is still unclear what the molecular underpinnings are for the development of ECM symbioses, and it might be lucrative to extend knowledge about AM symbiosis to ECM mechanisms given that there are some commonalities between certain mechanisms in the two kinds of mycorrhizal relationships. But a lot of mechanisms appear to be unique to both individual symbioses. Understanding how mycorrhizal fungi interact with other members of thesubterranean food web, and how these associations with soil biota shape ecosystem functioning, remains a keychallenge. This includes clarifying the flow of energy, metabolites, signals, and nutrients mediated throughmycorrhizal networks. The development of biological models for explaining plant-fungi associations needs to be carried out at a faster pace, and their coevolution with hosts also needs to be studied.

References

Agledal L, Niere M, Ziegler M (2010) The phosphate makes a difference: cellular functions of NADP. Redox Rep 15:2–10. https://doi.org/10.1179/174329210X12650506623122

Alves Cardoso Filho J (2023) Mycorrhizal association and plant disease protection: new perspectives. INTECH. https://doi.org/10.5772/intechopen.108538

Augé RM (2001) Water relations, drought and vesicular-arbuscular mycorrhizal symbiosis. Mycorrhiza 11:3–42. https://doi.org/10.1007/s005720100097

Averill C, Turner BL, Finzi AC (2014) Mycorrhiza-mediated competition between plants and decomposers drives soil carbon storage. Nature 505:543–545. https://doi.org/10.1038/nature12901

Azeez MO, Adesanwo OO, Adepetu JA (2015) Effect of copper (cu) application on soil available nutrients and uptake. Afr J Agric Res 10:359–364. https://doi.org/10.5897/AJAR2014.9010

Banasiak J, Jamruszka T, Murray JD, Jasiński M (2021) A roadmap of plant membrane transporters in arbuscular mycorrhizal and legume–rhizobium symbioses. Plant Physiol 187(4):2071–2091. https://doi.org/10.1093/plphys/kiab280

Bender SF, Wagg C, van der Heijden MGA (2016) An underground revolution: biodiversity and soil ecological engineering for agricultural sustainability. Trends Ecol Evol 31:440–452. https://doi.org/10.1016/j.tree.2016.02.016

Besserer A, Puech-Pagès V, Kiefer P, Gomez-Roldan V, Jauneau A, Roy S, Portais JC, Roux C, Bécard G, Séjalon-Delmas N (2006) Strigolactones stimulate arbuscular mycorrhizal fungi by activating mitochondria. PLoS Biol 4(7):e226. https://doi.org/10.1371/journal.pbio.0040226

Bhantana P, Rana MS, Sun XC, Moussa MG, Saleem MH, Syaifudin M, Shah A, Poudel A, Pun AB, Bhat MA, Mandal DL (2021) Arbuscular mycorrhizal fungi and its major role in plant growth, zinc nutrition, phosphorous regulation and phytoremediation. Symbiosis 84. https://doi.org/10.1007/s13199-021-00756-6

Bitomský M, Duchoslav M, Lososová Z, Rūsiņa S, Schaefer H, Klimešová J, Pakeman RJ, Mládková P (2021) Mycorrhizal status is a poor predictor of the distribution of herbaceous species along the gradient of soil nutrient availability in coastal and grassland habitats. Mycorrhiza 31(5):577–587. https://doi.org/10.1007/s00572-021-01047-0

Bonfante P, Genre A (2010) Mechanisms underlying beneficial plant–fungus interactions in mycorrhizal symbiosis. Nat Commun 1(1):48. https://doi.org/10.1038/ncomms1046

Bueno CG, Meng Y, Neuenkamp L (2022) How can mycorrhizal symbiosis mediate multiple abiotic stresses in woody plants? Flora 295:152146. https://doi.org/10.1016/j.flora.2022.152146

Choi J, Summers W, Paszkowski U (2018) Mechanisms underlying establishment of arbuscular mycorrhizal symbioses. Annu Rev Phytopathol 56(1):135–160. https://doi.org/10.1146/annurev-phyto-080516-035521

Clark RB, Zeto SK (2000) Mineral acquisition by arbuscular mycorrhizal plants. J Plant Nutr 23:867–902. https://doi.org/10.1080/01904160009382068

Cope KR, Bascaules A, Irving TB, Venkateshwaran M, Maeda J, Garcia K, Rush TA, Ma C, Labbé J, Jawdy S, Steigerwald E (2019) The ectomycorrhizal fungus *Laccaria bicolor* produces lipochitooligosaccharides and uses the common symbiosis pathway to colonize *Populus* roots. Plant Cell 31(10):2386–2410. https://doi.org/10.1105/tpc.18.00676

Dong Y, Wang Z, Sun H, Yang W, Xu H (2018) The response patterns of Arbuscular Mycorrhizal and ectomycorrhizal symbionts under elevated CO_2: a meta-analysis. Front Microbiol 9:1248. https://doi.org/10.3389/fmicb.2018.01248

Duponnois R, Prin Y, Galiana A (2008) The mycorrhizosphere effect: a multitrophic interaction complex improves mycorrhizal symbiosis and plant growth, pp 227–240. https://doi.org/10.1007/978-1-4020-8770-7_10

Ellouze W, Depauw RM, Singh AK, Cuthbert RD, Knox RE, Hamel C (2015) Potential to breed for mycorrhizal association in durum wheat. Can J Microbiol 62(3):263–271. https://doi.org/10.1139/cjm-2014-0598

Escudero-Martinez C, Bulgarelli D (2019) Tracing the evolutionary routes of plant–microbiota interactions. Curr Opin Microbiol 49:34–40. https://doi.org/10.1016/j.mib.2019.09.013

Fall AF, Nakabonge G, Ssekandi J, Founoune-Mboup H, Apori SO, Ndiaye A, Badji A, Ngom K (2022) Roles of Arbuscular Mycorrhizal fungi on soil fertility: contribution in the improvement of physical, chemical, and biological properties of the soil. Front Fungal Biol 3:1–11. https://doi.org/10.3389/ffunb.2022.723892

Garcia K, HaiderMZ DA, Corratgé-Faillie C, Conéjero G, Tatry MV, Becquer A, Amenc L, Sentenac H, Plassard C, Zimmermann S (2013) Promoter-dependent expression of the fungal transporter HcPT1. 1 under pi shortage and its spatial localization in ectomycorrhiza. Fungal Genet Biol 58:53–61. https://doi.org/10.1016/j.fgb.2013.06.007

Garcia K, Delaux PM, Cope KR, Ané JM (2015) Molecular signals required for the establishment and maintenance of ectomycorrhizal symbioses. New Phytol 208(1):79–87. https://doi.org/10.1111/nph.13423

Garcia K, Doidy J, Zimmermann SD, Wipf D, Courty PE (2016) Take a trip through the plant and fungal transportome of mycorrhiza. Trends Plant Sci 21(11):937–950. https://doi.org/10.1016/j.tplants.2016.07.010

Genre A, Chabaud M, Balzergue C, Puech-Pagès V, Novero M, Rey T, Fournier J, Rochange S, Bécard G, Bonfante P, Barker DG (2013) Short-chain chitin oligomers from arbuscular mycorrhizal fungi trigger nuclear Ca^{2+} spiking in *Medicago truncatula* roots and their production is enhanced by strigolactone. New Phytol 198(1):190–202. https://doi.org/10.1111/nph.12146

Harrison MJ, Buuren MLV (1995) A phosphate transporter from the mycorrhizal fungus *glomus versiforme*. Nature 378(6557):626–629. https://doi.org/10.1038/378626a0

Hildebrandt U, Janetta K, Bothe H (2002) Towards growth of arbuscular mycorrhizal fungi independent of a plant host. Appl Environ Microbiol 68(4):1919–1924. https://doi.org/10.1128/AEM.68.4.1919-1924.2002

Hobbie EA (2006) Carbon allocation to ectomycorrhizal fungi correlates with belowground allocation in culture studies. Ecology 87:563–569. https://doi.org/10.1890/05-0755

Hodge A, Storer K (2015) Arbuscular mycorrhiza and nitrogen: implications for individual plants through to ecosystems. Plant Soil 386:1–19. https://doi.org/10.1007/s11104-014-2162-1

Hodge A, Campbell C, Fitter A (2001) An arbuscular mycorrhizal fungus accelerates decomposition and acquires nitrogen directly from organic material. Nature 413:297–299. https://doi.org/10.1038/35095041

Horan DP, Chilvers GA, Lapeyrie FF (1988) Time sequence of the infection process eucalypt ectomycorrhizas. New Phytol 109(4):451–458. https://doi.org/10.1111/j.1469-8137.1988.tb03720.x

Jeffries P, Gianinazzi S, Perotto S, Turnau K, Barea JM (2003) The contribution of arbuscular mycorrhizal fungi in sustainable maintenance of plant health and soil fertility. Biol Fertil Soils 37(1):1–16. https://doi.org/10.1007/s00374-002-0546-5

Koide RT, Kabir Z (2000) Extraradical hyphae of the mycorrhizal fungus *glomus intraradices* can hydrolyse organic phosphate. New Phytol 148:511–517. https://doi.org/10.1046/j.1469-8137.2000.00776.x

Koide RT, Mosse B (2004) A history of research on arbuscular mycorrhiza. Mycorrhiza 14:145–163. https://doi.org/10.1007/s00572-004-0307-4

Kretzschmar T, Kohlen W, Sasse J, Borghi L, Schlegel M, Bachelier JB, Reinhardt D, Bours R, Bouwmeester HJ, Martinoia E (2012) A petunia ABC protein controls strigolactone-dependent symbiotic signalling and branching. Nature 483(7389):341–344. https://doi.org/10.1038/nature10873

Kumar A, Verma JP (2017) Does plant—microbe interaction confer stress tolerance in plants: a review? Microbiol Res 207:41–52. https://doi.org/10.1016/j.micres.2017.11.004

Kumar M, Varma A, Prasad R, Tuteja N, Kumar V (2017) Mycorrhizal fungi under biotic and abiotic stress. Springer, pp 57–69. https://doi.org/10.1007/978-3-319-57849-1_4

Leake JR, Ostle NJ, Rangel-Castro JI, Johnson D (2006) Carbon fluxes from plants through soil organisms determined by field $13CO_2$ pulse-labelling in an upland grassland. Appl Soil Ecol 33:152–175. https://doi.org/10.1016/j.apsoil.2006.03.001

Leifheit EF, Veresoglou SD, Lehmann A, Morris EK, Rillig MC (2014) Multiple factors influence the role of arbuscular mycorrhizal fungi in soil aggregation—a meta-analysis. Plant Soil 374:523–537. https://doi.org/10.1007/s11104-013-1899-2

Ling-Lee M, Chilvers GA, Ashford AE (1977) A histochemical study of phenolic materials in mycorrhizaland uninfected roots of Eucalyptus fastigata Deane and Maiden. New Phytologist, 78(2):313–328

Lone R, Koul KK, Ahmad J, Shuab R, Khan S (2017) Arbuscular mycorrhizal fungi for sustainable agriculture. Springer, Singapore, pp 553–577. https://doi.org/10.1007/978-981-10-3473-2_25

Luginbuehl LH, Oldroyd GE (2017) Understanding the arbuscule at the heart of endomycorrhizal symbioses in plants. Curr Biol 27(17):R952–R963. https://doi.org/10.1016/j.cub.2017.06.042

Maillet F, Poinsot V, André O, Puech-Pagès V, Haouy A, Gueunier M, Cromer L, Giraudet D, Formey D, Niebel A, Martinez EA (2011) Fungal lipochitooligosaccharide symbiotic signals in arbuscular mycorrhiza. Nature 469(7328):58–63

Martin F, Kohler A, Murat C, Veneault-Fourrey C, Hibbett DS (2016) Unearthing the roots of ectomycorrhizal symbioses. Nat Rev Microbiol 14(12):760–773. https://doi.org/10.1038/nrmicro.2016.149

Massicotte HB, Peterson RL, Melville LH (1989) Hartig net structure of ectomycorrhizae synthesized between *Laccaria bicolor* (Tricholomataceae) and two hosts: *Betula alleghaniensis* (Betulaceae) and *Pinus resinosa* (Pinaceae). Am J Bot 76(11):1654–1667. https://doi.org/10.1002/j.1537-2197.1989.tb15150.x

Melin E (1954) Growth factor requirements of mycorrhizal fungi of forest trees. Svensk Bot Tidskr 48:86–94

Mitra D, Djebaili R, Pellegrini M, Mahakur B, Sarker A, Chaudhary P, Khoshru B, Gallo MD, Kitouni M, Barik DP, Panneerselvam P, Mohapatra PKD (2021) Arbuscular mycorrhizal symbiosis: plant growth improvement and induction of resistance under stressful conditions. J Plant Nutr 44(13):1993–2028. https://doi.org/10.1080/01904167.2021.1881552

Miyata K, Kozaki T, Kouzai Y, Ozawa K, Ishii K, Asamizu E, Okabe Y, Umehara Y, Miyamoto A, Kobae Y, Akiyama K, Kaku H, Nishizawa Y, Shibuya N, Nakagawa T (2014) The bifunctional plant receptor, OsCERK1, regulates both chitin-triggered immunity and arbuscular mycorrhizal symbiosis in rice. Plant Cell Physiol 55(11):1864–1872. https://doi.org/10.1093/pcp/pcu129

Miyata K, Hayafune M, Kobae Y, Kaku H, Nishizawa Y, Masuda Y, Shibuya N, Nakagawa T (2016) Evaluation of the role of the LysM receptor-like kinase, OsNFR5/OsRLK2 for AM symbiosis in rice. Plant Cell Physiol 57(11):2283–2290. https://doi.org/10.1093/pcp/pcw144

Mukherjee A, Ané JM (2011) Germinating spore exudates from arbuscular mycorrhizal fungi: molecular and developmental responses in plants and their regulation by ethylene. Mol Plant-Microbe Interact 24(2):260–270. https://doi.org/10.1094/MPMI-06-10-0146

Nadal M, Sawers R, Naseem S, Bassin B, Kulicke C, Sharman A, An G, An K, Ahern KR, Romag A, Brutnell TP (2017) An N-acetylglucosamine transporter required for arbuscular mycorrhizal symbioses in rice and maize. Nat Plants 3(6):1–7. https://doi.org/10.1038/nplants.2017

Op den Camp R, Streng A, De Mita S, Cao Q, Polone E, Liu W, Ammiraju JS, Kudrna D, Wing R, Untergasser A, Bisseling T (2011) LysM-type mycorrhizal receptor recruited for rhizobium symbiosis in nonlegume Parasponia. Science 331(6019):909–912. https://doi.org/10.1126/science.1198181

Parniske M (2008) Arbuscular mycorrhiza: the mother of plant root endosymbioses. Nat Rev Microbiol 6(10):763–775. https://doi.org/10.1038/nrmicro1987

Pauwels R, Graefe J, Bitterlich M (2023) An arbuscular mycorrhizal fungus alters soil water retention and hydraulic conductivity in a soil texture specific way. Mycorrhiza 33:165–179. https://doi.org/10.1007/s00572-023-01106-8

Phillips RP, Meier IC, Bernhardt ES, Grandy AS, Wickings K, Finzi AC (2012) Roots and fungi accelerate carbon and nitrogen cycling in forests exposed to elevated CO_2. Ecol Lett 15:1042–1049. https://doi.org/10.1111/j.1461-0248.2012.01827.x

Plett JM, Martin F (2012) Poplar root exudates contain compounds that induce the expression of MiSSP7 in *Laccaria bicolor*. Plant Signal Behav 7(1):12–15. https://doi.org/10.4161/psb.7.1.18357

Plett JM, Kemppainen M, Kale SD, Kohler A, Legué V, Brun A, Tyler BM, Pardo AG, Martin F (2011) A secreted effector protein of *Laccaria bicolor* is required for symbiosis development. Curr Biol 21(14):1197–1203. https://doi.org/10.1016/j.cub.2011.05.033

Rajapitamahuni S, Kang BR, Lee TK (2023) Exploring the roles of arbuscular mycorrhizal fungi in plant–iron homeostasis. Agriculture 13:1918. https://doi.org/10.3390/agriculture13101918

Ramasamy K, Joe MM, Sa TM, Islam MR, Shagol C, Rangasamy A, Kim KY, Chung JB, Lee SM (2011) Synergistic effects of arbuscular mycorrhizal fungi and plant growth promoting rhizobacteria for sustainable agricultural production. Korean J Soil Sci Fertil 44(4):637–649. https://doi.org/10.7745/kjssf.2011.44.4.637

Rasmussen SR, Füchtbauer W, Novero M, Volpe V, Malkov N, Genre A, Bonfante P, Stougaard J, Radutoiu S (2016) Intraradical colonization by arbuscular mycorrhizal fungi triggers induction of a lipochitooligosaccharide receptor. Sci Rep 6(1):29733. https://doi.org/10.1038/srep29733

Rillig MC (2004) Arbuscular mycorrhizae and terrestrial ecosystem processes. Ecol Lett 7:740–754. https://doi.org/10.1111/j.1461-0248.2004.00620.x

Rillig MC, Mummey DL (2006) Mycorrhizas and soil structure. New Phytol 171:41–53. https://doi.org/10.1111/j.1469-8137.2006.01750.x

Rillig MC, Wright SF, Nichols KA, Schmidt WF, Torn MS (2001) Large contribution of arbuscular mycorrhizal fungi to soil carbon pools in tropical forest soils. Plant Soil 233:167–177. https://doi.org/10.1023/A:1010364221169

Six J, Bossuyt H, Degryze S, Denef K (2004) A history of research on the link between (micro) aggregates, soil biota, and soil organic matter dynamics. Soil Till Res 79:7–31. https://doi.org/10.1016/j.still.2004.03.008

Slimani A, Ait-El-Mokhtar M, Ben-Laouane R, Boutasknit A, Anli M, Abouraicha EF, Oufdou K, Meddich A, Baslam M (2024) Molecular and systems biology approaches for harnessing the symbiotic interaction in Mycorrhizal Symbiosis for grain and oil crop cultivation. Int J Mol Sci 25(2):912. https://doi.org/10.3390/ijms25020912

Smith SE, Read D (2008) Introduction. In: Smith SE, Read D (eds) Mycorrhizal symbiosis, 3rd edn. Academic, London, pp 1–9

Stracke S, Kistner C, Yoshida S, Mulder L, Sato S, Kaneko T, Tabata S, Sandal N, Stougaard J, Szczyglowski K, Parniske M (2002) A plant receptor-like kinase required for both bacterial and fungal symbiosis. Nature 417(6892):959–962

Tsuzuki S, Handa Y, Takeda N, Kawaguchi M (2016) Strigolactone-induced putative secreted protein 1 is required for the establishment of symbiosis by the arbuscular mycorrhizal fungus *Rhizophagus irregularis*. Mol Plant-Microbe Interact 29(4):277–286. https://doi.org/10.1094/MPMI-10-15-0234-R

Umehara M, Hanada A, Yoshida S, Akiyama K, Arite T, Takeda-Kamiya N, MagomeH KY, Shirasu K, Yoneyama K, Kyozuka J (2008) Inhibition of shoot branching by new terpenoid plant hormones. Nature 455(7210):195–200

van der Heijden MGA, Martin FM, Selosse M-A, Sanders IR (2015) Mycorrhizal ecology and evolution: the past, the present, and the future. New Phytol 205:1406–1423. https://doi.org/10.1111/nph.13288

Venkateshwaran M, Volkening JD, Sussman MR, Ané JM (2013) Symbiosis and the social network of higher plants. Curr Opin Plant Biol 16(1):118–127

Voiblet C, Duplessis S, Encelot N, Martin F (2001) Identification of symbiosis-regulated genes in *Eucalyptus globulus–Pisolithus tinctorius* ectomycorrhiza by differential hybridization of arrayed cDNAs. Plant J 25(2):181–191. https://doi.org/10.1046/j.1365-313x.2001.00953.x

Wang E, Schornack S, Marsh JF, Gobbato E, Schwessinger B, Eastmond P, Schultze M, Kamoun S,Oldroyd GED, & Zipfel C (2012) A Common Signaling Process that Promotes Mycorrhizal and OomyceteColonization of Plants. Current Biology, 22(23):2242–2246. https://doi.org/10.1016/j.cub.2012.09.044

Watkinson SC (2016) Mutualistic symbiosis between fungi and autotrophs. In: The fungi. Academic, pp 205–243. https://doi.org/10.1016/B978-0-12-382034-1.00007-4

Wilson GW, Rice CW, Rillig MC, Springer A, Hartnett DC (2009) Soil aggregation and carbon sequestration are tightly correlated with the abundance of arbuscular mycorrhizal fungi: results from long-term field experiments. Ecol Lett 12:452–461. https://doi.org/10.1111/j.1461-0248.2009.01303.x

Wu Y, Deng M, Huang J, Yang S, Guo L, Yang L, Ahirwal J, Peng Z, Liu W, Liu L (2022) Global patterns in mycorrhizal mediation of soil carbon storage, stability, and nitrogen demand: a meta-analysis. Soil Biol Biochem 166:108578. https://doi.org/10.1016/j.soilbio.2022.108578

Xie H, Li J, Zhang B, Wang L, Wang J, He H, Zhang X (2015) Long-term manure amendments reduced soil aggregate stability via redistribution of the glomalin-related soil protein in macro-aggregates. Sci Rep 5:14687. https://doi.org/10.1038/srep14687

Xie X, Lin H, Peng X, Xu C, Sun Z, Jiang K, Huang A, Wu X, Tang N, Salvioli A, Bonfante P (2016) Arbuscular mycorrhizal symbiosis requires a phosphate transceptor in the *Gigaspora margarita* fungal symbiont. Mol Plant 9(12):1583–1608

Zhang X, Dong W, Sun J, Feng F, Deng Y, He Z et al (2015) The receptor kinase CERK 1 has dual functions in symbiosis and immunity signalling. Plant J 81(2):258–267. https://doi.org/10.1111/tpj.12723

Zhang H, Powell JR, Plett JM, Churchill AC, Power SA, Macdonald CA, Jacob V, Kim GW, Pendall E, Tissue D, Catunda KM (2021) Climate warming negates arbuscular mycorrhizal fungal reductions in soil phosphorus leaching with tall fescue but not lucerne. Soil Biol Biochem 152:108075. https://doi.org/10.1016/j.soilbio.2020.108075

Zhu B, Li C, Gao T, Ding K, Zhang D, Ma F (2022) Functions of arbuscular mycorrhizal fungi in horticultural crops. Sci Hortic 303:111219. https://doi.org/10.1016/j.scienta.2022.111219

Zipfel C, Oldroyd GE (2017) Plant signalling in symbiosis and immunity. Nature 543(7645):328–336. https://doi.org/10.1038/nature22009

Economic Analysis of Mycorrhizal Applications and Their Potential for Enhancing Farm Profitability and Livelihoods

Khomdram Niren Singh, Nitesh Boro, Indrani Baruah, Shibani Gogoi, Surmani Huidrom, Mahamanikya Brahma, and Khomdram Bijoya Devi

Abstract

The incorporation of arbuscular mycorrhizal fungi (AMF) in agricultural practices presents a promising approach for improving agricultural profitability and sustainability. AMF plays a crucial role in improving nutrient uptake, water absorption, and plant resistance to different abiotic and biotic stresses. This chapter provides an economic analysis of AMF applications, emphasizing their potential to reduce input costs, enhance crop yield having better quality, and contribute to sustainable agricultural practices. The adoption of AMF in agriculture significantly enhances soil health, reduces dependency on synthetic fertilizers, and also mitigates environmental degradation. AMF inoculation has been shown to improve major mineral uptakes like phosphorus and nitrogen, which led to increased crop productivity by reducing nutrient runoff. Case studies from several agricultural settings validate the practical benefits of AMF, unveiling increased yield in crops such as maize, wheat, citrus fruits, and legumes. Besides, the application of AMF contributes to climate resilience by enhancing the sequestration of carbon, reducing the emission of greenhouse gases, and restoring degraded soils. However, some challenges like high costs, market availability, and lack of awareness among farmers hamper widespread adoption of the use of

K. N. Singh (✉) · N. Boro · I. Baruah · M. Brahma · K. B. Devi
Department of Botany, Gauhati University, Guwahati, Assam, India

S. Gogoi
Department of Botany, Gauhati University, Guwahati, Assam, India

Department of Biotechnology, Mizoram University, Aizawl, Mizoram, India

S. Huidrom
Microbial Resource Division, Institute of Bioresources and Sustainable Development (IBSD), Imphal, Manipur, India

Y. Siddiqui et al. (eds.), *Sustainable Mycorrhizal Cultivation*,
https://doi.org/10.1007/978-981-95-1778-7_13

AMF. Farmer education, research investments, and policy support are obligatory to facilitate the integration and use of AMF into typical agricultural practices. This chapter emphasizes the potential of mycorrhizal technology as a keystone for sustainable and economically viable agriculture.

Keywords

Mycorrhizal fungi · Crop plants · Sustainable agriculture · Benefits · Profitability

Abbreviations

ABA	Abscisic Acid
AM	Arbuscular Mycorrhiza
AMF	Arbuscular Mycorrhizal Fungi
CMV	Cucumber Mosaic Virus
GHG	Greenhouse Gases
GRSP	Glomalin-Related Soil Protein
LAR	Leaf Area Ratio
LDN	Land Degradation Neutrality
PGP	Plant Growth Promoting
PGPM	Plant Growth Promoting Microorganisms
Pi	Inorganic Phosphorus
PVY	Potato Virus Y
RGR	Relative Growth Rate
RWC	Relative Water Content
SLA	Specific Leaf Area
TBSV	Tomato Bush Stunting Virus

1 Introduction

Mycorrhiza is a beneficial symbiotic association between plant roots and fungi. Fungi extend to the root system of plants where they improve plants' water and nutrients access while receiving the organic carbon required for their growth in return (Goltapeh et al. 2008). Mycorrhizal fungi are one of the critical components in agricultural systems as they improve plant growth by improving nutrient absorption, water absorption, and help to tolerate drought stress, heavy metal stress, salt stress, and soil pathogens while improving the soil structure and soil carbon stability (Goltapeh et al. 2008). These fungi can release certain enzymes in the soil which mobilize tightly bound nutrients such as nitrogen, phosphorus, calcium, copper, magnesium, manganese, iron, zinc, and sulfur, thereby helping plants for their absorption (Begum et al. 2019). Mycorrhizal associations with plants not only improve nutrient absorption and accumulation in plant tissue, but they also interact with other soil microbes involved in important nutrient cycles such as the nitrogen cycle (Goltapeh et al. 2008). Such ecological roles, which rely on natural nutrient cycles, are important for managing low-input agricultural farmland to provide adequate nutrients for plant production. Figure 1 depicts an overview of the influence of mycorrhizal application in crop plants. Mycorrhiza can be classified into two types based on the location of fungi in the plant root tissue: ectomycorrhiza and endomycorrhiza. Ectomycorrhizae are a type of symbiotic association between fungi and plant roots, where fungi do not penetrate the plant roots. They are mostly associated with trees (pines, oaks, eucalyptus) and form a dense network of hyphae around the root tips called the mantle for accumulating nutrients and extend to the root cortex, forming a network of hyphae structure called the Hartig net for mutual exchange of carbon, nutrients, and water (Chilvers et al. 1987). Unlike ectomycorrhizae, endomycorrhizae are type of symbiotic association between fungi and plant roots, where fungi penetrate deep in the plant root tissue. They are associated with a wide variety of plants (agricultural crops, grass, herbaceous plants) where they form a highly branched structure called arbuscules for nutrient exchange and ballon-shaped swellings in the root cortex called vesicles for nutrient storage in the roots (Chilvers et al. 1987).

Agriculture is still a cornerstone of global food security and economic development, especially in developing countries where livelihood is closely linked to agriculture. However, the agricultural sector remains challenging due to various factors such as climate change, soil degradation due to the use of chemicals in farming, and rising cost of inputs, which ultimately threaten the productivity and profitability in this sector. The wide use of chemical fertilizers, herbicides, and fungicides on agricultural

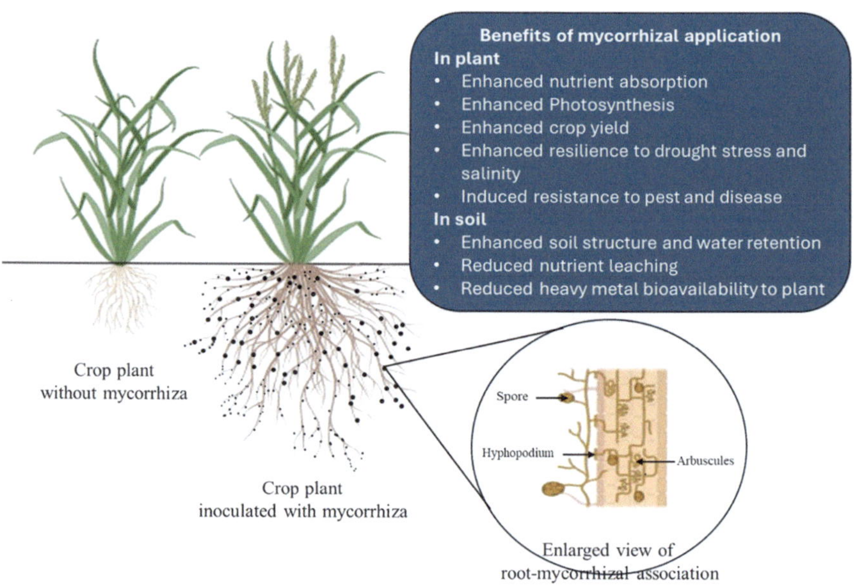

Fig. 1 Overview of the influence of mycorrhizal application on crop plants

land continuously for a long time has a negative impact on soil, plants, and human health, and on air and water systems (Yang et al. 2004). Therefore, an innovative alternative approach is required that can solve these challenges. The application of mycorrhiza is one such promising eco-friendly approach which can enhance the crop yields, improve soil quality, and reduce dependency on the use of chemicals in agricultural land. During the past decades, the use of AMF has increased significantly, showing improvement in soil health and crop productivity (Begum et al. 2019). Therefore, AMF is considered one of the best alternatives to considerably reduce the use of chemical fertilizers in agricultural land. The use of AMF not only increases the nutrient uptake in plants, but they also are found to enhance the quality of crops produced such as increased content of vitamin C, minerals, organic acids, and flavonoids in citrus fruits (Cheng et al. 2022). Many studies have confirmed the positive results of AMF applications in agricultural land (Sabia et al. 2015; Lu et al. 2015; Hijri 2016; Cao et al. 2021), their use as biofertilizers can reduce or completely replace the use of chemicals in agriculture. However, the practical use of this innovative approach by farmers depends not only on their effectiveness but also on their economic viability. Economic analysis to use AMF as biofertilizers in agricultural land is therefore very crucial as it will help farmers, policymakers, and stakeholders to evaluate the cost benefits and risks associated with using AMF for practical applications. This chapter explores the economic dimensions of mycorrhizal applications in agricultural land, highlighting the potential to transform agricultural practices while addressing the broader implications for agricultural economics and rural development.

2 Benefits of Mycorrhizal Applications for Agronomy

Our ignorance of malpractice in agriculture by using synthetic fertilizers and pesticides has stretched the effects of abiotic stresses on soil fertility, productivity, and degradation. About 6% of global agricultural land degradation is caused due to use of harmful chemicals, while the remaining are caused due to physical degradation (2%), water erosion (24%), and wind erosion (12%) (Saik et al. 2024). Several steps are taken for reversing soil degradation in the land degradation neutrality (LDN) program, of which the mycorrhizal approach is among one considered in agronomy. The use of mycorrhizal fungi is considered to support plants against various stresses, enhance nutrient uptake in plants, growth in plants, and increase resistance to pests and diseases (Jung et al. 2012; Sun et al. 2018; Zhang et al. 2022). The introduction of mycorrhizal application also helps in the aggregation of soil and maintenance of organic matter and thus influences conservation of soils' moisture content and permeability (Holatko et al. 2021). The importance of mycorrhizal fungi and their research on agricultural applications represents the link between the plants and soil nutrients, and thus, the current interests are focused on their role as biofertilizers and bio-stimulants. Table 1 lists the benefits of the mycorrhizal application in different crop plants.

Table 1 Benefits of the application of mycorrhizal fungi in different crop plants

Mycorrhizae	Crops	Benefits	References
Rhizophagus intraradices	*Solidago canadensis*	Acquisition of phosphorus from soil	Qi et al. (2022)
	Trifolium repens	GRSP production and soil aggregation	Zhang et al. (2019a)
	Medicago truncatula	Inhibit *Xanthomonas campestris*	Liu et al. (2007)
	Soyabean, tomato, *Cymbopogon citratus*	Induce defense against *Meloidogyne incognita*, *Pratylenchus brachyurus*	Dey and Ghosh (2022)
	Strawberry	Inhibit *Fusarium oxysporum* f.sp. *fragariae*	Yanan et al. (2015)
R. intraradices and *Funneliformis mosseae*	–	GRSP production and soil aggregation	Bedini et al. (2009)
F. mosseae	*Trifolium repens*	GRSP production and soil aggregation	Zhang et al. (2019a)
Glomus versiforme	*Trifolium repens*	GRSP production and soil aggregation	Zhang et al. (2019a)
	Grapevine	Induce defense against *Meloidogyne incognita*	Li et al. (2006)
Rhizophagus clarus	Soyabean, *Cymbopogon citratus*	Induce defense against *Meloidogyne incognita* and *Pratylenchus brachyurus*	Dey and Ghosh (2022)
Glomus fasciculatum	Tomato plants	Systemic induced resistance to *Alternaria alternata*	Nair et al. (2014)
Claroideoglomus etunicatum	Soyabean, *Cymbopogon citratus*	Induce defense against *Meloidogyne incognita* and *Pratylenchus brachyurus*	Dey and Ghosh (2022)

2.1 Better and Improved Uptake of Nutrients (Phosphorus, Nitrogen)

The demand for food has significantly increased globally in the past decades, resulting in the increased and repeated use of chemical fertilizers and pesticides in agricultural soil, reducing the soil quality and fertility. The inorganic or chemical fertilizers are present in the form of nitrogen, phosphorus, and potassium which are found in mixed composition or nowadays used as individual fertilizers. Among the nutrients, phosphorus and nitrogen are critical for plant growth, development, and yield, but their bioavailability in soil is insufficient for the plants. In acidic soils, the inorganic phosphorus anions are adsorbed to iron (Fe) and aluminum (Al), whereas in alkaline soils, it is adsorbed to calcium (Ca), and thus less than 10 μM concentrations are directly available to the plants (Smith and Smith 2011). For their bioavailability to plants, the AMF play a significant role in the uptake of phosphorus and other nutrients. About 50 percent of the soil microbial biomass accounts for the AMF (Ryan and Graham 2002). The fungus comprising the mycelium acts as a network extension of the root system and colonizes multiple hosts through a common network of mycorrhiza (Walder et al. 2012). These common mycorrhizal networks help in the transport of nutrients, water, and stress chemicals, as well as communication between the connected plants, and also form the pathway to transport of nitrogen from donor to recipient plants (He et al. 2009; Barto et al. 2011; Babikova et al. 2013; Bucking and Kafle 2015). Moreover, the AMF grows in two different environments, i.e., intraradical mycelium, which grows inside the plant roots, forming arbuscules, and extraradical mycelium, which grows in soil encountering soil variations. The intraradical mycelium involves the formation of intracellular mycelium growth, which later develops into symbiotic interfaces between the plant plasma membrane and peri-arbuscular membrane for the transport of potassium to plant cells. The extraradical mycelium forms a rapid network of mycelium outside the root in locating, uptake, and translocating the nutrients to the intraradical mycelium.

The plants in association with mycorrhiza express the H^+-ATPase gene along with the inorganic phosphorus transporter gene for the transport of phosphorus from soil to the plants. The phosphorus in soil is upregulated inside the extraradical mycelium as polyphosphate due to Pi influx generated by the H^+-ATPase and is later transported to the intraradical mycelium and then to the interfacial apoplast by Pi efflux (Smith and Smith 2011). It was also reported that the H^+ influx was generated at the tip and efflux was generated at the basal region of the root hairs for the transfer of phosphorus from soil to the plants (Kochian et al. 1994). The transfer of negatively charged polyphosphate is balanced by the presence of potassium (K^+), magnesium (Mg^+), or arginine (Arg^+) inside the mycelium. The Pi efflux in transferring phosphorus from the intraradical mycelium to the plant plasma membrane may be due to the higher ionic charge balance (Smith and Smith 2011). According to Qi et al. (2022), AMF *Glomus intraradices* induces a rapid colonization rate, abundance, spore production, and acquisition of phosphorus in plants growing in phosphate-deficient environments. It is also reported that plants inoculated with AMF increase the concentrations and storage of phosphorus in them (Wu et al. 2024).

As mycorrhizal fungi are nonsaprophytic in nature, the uptake of mineral nitrogen from organic matter is limited (Hodge 2014). Mycorrhizal fungi can acquire the inorganic nitrogen in the form of nitrates (NO_3^-) and ammonium (NH_4^+) from soil to plants, which is more feasible as compared to organic matter. In most agricultural soil, nitrogen is easily available to the plants in the form of nitrates, but in acidic or undisturbed soils, ammonium predominates in the soil, and nitrate availability is limited. Similar to the phosphorus uptake, the ammonium or nitrates are absorbed through the extraradical mycelium and are rapidly assimilated into amino acids, glutamine, and then to arginine via various metabolic pathways. The arginine is then transported from the extraradical mycelium to the intraradical mycelium via the same phosphate uptake pathway. Later, arginine is broken down in the intraradical mycelium and forms carbon dioxide, urea, and ammonium, from which ammonium is transported to the plant plasma membrane (Jin et al. 2005). According to Wu et al. (2024), AMF-inoculated plants showed an increase in nitrogen concentration (about 16%) and nitrogen uptake (about 67%) in plants. Overall, in nutrient-limited environments, the AMF proliferate extensively, and the plants prefer the mycorrhizal uptake pathway in response to nutrient foraging (Sun et al. 2024). It is also reported that nitrogen acquisition from organic matter increases many fold due to the implication of AMF and soil microbial communities together (Hestrin et al. 2019).

2.2 Enhanced Soil Structure and Water Retention

AMF play a major role in enhancing the soil structure and water retention by the process of soil aggregation. The characteristics of soil aggregation, such as arrangement of soil particles, pore size, and stability, evaluate the quality of soil. The AMF regulate the root exudation and, along with its hyphal network, entrap the soil particles, organic matter, and form small aggregates (Lu et al. 2019). Among the various factors in the formation of soil structure, the AMF produce a hydrophobic and thermos-stable glycoprotein substance known as glomalin- or glomalin-related soil protein (GRSP), which also helps in soil aggregation and increases its stability (Muneer et al. 2020). The presence of GRSP in soil prolongs the soil carbon presence and its stabilization during carbon flux, which enables organic matter stabilization, water-stable soil aggregation, and thereby prevents soil degradation (Agnihotri et al. 2022). According to Bedini et al. (2009), the implication of plants with AMF species *Glomus mosseae* and *Glomus intraradices* showed a significant increase in GRSP and soil aggregation as compared to nonmycorrhizal treated plants. Similar observations of increased GRSP and soil aggregation were also reported when plants were treated with three other AMF species, such as *Funneliformis mosseae, Glomus versiforme*, and *Rhizophagus intraradices* (Zhang et al. 2019a). Moreover, the AMF with glomalin production reduce the heavy metal toxicity in soil by adsorbing the heavy metals on the surface of mycelium (Vilela and Barbosa 2019). Wu et al. (2014b) also reported the adsorption of 4% total Ni and Pb from soil by AMF produced glomalin. The application of AMF in heavy metal-polluted soil may be one of the effective ways to recover the heavy metals and improve the quality of soil.

2.3 Increased Resistance to Pests and Diseases

The AMF induce resistance to diseases through various ways, such as enhancing plant vigor through nutrient supply, activation of plant hormones, and production of toxins or bioactive compounds against pathogenic bacteria and fungi (Song et al. 2020; Bencherif et al. 2019; Kaur and Suseela 2020; Wang et al. 2017). The mycorrhizal fungi form a synergistic relation with other rhizobacteria, which act as a helper agent in antagonizing the pathogens (Perez-de-Luque et al. 2017). The colonization of AMF *Funneliformis mosseae* in tomato reduces the diseases caused by the fungi *Fusarium oxysporum* and *Alternaria solani* (Song et al. 2015). Similarly, the AMF *G. intraradices* colonized in *M. truncatula* reduce the shoot disease caused by the bacteria *Xanthomonas campestris* (Liu et al. 2007). Also, plants incorporated with AMF show better resistance to nematode infection, mostly against the

Meloidogyne nematodes (da Silva Campos 2020). The AMF *Funneliformis mosseae*, *Rhizophagus clarus*, and *Claroideoglomus etunicatum* were also reported to inhibit the nematodal infection caused by *M. incognita* in soybean and *Pratylenchus brachyurus* in *Cymbopogon citratus* (Dey and Ghosh 2022). Not many reports were available on the impact of AMF against root-feeding insects. Plants associated with AMF increase the enzymatic activities of phenol oxidase, phenylalanine ammonia lyase, and lysyl oxidase, producing toxic compounds such as quinone or hydroperoxides which result in resistance to insects (Selvaraj et al. 2020; Bhonwong et al. 2009).

3 Cost-Benefit Analysis of Mycorrhizal Application in Agriculture

In recent last two decades, the production and use of mycorrhiza-based biofertilizer in the market have surged to a new height and emerged as a potential alternative to chemical fertilizers (Basiru et al. 2020). Mycorrhizal application in agriculture, having its potential to produce high yield and good-quality crop products, while maintaining the soil health for longer periods, has gained interest in most of the modern farmers. Their use as biofertilizers alone or in combination with other plant growth-promoting microorganisms (PGPM) or with compost can significantly improve crop production, improving the livelihood of people. Based on estimates, the global biofertilizer market was 3.72 billion USD in 2025 (Mordor Intelligence 2025), while it is estimated to reach 5.2 billion USD by 2028 (Joshi and Gauraha 2022). Therefore, there is a chance that biofertilizers such as mycorrhiza based, which is a sustainable approach, will completely replace chemical-based fertilizers in the future. To boost this conversion, an awareness program about the importance and benefits of mycorrhizal biofertilizer targeting local farmers across every corner of the world is required. A systematic cost-benefit analysis will ensure and encourage farmers to adopt and implement this technology at a higher pace. To address this, careful consideration of both the direct and indirect factors involved in mycorrhizal application is required.

3.1 Cost Assessment

It involves assessing all the costs required for AMF applications. Determining the exact cost is complex; however, the analysis of various factors that influence the cost can be an important aspect. The various costs and the factors affecting them are as follows.

3.1.1 Product Cost

This includes the cost of purchasing the AMF inoculum. The inoculum cost may depend on the producer, the type of AMF species used, and the concentration of the product. Inoculum comes in different forms, such as granulated, powdered, or liquid, which influence the product cost. According to a study by Basiru et al. (2020), based on the data availability across Asia, Europe, and America, 90% of the mycorrhizal products were in solid form (powdered: 65%, granulated: 25%), while the rest 10% in liquid form. However, recently, the liquid form of the product is growing very fast in the market (Basiru et al. 2020).

3.1.2 Application Cost

This includes labor costs for applying the inoculum in the field. The inoculum can be applied through seed coating, soil incorporation, or the root dipping method. Each method comes with varying labor costs. Soil types and preexisting soil conditions (acidic, drought conditions, initial mycorrhizal population, nutrient availability, etc.) may also influence the amount of AMF required for application directly, influencing the cost of application. Mycorrhizal application is cost effective in soil where the phosphorus level is very low. In India, mycorrhizal application is widely being used for the cultivation of rice in regions where phosphorus level is low, as this allows to cut down application costs to a significant level than using chemical fertilizers (Berruti et al. 2016). Depending on the crop type, different crops require varying quantities of AMF inoculum, influencing the cost of application.

3.1.3 Equipment Costs

Equipment costs, if any, are associated with the application method.

3.2 Benefit Assessment

While the initial cost is involved for the application of AMF in agriculture, it is important to understand the potential long-term benefits of AMF.

3.2.1 Increased Crop Yield

AMF increases the crop yield by improving the plant's health and growth, which is a direct effect of enhanced nutrient uptake capacity. Uptake of nutrients, viz., N, P, and K, which are vital for the growth of plant, is improved as a result of AMF application.

3.2.2 Reduced Input of Chemical Fertilizers

AMF increases the plant's nutrient uptake capacity, which reduces the dependency on chemical fertilizers. Research studies have already demonstrated that the inoculation of AMF led to an increase in phosphorus uptake, thereby reducing the necessity for conventional synthetic fertilizers prepared from harmful chemicals and hence encouraging sustainable agricultural practices (Begum et al. 2019).

3.2.3 Enhancement of Soil Health

Soil health is an essential pillar of sustainable agriculture, and the application of AMF led to improvement in soil structure, enhancement of nutrient dynamics, and promotion of long-term soil fertility. AMF contribute to soil aggregation by forming symbiotic relationships with plant roots, via the secretion of glomalin, a glycoprotein that stabilizes soil particles, thus enhancing aeration and water retention capability of soil (Rillig et al. 2015).

3.2.4 Enhanced Plant Resilience

AMF shows a pivotal role in improving plant resilience to various environmental stresses, such as drought, salinity, and pathogenic infections. They improve water and nutrient uptake efficiency, leading to greater drought tolerance through improved root hydraulic conductivity and osmotic adjustment (Augé 2001). Besides, AMF have been demonstrated to relieve salt stress by different processes like regulation of ion balance, reduction of sodium accumulation, and increasing mineral uptake like potassium and calcium, thus enhancing plant growth in saline environments (Evelin et al. 2009).

3.3 Calculation and Analysis

As many factors are involved in the real-world application of mycorrhiza in the field, providing an accurate value of cost benefit is actually hard, which depends on an accurate measurement of all the costs involved during mycorrhizal application to calculate the actual economic gain from the increased yields and reduced inputs. However, considering the long-term effect of mycorrhiza in soil, from the environmental sustainability perspective, mycorrhizal application in agriculture can be considered highly beneficial. On the other hand, despite the claim that AMF has, some commercially available AMF products fail to give results that significantly impact the benefit. The effectiveness of AMF species to different crop types and soil types will also greatly influence the cost benefit. For example, the application of *Rhizophagus irregularis* showed good results in potato, cassava, sweet potato, malanga, pepper, cucumber, tomato, and banana, also outperforming other species such as *Gigaspora magarita* and *Funneliformis mosseae*, in terms of phosphorus absorption in plants (Basiru et al. 2020).

4 Impact on Crops Yield and Quality

AMF act as an important factor in plant growth and nutrition, owing to their symbiotic relationship with higher plants, which help them to contribute to plant mineral uptake (Smith and Read 2010). The symbiosis of AMF is extremely indispensable for amplifying the absorption of the corresponding static and insoluble phosphates present in the soil (Fitter et al. 2011). The reason behind improved crop productivity even during phosphorus deficiency is due to the capability of AMF in secreting phosphatases, which hydrolyze organic phosphate compounds, releasing phosphate and making it available for plant absorption (Marschner 2011). AMF not only improve plant nutrition but also intervene with the phytohormones present in the plants, hence controlling plant development and relieving the consequences of environmental stresses. This not only increases the biomass and yield of plants but also helps in improving various other qualities of plants (Antunes et al. 2012).

4.1 Impact of AMF on Different Agricultural Practices

Seed inoculation and application of mycorrhiza in soil is one of the principal agricultural practices that contribute to vigorous agricultural ecosystems and expansion of crop productivity (Deveautour et al. 2021). To date, successful AMF inoculation is seen in numerous commercial crops like cereals (rice, wheat, etc.), legumes, and various horticultural crops (citrus, tomato, potato, etc.).

4.1.1 Impact on Fruits and Other Horticultural Crops

Inoculation of AMF in horticultural crops enhances water utilization and nutrient assimilation of host plants, amplifies photosynthesis, and boosts reproductive growth thereby increasing the yield and quality of horticultural crops. Moreover, this can help in meeting the rising crop demand related to population growth. Due to the use of AMF in agriculture, the use of synthetic fertilizers and other chemicals has been greatly reduced, which is truly essential for the development of sustainable modern agriculture (Zhu et al. 2022). Navarro and Morte (2024), in their recent experiment with sour-orange seedlings, conveyed that inoculation of two AMF species, viz. *Rhizophagus irregularis* and *F. mosseae* in the seedlings, exhibited high colonization percentages, notable growth improvements, increased phosphorus nutrition, and chlorophyll content, which remarkably intensified the performance of AMF-inoculated plants.

In *Citrus reticulata*, size, weight, coloration value, quality, and soluble solids of the fruit were significantly improved when field inoculated with a mixture of exotic AMF (*D. versiformis*, *R. intraradices,* and *F. mosseae*) (Cao et al. 2021). Similarly, in *Camellia sinensis,* metabolic activity and other various qualities related to the plant's growth were upgraded with AMF mixed inoculation (Singh et al. 2010).

A noteworthy growth can be seen in melon (*Cucumis melo*), besides an enhancement in the concentration of sugars and carotenoids in the pulp when inoculated with AMF. Remarkably, it also has a great effect on the growth, flowering, yield, and quality of melon (Copetta et al. 2021). Grapevines inoculated with AMF showed improved berry quality and improved production of anthocyanins, flavonols, and resveratrol (Torres et al. 2018; Torres et al. 2016).

4.1.2 Impact on Vegetables

A significant increase in the yield of eggplant (1.3–1.5 average increase in yield per plant) was reported with the application of farm-produced AMF in the soil (*C. claroideum, F. mosseae*) (Douds et al. 2017). In another study conducted by Felföldi et al. (2022), AMF inoculation in seedling roots of tomato showed vigorous growth in plants, superiorly developed root system, bigger fruits, and a large-scale production compared to other noninoculated plants.

On AMF (*Glomus intraradices* or *Glomus versiforme*) inoculation, onion (*Allium cepa*) plants grown with mineral soil in pots had a larger biomass than noninoculated plants and attained marketable size (>25 mm bulb diameter) at a much earlier time (Charron et al. 2001). Interestingly, an increase in the yield quality and quantity of strawberries at a large scale was reported with the use of AMF in coconut shell substrate (Robinson Boyer et al. 2016). In another study, a 13% and 5% increase in the Ca content of pea pods and okra fruits was reported with AMF inoculation (Kumar et al. 2017).

The quality of tomato is enhanced with increased flowering, better color and shape, increased size and weight with elevated content of carotenoids, vitamin C, lycopene, and nitrate when inoculated with mixed AMF in addition to magnified accumulation of Ca, Cl, K, Mg, Na, and citrate (Ramona et al. 2020; Ziane et al. 2021; Chouyia et al. 2022).

4.1.3 Impact on Cereals and Legumes

The role of AM fungi for nutrient absorption and quality yield in cereals and legumes is extensively studied and recognized. A study conducted by Zhang et al. (2015) showed that with the inoculation of AMF (*Rhizophagus intraradices*) in rice (*Oryza sativa* L.) during the seedling stage, N accumulation in vegetative tissues increased significantly. Consequently, the distribution of N was more strongly impacted in seeds than C, where the C:N ratio was reduced notably, and protein content was increased by 7.4%. Furthermore, grain yield was remarkably increased by 28.2% where the ratio of grain to straw increased by 18.4%. Additionally, rice roots inoculation showed transformation in C arrangements with elevated C accumulations and concentrations. Research carried out by Beslemes et al. (2023) in a two-year field trial with a two-rowed barley crop showed taller plants and a higher leaf area index, with enhanced biomass and seed yield. Moreover, the AMF-inoculated plants showed higher values of nitrogen and phosphorus uptake with higher utilization indices. A meta-analysis of wheat was carried out with 38 field trials and 333 observations, where the results clearly showed an increase in aerial biomass, crop yield efficiency, harvest production, and nutrient content, as well as an interdependency between AMF colonization of plant root and grain yield (Pellegrino et al. 2015). A field trial carried out by Amirnia et al. (2019) with lentil (*Lens culinaris* Medik) applied with nitrogen-fixing soil bacteria (*Azotobacter*) and mycorrhizal fungi (*Glomus intraradices*) showed improved leaf moisture content, high chl-a and total chlorophyll concentration, with lessened proline content, elevated biomass yield, seed yield, and seed protein.

4.2 Importance of AMF in Mitigating Stress

Along with better plant yield and quality, AMF help plants during stress conditions by stimulating their tolerance to biotic and abiotic constraints (Zhu et al. 2010), because of modification in plant physiology (Miransari et al. 2008). Mycorrhiza-associated plants display good resistance to extreme conditions and phytopathogens, which leads to a reduction in crop loss due to environmental stress (Liu et al. 2007; Deja-Sikora et al. 2019). An increase in bioavailability of soil nutrients during Cd stress is seen in red gram (*Cajanus cajan* L.) when AMF colonize the plant roots, which effectively modulates soil enzyme activity and imparts tolerance of red gram to Cd stress (Bisht and Garg 2022). As it is known, AMF form a mutually beneficial relationship with tomato plants and promote vigorous plant growth and yield; they also optimize the physiological processes and mitigate the oxidative stress in tomato plants under salt/saline stress, improving the plants' tolerance to salt (Hajiboland et al. 2010; Kong et al. 2020). One of the most destructive abiotic stresses is drought stress, as it adversely affects crop growth, development, and yield (Zhu et al. 2022). During drought stress in apple (*M. hupehensis*), interaction with AMF (*R. irregularis*) resulted in enhanced capacity of gas exchange, increased parameters of chlorophyll fluorescence, increased osmoregulation, free radical scavenging, and Mitogen-Activated Protein Kinase (MAPK) signaling, which resulted in increased tolerance to drought in apple (Huang et al. 2020). Furthermore, under drought stress conditions in sesame (*Sesamum indicum* L.), AMF application remarkably increases their chlorophyll content, unsaturated fatty acid content, and oil yield (Gholinezhad and Darvishzadeh 2021). Another important stress affecting crops, which is cold stress, includes chilling and freezing damage (Li et al. 2017). The symbiosis of AMF with watermelon (*Citrullus lanatus*) seedlings alters chlorophyll content and improves photosynthesis and tolerance to oxidative stress induced by low temperature, which helps watermelon seedlings to tolerate cold stress (Shirani Bidabadi and Mehralian 2020). Mycorrhiza-associated plants display good water conservation, efficient water utilization, improved plant water level, improved regulation of osmotic pressure, and gas exchange, in addition to better PSII regulation and nutrient absorption under cold stress (Latef et al. 2016).

Symbiotic association of plants with mycorrhizal fungi also improves their tolerance to biotic stress. The tolerance of host plants to various biotic stresses can be improved by inoculation of AMF (Dowarah et al. 2022). The disease Fusarium wilt of pepper (*L. grossum*) is caused by the mutual action of bacterial pathogen *R. solanacearum*, fungal pathogen *P. capsici,* and root-knot disease-causing nematode *M. incognita*. Inoculation of pepper plants with AM fungi showed decreased (50.20%) disease index as compared to the standard group (82.14%), signifying effective control of AMF on the spreading of the pathogens (Raghavendra Kumar et al. 2018). Bacterial wilt caused by *Ralstonia solanacearum* on tomato plants generates major economic losses. Both bacterial wilt of tomato and Fusarium wilt symptoms can be alleviated by AMF (*R. irregularis* MUCL 41833) (Chave et al. 2017). In potatoes, the proliferation of a fatal fungus (*F. sambucinum*) can be controlled by the inoculation of AMF, consequently diminishing its production of toxin (Deja-Sikora et al. 2020). Viral infections are known to cause diseases in plants, which can lead to greater damage in plants. Some of the studies demonstrated the protective effect of AMF on virus-infected plants. AMF *(R. irregualris*), when inoculated both before and after viral infection, boost the tolerating capacity of plants to Lycopersicon virus 4 and cucumber mosaic virus (CMV) (Miozzi et al. 2020; Khoshkhatti et al. 2020). The effects of potato virus Y (PVY) infection can be seen to be controlled and mitigated by AMF (*G. irregulare*) vaccination (Deja-Sikora et al. 2020). Additionally, AMF can also protect plants from nematode infestation and pests. *Psidium guajava* L. seedlings inoculated with AMF provide prominent resilience to invasion by nematodes for a prolonged duration in the field (De Sá and Campos 2020). At the sowing stage, AMF inoculation in peppers exhibits stronger tolerance to nematodes (*N. aberrans*), which results in the decline of nematode population (Bernardo et al. 2021). He et al. (2017) stated that AMF-associated plants alter the ratio of their hormonal and photosynthetic properties under infestation by *S. exigua*, a beet armyworm. Also, the root colonization of AMF in tomato is stimulated by the feeding activity of *S. exigua* and significantly enhances the net carbon fixation rate and stomatal conductance rate in tomato. Additionally, Kaur et al. (2020) demonstrated that the symbiotic association of plants with AMF helps in attracting insects and deterring herbivores.

5 Reduction in Input Costs and Sustainability

The human population has been growing at an unprecedented rate, and this exerts pressure on crop productivity to sustain life, leading to the utilization of more chemical fertilizers and pesticides for surplus crop yield and to maximize the agricultural output. However, the use of synthetic fertilizers has caused a financial burden upon farmers, harming economic stability. The need of the hour is the adoption of sustainable farming approaches to reduce input costs while maintaining crop productivity that can optimize resource utilization and allocation, reaching our goals of economic and environmental sustainability. Partially reduced reliance on conventional inorganic chemical pesticides and fertilizers prompts the foundation of sustainable agricultural development (Alori et al. 2017). The use of these agrochemicals cannot be denied; they have helped in meeting agricultural productivity demand, on the other hand, causing massive detrimental effects on the environment.

The application of AMF can cut down the energy inputs and drastically reduce the use of fertilizers, promoting plant and soil health. Reaching out to farmers and encouraging them to understand the prospect of sustainable agronomic practices through the usage of AM fungi could be a step toward economic recovery. The major mechanism of the AMF symbiosis is fast absorption from beyond the root depletion zone through an expansive network of mycelia (Johri et al. 2015). The biological role played by AMF in soil nutrient cycling ensures the availability of nutrients in soil with less fertility (Johnson et al. 2015). Thirkell et al. (2016), through their extensive work, observed an evident increase in plant biomass due to the transfer of nitrogen from organic matter to the plant tissue via AM fungal hyphae.

The AMF form symbiosis with a vast number of known plants that highlights them from an ecological and agricultural standpoint (Tisserant et al. 2013). Critical analysis and comprehending the role of the plant–microbe interactions in the nutrition of plants and biogeochemical cycles are pivotal for the onset of development of a sustainable agricultural system that is based largely on the augmentation of naturally occurring ecological processes (Lazcano et al. 2014). The AM interaction with the host plant stimulates the expression of genes for inorganic phosphorus (Pi) transporter proteins, enhancing Pi uptake by plants (Xie et al. 2013). Various works have given prominence to the potential of AMF in nutrient absorption for the development and proper growth of the plant, mainly through phosphorus assimilation (Vosátka et al. 2012). AMF symbiosis displays major beneficial functions that include increased water uptake and water-holding capacity, building up the plant tolerance to drought, increased tolerance to heavy metal toxicity and soil salinity, overcoming the biotic stress, offering bio-protection against pathogen, improving soil quality, and boosting plant vigor and yield, among others (Garnett et al. 2013). With the application of AM in water-stressed or mined areas, it may be possible to reclaim the agricultural land as AM fungi capture the hazardous elements within their mycelial mat (Lopes Leal et al. 2016). Regulating the plant water stress is one of the key features executed by this symbiotic association, which can be through various ways, one of them is by influencing the thresholds for stomatal closure (Ruiz-Lozano et al. 2012). AMF-associated plants are better at regulating their ABA (abscisic acid) levels, which allows a more adequate balance between transpiration via leaf and root movement during the periods of drought (Ruiz-Lozano et al. 2006).

Organically managed systems invite ways that can reduce the usage of fungicides, and here the role of AM symbiosis intensifies as they impact the plant, soil, and greenhouse gas (GHG) emissions. The dependence on slow release of organic fertilizers in this system shows dependence on biological interactions in soil for the release and uptake of nutrients, which is found to be elevated than conventional cropping systems (Drinkwater et al. 1995). The application of AMF enhances the quality of food crops, particularly enhancement of plant sugar content along with the enhancement of antioxidants and some inorganic elements (Albrechtova et al. 2012). Mycorrhizal presence in the soil modifies the host plant defense system, leading to plant immune responses to diseases through a local and systemic approach (Jung et al. 2012; Philippot et al. 2013). The minimization of heavy metal content in soil can be achieved by AM fungal application that assures the quality of edible food crops; other AM fungi enhance phytoextraction, facilitating remediation of contaminated soil (Igiehon and Babalola 2007). The bioavailability of phosphorus is an emerging environmental issue; the application of combined forms of AMF and PGP microbes could be a solution to this situation. These two groups work synergistically to benefit the plant and the environment broadly. The presence of AMF in the vicinity of the crop land improves the structure of soil through the discharge of proteinaceous and nonproteinaceous organic compounds in the form of aggregates which remain stable for at least 6 months even after the hyphal network disintegrates, helping to maintain and conserve the soil ecosystem (Bitterlich et al. 2018; Zhang et al. 2019b). The leaching of the nitrate could also be reduced through this symbiotic association, enhancing the N uptake and assimilation, conserving the loss of nitrogen from the ecosystem as nitric oxide (Asghari and Cavagnaro 2012). The AM symbiosis has shown great promise to increase not only the water-holding capacity but also the efficient utilization of nutrients, improving the environmental sustainability in agrosystems (Gianinazzi et al. 2010; Jackson et al. 2012). Proper harnessing of these mutualistic fungi can possibly diminish the excessive usage of agrochemicals, which lead to hazardous outcomes for the environment. Taking up small-scale production of AMF is feasible and cheaper than synthetic fertilizers and gives a more credible approach to farmers about the cost effectiveness of the AMF, having a positive impact on plant health. Boosting the abundance of AMF in the plant's natural rhizosphere evinces a more improved strategy to practices incorporating chemical fertilizers (Igiehon and Babalola 2007).

6 Market Potential and Adoption Challenges

6.1 Availability and Cost of Commercial Mycorrhizal Products

The mycorrhizal product market is witnessing substantial growth, fueled by heightened awareness of sustainable agriculture and the benefits it provides for plant health and productivity. This surge can be attributed to the increasing adoption of organic farming practices, the imperative to enhance crop yields, and the heightened emphasis on soil health. Although there is a wide variety of mycorrhiza-based products on the market, their revenue remains relatively low, even in highly developed countries. The global revenue of mycorrhizal products in 2022 was just $558 million, representing only a fraction of the total $2 billion spent on biofertilizers worldwide (Veresoglou and Agathokleous 2023). By 2029, the estimated market size of mycorrhizal products will be $1.87 billion (Sharma et al. 2023).

The mycorrhizal product industry exhibits a fragmented structure with a high degree of regional specialization. Leading companies have developed strong market positions within their respective geographical areas through extensive distribution networks and tailored product ranges. The product is characterized by a combination of multinational enterprises and regional specialists, encompassing both dedicated biofertilizer firms and diversified agricultural input providers (Aloo et al. 2021). Some of the major key players that have a role in mycorrhizal application products in the markets are Gujarat State Fertilizers & Chemicals Ltd., Inc., Valent Biosciences, Atlantica Biolchim SPA, Koppert Biological Systems Inc., Symborg company, and LLCEzzy Bioscience Pvt. Ltd. (Mordor Intelligence 2025). Most of them are engaged in innovating and developing new products to meet the growing demand for sustainable agricultural solutions.

Mycorrhizal products provide numerous advantages for plant health and productivity, but the cost can be a significant barrier for resource-constrained farmers (Mishra et al. 2018). Mycorrhizal inoculants, particularly high-quality ones with diverse fungal strains, can be expensive, which restricts their use in low-income areas or by farmers with limited budgets. An extra cost may also be associated, depending on the product type and the method of application. Additionally, certain regions may not have readily available mycorrhizal products, and in some instances, products are not easily accessible, leading to higher costs of buying. On the other hand, many farmers may be unaware of the benefits of mycorrhizal products or lack knowledge of their effective uses, which can hinder adoption even when cost is not a primary concern (Abbott and Lumley

2014). Overcoming these multifaceted problems requires teamwork between researchers, regulators, and key industry stakeholders. Further, increasing the funding for research activity can help understand more about mycorrhiza, which can be helpful in developing mycorrhizal inoculum for effective use in agriculture (Basiru et al. 2020).

6.2 Farmers' Awareness and Perception

Farmers often lack the technical knowledge to understand mycorrhizal products, leading to knowledge gaps about their benefits for plants and soil. This lack of knowledge creates a barrier to learning the proper techniques for utilizing these products. Additionally, limited access to information, particularly in remote areas, further constrains farmers' ability to understand and capitalize on the potential advantages of mycorrhizal applications (Abbott and Lumley 2014). A common misconception among farmers is that fertilizers are limited to chemical fertilizers and that mycorrhizae are solely relevant to organic farming. Additionally, they assume that it requires specialized equipment and are unsure about the results of the investment, and that the cost of mycorrhizal takes time to manifest. So, it discouraged farmers because of their expected immediate results. It leads to being misperceived as being overestimated, and the potential investment returns may be underappreciated (Vosátka et al. 2012). However, this misconception and gap can be addressed by providing awareness programs among the farmers for the benefit of mycorrhizae. These programs may include explaining the benefits of understanding the farmer's needs, delivering information through local language, visual aids, conducting field demonstrations, and providing the farmer with hands-on experience. This can help the farmers to how mycorrhizal can improve crop yields and soil health (Tawaha et al. 2021).

Furthermore, encouraging the networking of farmer-to-farmer interaction where they can share their experience and information and collaborate with agricultural services expertise by fostering educational initiatives through conducting outreach programs and training will develop a positive impact and trust in the mycorrhizal products. Utilizing media outreaches such as local channels, newspapers, and awareness campaigns can also raise awareness about the benefits and advantages of mycorrhizae. It also requires the role of research and development to improve the efficacy of mycorrhizal products for easy use and cost-effectiveness, which will empower the farmers to build trust in the use of mycorrhizal products and unlock their full potential for sustainable agriculture (Basiru et al. 2020).

6.3 Policy and Institutional Support

It is essential for the widespread adoption and successful implementation of policy and institutional support for mycorrhizal products. Establishing standard guidelines for the production, quality control, and labeling of mycorrhizal products will ensure effective and safe products. Providing financial incentives and subsidies to farmers can encourage them to adopt mycorrhizal products. Additionally, allocating funds for research on mycorrhizal fungi for the development and improvement of mycorrhizal products and understanding the crop varieties in different environments can be a fruitful step to promote sustainable agricultural practices (Osemwegie et al. 2021; Rillig et al. 2016).

The extension services, such as institutional support, can provide the farmers with technical assistance and educate them on the benefits and use of mycorrhizal products. Research institutions can support the conduct of experiments and the development of products under convenient conditions in the field. It is also necessary to encourage seed input suppliers to stock and promote the products to support organizations and cooperatives in producing and distributing the mycorrhizal products (Aloo et al. 2021). Many countries have their own policies and institutional support for encouraging sustainable agriculture practices. For instance, the Indian government has many schemes to encourage and support farmers to use biofertilizers, European nations also have schemes for supporting research and activities related to biofertilizers and sustainable agricultural practices, and some African countries are providing training to farmers and adopting biofertilizers in their policies as well (Osemwegie et al. 2021). Policies and institutional support help create widespread adoption of mycorrhizal products, contributing to sustainable agriculture practice and food security.

7 Environmental and Ecological Benefits

7.1 Reduction in Chemical Runoff

For proper plant growth and development, essential macronutrients like nitrogen and phosphorus are required (Ye et al. 2022; Khan et al. 2023). They are the major components for the synthesis of proteins and nucleic acids, making them regulate cellular processes and establishing optimal plant health (Wang et al. 2021). Only 55% of the available nitrogen is utilized by the global pooled crops efficiently; the remaining percentage of nitrogen is lost by various mechanisms from the soil (Martínez-Dalmau et al. 2021). Volatilization is one of the pathways through which nitrogen is lost in the form of ammonia (NH_3) and nitrous oxide (N_2O) from agricultural fields (Ramalingappa et al. 2023). Furthermore, surface runoff from agricultural lands impairs nitrogen loss, and this surplus nitrogen not only decreases the effectiveness of chemical fertilizer but also causes ecological imbalances, which negatively impact the aquatic ecosystems by eutrophication and in soil fertility (Wang and Huang 2021).

7.2 Nitrogen and Phosphorus Uptake by AMF Reduces Sustainable Agricultural Practices

Mycorrhizal networks with the plant's roots hold significant value in nutrient uptake and soil health as they form mutualistic relationships with more than 80% of the terrestrial plants (Chowdhury et al. 2022). They get carbon from their host plants, and they extend their hyphal network into the soil and facilitate inorganic nitrogen (NO_3^- or NH_4^+) from the soil to their host plants (Govindarajulu et al. 2005). Thus, enhances nitrogen use efficiency for the host plants, ensuring that a greater proportion of applied nitrogen is absorbed by plants rather than being lost in the environment (Verzeaux et al. 2017). This process not only increases crop productivity but also reduces the negative environmental effects of excessive nitrogen fertilizer use, groundwater contamination, GHG emissions, and in soil health, subsequently promoting this symbiotic interaction contributes to more sustainable agricultural practices by reducing reliance on synthetic fertilizers and improving soil health and ecosystem stability (Fall et al. 2022; Wahab et al. 2023).

The fundamental relationship between arbuscular mycorrhizae and the plant roots plays a vital role in soil phosphorus cycling which also significantly influences phosphorus dynamics in agricultural fields and agroccosystems (Jansa et al. 2011). AMF extend their hyphal networks into the soil, increasing the surface area available for phosphorus uptake beyond the depletion zones around roots (Jiang et al. 2021). The AMF network facilitates the uptake and assimilation of phosphorus which are not accessible by the plants through their mobilization and transportation mechanism that ensures optimal utilization of phosphorus by the associated plants (Schnepf et al. 2008). Thus, the environmental issues like eutrophication and phosphorus leaching could be minimized by establishing an AMF symbiosis system that could strengthen eco-friendly agricultural practices by reducing reliance on synthetic phosphorus fertilizer and sustainably improving soil fertility and crop production (Zhang et al. 2015; El-Sherbeny et al. 2022; Fall et al. 2022); therefore, this mutualistic relationship reduces N and P loss from a wide range of crop systems by preventing leaching off, surface runoff, and volatilization (Demenois et al. 2018; Parihar et al. 2020; Fall et al. 2022). Along with the facilitation of minerals, AMF exudate glycoprotein glomalin, which contributes to the formation of soil aggregates, this compound enhances the soil structure and stability by binding the soil particles together, which in turn helps stabilize organic matter and sequester carbon (Rillig et al. 2002; Lutgen et al. 2003; Wu et al. 2014a, b). These aggregates markedly enhance the soil fertility and water retention capacity, aiding in improving the agroecosystem.

7.3 Enhancement of Biodiversity and Carbon Sequestration

CO_2 emissions are a major contributor to global warming and climate change, which directly impact agricultural fields by altering precipitation patterns that disturb the environment, causing loss of biodiversity and shifts in the ecosystem (Muluneh

2021). It has become crucial to mitigate these effects and safeguard food security, natural habitats, and the climate by taking an innovative strategy to reduce carbon emissions (Saleem et al. 2024).

Various factors like temperature, moisture, nutrient availability, and composition of the ecosystem affect the net carbon balance in the environment (Xia et al. 2023). However, the stored carbon is continuously subjected to loss via autotrophic and heterotrophic respiration, which releases carbon back to the environment. Comprehending this process is crucial for monitoring carbon fluxes and generating constructive procedures to mitigate climate change. Photosynthetic plants assimilate the carbon present in the atmosphere through the conversion of organic carbon contribution to biomass accumulation and carbon pools in soil (Wang et al. 2022). Mycorrhizal fungi are essential mediators of carbon and nutrient cycling within terrestrial ecosystems, influencing biogeochemical processes and ecosystem stability (Read and Perez-Moreno 2003).

Mycorrhizal fungi establish complex networks inside the soil with the plant roots, extending beyond the vicinity of the root zone to access and provide nutrients like phosphorus and nitrogen to the host plants in exchange for carbon from plant photosynthesis (Thirkell et al. 2020). The role of mycorrhiza in carbon sequestration is via increasing plant productivity (Wang et al. 2016). This fungus enhances the availability of water and minerals, allowing plants to thrive, and capture higher amounts of CO_2 from the atmosphere through photosynthesis, and this carbon assimilated by the plants diffuses into the soil, which increases soil organic matter (Begum et al. 2019). Mycorrhiza exudates glycoproteins such as glomalin, which can bind to soil particles, increasing soil stability (Rillig et al. 2002). This mechanism improves soil structure and promotes carbon storage, contributing to carbon sequestration (Rillig et al. 2002; Hossain 2021). Furthermore, the mycorrhizal relationship with the plant roots also helps in reducing soil respiration by shaping microbial communities and decelerating the breakdown of organic matter (Berg and Smalla 2009; Frey 2019; Zheng et al. 2022).

Mycorrhizal fungi are known to play a vital role in the carbon dynamic cycle, which contributes to climate change mitigation (Treseder 2016). They are responsible for soil carbon storage by improving organic carbon matter and promoting plant growth and development (Ronsheim 2012). These mutualistic relations between the plant roots and mycorrhiza can be utilized to improve carbon sequestration by sustainable land management strategies, afforestation, and by reducing soil disturbance, which subsequently enhance carbon sequestration by extending their hyphal network and help in retaining carbon in the soil for longer periods of time, reducing atmospheric CO_2 concentration. Integrating mycorrhizal fungi into conservation strategies can increase soil fertility, promote healthy biodiversity, and contribute to climate change mitigation.

8 Case Studies of Mycorrhizal Applications in Agriculture

Mycorrhizal application in agricultural systems has become a cornerstone for sustainable farming practice as this offers a solution to challenges such as soil degradation and resource inefficiency. By harnessing the benefits of the symbiotic association between Mycorrhiza and plants, scientists and farmers can now reimagine crop productivity and sustainability. There are several studies on the positive impact of the use of mycorrhiza in agricultural systems worldwide. Some of the relevant findings can be discussed below. The reduction of the use of chemical fertilizers, mainly N, P, and K in agriculture is a critical success. It was estimated that the use of chemical fertilizers can be reduced by up to 50% by using AMF (Begum et al. 2019). In Uganda, a 50% reduction in NPK fertilizer by using AMF in field conditions achieved a comparable maize production. This approach not only lowered the cost but also improved soil quality for the long-term by reducing nutrient runoff (Fall et al. 2023). In China, Qian et al. (2024) reported a 30% reduction of chemical fertilizers by using AMF for maize cultivation in a field trial, while maintaining their growth and production. In another study, a meta-analysis conducted on the use of AMF in hydroponic farm systems concluded that the use of AMF significantly increases the growth and production of leafy vegetables, which makes them a viable and valuable addition for creating a resilient and healthy crop in hydroponic farm systems (Othman et al. 2024). Such strategies align with global efforts to reduce, and to some extent replace, the chemical inputs in agriculture. In another case, the implementation of AMF in regions having water scarcity has proven indispensable for mitigating drought impact. For example, in a case study conducted in India, the application of AMF (*Rhizophagus intraradices*, *Funneliformis mosseae*) enhances specific leaf area (SLA), relative growth rate (RGR), photosynthetic yield, and leaf area ratio (LAR), in tomato under drought as well as watered conditions. AMF also enhance the status of nutrients, relative water content (RWC), and translocation of minerals under water stress, promoting tolerance to drought conditions (Singh and Singh 2024). Similarly, in Brazil, Oliveira et al. (2022) demonstrated increased drought stress tolerance by two cultivars of soybean (Anta82 and Desafio) with the application of AMF (*Rhizophagus clarus*). These findings highlight the importance of AMF as a biological buffer to the changing extreme climatic conditions. Mycorrhizae also help in soil health restoration by increasing the soil organic matter content, improving soil aggregation, and improving water retention capacity (Asmelash et al. 2016). Zhou et al. (2024) demonstrated the restoration of degraded grassland by inoculating the soil with AMF. AMF inoculation showed a 190% increase in plant community productivity by maintaining the microbial community structure of the grassland, which helps in breaking down the soil organic matter. AMF help stabilize the soil even after 5 months of the death of the host plant, acting as an important component for soil restoration. A recent study also highlights novel applications of AMF beyond traditional applications in the field. A study by Belval et al. (2024) in France confirmed the prevention of a common grapevine trunk disease called Botryosphaeria dieback in grapevine caused by the fungus *Neofusicoccum parvum*. The underlying immune response was the upregulation of lipid metabolism, such as fatty acids, glycerophospholipids, and oxylipins in the grapevine plant as a result of AMF introduction (Belval et al. 2024). This allowed farmers to slash the use of chemical pesticides while maintaining the yield. Similarly, a study conducted by Wen et al. (2024) introduced AMF (*Rhizophagus irregularis*) amended biochar to soil under salt stress conditions and reported improvement in growth and yield of the sorghum plant while improving soil quality. Despite many successful reports on the application of AMF in agriculture, scalability remains a hurdle. A customized AMF consortium is required as efficacy varies depending on crop type, soil type, and microbial competition (Herrera-Peraza et al. 2011). For example, application of native AMF consortia in sunflower field improves the grain yield and oil yield (Arcidiacono et al. 2025), which may not be as effective in other soil types such as soil rich in clay. Therefore, AMF application in agricultural land requires prior knowledge of AMF–soil compatibility, and for this purpose, providing proper education and awareness to the farmers regarding AMF and convincing them to use it is needed.

9 Conclusion and Future Prospects

The application of AMF in agriculture renders a sustainable, cost-effective, and economically viable alternative to the conventional chemical fertilizers, boosting nutrient uptake from soil, soil health, and crop resilience. AMF play a major role in maintaining sustainable agriculture by reducing dependency on chemical fertilizers, mitigating environmental impact, and improving drought tolerance. However, for the widespread adoption of AMF by the farmers, several challenges such as farmer awareness, economic feasibility, and soil compatibility must be focused on and addressed.

Further future research should focus on impactful areas such as improving AMF inoculum generation, optimizing AMF formulations to be utilized for different soil types and crops, and ultimately integrating AMF with precision agriculture. Heightened policy support, farmer awareness regarding AMF, and technological innovations will further enhance AMF's role in inducing food security as well as long-term agricultural sustainability.

References

Abbott LK, Lumley S (2014) Assessing economic benefits of arbuscular mycorrhizal fungi as a potential indicator of soil health. In: Solaiman Z, Abbott L, Varma A (eds) Mycorrhizal fungi: use in sustainable agriculture and land restoration. Springer, Berlin/Heidelberg, pp 17–31

Agnihotri R, Sharma MP, Prakash A et al (2022) Glycoproteins of arbuscular mycorrhiza for soil carbon sequestration: review of mechanisms and controls. Sci Total Environ 806:150571

Albrechtova J, Latr A, Nedorost L et al (2012) Dual inoculation with mycorrhizal and saprotrophic fungi applicable in sustainable cultivation improves the yield and nutritive value of onion. Sci World J 2012:374091

Aloo BN, Mbega ER, Tumuhairwe JB, Makumba BA (2021) Advancement and practical applications of rhizobacterial biofertilizers for sustainable crop production in sub-Saharan Africa. Agric Food Secur 10(1):57

Alori ET, Dare MO, Babalola OO (2017) Microbial inoculants for soil quality and plant health. In: Lichtfouse E (ed) Sustainable agriculture reviews, vol 22. Springer, Cham, pp 281–307

Amirnia R, Ghiyasi M, Siavash Moghaddam S et al (2019) Nitrogen-fixing soil bacteria plus mycorrhizal fungi improve seed yield and quality traits of lentil (*Lens culinaris* Medik). J Soil Sci Plant Nutr 19:592–602

Antunes PM, Franken P, Schwarz D et al (2012) Linking soil biodiversity and human health: do arbuscular mycorrhizal fungi contribute to food nutrition. In: Soil ecology and ecosystem services, pp 153–172

Arcidiacono M, Ercoli L, Piazza G, Pellegrino E (2025) Field inoculation with a local arbuscular mycorrhizal (AM) fungal consortium promotes sunflower agronomic traits without changing the composition of AM fungi coexisting inside the crop roots. Appl Soil Ecol 206:105830

Asghari HR, Cavagnaro TR (2012) Arbuscular mycorrhizas reduce nitrogen loss via leaching. PLoS One 7(1):e29825

Asmelash F, Bekele T, Birhane E (2016) The potential role of arbuscular mycorrhizal fungi in the restoration of degraded lands. Front Microbiol 7:1095

Augé RM (2001) Water relations, drought and vesicular-arbuscular mycorrhizal symbiosis. Mycorrhiza 11(1):3–42

Babikova Z, Gilbert L, Bruce TJA et al (2013) Underground signals carried through common mycelial networks warn neighboring plants of aphid attack. Ecol Lett 16(7):835–843

Barto EK, Hilker M, Muller F et al (2011) The fungal fast lane: common mycorrhizal networks extend bioactive zones of allelochemicals in soils. PLoS One 6:e27195

Basiru S, Mwanza HP, Hijri M (2020) Analysis of arbuscular mycorrhizal fungal inoculant benchmarks. Microorganisms 9(1):81

Bedini S, Pellegrino E, Avio L et al (2009) Changes in soil aggregation and glomalin-related soil protein content as affected by the arbuscular mycorrhizal fungal species *Glomus mosseae* and *Glomus intraradices*. Soil Biol Biochem 41:1491–1496

Begum N, Qin C, Ahanger MA et al (2019) Role of arbuscular mycorrhizal fungi in plant growth regulation: implications in abiotic stress tolerance. Front Plant Sci 10:1068

Belval L, Roth L, Martin IR et al (2024) Effect of arbuscular mycorrhizal symbiosis on grapevine response to *Neofusicoccum parvum*, a major trunk disease fungus. Plant Stress 14:100582

Bencherif K, Djaballah Z, Brahimi F et al (2019) Arbuscular mycorrhizal fungi affect total phenolic content and antimicrobial activity of *Tamarix gallica* in natural semi-arid Algerian areas. S Afr J Bot 125:39–45

Berg G, Smalla K (2009) Plant species and soil type cooperatively shape the structure and function of microbial communities in the rhizosphere. FEMS Microbiol Ecol 68(1):1–3

Bernardo VF, Garita SA, Arango MC et al (2021) Arbuscular mycorrhizal fungi against the false root-knot nematode activity in *Capsicum annuum*: physiological responses in plants. Biocontrol Sci Tech 31(2):119–131

Berruti A, Lumini E, Balestrini R, Bianciotto V (2016) Arbuscular mycorrhizal fungi as natural biofertilizers: let's benefit from past successes. Front Microbiol 6:1559

Beslemes D, Tigka E, Roussis I et al (2023) Effect of arbuscular mycorrhizal fungi on nitrogen and phosphorus uptake efficiency and crop productivity of two-rowed barley under different crop production systems. Plants 12(9):1908

Bhonwong A, Stout MJ, Attajarusit J et al (2009) Defensive role of tomato polyphenol oxidases against cotton bollworm (*Helicoverpa armigera*) and beet armyworm (*Spodoptera exigua*). J Chem Ecol 35:28–38

Bisht A, Garg N (2022) AMF modulated rhizospheric microbial enzyme activities and their impact on sulphur assimilation along with thiol metabolism in pigeonpea under Cd stress. Rhizosphere 21:100478

Bitterlich M, Franken P, Graefe J (2018) Arbuscular mycorrhiza improves substrate hydraulic conductivity in the plant available moisture range under root growth exclusion. Front Plant Sci 9:301

Bucking H, Kafle A (2015) Role of arbuscular mycorrhizal fungi in the nitrogen uptake of plants: current knowledge and research gaps. Agronomy 5(4):587–612

Cao MA, Wang P, Hashem A et al (2021) Field inoculation of arbuscular mycorrhizal fungi improves fruit quality and root physiological activity of citrus. Agriculture 11(12):1297

Charron G, Furlan V et al (2001) Response of onion plants to arbuscular mycorrhizae: 1. Effects of inoculation method and phosphorus fertilization on biomass and bulb firmness. Mycorrhiza 11:187–197

Chave M, Crozilhac P, Deberdt P et al (2017) *Rhizophagus irregularis* MUCL 41833 transitorily reduces tomato bacterial wilt incidence caused by *Ralstonia solanacearum* under in vitro conditions. Mycorrhiza 27:719–723

Cheng XF, Xie MM, Li Y et al (2022) Effects of field inoculation with arbuscular mycorrhizal fungi and endophytic fungi on fruit quality and soil properties of Newhall navel orange. Appl Soil Ecol 170:104308

Chilvers GA, Lapeyrie FF, Horan DP (1987) Ectomycorrhizal vs endomycorrhizal fungi within the same root system. New Phytol 107(2):441–448

Chouyia FE, Fiorentino N, Rouphael Y et al (2022) Assessing the effect of P-solubilizing bacteria and mycorrhizal fungi on tomato yield and quality under different crop rotations. Sci Hortic 293:110740

Chowdhury S, Lange M, Malik AA et al (2022) Plants with arbuscular mycorrhizal fungi efficiently acquire nitrogen from substrate additions by shaping the decomposer community composition and their net plant carbon demand. Plant Soil 475(1):473–490

Copetta A, Todeschini V, Massa N et al (2021) Inoculation with arbuscular mycorrhizal fungi improves melon (*Cucumis melo*) fruit quality under field conditions and plant performance in both field and greenhouse. Plant Biosyst 155(5):1063–1074

da Silva Campos MA (2020) Bioprotection by arbuscular mycorrhizal fungi in plants infected with *Meloidogyne* nematodes: a sustainable alternative. Crop Prot 135:105203

De Sá CS, Campos MA (2020) Arbuscular mycorrhizal fungi decrease *Meloidogyne enterolobii* infection of Guava seedlings. J Helminthol 94:e183

Deja-Sikora E, Mercy L, Baum C et al (2019) The contribution of endomycorrhiza to the performance of potato virus Y-infected solanaceous plants: disease alleviation or exacerbation? Front Microbiol 10:516

Deja-Sikora E, Kowalczyk A, Trejgell A et al (2020) Arbuscular mycorrhiza changes the impact of potato virus Y on growth and stress tolerance of *Solanum tuberosum* L. *in vitro*. Front Microbiol 10:2971

Demenois J, Rey F, Ibanez T et al (2018) Linkages between root traits, soil fungi and aggregate stability in tropical plant communities along a successional vegetation gradient. Plant Soil 424:319–334

Deveautour C, Donn S, Bennett AE et al (2021) Variability of arbuscular mycorrhizal fungal communities within the root systems of individual plants is high and influenced by host species and root phosphorus. Pedobiologia 84:150691

Dey M, Ghosh S (2022) Arbuscular mycorrhizae in plant immunity and crop pathogen control. Rhizosphere 22:100524

Douds DD Jr, Carr ER, Shenk JE et al (2017) Positive yield response of eggplant (*Solanum melongena* L.) to inoculation with AM fungi produced on-farm. Sci Hortic 224:48–52

Dowarah B, Gill SS et al (2022) Arbuscular mycorrhizal fungi in conferring tolerance to biotic stresses in plants. J Plant Growth Regul 41(4):1429–1444

Drinkwater LE, Letourneau DK, Workneh FA et al (1995) Fundamental differences between conventional and organic tomato agroecosystems in California. Ecol Appl 5(4):1098–1112

El-Sherbeny TM, Mousa AM et al (2022) Use of mycorrhizal fungi and phosphorus fertilization to improve the yield of onion (*Allium cepa* L.) plant. J Biol Sci 29(1):331–338

Evelin H, Kapoor R, Giri B (2009) Arbuscular mycorrhizal fungi in alleviation of salt stress: a review. Ann Bot 104(7):1263–1280
Fall AF, Nakabonge G, Ssekandi J et al (2022) Roles of arbuscular mycorrhizal fungi on soil fertility: contribution in the improvement of physical, chemical, and biological properties of the soil. Front Fungal Biol 3:723892
Fall AF, Nakabonge G, Ssekandi J et al (2023) Combined effects of indigenous arbuscular mycorrhizal fungi (AMF) and NPK fertilizer on growth and yields of maize and soil nutrient availability. Sustainability 15(3):2243
Felföldi Z, Vidican R, Stoian V et al (2022) Arbuscular mycorrhizal fungi and fertilization influence yield, growth and root colonization of different tomato genotype. Plants 11(13):1743
Fitter AH, Helgason T et al (2011) Nutritional exchanges in the arbuscular mycorrhizal symbiosis: implications for sustainable agriculture. Fungal Biol Rev 25(1):68–72
Frey SD (2019) Mycorrhizal fungi as mediators of soil organic matter dynamics. Annu Rev Ecol Evol Syst 50(1):237–259
Garnett T, Appleby MC, Balmford A (2013) Sustainable intensification in agriculture: premises and policies. Science 341(6141):33–34
Gholinezhad E, Darvishzadeh R (2021) Influence of arbuscular mycorrhiza fungi and drought stress on fatty acids profile of sesame (*Sesamum indicum* L.). Field Crop Res 262:108035
Gianinazzi S, Gollotte A, Binet MN (2010) Agroecology: the key role of arbuscular mycorrhizas in ecosystem services. Mycorrhiza 20(8):519–530
Goltapeh EM, Danesh YR, Prasad R, Varma A (2008) Mycorrhizal fungi: what we know and what should we know? In: Varma A (ed) Mycorrhiza. Springer, Berlin/Heidelberg, pp 3–27
Govindarajulu M, Pfeffer PE, Jin H et al (2005) Nitrogen transfer in the arbuscular mycorrhizal symbiosis. Underw Nat 435(7043):819–823
Hajiboland R, Aliasgharzadeh N, Laiegh SF et al (2010) Colonization with arbuscular mycorrhizal fungi improves salinity tolerance of tomato (*Solanum lycopersicum* L.) plants. Plant Soil 331:313–327
He X, Xu M, Qiu GY et al (2009) Use of ^{15}N stable isotope to quantify nitrogen transfer between mycorrhizal plants. J Plant Ecol 2(3):107–118
He L, Li C et al (2017) Indirect interactions between arbuscular mycorrhizal fungi and *Spodoptera exigua* alter photosynthesis and plant endogenous hormones. Mycorrhiza 27:525–535
Herrera-Peraza RA, Hamel C, Fernández F et al (2011) Soil–strain compatibility: the key to effective use of arbuscular mycorrhizal inoculants? Mycorrhiza 21:183–193
Hestrin R, Hammer EC, Mueller CW et al (2019) Synergies between mycorrhizal fungi and soil microbial communities increase plant nitrogen acquisition. Commun Biol 2:233
Hijri M (2016) Analysis of a large dataset of mycorrhiza inoculation field trials on potato shows highly significant increases in yield. Mycorrhiza 26(3):209–214
Hodge A (2014) Interaction between arbuscular mycorrhizal fungi and organic material substrates. In: Sariaslani S, Gadd GM (eds) Advances in applied microbiology, pp 47–99
Holatko J, Brtnicky M, Kucerik J et al (2021) Gomalin—truths, myths, and the future of this elusive soil glycoprotein. Soil Biol Biochem 153:108116
Hossain MB (2021) Glomalin and contribution of glomalin to carbon sequestration in soil: a review. TURJAF 9(1):191–196
Huang D, Ma M, Wang Q et al (2020) Arbuscular mycorrhizal fungi enhanced drought resistance in apple by regulating genes in the MAPK pathway. Plant Physiol Biochem 149:245–255
Igiehon NO, Babalola OO (2007) Biofertilizers and sustainable agriculture: exploring arbuscular mycorrhizal fungi. Appl Microbiol Biotechnol 101:4871–4881
Jackson LE, Bowles TM, Hodson AK et al (2012) Soil microbial-root and microbial-rhizosphere processes to increase nitrogen availability and retention in agroecosystems. Curr Opin Environ Sustain 4(5):517–522
Jansa J, Finlay R, Wallander H et al (2011) Role of mycorrhizal symbioses in phosphorus cycling. In: Phosphorus in action: biological processes in soil phosphorus cycling, pp 137–168

Jiang F, Zhang L, Zhou J et al (2021) Arbuscular mycorrhizal fungi enhance mineralisation of organic phosphorus by carrying bacteria along their extraradical hyphae. New Phytol 230(1):304–315

Jin H, Pfeffer P, Douds D et al (2005) The uptake, metabolism, transport and transfer of nitrogen in an arbuscular mycorrhizal fungi. New Phytol 168:687–696

Johnson NC, Wilson GW, Wilson JA (2015) Mycorrhizal phenotypes and the l aw of the minimum. New Phytol 205(4):1473–1484

Johri AK, Oelmüller R, Dua M (2015) Fungal association and utilization of phosphate by plants: success, limitations, and future prospects. Front Microbiol 6:984

Joshi SK, Gauraha AK (2022) Global biofertilizer market: Emerging trends and opportunities. Trends Appl Microbiol Sustain Econ 689–697

Jung SC, Martinez-Medina A, Lopez-Raez JA et al (2012) Mycorrhiza-induced resistance and priming of plant defenses. J Chem Ecol 38:423–441

Kaur S, Suseela V (2020) Unravelling arbuscular mycorrhiza-induced changes in plant primary and secondary metabolome. Meta 10:335

Kaur J, Chavana J, Soti P et al (2020) Arbuscular mycorrhizal fungi (AMF) influences growth and insect community dynamics in Sorghum-sudangrass (*Sorghum* x *drummondii*). Arthropod-Plant Inte 14(3):301–315

Khan F, Siddique AB, Shabala S et al (2023) Phosphorus plays key roles in regulating plants' physiological responses to abiotic stresses. Plants 12(15):2861

Khoshkhatti N, Eini O et al (2020) Differential response of mycorrhizal plants to tomato bushy stunt virus and tomato mosaic virus infection. Microorganisms 8(12):2038

Kochian LV, Jones DL, Shaff JE (1994) The role of ion transport process in root hair tip growth in *Limnobium stoloniferum*. Curr Top Plant Physiol 12:150–160

Kong L, Gong X, Zhang X et al (2020) Effects of arbuscular mycorrhizal fungi on photosynthesis, ion balance of tomato plants under saline-alkali soil condition. J Plant Nutr 43(5):682–698

Kumar A, Choudhary AK et al (2017) Agronomic bio-fortification and quality enhancement in okra–pea cropping system through arbuscular mycorrhizal fungi at varying phosphorus and irrigation regimes in Himalayan acid alfisol. J Plant Nutr 40(8):1213–1229

Latef AA, Hashem A, Rasool S (2016) Arbuscular mycorrhizal symbiosis and abiotic stress in plants: a review. J Plant Biol 59:407–426

Lazcano C, Barrios-Masias FH et al (2014) Arbuscular mycorrhizal effects on plant water relations and soil greenhouse gas emissions under changing moisture regimes. Soil Biol Biochem 74:184–192

Li HY, Yang GD, Shu HR et al (2006) Colonization by the arbuscular mycorrhizal fungus *Glomus versiforme* induces a defense response against the root-knot nematode *Meloidogyne incognita* in the grapevine (*Vitis amurensis* Rupr.), which includes transcriptional activation of the class III chitinase gene VCH3. Plant Cell Physiol 47(1):154–163

Li H, Chang J, Zheng J et al (2017) Local melatonin application induces cold tolerance in distant organs of *Citrullus lanatus* L. via long distance transport. Sci Rep 7(1):40858

Liu J, Maldonado-Mendoza I, Lopez-Meyer M et al (2007) Arbuscular mycorrhizal symbiosis is accompanied by local and systemic alterations in gene expression and an increase in disease resistance in the shoots. Plant J 50:29–544

Lopes Leal P, Varón-López M, de Oliveira G, Prado I et al (2016) Enrichment of arbuscular mycorrhizal fungi in a contaminated soil after rehabilitation. Braz J Microbiol 47(4):853–862

Lu FC, Lee CY, Wang CL (2015) The influence of arbuscular mycorrhizal fungi inoculation on yam (*Dioscorea* spp.) tuber weights and secondary metabolite content. Peer J 3:e1266

Lu LH, Zou YN, Wu QS (2019) Mycorrhizas mitigate soil replant diseases of peach through regulating root exudates, soil microbial communities, and soil aggregate stability. Commun Soil Sci Plant Anal 50:909–921

Lutgen ER, Muir-Clairmont D, Graham J et al (2003) Seasonality of arbuscular mycorrhizal hyphae and glomalin in a western Montana grassland. Plant Soil 257:71–83

Marschner H (2011) Marschner's mineral nutrition of higher plants. Academic

Martínez-Dalmau J, Berbel J et al (2021) Nitrogen fertilization. A review of the risks associated with the inefficiency of its use and policy responses. Sustainability 13(10):5625

Miozzi L, Vaira AM et al (2020) Arbuscular mycorrhizal symbiosis primes tolerance to cucumber mosaic virus in tomato. Viruses 12(6):675

Miransari M, Bahrami HA, Rejali F et al (2008) Using arbuscular mycorrhiza to alleviate the stress of soil compaction on wheat (*Triticum aestivum* L.) growth. Soil Biol Biochem 40(5):1197–1206

Mishra V, Ellouze W, Howard RJ (2018) Utility of arbuscular mycorrhizal fungi for improved production and disease mitigation in organic and hydroponic greenhouse crops. J Hortic 5(03):237

Mordor intelligence (2025) Mycorrhiza market size & share analysis – growth trends & forecasts up to 2030. https://www.mordorintelligence.com/industry-reports/mycorrhiza-based-biofertilizer-market. Accessed 23 Feb 2025

Muluneh MG (2021) Impact of climate change on biodiversity and food security: a global perspective—a review article. Agric Food Secur 10(1):1–25

Muneer MA, Wang P, Zaib-un-Nisa V et al (2020) Potential role of common mycorrhizal networks in improving plant growth and soil physicochemical properties under varying nitrogen levels in a grassland ecosystem. Glob Ecol Conserv 24:e01352

Nair A, Kolet SP, Thulasiram HV et al (2014) Systemic jasmonic acid modulation in mycorrhizal tomato plants and its role in induced resistance against *Alternaria alternata*. Plant Biol 17:625–631

Navarro JM, Morte A (2024) Arbuscular mycorrhizal fungi as biofertilizers to increase the plant quality of sour-orange seedlings. Agronomy 14(1):230

Oliveira TC, Cabral JSR, Santana LR et al (2022) The arbuscular mycorrhizal fungus *Rhizophagus clarus* improves physiological tolerance to drought stress in soybean plants. Sci Rep 12(1):9044

Osemwegie OO, Adetunji CO, Oghenekaro AO et al (2021) Arbuscular mycorrhizae: under-tapped potential benefits and perspective on Africa. OnLine J Biol Sci 21(1):12–25

Othman YA, Alananbeh KM, Tahat MM (2024) Can arbuscular mycorrhizal fungi enhance crop productivity and quality in hydroponics? A meta-analysis. Sustainability 16(9):3662

Parihar M, Rakshit A, Meena VS et al (2020) The potential of arbuscular mycorrhizal fungi in C cycling: a review. Arch Microbiol 202:1581–1596

Pellegrino E, Öpik M et al (2015) Responses of wheat to arbuscular mycorrhizal fungi: a meta-analysis of field studies from 1975 to 2013. Soil Biol Biochem 84:210–217

Perez-de-Luque A, Tille S, Johnson I et al (2017) The interactive effects of arbuscular mycorrhiza and plant growth-promoting rhizobacteria synergistically enhance host plant defences against pathogens. Sci Rep 7:16409

Philippot L, Raaijmakers JM, Lemanceau P et al (2013) Going back to the roots: the microbial ecology of the rhizosphere. Nat Rev Microbiol 11(11):789–799

Qi S, Wang J, Wan L et al (2022) Arbuscular mycorrhiza fungi contribute to phosphorus uptake and allocation strategies of *Solidago canadensis* in a phosphorus-deficient environment. Front Plant Sci 13:831654

Qian S, Xu Y, Zhang Y et al (2024) Effect of AMF inoculation on reducing excessive fertilizer use. Microorganisms 12(8):1550

Raghavendra Kumar M, Ashwin R, Bagyaraj DJ (2018) Screening arbuscular mycorrhizal fungi in order to select the best for alleviating wilt disease complex of capsicum. Proc Natl Acad Sci India Sect B Biol Sci 88:679–684

Ramalingappa PL, Shrivastava M, Dhar S et al (2023) Reducing options of ammonia volatilization and improving nitrogen use efficiency via organic and inorganic amendments in wheat (Triticum aestivum L.). PeerJ 11:e14965

Ramona S, Stephanie W et al (2020) Effects of arbuscular mycorrhization on fruit quality in industrialized tomato production. Int J Mol Sci 21:7029

Read DJ, Perez-Moreno J (2003) Mycorrhizas and nutrient cycling in ecosystems–a journey towards relevance? New Phytol 157(3):475–492

Rillig MC, Wright SF et al (2002) The role of arbuscular mycorrhizal fungi and glomalin in soil aggregation: comparing effects of five plant species. Plant Soil 238:325–333

Rillig MC, Aguilar-Trigueros CA et al (2015) Plant root and mycorrhizal fungal traits for understanding soil aggregation. New Phytol 205(4):1385–1388

Rillig MC, Sosa-Hernández MA, Roy J et al (2016) Towards an integrated mycorrhizal technology: harnessing mycorrhiza for sustainable intensification in agriculture. Front Plant Sci 7:1625

Robinson Boyer L, Feng W, Gulbis N et al (2016) The use of arbuscular mycorrhizal fungi to improve strawberry production in coir substrate. Front Plant Sci 7:1237

Ronsheim ML (2012) The effect of mycorrhizae on plant growth and reproduction varies with soil phosphorus and developmental stage. Am Midl Nat 167(1):28–39

Ruiz-Lozano JM, Porcel R et al (2006) Does the enhanced tolerance of arbuscular mycorrhizal plants to water deficit involve modulation of drought-induced plant genes? New Phytol 1:693–698

Ruiz-Lozano JM, Porcel R, Bárzana G (2012) Contribution of arbuscular mycorrhizal symbiosis to plant drought tolerance: state of the art. In: Aroca R (ed) Plant responses to drought stress: from morphological to molecular features. Springer, pp 335–362

Ryan MH, Graham JH (2002) Is there a role for arbuscular mycorrhizal fungi in production agriculture? Plant Soil 244:263–271

Sabia E, Claps S, Morone G et al (2015) Field inoculation of arbuscular mycorrhiza on maize (*Zea mays* L.) under low inputs: preliminary study on quantitative and qualitative aspects. Ital J Agron 10(1):30–33

Saik P, Koshkalda I, Bezuhla L et al (2024) Achieving land degradation neutrality: land-use planning and ecosystem approach. Front Environ Sci 12:1446056

Saleem A, Anwar S, Nawaz T et al (2024) Securing a sustainable future: the climate change threat to agriculture, food security, and sustainable development goals. JAS 11:1–7

Schnepf A, Roose T et al (2008) Impact of growth and uptake patterns of arbuscular mycorrhizal fungi on plant phosphorus uptake—a modelling study. Plant Soil 312:85–99

Selvaraj A, Thangavel K, Uthandi S (2020) Arbuscular mycorrhizal fungi (*Glomus intraradices*) and diazotrophic bacterium (*Rhizobium* BMBS) primed defense in blackgram against herbivorous insect (*Spodoptera litura*) infestation. Micriobiol Res 231:126355

Sharma K, Thokchom SD, Gupta S, Rupam K (2023) Fungi and fungal products in human welfare and biotechnology. Springer, Singapore

Shirani Bidabadi S, Mehralian M (2020) Arbuscular mycorrhizal fungi inoculation to enhance chilling stress tolerance of watermelon. Gesunde Pflanzen 72(2):171–179

Singh M, Singh PK (2024) Enhancing growth and drought tolerance in tomato through arbuscular mycorrhizal symbiosis. Rodriguésia 75:e00482024

Singh S, Pandey A, Kumar B et al (2010) Enhancement in growth and quality parameters of tea [*Camellia sinensis* (L.) O. Kuntze] through inoculation with arbuscular mycorrhizal fungi in an acid soil. Biol Fertil Soils 46:427–433

Smith SE, Read DJ (2010) Mycorrhizal symbiosis, 3rd edn. Academic press

Smith SE, Smith FA (2011) Roles of arbuscular mycorrhizas in plant nutrition and growth: new paradigms from cellular to ecosystem scales. Ann Rev Plant Biol 62:227–250

Song Y, Chen D, Lu K et al (2015) Enhanced tomato disease resistance primed by arbuscular mycorrhizal fungus. Front Plant Sci 6:786

Song Z, Bi Y, Zhang J et al (2020) Arbuscular mycorrhizal fungi promote the growth of plants in the mining associated clay. Sci Rep 10:2663

Sun Z, Song J, Xin X et al (2018) Arbuscular mycorrhizal fungal 14-3-3 proteins are involved in arbuscule formation and responses to abiotic stresses during AM symbiosis. Front Microbiol 9:91

Sun J, Tang Y, Chen K et al (2024) Nitrogen level determines arbuscular mycorrhizal fungi nitrogen uptake rate of *Stipa purpurea* in alpine steppe. Plant Soil 512:677–693

Tawaha ARM, Khanum S, Al Tawaha AR et al (2021) Use of mycorrhiza in organic farming. IOP Conf Ser Earth Environ Sci 788(1):012167

Thirkell TJ, Cameron DD et al (2016) Resolving the 'nitrogen paradox' of arbuscular mycorrhizas: fertilization with organic matter brings considerable benefits for plant nutrition and growth. Plant Cell Environ 39(8):1683–1690

Thirkell TJ, Pastok D et al (2020) Carbon for nutrient exchange between arbuscular mycorrhizal fungi and wheat varies according to cultivar and changes in atmospheric carbon dioxide concentration. Glob Chang Biol 26:1725–1738

Tisserant E, Malbreil M, Kuo A (2013) Genome of an arbuscular mycorrhizal fungus provides insight into the oldest plant symbiosis. PNAS 110(50):20117–20122

Torres N, Goicoechea N, Morales F et al (2016) Berry quality and antioxidant properties in *Vitis vinifera* cv. Tempranillo as affected by clonal variability, mycorrhizal inoculation and temperature. Crop Pasture Sci 67(9):961–977

Torres N, Antolín MC et al (2018) Arbuscular mycorrhizal symbiosis as a promising resource for improving berry quality in grapevines under changing environments. Front Plant Sci 9:897

Treseder KK (2016) Model behavior of arbuscular mycorrhizal fungi: predicting soil carbon dynamics under climate change. Botany 94(6):417–423

Veresoglou SD, Agathokleous E (2023) Pertinent challenges in mycorrhizal ecopreneurship for sustainable food production. Sustain Horiz 8:100064

Verzeaux J, Hirel B, Dubois F et al (2017) Agricultural practices to improve nitrogen use efficiency through the use of arbuscular mycorrhizae: basic and agronomic aspects. Plant Sci 264:48–56

Vilela LAF, Barbosa MV (2019) Contribution of arbuscular mycorrhizal fungi in promoting cadmium tolerance in plants. In: Hasanuzzaman M, Prasad MNV, Nahar K (eds) Cadmiun tolerance in plants: agronomic. Academic, Molecular, signaling and omics approaches, pp 553–586

Vosátka M, Látr A, Gianinazzi S, Albrechtová J (2012) Development of arbuscular mycorrhizal biotechnology and industry: current achievements and bottlenecks. Symbiosis 58:29–37

Wahab A, Muhammad M, Munir A et al (2023) Role of arbuscular mycorrhizal fungi in regulating growth, enhancing productivity, and potentially influencing ecosystems under abiotic and biotic stresses. Plants 12(17):3102

Walder F, Niemann H, Natarajan M et al (2012) Mycorrhizal networks: common goods of plants shared under unequal terms of trade. Plant Physiol 159(2):789–797

Wang L, Huang D (2021) Nitrogen and phosphorus losses by surface runoff and soil microbial communities in a paddy field with different irrigation and fertilization managements. PLoS One 16(7):e0254227

Wang ZG, Bi YL, Jiang B et al (2016) Arbuscular mycorrhizal fungi enhance soil carbon sequestration in the coalfields, Northwest China. Sci Rep 6(1):34336

Wang YY, Yin QS, Qu Y et al (2017) Arbuscular mycorrhiza-mediated resistance in tomato against *Cladosporium fulvum*-induced mould disease. J Phytopathol 166:67–74

Wang W, Li A, Zhang Z et al (2021) Posttranslational modifications: regulation of nitrogen utilization and signaling. PCP 62(4):543–552

Wang XX, Li J, Wang D et al (2022) Straw incorporation effects on net photosynthetic carbon assimilation and maize growth. Front Agron 4:805320

Wen Y, Wu R, Qi D et al (2024) The effect of AMF combined with biochar on plant growth and soil quality under saline-alkali stress: insights from microbial community analysis. Ecotoxicol Environ Saf 281:116592

Wu QS, Cao MQ, Zou YN et al (2014a) Direct and indirect effects of glomalin, mycorrhizal hyphae and roots on aggregate stability in rhizosphere of trifoliate orange. Sci Rep 4(1):5823

Wu Z, McGrouther K, Huang J et al (2014b) Decomposition and the contribution of glomalin-related soil protein (GRSP) in heavy metal sequestration: field experiment. Soil Biol Biochem 68:283–290

Wu Y, Chen C, Wang G (2024) Inoculation with arbuscular mycorrhizal fungi improves plant biomass and nitrogen and phosphorus nutrients: a meta-analysis. BMC Plant Biol 24:960

Xia J, Chen Y, Yuan W et al (2023) The effects of multiple environmental factors on global carbon allocation. Ecol Process 12(1):60

Xie X, Huang W, Liu F (2013) Functional analysis of the novel mycorrhiza-specific phosphate transporter AsPT1 and PHT1 family from *Astragalus sinicus* during the arbuscular mycorrhizal symbiosis. New Phytol 198(3):836–852

Yanan W, Xusheng Z, Baozhong Y et al (2015) Biochemical defences induced by mycorrhiza fungi *Glomus mosseae* in controlling strawberry *Fusarium* wilt. Open Biomed Eng J 9:301–304

Yang S, Li F, Malhi SS et al (2004) Long-term fertilization effects on crop yield and nitrate nitrogen accumulation in soil in northwestern China. Agron J 96(4):1039–1049
Ye JY, Tian WH et al (2022) Nitrogen in plants: from nutrition to the modulation of abiotic stress adaptation. Stress Biol 2(1):4
Zhang S, Wang L, Ma F et al (2015) Can arbuscular mycorrhiza and fertilizer management reduce phosphorus runoff from paddy fields? J Environ Sci Int 33:211–218
Zhang Z, Mallik A, Zhang J et al (2019a) Effects of arbuscular mycorrhizal fungi on inoculated seedling growth and rhizosphere soil aggregates. Soil Tillage Res 194:104340
Zhang Q, Zhu D, Ding J et al (2019b) The fungicide azoxystrobin perturbs the gut microbiota community and enriches antibiotic resistance genes in *Enchytraeus crypticus*. Environ Int 131:104965
Zhang W, Yu L, Han B et al (2022) Mycorrhizal inoculation enhances nutrient absorption and induces insect-resistant defense of *Elymus nutans*. Front Plant Sci 13:898969
Zheng X, An Z, Cao M et al (2022) Arbuscular mycorrhizal hyphal respiration makes a large contribution to soil respiration in a subtropical forest under various N input rates. Sci Total Environ 852:158309
Zhou J, Su Y, Li X et al (2024) Arbuscular mycorrhizae mitigate negative impacts of soil biodiversity loss on grassland productivity. J Environ Manag 349:119509
Zhu X, Song F et al (2010) Influence of arbuscular mycorrhiza on lipid peroxidation and antioxidant enzyme activity of maize plants under temperature stress. Mycorrhiza 20:325–332
Zhu B, Gao T, Zhang D et al (2022) Functions of arbuscular mycorrhizal fungi in horticultural crops. Sci Hortic 303:111219
Ziane H, Hamza N et al (2021) Arbuscular mycorrhizal fungi and fertilization rates optimize tomato (*Solanum lycopersicum* L.) growth and yield in a Mediterranean agroecosystem. J Saudi Soc Agric Sci 20(7):454–458

Environmental Benefits of Mycorrhizal Symbiosis: Soil Conservation, Biodiversity Conservation, and Climate Change Mitigation

Nur Maizatul Idayu Othman, Nur Qursyna Boll Kassim,
Nuraini Mohd Noor, Nur Hafizah Zakaria,
Nurul Wahida Ramli, Nur' Amira Hamid,
and Fadzilah Adibah Abdul Majid

Abstract

The environmental benefits of mycorrhizal symbiosis, particularly in the context of soil conservation, biodiversity conservation, and climate change mitigation, are increasingly gaining attention, especially in the ecological research area. Mycorrhizal fungi, especially arbuscular mycorrhizal fungi, usually form a symbiotic relationship with approximately 80% of terrestrial plant families. This symbiotic relationship enhanced nutrient uptake, particularly phosphorus, and improved plant resilience to abiotic stresses such as salinity, flooding, and drought due to climate change. This symbiosis not only supports plant health but also promotes soil organic matter formation and their aggregate stability, which are critical for soil conservation and carbon sequestration in agroecosystems. Additionally, mycorrhizal networks also facilitate nutrient cycling and carbon sequestration, with studies indicating that mycorrhizal colonization can lead to

N. M. I. Othman (✉) · N. Q. B. Kassim
Faculty of Plantation and Agrotechnology, Universiti Teknologi MARA (UiTM), Malacca, Malaysia

Soil Conservation and Management Research Interest Group (RIG), Universiti Teknologi MARA (UiTM), Shah Alam, Selangor, Malaysia
e-mail: nurmaizatul@uitm.edu.my

N. M. Noor · N. W. Ramli · N. A. Hamid
Faculty of Plantation and Agrotechnology, Universiti Teknologi MARA (UiTM), Malacca, Malaysia

N. H. Zakaria · F. A. A. Majid
Institute of Climate Adaptation and Marine Biotechnology (ICAMB), Universiti Malaysia Terengganu, Kuala Nerus, Terengganu, Malaysia

Y. Siddiqui et al. (eds.), *Sustainable Mycorrhizal Cultivation*,
https://doi.org/10.1007/978-981-95-1778-7_14

an average increase of about 25% in soil organic carbon content compared to a non-mycorrhizal soil system. Recent research has shown that mycorrhizal networks can transfer up to 40% of nitrogen and 50% of the phosphorus required by plants, significantly reducing the need for chemical fertilizer. Moreover, AMF-inoculated plants have demonstrated up to 80% higher drought tolerance compared to non-inoculated plants. Furthermore, mycorrhizal fungi play an important role in enhancing biodiversity through supporting the diverse range of plant species that lead to various ecosystems and new habitats. The presence of mycorrhizal fungi can influence plant community dynamics and niche differentiation, promoting a more resilient environment capable of any climate changes. This book chapter explores the multifaceted roles of mycorrhizal symbiosis extending beyond the individual plant and encompassing its environmental benefits on soil conservation, biodiversity conservation, and climate change.

Keywords

Mycorrhizal symbiosis · Soil conservation · Biodiversity · Arbuscular mycorrhizal fungi · Abiotic stress

1 Introduction

Mycorrhizal symbiosis is a well-recognized mutualistic association, where plants and mycorrhizal fungi form a partnership that benefits both organisms (Genre et al. 2020). This mutually beneficial relationship between two organisms was discovered in 1866 by Heinrich Anton de Bary, who observed a close relationship between a fungus and an alga (Shi et al. 2023). Later, in 1885, Albert Bernhard Frank created the term "mycorrhiza" (derived from mycor, meaning fungus, and rhiza meaning root) to describe the unique interaction between plant roots and ectomycorrhizal fungi. Two years after that, in 1887, Frank identified a differentiation between eutrophic and endotrophic mycorrhizae, which then comprised solely ericaceous and orchid mycorrhizae (Shi et al. 2023). Following over 150 years of investigation into this subject, it is now established that mycorrhizae signify one of the most significant biological interkingdom interactions, encompassing approximately 50,000 taxa of soil fungi and around 340,000 terrestrial plants, with the sole exceptions being Brassicaceae and Chenopodiaceae, which seemingly do not engage in such interaction (Genre et al. 2020; Shi et al. 2023). According to Martin and van der Heijden (2024), mycorrhizal symbiosis is recognized as ubiquitous in nearly all terrestrial ecosystems, including deserts, tropical forests, savannahs, and agricultural land.

Mycorrhizal fungi are pivotal in ecological and environmental research, largely due to the significant impact on soil health, plant development, and overall ecosystem dynamics (Ma et al. 2022). These fungi inhabit various regions of plant roots, including within the root cortex, on root surfaces, and around root epidermal cells (van der Heijden et al. 2015; Usman et al. 2021; Abdalla et al. 2023). Researchers have identified four primary mycorrhizal symbiosis types, each with distinct structures and functions: arbuscular mycorrhiza, ectomycorrhiza, orchid mycorrhiza, and ericoid mycorrhiza (Martin and van der Heijden 2024). Among the various types of mycorrhiza, the arbuscular mycorrhiza, belonging to the phylum Glomeromycota, stands out as one of the most prevalent, ancient, and extensively distributed, forming associations with approximately 80% of all terrestrial plant species (Figueiredo et al. 2021). This type of fungi is especially common in warm climates and diverse ecosystems like tropical forests. Ectomycorrhizal fungi, meanwhile, are the second most widespread type found in nature. While fewer plant species are known to engage in symbiosis with ectomycorrhizal fungi compared to arbuscular mycorrhiza fungi (AMF), the hosts of ectomycorrhiza fungi are often widely distributed, plentiful, and prominent members of their respective groups (Figueiredo et al. 2021). In contrast to AMF, ectomycorrhiza fungi predominantly inhabit colder regions and ecosystems characterized by a lower diversity of host species, such as temperate and boreal forests (Figueiredo et al. 2021).

Through close cellular interactions between the symbiotic partners, mycorrhizal fungi benefit from the carbon produced during photosynthesis by the host plant, receiving essential compounds such as sugars and fatty acids. In return, these fungi significantly enhance the access of the plants to water and vital mineral nutrients, particularly phosphorus (P) and nitrogen (N). This nutrient exchange is not limited to these micronutrients; mycorrhizal fungi also facilitate the uptake of a variety of micronutrients, including copper (Cu), iron (Fe), manganese (Mn), and zinc (Zn) (van der Heijden et al. 2015; Genre et al. 2020; Shi et al. 2021; Rui et al. 2022; Martin and van der Heijden 2024). This mutually beneficial relationship not only promotes the growth and health of the plant but also supports the overall vitality of the ecosystem, as healthy plants contribute to soil stability and biodiversity. Additionally, during the symbiotic relationship, the hyphae of mycorrhizal fungi, along with the roots of plants, attract a unique community of microorganisms into their mycorrhizosphere. Mycorrhizal fungi have the ability to facilitate plant interactions with this mycorrhizosphere microbiome, which contributes to enhanced plant growth and improved nutrient absorption (van der Heijden et al. 2015; Usman et al. 2021; Ma et al. 2022; Shi et al. 2023). Furthermore, this symbiotic relationship boosts the host plant's resilience to various stresses. Specifically, it enhances their tolerance to biotic stresses (e.g., nematodes, insects, bacteria, and viruses) and abiotic stresses (e.g., salinity, flooding, extreme temperature, and drought) (Mitra et al. 2021; Ledford et al. 2023). Beyond these benefits to individual plants, mycorrhizal fungi deliver critical ecosystem services across both natural and agricultural environments. According to Genre et al. (2021), these mycorrhizal fungi are essential drivers of carbon sequestration, as they contribute to soil carbon storage through organic matter accumulation and glomalin production, a protein that binds soil particles. Moreover, mycorrhizal fungi are essential for soil aggregation, improving retention of water, enhancing the structure of the soil, and facilitating the cycling of nutrient. Their influence on soil composition extends to shaping the diversity and composition of microbial and plant communities. As a result, mycorrhizal fungi contribute significantly to promoting biodiversity within ecosystems (Genre et al. 2021). Given their impact on these ecosystem functions, mycorrhizas have likely influenced the evolution of the biosphere since the earliest stages of plant colonization on land.

In the context of today's escalating environmental challenges, mycorrhizal fungi are increasingly recognized as essential contributors to ecosystem stability and resilience, highlighting their importance in sustainable land management and climate change mitigation efforts (Genre et al. 2020). Hence, this chapter intends to thoroughly examine mycorrhizal fungi, highlighting their ecological functions and environmental importance. This review focuses on mycorrhizal symbiosis, defining its forms and examining its effects on nutrient intake, soil conservation, and resistance under stress. Furthermore, it analyzes their function in nutrient cycle, drought resilience, biodiversity preservation, and climate change mitigation, emphasizing their applications in ecological, agricultural, and environmental domains.

2 Types of Mycorrhizal Fungi and Their Ecological Roles

Mycorrhiza refers to the mutualism between certain fungi and higher plant roots. These soil fungi are common symbioses that serve as channels for the movement of materials and energy between plants and soil (Cardon and Whitbeck 2007). Mycorrhizal fungi offer protection against pathogens, enhancing plant nutrition uptake, production of plant growth-promoting substances, and increased resistance to root infections and water stress. Mycorrhizal fungi commonly encourage plants to decrease root biomass while also improving nutrient uptake capabilities by growing far beyond root surfaces and proliferating in soil pores too small for root hair to reach. Mycelial networks of mycorrhizal fungi frequently connect plant root systems over broad areas. These fungi usually make up the majority of the biomass of soil microbes (Mohammadi et al. 2011). Besides, mycorrhizosphere affects microbial alterations that result in plant growth and health (Azcón-Aguilar et al. 2002).

2.1 Mycorrhizal Fungi

There are seven identified categories of mycorrhizal fungal associations known to exist, including Arbuscular mycorrhiza, Ectomycorrhiza, Ectendomycorrhiza, Arbutoid mycorrhiza, Monotropoid mycorrhiza, Ericoid mycorrhiza, and Orchid mycorrhiza (Smith and Read 2008). The most common mycorrhizal type is AMF, which is also referred to as endomycorrhizas, due to the fact that their fungal companion creates structures inside plant roots. Most AMF species belong to the subphylum Glomeromycotina of the phylum Mucoromycota (Spatafora et al. 2016). The classification of AMF is categorized according to their structure of soil-borne resting spore, biochemical properties, and molecular studies. They can be found in most ecosystems worldwide and in numerous significant crop species, such as wheat, maize, rice, grape, soybean, and cotton (Peterson et al. 2004). AMF are obligate biotrophs that consume the byproducts of their live plant host and are not specific to their potential host (Tahat et al. 2010). It is known that AMF symbionts play a significant role as microorganisms in the development of the primary soil biogeochemical cycles (P, N, and C). Ectomycorrhizal (ECM) represents a group of fungi that only live surround the host roots rather than penetrating inside. According to Shalini et al.(2007), ectomycorrhizae are most prevalent in species of ornamental and forest trees. The presence of the mantle and Hartig net helps in distinguishing ectomycorrhizal. The branch structures of the Hartig net allow for extensive surface contact between cells of the two symbionts (Peterson et al. 2004).

2.2 Role of Arbuscular Mycorrhizal Fungi in Plant Health and Soil Ecosystems

AMF are a type of fungus that is associated with most agricultural crops and offers biological defence against soil-borne illness (Smith and Read 2008). One of the distinctive features of AMF, when in symbiotic relationship with plant roots, is the ability to expand surface area through the development of extended hyphae, which helps plants grow in very severe environments like drought stress and nutritional deficiencies. Supplementing plants with beneficial microbes (antagonistic bacteria) and AMF improves their defence against diseases by competing for nutrients and space, generating antibiotics, parasitizing pathogens, or inducing host plant resistance. Consequently, these microorganisms have been extensively employed for biocontrol of pathogens (Berg et al. 2007). Disease reduction in AMF-colonized host plants is the outcome of the intricate relationship between pathogens, AMF, and plants (Harrier and Watson 2004). AMF symbiosis has been demonstrated to lessen the damage that is inflicted by soil-borne pathogens (Azcón-Aguilar et al. 2002). AMF were found to effectively delay the time required by *Ganoderma boninense* to infect and destroy oil palm plants, and the seedlings showed greater resistance (Rini 2001). AMF improve the intake of mineral nutrients, particularly phosphorus, which promotes plant growth and resistance to diseases. Symbiotic interaction between AMF and plant roots may prevent disease spore development by changing the patterns of root exudation (Strack et al. 2003). AMF produce enormous, thick-walled resting spores in the soil that can withstand unfavorable conditions and germinate under good circumstances (Bagyaraj 2014). In the rhizosphere, AMF increase the number and activity of phosphate-solubilizing and nitrogen-fixating bacteria (Reverkar et al. 2005). The hyphae of AMF allow for the exchange of nutrients by penetrating plant roots without damaging them. The hyphae help in local transportation of nitrogen availability, and it has been shown that mycorrhizal associations may be important for the decomposition and mineralization of plant organic matter and mobilize nutrients, especially nitrogen, for the benefit of the host plant (Lambers et al. 2008). The primary advantages of AMF include improving the plant's uptake of mineral nutrients and boosting the host plant's resistance to reduce in uptake of growth-inhibiting components. AMF benefit the soil by generating glomalin and organic acids, which prevent soil erosion, chelate heavy metals, increase carbon sequestration, and stabilize soil macro-aggregation. Additionally, AMF attract bacteria that produce alkaline phosphatase, an enzyme linked to the availability of organic phosphorus in mineralization soil. AMF also affect the diversity, composition, and activity of microbial communities in the soil through mechanisms of antagonism. These AMF activities help to increase soil fertility (Fall et al. 2022). Besides, AMF increase soil volume and helps to overcome water and nutrient depletion areas around root surfaces. According to, AMF are utilized as bioinoculants and as key biofertilizers in sustained crop yield. The significance of incorporating AMF in sustainable agriculture is demonstrated by the fact that they encourage stronger plant development and greater resilience to disease attacks (Linderman 1994). In conclusion, AMF establish symbiotic connections with most plants and play a crucial role in plant community development, nutrient absorption, water management, and general productivity. They also act as bioprotectants against pathogens and toxic stresses (Van der Heijden et al. 1998). However, to optimize their benefits, it is essential to implement management practices including minimum tillage, reduced chemical fertilizers, practices appropriate crop rotation with minimal fallow, and careful pesticide use.

3 Mycorrhizal Fungi and Nutrient Uptake

Mycorrhizal fungi play a significant role in nutrient uptake in plants by establishing symbiotic relationships with the plant that promote both plant growth and soil health. With this interaction, plants have an enhanced ability to access and absorb nutrients from the soil as the fungi could extend their hyphae far beyond the reach of the plant's roots. These symbiotic relationships between plants and fungi not only improve nutrient acquisition but also help plants to cope with unfavorable environmental conditions.

3.1 Enhanced Phosphorus Uptake Through Mycorrhizal Associations

Phosphorus is vital for plants to perform their normal physiological functions, such as energy transfer, photosynthesis, intracellular signaling, as well as gene replication and expression (Bhantana et al. 2021). The demand for phosphorus is high during the early stages of plant growth, which cannot be sufficiently absorbed by the plant roots (Saboor et al. 2021). Phosphorus is also one of the least accessible macronutrients in soils due to its tendency to form insoluble compounds (Wang et al. 2020). Many studies have reported that arbuscular mycorrhizae fungi (AMF), an obligate biotroph, could facilitate the plant to uptake phosphorus through the mycorrhizal uptake pathway (MUP) (Zhang et al. 2021; Watts-Williams 2022). In this pathway, Mycorrhizal fungi have developed extensive networks of hyphae that extend deep further into the soil to absorb the unreachable phosphorus. The AMF also produced low molecular weight organic acids (LMWOAs) such as oxalate and citrate (Andrino et al. 2021). These organic acids significantly contribute to phosphorus solubilization through mechanisms such as metal ion chelation and pH modification. Additionally, the fungi possess specialized high-affinity phosphate transporters that enable efficient uptake of inorganic phosphorus from the soil (Rui et al. 2022). Along the pathway, the phosphorus absorbed by the fungal hyphae will be transported through the fungal network and delivered to the arbuscules (Ferrol et al. 2019). Arbuscules are specialized mycorrhizae structures that form within the cortical cells of the plant root. These structures are branched, microscopic, and tailored to optimize the nutrient exchange between plants and fungi. Each branch of the arbuscule is covered by the plant plasma membrane, forming a unique structure known as the periarbuscular membrane (Ivanov et al. 2019). This specialized membrane contains protein transporters that facilitate the active transfer of phosphorus and other vital nutrients across the plant–fungal interface, which allows the plant to receive phosphorus directly from the fungus into its root cells (Di Martino and Crawford 2021). Ultimately, this symbiotic relationship with fungi allows plants to thrive in nutrient-poor soil conditions and bypass the challenges of dealing with limited availability of phosphorus in the soil.

3.2 Mycorrhizal Contributions to Plant Resilience Against Abiotic Stresses

Abiotic stresses pose substantial challenges for growers to produce plants with optimum development and yield formation. As environmental pressures continue to intensify, it has become crucial to understand the mechanisms that could enhance plant resilience. In addition to supplying nutrients, mycorrhizae also enhance plant growth and development by strengthening the plant's immune system to resist biotic and abiotic stress more effectively. Drought stress is one of the most serious abiotic stresses that impaired plant morphology, biomass distribution, seed germination, nutrient availability, and photosynthesis process (Chen et al. 2020). The symbiotic relationship between AMF and plants plays a crucial role in drought adaptation and enhances plant resilience. Under water-stress conditions, plants benefit from the extensive extraradical hyphae network formed in the soil during AMF colonization. This hyphae provides a continuous supply of water and nutrients, especially phosphorus and nitrogen to the plant (Cheng et al. 2021). One mechanism for plants to cope with drought stress is through the formation of phytohormones known as abscisic acid (ABA).

Under water-stress conditions, this hormone will encourage stomata to remain closed for a longer period in order to reduce water loss from the cell. Interestingly, plants with AMF colonization have lower ABA levels in their roots and leaves compared to non-AMF plants. This mechanism will allow plants to have normal stomata function even under drought conditions (Diagne et al. 2020). It also demonstrated that AMF will induce the formation of jasmonic acid (JA), a hormone that is involved in the regulation of aquaporins, which plays an important function in water uptake (Fresno et al. 2023). By adjusting aquaporins' activity, plants can have better management of water uptake and distribution while overcoming drought stress. Besides drought, salinity stress or high salt levels also have a major impact on plants, such as osmotic stress, ion toxicity, nutrient imbalance, and photosynthesis reduction (Zong et al. 2023). Under high salinity conditions, AMF will selectively absorb more ions of potassium (K^+) while reducing the absorption of sodium (Na^+) from the soil. Through this mechanism, AMF help the plant to maintain a favorable K^+/Na^+ ratio in the cell and reduces the ion toxicity level as saline soils contain a high concentration of Na^+ in the rhizosphere (Klinsukon et al. 2021). Besides, AMF also help plants to maintain osmotic balance by improving water uptake, thus reducing the osmotic stress caused by high salt concentrations in the soil. Overall, AMF play a significant role in developing plants' resilience against abiotic stresses through various strategic mechanisms that have a positive impact on plant health and development (Table 1).

4 Mycorrhizal Fungi and Soil Conservation

Soil conservation is crucial to human beings due to the rising issues of food security to meet the needs of high food production for an increasing world population (Blanco and Lal 2008). Despite the extensive research and abundant literature on soil conservation strategies, little is known about the contribution of soil microorganisms, specifically the mycorrhizal fungi. The mycorrhizal fungi not only influence the cycling of nutrients in the soil system through organic matter decomposition but also modify various biochemical and biophysical mechanisms during the process. Thus, this chapter briefly discusses the contribution of mycorrhizal fungi to soil conservation, focusing on organic matter formation, soil aggregate stability, and soil carbon sequestration.

Table 1 AMF improves plant drought tolerance via water uptake and stomatal regulation, and salinity tolerance via ionic balance and osmotic adjustment to mitigate the effects of drought and salinity stress

Abiotic stress	AMF strategies	Benefits to plant
Drought	**Water uptake**: AMF extraradical hyphae extend the root's reach, increasing water absorption**Hormone regulation**: reduces abscisic acid (ABA) levels for normal stomata function; induces jasmonic acid (JA) for regulating aquaporins	Maintains water balance and conserves water Improves stomatal function under drought conditions
Salinity	**Ion regulation**: relectively absorbs potassium (K^+) over sodium (Na^+) to maintain a favorable K^+/Na^+ ratio **Osmotic balance**: enhances water uptake, reducing osmotic stress	Reduces ion toxicity and improves ionic balance Maintains cell function and reduces salt-induced damage

4.1 Mycorrhizal Role in Soil Organic Matter Formation

Soil organic matter holds a significant amount of the world's carbon and is crucial for sustaining productive soils and ensuring water quality (Lehmann and Kleber 2015). The compounds that contribute most significantly to stable soil organic matter pools are derived from microbes, formed during the decomposition and assimilation of plant exudates, litter, and roots (Cotrufo et al. 2013). The graphical illustration of mycorrhizal contribution toward organic matter formation is shown in Fig. 1.

Mycorrhizal fungi contribute to soil organic matter formation in several ways, including root exudation, fungal biomass, soil particle stabilization, and enhanced nutrient cycling (Phillips 2017). Root exudates refer to various organic compounds released by plants through their roots. It is a range of substances secreted by the roots of living plants, along with microbially modified products derived from these substances. They consist of low molecular weight organic compounds that are passively released from root cells, as well as mucilage associated with the roots (Catherine and Josep 2020). These can include sugars, amino acids, and organic acids. Mycorrhizal fungi colonize the plant roots and enhance this exudation. They absorb these root exudates and use them as a food source, and in return, they will provide the plants with improved access to nutrients (Wahab et al. 2023). On the other hand, the organic compounds released by mycorrhizal fungi and root exudates will stimulate the growth of beneficial soil microbes. These microbes, in turn, decompose organic matter in the soil, breaking it down into simpler compounds (Brimecombe et al. 2000). During the decomposition process of organic materials such as plant litter and decaying roots, the soil microbes will produce byproducts, which then will form humus (Condron et al. 2010). Humus is a stable form of organic matter that can improve the soil properties, including the soil structure, soil

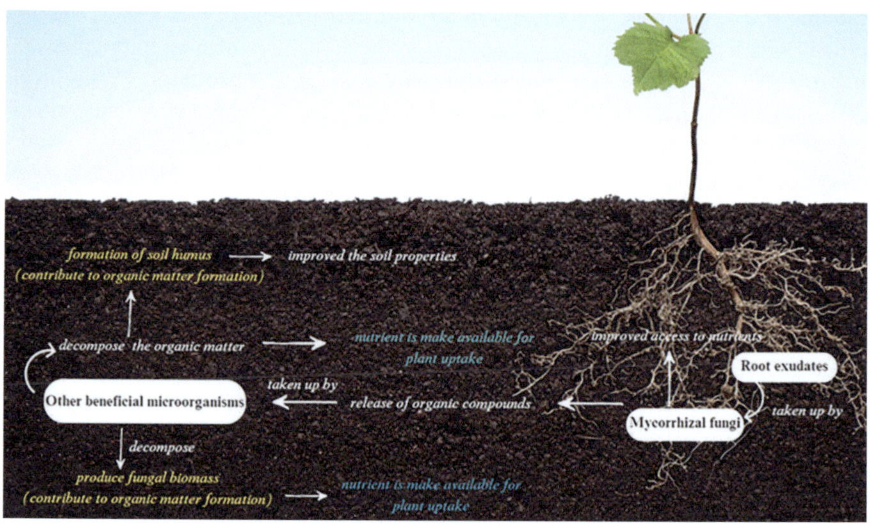

Fig. 1 Graphical illustration on mycorrhizal contribution toward organic matter formation

water-holding capacity, and enhance the availability of soil nutrients (Khatoon et al. 2017). Throughout the cycle, the soil microbial activities, which are supported by mycorrhizal fungi, will help to recycle the nutrients back into the soil and make these nutrients available for plant uptake. As such, it creates a dynamic nutrient cycle that enhances soil fertility and plant growth, which is vital in the soil conservation process (Prasad et al. 2021).

4.2 Impact of Mycorrhizal Fungi on Soil Aggregate Stability

Soil structure is one of the soil's physical properties that plays a vital role in the retention, transformation, and transfer efficiency of water, air, and nutrients in the soil ecosystem. The soil structure is influenced by the composition of soil aggregate and its stability (Li et al. 2022). One of the contributors to the development of protective vegetation cover is mycorrhizal fungi. Many findings have shown that mycorrhizal fungi can enhance plant growth and play a significant role in soil aggregate stability (Graf and Frei 2013). This is supported by the findings from Bearden and Petersen (2000), where their findings revealed that mycorrhizal fungi play a key role in vertisols, where a significant effect on soil aggregate stability is observed after just one growing season. The impact is suspected to be due to the role of both hyphae and the promotion of root growth by the mycorrhizal fungi. It has been demonstrated that the mycorrhizal fungi hyphae can bind soil microaggregates with a size of less than 250 μm in diameter into macroaggregates, which have a size of greater than 250 μm in diameter (Tisdall et al. 1997). However, Rillig et al. (2003) in their study revealed that glomalin has a significant direct impact, which is greater than mycorrhizal fungi hyphae alone. This showed that the glomalin-related soil protein (GRSP) plays an important role in the hypha-mediated mechanism for stabilizing soil aggregates, specifically in the 1–2 mm size class. It is known that the GRSP, which is the common glycoprotein produced by mycorrhizal fungi, plays an important role in ecosystem functioning and ecological restoration (Yang et al. 2017). In addition, the GRSP has also been found to be the main factor that contributes to soil aggregate stability in the rhizosphere (Wu et al. 2014).

4.3 Mycorrhizal Involvement in Soil Conservation and Carbon Sequestration

The soil organic carbon sequestration (SOC) includes the reduction of anthropogenic CO_2 emissions through the process of capturing and storing carbon in both biological and soil carbon pools (Olson et al. 2014). A symbiotic relationship between most terrestrial plants and various soil fungi from the phylum Glomeromycota is essential in transferring nutrients and enhancing soil carbon sequestration (Solaiman 2014). In this case, the mycorrhizal fungi act as an important link between the plant and the soil. Their main role is to move the carbon from the plants into the soil. Further, this will help to improve the carbon sequestration in soil (Francis and Read 1984). Apart from that, the mycorrhizal fungi transport the carbon away from the plant roots zone, where respiration is high, moving it into more stable soil aggregates (Frey 2019). A study by Thirkell et al. (2020) reveals that when the atmospheric CO_2 levels rise due to the plant's photosynthesis process, the plants will send more carbon to mycorrhizal fungi. This will further encourage the growth of those mycorrhizal fungi. In addition, the mycorrhizal fungi also contribute significantly to the addition of soil microbial biomass. Their structures account for approximately 15% of soil organic carbon and can make up between 20% and 80% of the soil biomass. The production of protein known as glomalin by the mycorrhizal fungi helps in protecting the organic matter from breaking down. Thus, this helps in maintaining soil structure by making the soil more stable and less prone to loose soil carbon. Further, it will effectively increase the soil carbon storage and sequestration (Bhattacharyya et al. 2022).

5 Mycorrhizal Nutrient Cycling and Networks

Under the soil's surface lies a long-time mutual relationship, which is the symbiosis between plants and mycorrhizal fungi. These microorganisms are active soil components but are also important facilitators of plant health, nutrient cycling, and to make ecosystem functioning. Through the evolutionary period, mycorrhizal fungi, for instance, AMF, have developed complex interactions with plants. It is reshaping how the plants access and utilize nutrients from its surroundings and environment. AMF colonize plant roots and extend their hyphae, which are microscopic filaments, into the surrounding soil, creating extensive networks. The extensive networks provide a functional extension of the plant root system (Çığ et al. 2023). These hyphal networks significantly increase the plant's ability to access important nutrients, for example, phosphorus and nitrogen. Both macronutrients are limited in the soil. Previous studies showed that AMF networks can extend to several meters away from the initial root zone. It is estimated that the enlargement of the area that a plant can absorb nutrients to grow (Rosliani et al. 2023). This mutual benefit consists of an exchange for both the plant and the fungi. The fungi benefit from sugars and other organic compounds produced by the plant through photosynthesis. The plant receives benefits from improved nutrient uptakes, growth, and more resilience toward climate change and environmental stress. Beyond the plant healthiness level, these fungal networks are also contributing to the stability of ecosystems through a way of promoting nutrient transport among adjacent plants. In addition, it also helps to improve the structure of the soil, which leads to the continuation of ecosystem stability.

5.1 Mycorrhizal Facilitates Nutrient Cycling in Ecosystems

The fungi's ability to enhance nutrient uptake is specifically pronounced in the case of phosphorus, nitrogen, and potassium. These are the essential macronutrients for plant growth. Under high salt conditions, for example, mycorrhizal inoculation has been proven to increase phosphorus concentration in plant leaves by up to 45% as compared to non-inoculated plants (Çığ et al. 2023). This significant finding of efficacy stems from the fungi's specialized structures called arbuscules. Arbuscules form within plant root cells and serve as the main site for nutrient translocation and transfer. The relationship involving the fungi and plants represents one of nature's most complicated productions through photosynthesis. Previous findings have shown that plants can distribute about 20% of their fixed carbon to their fungal companions or partners (Martin and van der Heijden 2024). The investment of carbon pays significant results where mycorrhizal fungi can improve nutrient efficiency by several hundred percent in comparison to non-mycorrhizal plants (Jo et al. 2019).

Beyond their role in nutrient exchange, mycorrhizal fungi serve as a main component of soil structure. Their hyphal networks produce a glycoprotein which can be called as glomalin. Glomalin acts as a natural glue, binding soil particles to each other. This process creates stable soil aggregates that enhance soil aeration and improve water retention (Bayona 2020). Research on coffee plantations has proven that areas with abundant mycorrhizal fungi comprise up to 30% of soil organic carbon content in comparison to areas without any presence of fungal (Szentpéteri et al. 2023). The influence of fungi on soil structure extends to their role in organic matter decomposition. Through their widespread networks, they help the degradation process of complex organic compounds. Then, releasing nutrients that remain locked in the plant or animal debris. The process or workflow is essentially important in the ecosystem of the forest. Mycorrhizal fungi can increase the speed of leaf and litter decomposition by up to 40% (Tedersoo et al. 2020).

The effectiveness of mycorrhizal fungi varies in its importance based on its environmental conditions, the surroundings, and soil properties. Soil that has nutrient depletion causes fungi to become more important for plant survival. Studies have shown that yield increases by 50% when inoculated with fungi (Gilbert et al. 2023). However, their benefits might not be affected in high-nutrient soil. This is because high-nutrient soil may provide nutrients to the plant directly. Modern agriculture practices usually inadvertently disrupt this crucial fungal spread of network. Intensive farming techniques such as heavy tillage and chemical fertilization can contribute to the reduction of fungal diversity by about 70% (Banerjee et al. 2019). This reduction caused negative implications for soil health level and crop yield production. This emphasizes the requirement for more sustainable agriculture that preserves these beneficial mutual relationships.

The role of mycorrhizal fungi in ecosystem resilience cannot be overstated. These fungi create underground networks that connect different plant species. This might facilitate nutrient sharing and improve overall ecosystem stability. Research has proven that plant communities with high mycorrhizal networks can withstand environmental stress better. For example, survival rates of mycorrhizal fungi are about 40% during drought and high water-stress conditions (Tedersoo et al. 2020; Eisenhauer et al. 2022). In the ecosystem of soil restoration works, mycorrhizal fungi have emerged as a highly beneficial partner. Recent studies have demonstrated that mycorrhizal fungi inoculation can increase plant establishment success rate by

60% during the remediation process (Reddy et al. 2024). This has shown effective effects of remediation or restoration for degraded landscapes and to maintain biodiversity.

5.2 Role of Mycorrhizal Fungi in Reducing Chemical Fertilizer Dependence

The agricultural sectors show high usage of chemical fertilizer, which has raised concerns regarding environmental sustainability, soil degradation, and long-term agricultural viability. AMF have been established as a promising biological alternative, demonstrating an essential role in reducing the high usage of chemical fertilizers, and at the same time, maintaining the yield production of crops. AMF significantly enhance plant nutrient acquisition through their extensive hyphal networks, which effectively increase the root surface area available for the absorption of nutrients. These widespread networks also enable plants to access soil macro- and micronutrients that would remain inaccessible through conventional root systems (Alsunuse et al. 2021). Studies have demonstrated substantial enhancements in phosphorus uptake efficiency in major crops, for example, maize and tomato, especially under phosphorus-depleted soil conditions (Chafai et al. 2023). Quantitative studies also revealed that AMF integration with chemical fertilizers can help to reduce by about 25% chemical fertilizer application without compromising crop yields. A comparative study analysis on tomato cultivation showed that higher yield was observed in mycorrhizal-inoculated plants as compared to other treatments with chemical fertilizer application (Torre et al. 2022). This reduction of chemical inputs not only provides less cost of production through decreased fertilizer costs but also helps to mitigate the environmental impacts caused by chemical fertilizers, especially under excessive application.

The contributions of mycorrhizal fungi extend beyond nutrient acquisition to basic improvements in soil functionality and structure. AMF hyphal networks help better soil aggregation, structure, stability, and formation (Hu et al. 2024). These structural modifications create optimal and suitable conditions for root development and nutrient retention. In addition, mycorrhizal presence also stimulates the other beneficial soil microorganisms activity. This contributes to the nutrient cycle and organic matter composition (Reddy et al. 2024). Mycorrhizal-colonized plants showed better crop resilience toward environmental stress, including drought and nutrient deficiencies. This capability has increased stress tolerance due to the improvement of root functionality and increased nutrient acquisition mechanisms (Jia et al. 2020).

The integration of mycorrhizal fungi into agricultural system showed a better strategy to reduce chemical fertilizer dependency. Mycorrhizal inoculation has also shown efficacy in organic farming systems, where chemical input restrictions require alternative fertility management approaches (Bencherif and Samia 2021). The application of AMF as organic and biofertilizers demonstrated the potential for enhancing crop yield production while maintaining soil health, and at the same time reducing the effect of environmental impact for a long time. Soil structure enhancements through mycorrhizal colonization have significantly contributed to the overall ecosystem functioning (Cifizzari et al. 2023). Colonization also enhanced soil fertility parameters such as cation exchange capacity and improved the tolerance of crop stress. Further research should prioritize the optimization of methodologies, investigation of plant–fungi interaction, and identification of optimal AMF strains for various crop species. The identification and advancement of effective AMF strains for specific crop and soil combinations remain crucial for maximizing the benefits of mycorrhizal interactions (Li et al. 2019a, b; Cifizzari et al. 2023).

5.3 Mycorrhizal Contributions to Nitrogen and Phosphorus Transfer in Plants

Mycorrhizal fungi, specifically AMF, play a significant role in enhancing the availability and transferring the essential nutrients such as nitrogen and phosphorus to plants. These symbiotic relationships between fungi and plant roots are important to plant nutrition and ecosystems health. Mycorrhizal fungi extend their hyphal networks into the soil, significantly increasing the surface that is accessible for nutrient absorption. This extension allows fungi to access soil nutrients that are frequently beyond the reach of plant roots. Research has shown that AMF can contribute from 30% to 50% of the nitrogen that is required by the host plants (Zhou et al. 2022; Beltayef et al. 2020). This contribution is specifically crucial in low-nutrient soils, where the availability of nitrogen is very low. The hyphal networks of mycorrhizal fungi not only enhance nitrogen uptake but also increase the breakdown of organic matter. This will then help to release nitrogen into the soil (Beltayef et al. 2020). In addition, mycorrhizal fungi are also well-known for their role in the acquisition of phosphorus. A study has indicated that 80% of the phosphorus absorbed by mycorrhizal plants is provided by the fungi (Akpodé et al. 2023). The mycorrhizal fungi can absorb the phosphorus in the form that is unavailable to plants, such as organic phosphorus compounds. The mycorrhizal fungi can absorb phosphorus in the forms that is unavailable to plants, such as organic phosphorus compounds. This capability is particularly important in soils, where phosphorus is bound to soil particles and iron (Fe). Thus, it makes it unavailable for plant uptake (Zhang 2023). The exudation of organic acids aided by mycorrhizal fungi further enhances the solubilization of phosphorus, which makes it more available to plants.

The enhanced nutrient uptake facilitated by mycorrhizal fungi leads to significant improvements in plant growth and its growth. Inoculation with mycorrhizal fungi has been proven to increase biomass and other related parameters to plant health, a fact referred to as the positive mycorrhizal growth response (MGR) (Zhang et al. 2022; Jiang et al. 2021). For example, studies have demonstrated that plants inoculated with AMF exhibited a widespread network of root systems, which improve the availability of water and other essential nutrients, including nitrogen and phosphorus (Rostami et al. 2024). This growth response is particularly beneficial in agricultural systems, where maximizing nutrient uptake is important for crop yield. By enhancing nutrient uptake, mycorrhizal fungi help plants maintain physiological functions under adverse conditions, thereby promoting productivity and increasing their survival rate (Salim 2023; Boutaskni et al. 2020).

6 Mycorrhizal Fungi and Drought Tolerance

Drought stress is a formidable environmental challenge that significantly impacts agricultural productivity and plant growth. The consequence is particularly severe during climate change, which is expected to increase the intensity and frequency of drought conditions. There is an urgency for sustainable agricultural practices that can bolster drought resilience in crops. Among the potential solutions, mycorrhizal fungi, especially AMF, have emerged as essential partners in promoting plant tolerance to drought (Raklami et al. 2019). These fungi establish mutual symbiosis relationships with the roots of most terrestrial plants, thereby enhancing their capacity to endure drought through various physiological and biochemical pathways.

6.1 Mycorrhizal Mechanisms for Enhancing Drought Tolerance in Plants

One of the main functions of mycorrhizal fungi in supporting drought resilience is their ability to enhance water uptake. The widespread or extensive networks of hyphae produced by AMF extend well beyond the root zone of plants. This is enabling them to tap into a larger soil volume for any water resources. This adaptation is specifically critical during dry seasons, as AMF can help to facilitate the transfer of water to the host plants. This might help to alleviate the stress associated with drought conditions (Kakouridis et al. 2022). Numerous studies have indicated that plants colonized by AMF possess significantly higher water content as compared to non-mycorrhizal plants during low water levels or drought periods (Oliveira et al. 2022; Sachsenmaier et al. 2024). This enhancement in water retention not only aids in maintaining turgor pressure within plant cells but also supports crucial physiological functions such as photosynthesis.

The ability of AMF to increase water uptake stems from their unique hyphal structures, which can infiltrate soil microenvironments that plant roots cannot reach. This expanded surface area significantly enhances the plants' capacity to absorb moisture, allowing them to forage effectively for water. During drought events, this ability to access deeper soil moisture can be critical for plant survival. Beyond enhancing water uptake, mycorrhizal fungi also play a pivotal role in nutrient acquisition, particularly with respect to nitrogen (N) and phosphorus (P)—two essential nutrients that support plant growth and resilience during drought. AMF assist in the absorption of these nutrients via their hyphal networks, which can access soil resources that are otherwise out of reach for plant roots (Behrooz et al. 2019; Rasouli et al. 2023). This increase in nutrient availability leads to enhanced plant growth and metabolic functions, enabling plants to better withstand the physiological challenges imposed by drought conditions (Metwally 2019). The contribution of AMF to nutrient uptake extends to micronutrients like zinc and copper, which are vital for various biochemical processes in plants. Improved nutrient accessibility translates into better root development and overall plant vigor, which can ultimately result in higher crop yields. Moreover, mycorrhizal fungi are known to promote osmotic adjustment in plants. They facilitate the accumulation of low molecular weight solutes, such as proline and sugars, which help stabilize cellular structures and maintain turgor during periods of drought (Behrooz et al. 2019; Wang et al. 2023). This osmotic adjustment is crucial for plant survival in water-limited environments, as it allows for the retention of water and the maintenance of essential metabolic functions.

The presence of mycorrhizal fungi also induces specific physiological and molecular adaptations in plants facing drought stress. For example, AMF colonization can enhance the expression of genes that are linked to drought tolerance, particularly those involved in ABA signaling pathways (Huang et al. 2021). ABA, a plant hormone, plays a significant role in regulating stomatal closure, which minimizes water loss through transpiration during drought episodes. Elevated ABA levels in mycorrhizal plants contribute to improved water management, allowing them to endure longer periods of drought (Huang et al. 2021). In addition, mycorrhizal fungi have been shown to boost the antioxidant capacity of plants, which helps mitigate oxidative stress caused by reactive oxygen species (ROS) during drought (Wang et al. 2023; Bahadur et al. 2019). This protective mechanism is essential for preserving cellular integrity and functionality under stressful conditions, further enhancing the resilience of mycorrhizal plants. The synergistic relationship between AMF and plant physiology extends to improved photosynthetic performance. Research has shown that AMF colonization can enhance chlorophyll content and increase photosynthetic rates, further supporting plant growth during drought conditions.

Incorporating mycorrhizal fungi into agricultural practices offers a sustainable strategy for enhancing drought resilience in crops. By fostering mycorrhizal relationships, farmers can improve soil health, increase nutrient and water uptake, and bolster crop resilience to drought stress (Cosme 2023). The application of mycorrhizal inoculants as biofertilizers is gaining momentum as a practical approach to boosting crop productivity, particularly in regions vulnerable to water shortages (Oliveira et al. 2022; Sachsenmaier et al. 2024). Utilizing mycorrhizal fungi in agricultural systems can lead to reduced dependency on chemical fertilizers and irrigation practices, promoting more sustainable farming methods. A healthy soil microbiome, supported by mycorrhizal associations, not only enhances drought resilience but also contributes to long-term soil health, benefiting crop production and environmental sustainability.

6.2 Comparison of Drought Tolerance Between Mycorrhizal and Non-mycorrhizal Plants

Mycorrhizal fungi play a crucial role in improving drought tolerance in plants, primarily by enhancing water uptake. AMF form symbiotic relationships with plant roots, extending their hyphal networks deep into the soil. This extension allows plants to access water beyond the reach of their roots alone, giving them a significant advantage under drought conditions. Research confirms that plants associated with AMF maintain higher water content compared to non-mycorrhizal plants during drought stress (Li et al. 2019a, b; Metwally 2019). For example, AMF-inoculated plants can extract water from deeper soil layers, which are typically inaccessible to non-mycorrhizal plants, helping them mitigate drought's negative impacts (Kebede et al. 2023; Cosme 2023). Beyond water uptake, mycorrhizal fungi enhance the absorption of essential nutrients, particularly nitrogen (N) and phosphorus (P), both critical for plant growth and stress resilience. AMF's hyphal networks tap into nutrient sources that roots alone cannot access, making these nutrients available even under water-limited conditions (Tao et al. 2022; Behrooz et al. 2019). This enhanced nutrient acquisition supports metabolic processes and promotes overall plant health, enabling better drought resistance. With improved nutrient status, mycorrhizal plants are more physiologically equipped to cope with drought stress, ensuring higher resilience and growth.

Plants colonized by AMF exhibit improved physiological responses to drought stress. For instance, AMF colonization increases the activity of antioxidant enzymes, which play a crucial role in minimizing oxidative stress caused by drought (Jabborova et al. 2021; Li et al. 2021). Drought conditions generate ROS that can damage plant cells, but the enhanced antioxidant capacity provided by AMF helps protect cellular structures, improving plant survival. Additionally, AMF influence gene expression related to drought tolerance. Studies indicate that AMF can regulate the expression of genes involved in the MAPK signaling pathway, which governs stress response mechanisms (Oliveira et al. 2022; Gao et al. 2022). These regulatory effects promote physiological adjustments, such as stomatal closure, which reduces water loss through transpiration and conserves moisture during drought (Rasouli et al. 2023). As a result, mycorrhizal plants can retain water and manage stress efficiently.

Numerous comparative studies have highlighted the advantages of mycorrhizal plants over non-mycorrhizal plants under drought stress. For instance, research on crops like wheat and soybean shows that AMF-inoculated plants exhibit better growth, higher chlorophyll content, and increased photosynthetic efficiency compared to non-mycorrhizal plants during drought conditions (Zhang et al. 2024). In contrast, non-mycorrhizal plants frequently show stunted growth, wilting, and reduced photosynthetic activity when exposed to similar water stress. Mycorrhizal fungi also improve root morphology, increasing root surface area and density. This structural adaptation enhances the plant's capacity to absorb water and nutrients, which is particularly beneficial under drought conditions (Cheng et al. 2021; Fernández et al. 2019). A larger and denser root network improves resource acquisition, ensuring that plants can continue functioning even with limited water availability.

Incorporating mycorrhizal fungi into agricultural practices offers significant benefits for sustainable farming, particularly in drought-prone regions. Mycorrhizal inoculation in crops can enhance drought resilience, reduce irrigation needs, and improve productivity under water-limited conditions (Wang et al. 2024). As climate change exacerbates water scarcity, leveraging mycorrhizal fungi will be critical for building resilient agricultural systems. Using AMF as biofertilizers can enhance soil health, promote sustainable crop production, and reduce reliance on chemical fertilizers. However, further research is needed to optimize inoculation techniques and identify the most effective AMF strains for specific crops and environmental conditions. Understanding plant–fungi interactions and tailoring AMF inoculation strategies will maximize the potential of mycorrhizal fungi in improving drought tolerance and supporting sustainable agriculture.

7 Mycorrhizal Fungi and Biodiversity Conservation

Mycorrhizal fungi, particularly AMF, play a crucial role in shaping plant community dynamics and supporting biodiversity conservation. These symbiotic fungi form intimate associations with the roots of many terrestrial plants, facilitating the uptake of nutrients and enhancing plant resilience to environmental stresses. This chapter examines the mechanisms through which mycorrhizal fungi influence plant community structure and function, with a focus on the implications for ecosystem health and biodiversity conservation.

7.1 Mycorrhizal Influence on Plant Community Dynamics

AMF contribute to plant community dynamics by improving nutrient acquisition and promoting plant health. Their hyphal networks extend far beyond the root zones of individual plants, significantly increasing the surface area available for nutrient absorption. Through these networks, AMF enhance the uptake of essential nutrients such as phosphorus and nitrogen—nutrients that are often limiting in natural ecosystems (Carteron et al. 2024; Fu et al. 2021). This improved nutrient availability not only benefits individual plant growth but also affects the competitive relationships between species. In competitive environments, mycorrhizal plants typically exhibit higher growth rates and greater biomass than non-mycorrhizal plants, often gaining a competitive edge (Yan et al. 2022; Sathiyadash et al. 2022). Additionally, AMF can influence the relative abundance and distribution of plant species within a community by selectively enhancing or suppressing plant performance. For instance, the presence of AMF may boost the growth of certain plant species, allowing them to outcompete others and shift the overall community composition (Chagnon et al. 2019; Nuske et al. 2019). Such selective impacts can promote plant diversity, as communities with a wider variety of species tend to be more stable and resilient to environmental fluctuations and disturbances.

The influence of mycorrhizal fungi extends beyond individual plants, impacting overall ecosystem processes and functioning. AMF associations contribute to soil health by improving soil structure, increasing water retention, and enhancing microbial diversity (Leopold et al. 2020; Mafa-Attoye et al. 2022). These improvements are critical for maintaining ecosystem stability, particularly in regions threatened by climate change, land-use changes, or other disturbances. The diversity of mycorrhizal fungi is closely linked to ecosystem productivity and nutrient cycling. Research has demonstrated that ecosystems with high mycorrhizal diversity exhibit more efficient carbon storage and nutrient cycling processes (Liang et al. 2019; Liang et al. 2022). For example, studies in subtropical forests reveal that diverse mycorrhizal networks are associated with higher biomass production and increased tree productivity (Rodriguez-Ramos et al. 2020; Kivlin 2020). These findings underscore the importance of conserving mycorrhizal diversity to maintain ecosystem productivity and support ecosystem services. Understanding the interactions between mycorrhizal fungi and plant communities is essential for developing effective biodiversity conservation strategies, especially in the context of climate change and habitat degradation. The ability of AMF to influence plant growth, community composition, and ecosystem function provides opportunities to enhance the resilience of both plants and ecosystems facing increasing environmental pressures (Dietrich et al. 2022; Barceló et al. 2019).

Conservation strategies that prioritize the preservation of mycorrhizal diversity can significantly improve the stability and resilience of ecosystems. For instance, restoring degraded ecosystems using AMF inoculants can promote the re-establishment of native plant species, improve soil structure, and enhance ecosystem functioning (Bennett and Classen 2020; Sathiyadash et al. 2022). Additionally, agricultural practices that foster mycorrhizal associations, such as reduced tillage, crop rotation, and organic farming, can improve both soil health and biodiversity, contributing to sustainable land management. Promoting these practices is especially crucial as ecosystems worldwide face increasing challenges from invasive species, land-use changes, and water scarcity. Ensuring that conservation and restoration efforts account for the role of mycorrhizal fungi will enhance the adaptive capacity of ecosystems, ultimately supporting long-term biodiversity conservation.

8 Mycorrhizal Fungi and Climate Change Mitigation

As the main source of basic foods, crop production is vulnerable to the climate change effects (Santos et al. 2023). A dramatic increase in atmospheric greenhouse gas concentrations has resulted from human activities, such as deforestation, animal farming, synthetic fertilizer usage, land use changes, waste disposal, and burning fossil fuels, which in turn threaten agricultural productivity and sustainability. Additionally, global warming may impact these essential crops' nutritional value (Filonchyk et al. 2024). According to one study, increased carbon dioxide (CO_2)

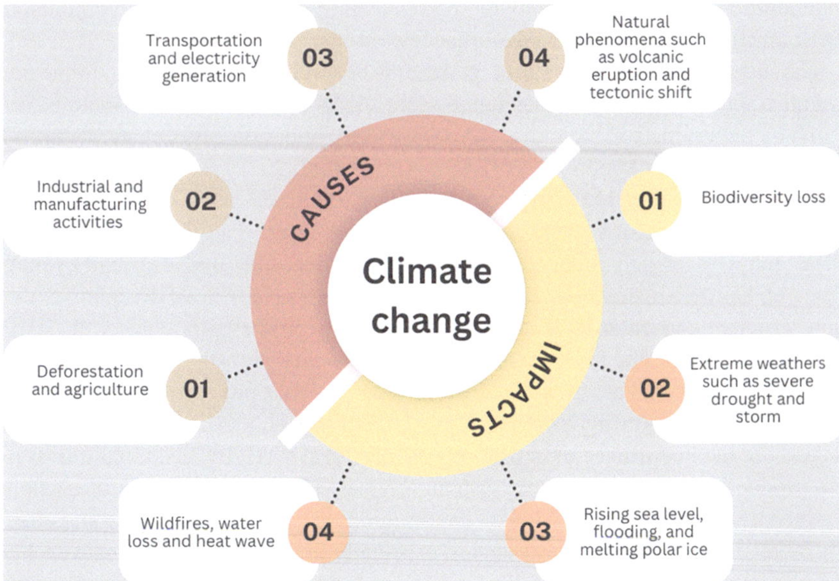

Fig. 2 Summary of causes and impacts of climate change

levels can dramatically reduce the concentrations of key mineral contents in crops such as wheat, rice, and soybeans by up to 8%. According to research, long-term climate change may result in significant losses in agricultural yields, notably for wheat, maize, rice, and soybeans (Wing et al. 2021). This emphasizes the demand for immediate action to limit the impact on agriculture worldwide, as climate change threatens both food security and the nutritional value of staple crops (Fig. 2).

8.1 Role of Mycorrhizal Symbiosis in Enhancing Climate Resilience

Mycorrhizal fungi contribute significantly to minimizing the negative impacts of climate change on agricultural production. The terrestrial carbon cycle and ecosystem resilience are significantly impacted by the mutualistic symbiosis between plants and mycorrhizal fungi, which dates back more than 400 million years (Wahab et al. 2023). In this interaction, mycorrhizal fungi accumulate vital soil nutrients like phosphorus and nitrogen, which plants exchange for carbon generated by photosynthetic activity. Currently, over 90% of terrestrial plants participate in this interaction, highlighting its ecological significance. According to a study, about 13 billion metric tons of CO_2 are transported from plants to mycorrhizal fungi each year, which accounts for 36% of the world's carbon emissions from fossil fuel burning (Hawkins et al. 2023).

Through the formation of huge networks of thin filaments called hyphae, mycorrhizal fungi play a critical role in mitigating climate change by enabling plants to absorb water and nutrients from a wider region beneath (Wahab et al. 2023). In addition to influencing soil ecosystems and promoting plant development, this living network serves as a major carbon storage, preserving the global carbon balance. By forming organic "glues," such as extracellular polysaccharides and glycoproteins like glomalin, the fungi improve the structure and porosity of the soil by binding soil particles into aggregates (Mummey et al. 2009; Agnihotri et al. 2022). In addition to preventing erosion, this increase in soil stability allows the soil to retain more organic carbon, which is crucial for lowering atmospheric CO_2 levels. Furthermore, soils with proper structure retain moisture better, which promotes plant development and increases the absorption of carbon through photosynthesis. Additionally, mycorrhizal fungi promote long-term carbon storage in biomass and soil organic material, improve plant health, and aid in the cycling of nutrients. Collectively, these mechanisms promote soil health and resilience, highlighting the crucial role mycorrhizal fungi play in mitigating climate change (Hawkins et al. 2023).

In addition to carbon sequestration, mycorrhizal fungi play an important role in increasing climate resilience in terrestrial ecosystems. Due to climate change, environmental stresses, including drought, soil salinity, and nutrient depletion, are becoming more common. These fungi help plants cope with these challenges (Wahab et al. 2023). Trees are generally credited with absorbing CO_2, although they do not do it alone. Forests feature a diverse microbiome of microorganisms that interact closely with the plants (Baldrian and Kohout 2023). All trees help absorb CO_2 and slow down global warming, but some are far more efficient than others. According to a recent study, identifying the mycorrhizal fungi that are present in a forest's microbiome is crucial for assessing their effectiveness as carbon sequesters. A variety of root fungi may develop associations with trees, but ectomycorrhizal fungi in particular have been shown to greatly increase plants' absorption of CO_2 (Xie et al. 2024). Ectomycorrhizal fungi not only facilitate the uptake of CO_2, but they also slow down the breakdown of organic matter, which normally releases carbon deposited in the soil back into the atmosphere. These fungi assist in keeping carbon trapped in trees and soils, preventing its release into the atmosphere and enhancing the forest's potential to operate as a carbon sink (Shao et al. 2023).

8.2 Mycorrhizal Fungi as a Tool for Mitigating the Impacts of Climate Change

Mycorrhizal fungi are essential for carbon sequestration and climatic resilience, making them a useful tool for reducing the effects of climate change (Hawkins et al. 2023). Applying mycorrhizal usage in agriculture and forestry provides a natural option to increase carbon storage and improve ecosystem stability. Mycorrhizal fungi, for example, can improve soil quality, encourage healthier plant development, and lessen the need for artificial fertilizers in agricultural techniques (Vassilev and Mendes 2024). This approach helps the Earth store more carbon while also increasing crop yields. Choosing plant species with robust mycorrhizal relationships in afforestation and reforestation programs contributes to the development of resilient ecosystems that can tolerate climatic stresses. These activities, which nurture mycorrhizal networks, can boost soil carbon storage and promote biodiversity, both of which are crucial for ecosystem resilience (Usman et al. 2021). Furthermore, the inoculation of mycorrhizal fungi into damaged areas can revive soil health and boost its carbon sequestration capability (Dal Cortivo et al. 2018). In areas where deforestation, erosion, or unsustainable farming methods are prevalent, these strategies are very beneficial.

In conclusion, mycorrhizal fungi play an important role in mitigating climate change. They are essential parts of strategies for climate mitigation and adaptation because they improve carbon storage and increase plant resilience. Carbon sequestration and vital ecosystem services like soil stability, nitrogen cycling, and water retention can be enhanced by promoting mycorrhizal interactions through sustainable land management techniques. These advantages are important in developing resilient agricultural systems that can thrive amidst changing climatic conditions.

9 Conclusion

As a conclusion, mycorrhizal symbiosis is very important in many aspects. Their roles in promoting environmental and ecological health are very crucial. By presenting the mycorrhizal symbiosis, the soil structure, nutrient uptake, and plant growth can be improved. These beneficial fungi establish a mutual relationship with the roots of plant, especially terrestrial plants by enhancing the nutrient exchange, such as phosphorus, and augmenting the soil water retention. Furthermore, the mycorrhizal networks can improve soil microbial diversity and activity that can promote a balanced ecosystem. Besides that, mycorrhizal fungi can protect the plant in terms of acting as a plant resilience against abiotic stress. A healthy and robust ecosystem is also facilitated by the mycorrhizal network, which is essential for increasing soil microbial diversity. Mycorrhizal organisms can also improve soil carbon sequestration. Mycorrhizal networks aid in lowering atmospheric CO_2 levels by promoting carbon storage. Additionally, these networks enhance plant resistance to abiotic conditions like drought, nutrient shortage, and high heat, in addition to supporting a range of advantageous bacteria. Mycorrhizal relationships contribute to the development of a more stable and healthier growing environment for plants by promoting nutrient exchange and interaction between plants and soil bacteria. The improvement of ecosystem functions and promotion of the general health of the environment by cultivating this relationship can be done. Thus, it is crucial for us to preserve mycorrhizal organisms to build a healthy ecosystem.

The future insights into the direction for research on mycorrhizal fungi and their applications. The focus on the molecular mechanism underlying mycorrhizal interactions, their potential of these fungi in bioremediation, and their applications in sustainable agriculture, especially in arid and nutrient-poor soils, needs to be more serious in order to sustain land management and food security for long-term purposes. The understanding of this variability in mycorrhizal partnership across different ecosystems will be crucial in leveraging these fungi for ecological restoration and sustainable land management practice. Additionally, exploring the impact of climate change on mycorrhizal functions and their adaptations could reveal new strategies for enhancing plant resilience and ecosystem services in changing environments. For instance, the molecular tool of model organisms can be used to forecast how mycorrhizae would react to environmental changes and to identify the genes that are crucial to our comprehension of the distribution and roles of mycorrhizae over time and place. However, the limited number of fungal and plant genomes currently available hinders mycorrhizae ecologists who are interested in employing genomic approaches. However, when more plant genome sequences become accessible, these technologies can be used more frequently in ecological research. Furthermore, new genetic engineering methods like Cas9/CRISPR will make it possible to clarify the role and ecological significance of one or more genes in mycorrhizal function in species that are not models. Utilizing this genetic method in conjunction with biological information will be essential to comprehending mycorrhizal function in the environment. Major advances have been made in the field of mycorrhizal research. It all started with the discovery that mycorrhizal associations are abundant and important for plant nutrition. Now, the ecological function of this symbiosis seems to be much better accepted and well understood. There are no longer black box mysteries after more than a century. This symbiosis's evolution and biodiversity are widely dispersed and up-to-date. Numerous mycorrhizal fungi have had their genomes sequenced, and the molecular interactions that underpin the symbiosis are beginning to be revealed.

References

Abdalla M, Bitterlich M, Jansa J, Püschel D, Ahmed MA (2023) The role of arbuscular mycorrhizal symbiosis in improving plant water status under drought. J Exp Bot 74(16):4808–4824. https://doi.org/10.1093/jxb/erad249

Agnihotri R, Sharma MP, Prakash A, Ramesh A, Bhattacharjya S, Patra AK, Manna MC, Kurganova I, Kuzyakov Y (2022) Glycoproteins of arbuscular mycorrhiza for soil carbon sequestration: review of mechanisms and controls. Sci Total Environ 806:150571. https://doi.org/10.1016/j.scitotenv.2021.150571

Akpodé C, Assogba SA, Hoteyi SM, Aguégué MR, Agbodjato NA, Adoko MY, Amogou O, Adjanohoun A, Baba-Moussa L (2023) Use of the biostimulant based on the mycorrhizae consortium of the glomeraceae family in the field to improve the production and nutritional status of maize (*Zea mays* L.) plants in Central Benin. Adv Microbiol 13(6):323–345. https://doi.org/10.4236/aim.2023.136021

Alsunuse BT, AL-Ani MAM, Faituri M, Ashilenje DS, Alawami AA, Stahl PD (2021) Effects of arbuscular mycorrhizal fungi on growth and phosphorus uptake of maize (*Zea mays* L.) at different levels of soil phosphorus and soil moisture. J Dryland Agric 7(3):22–33. https://doi.org/10.5897/joda2020.0063

Andrino A, Guggenberger G, Kernchen S, Mikutta R, Sauheitl L, Boy J (2021) Production of organic acids by arbuscular mycorrhizal fungi and their contribution in the mobilization of phosphorus bound to iron oxides. Front Plant Sci 12:661842

Azcón-Aguilar C, Jaizme-Vega MC, Calvet C (2002) The contribution of arbuscular mycorrhizal fungi to the control of soil-borne plant pathogens. In: Mycorrhizal technology in agriculture: From genes to bioproducts, pp 187–197

Bagyaraj DJ (2014) Mycorrhizal fungi. Proc Indian Natl Sci Acad 80(2):415–428

Bahadur A, Batool A, Nasir F, Jiang S, Qin M, Zhang Q, Pan J, Liu Y, Feng H (2019) Mechanistic insights into arbuscular mycorrhizal fungi-mediated drought stress tolerance in plants. Int J Mol Sci 20(17):4199. https://doi.org/10.3390/ijms20174199

Baldrian P, Kohout P (2023) Forest microbiome and global change. Nat Rev Microbiol 21(8):487–501. https://doi.org/10.1038/s41579-023-00876-4

Banerjee S, Walder F, Büchi L, Meyer M, Held AY, Gattinger A, Keller T, Charles R, van der Heijden MG (2019) Agricultural intensification reduces microbial network complexity and the abundance of keystone taxa in roots. ISME J 13(7):1722–1736. https://doi.org/10.1038/s41396-019-0383-2

Barceló M, Bodegom PMV, Soudzilovskaia NA (2019) Climate drives the spatial distribution of mycorrhizal host plants in terrestrial ecosystems. J Ecol 107(6):2564–2573. https://doi.org/10.1111/1365-2745.13275

Bayona CPA (2020) Soil quality indicators associated with the application of mycorrhizal fungi in coffee plantations. Ingeniería Solidaria 16(3). https://doi.org/10.16925/2357-6014.2020.03.05

Bearden BN, Petersen L (2000) Influence of arbuscular mycorrhizal fungi on soil structure and aggregate stability of a vertisol. Plant Soil 218(1):173–183

Behrooz A, Vahdati K, Rejali F, Lotfi M, Sarikhani S, Leslie CA (2019) Arbuscular mycorrhiza and plant growth-promoting bacteria alleviate drought stress in walnut. HortScience 54(6):1087–1092. https://doi.org/10.21273/hortsci13961-19

Beltayef H, Melki M, Saidi W, Hajri R, Cruz C, Muscolo A, ben Youness M (2020) Potential *Piriformospora indica* effect on growth and mineral nutrition of *Phaseolus vulgaris* crop under low phosphorus intake. J Plant Nutr 44(4):498–507. https://doi.org/10.1080/01904167.2020.1845366

Bencherif K, Samia T (2021) Native arbuscular mycorrhizal fungi and agro-industries in arid lands: productions, applications strategies and challenges, Mycorrhizal Fungi – Utilization in Agriculture and Forestry, vol 28, p 77. https://doi.org/10.5772/intechopen.94084

Bennett AE, Classen AT (2020) Climate change influences mycorrhizal fungal–plant interactions, but conclusions are limited by geographical study bias. Ecology 101(4). https://doi.org/10.1002/ecy.2978

Berg G, Grosch R, Scherwinski K (2007) Risk assessment for microbial antagonists: are there effects on non-target organisms. Gesunde Pflanzen 59:107–117

Bhantana P, Rana MS, Sun XC, Moussa MG, Saleem MH, Syaifudin M, Shah A, Poudel A, Pun AB, Bhat MA, Mandal DL (2021) Arbuscular mycorrhizal fungi and its major role in plant growth, zinc nutrition, phosphorous regulation and phytoremediation. Symbiosis 84:19–37

Bhattacharyya SS, Ros GH, Furtak K, Iqbal HM, Parra-Saldívar R (2022) Soil carbon sequestration–an interplay between soil microbial community and soil organic matter dynamics. Sci Total Environ 815:152928

Blanco H, Lal R (2008) Principles of soil conservation and management. Springer, New York

Boutaskni A, Baslam M, Ait-El-Mokhtar M, Anli M, Ben-Laouane R, Douira A, El MC, Mitsui T, Wahbi S, Meddich A (2020) Arbuscular mycorrhizal fungi mediate drought tolerance and recovery in two contrasting carob (*Ceratonia siliqua* L.) ecotypes by regulating stomatal, water relations, and (in)organic adjustments. Plants 9(1):80. https://doi.org/10.3390/plants9010080

Brimecombe MJ, De Leij FA, Lynch JM (2000) The effect of root exudates on rhizosphere microbial populations. In: The rhizosphere. CRC Press, pp 111–156

Cardon ZG, Whitbeck JL (2007) The rhizosphere. Elsevier Academic Press, p 235

Carteron A, Cantera I, Guerrieri A, Marta S, Bonin A, Ambrosini R, Anthelme F, Azzoni RS, Almond P, Alviz Gazitúa P, Ficetola GF (2024) Dynamics and drivers of mycorrhizal fungi after glacier retreat. New Phytol 242(4):1739–1752. https://doi.org/10.1111/nph.19682

Catherine P, Josep P (2020) A return to the wild: root exudates and food security. Trends Plant Sci 25(1):14–21. https://doi.org/10.1016/j.tplants.2019.09.010

Chafai W, Haddioui K, Serghini-Caid H, Labazi H, AlZain M, Noman O, Parvez MK, Addi M, Khalid A (2023) Impact of arbuscular mycorrhizal fungal strains isolated from soil on the growth, yield, and fruit quality of tomato plants under different fertilization regimens. Horticulturae 9(9):973. https://doi.org/10.3390/horticulturae9090973

Chagnon P, Bradley RL, Klironomos JN (2019) Mycorrhizal network assembly in a community context: the presence of neighbours matters. J Ecol 108(1):366–377. https://doi.org/10.1111/1365-2745.13230

Chen W, Meng P, Feng H, Wang C (2020) Effects of arbuscular mycorrhizal fungi on growth and physiological performance of *Catalpa bungei* CA Mey. Under drought stress. Forests 11(10):1117

Cheng S, Zou YN, Kuča K, Hashem A, Abd Allah EF, Wu QS (2021) Elucidating the mechanisms underlying enhanced drought tolerance in plants mediated by arbuscular mycorrhizal fungi. Front Microbiol 12:809473

Cifizzari KP, Moyer M, Cheeke TE (2023) Growth responses of potted vitis vinifera cultivars differ to a mycorrhizal inoculant and phosphorus fertilizer. HortScience 58(6):643–650. https://doi.org/10.21273/hortsci17114-23

Çığ A, Gülser F, Gülser E (2023) The responses of nutrient uptakes in different organs of *Narcissus tazetta* (l.) grown under saline conditions to mycorrhizal inoculation. Pol J Environ Stud 33(2):1631–1641. https://doi.org/10.15244/pjoes/173989

Condron L, Stark C, O'Callaghan M, Clinton P, Huang Z (2010) The role of microbial communities in the formation and decomposition of soil organic matter. In: Dixon G, Tilston E (eds) Soil microbiology and sustainable crop production. Springer, Dordrecht. https://doi.org/10.1007/978-90-481-9479-7_4

Cosme M (2023) Mycorrhizas drive the evolution of plant adaptation to drought. Commun Biol 6(1). https://doi.org/10.1038/s42003-023-04722-4

Cotrufo MF, Wallenstein MD, Boot CM, Denef K, Paul E (2013) The microbial efficiency-matrix stabilization (MEMS) framework integrates plant litter decomposition with soil organic matter stabilization: do labile plant inputs form stable soil organic matter? Glob Chang Biol 19(4):988–995

Dal Cortivo C, Barion G, Ferrari M, Visioli G, Dramis L, Panozzo A, Vamerali T (2018) Effects of field inoculation with VAM and bacteria consortia on root growth and nutrients uptake in common wheat. Sustainability 10:3286. https://doi.org/10.3390/su10093286

Di Martino C, Crawford TW (2021) Roles and implications of arbuscular mycorrhizas in plant nutrition. In: Handbook of plant and crop physiology. CRC Press, pp 321–341

Diagne N, Ngom M, Djighaly PI, Fall D, Hocher V, Svistoonoff S (2020) Roles of arbuscular mycorrhizal fungi on plant growth and performance: importance in biotic and abiotic stressed regulation. Diversity 12(10):370

Dietrich P, Ferlian O, Huang Y, Luo S, Quosh J, Eisenhauer N (2022) Tree diversity effects on productivity depend on mycorrhizae and life strategies in a temperate forest experiment. Ecology 104(2). https://doi.org/10.1002/ecy.3896

Eisenhauer N, Bonfante P, Buscot F, Cesarz S, Guerra C, Heintz-Buschart A, Hines J, Patoine G, Rillig M, Schmid B, Verheyen K (2022) Biotic interactions as mediators of context-dependent biodiversity-ecosystem functioning relationships. Res Ideas Outcomes 8:1–38. https://doi.org/10.3897/rio.8.e85873

Fall AF, Nakabonge G, Ssekandi J, Founoune-Mboup H, Apori SO, Ndiaye A, Badji A, Ngom K (2022) Roles of arbuscular mycorrhizal fungi on soil fertility: contribution in the improvement of physical, chemical, and biological properties of the soil. Front Fungal Biol 3:1–11

Fernández I, Cosme M, Stringlis IA, Yu K, De Jonge R, Van Wees SM, Pozo MJ, Pieterse CM, Van Der Heijden MG (2019) Molecular dialogue between arbuscular mycorrhizal fungi and the nonhost plant *Arabidopsis thaliana* switches from initial detection to antagonism. New Phytol 223(2):867–881. https://doi.org/10.1111/nph.15798

Ferrol N, Azcón-Aguilar C, Pérez-Tienda J (2019) Arbuscular mycorrhizas as key players in sustainable plant phosphorus acquisition: an overview on the mechanisms involved. Plant Sci 280:441–447

Figueiredo AF, Boy J, Guggenberger G (2021) Common mycorrhizae network: a review of the theories and mechanisms behind underground interactions. Front Fungal Biol 2:735299. https://doi.org/10.3389/ffunb.2021.735299

Filonchyk M, Peterson MP, Zhang L, Hurynovich V, He Y (2024) Greenhouse gases emissions and global climate change: examining the influence of CO_2, CH_4, and N_2O. Sci Total Environ 935:173359. https://doi.org/10.1016/j.scitotenv.2024.173359

Francis R, Read DJ (1984) Direct transfer of carbon between plants connected by vesicular–arbuscular mycorrhizal mycelium. Nature 307(5946):53–56

Fresno DH, Solé-Corbatón H, Munné-Bosch S (2023) Water stress protection by the arbuscular mycorrhizal fungus *Rhizoglomus irregulare* involves physiological and hormonal responses in an organ-specific manner. Physiol Plant 175(1):e13854

Frey SD (2019) Mycorrhizal fungi as mediators of soil organic matter dynamics. Annu Rev Ecol Evol Syst 50(1):237–259

Fu W, Chen B, Rillig MC, Jansa J, Ma W, Xu C, Luo W, Wu H, Hao Z, Wu H, Zhao A (2021) Community response of arbuscular mycorrhizal fungi to extreme drought in a cold-temperate grassland. New Phytol 234(6):2003–2017. https://doi.org/10.1111/nph.17692

Gao C, Xu L, Montoya L, Madera M, Hollingsworth J, Chen L, Purdom E, Singan V, Vogel J, Hutmacher RB, Dahlberg JA (2022) Co-occurrence networks reveal more complexity than community composition in resistance and resilience of microbial communities. Nat Commun 13(1):1–12. https://doi.org/10.1038/s41467-022-31343-y

Genre A, Lanfranco L, Perotto S, Bonfante P (2020) Unique and common traits in mycorrhizal symbiosis. Nat Rev Microbiol 18:649–660. https://doi.org/10.1038/s41579-020-0402-3

Gilbert L, Jeanniard-du-Dot T, Authier M, Chouvelon T, Spitz J (2023) Composition of cetacean communities worldwide shapes their contribution to ocean nutrient cycling. Nat Commun 14(1). https://doi.org/10.1038/s41467-023-41532-y

Graf F, Frei M (2013) Soil aggregate stability related to soil density, root length, and mycorrhiza using site-specific *Alnus incana* and *Melanogaster variegatus* sl. Ecol Eng 57:314–323

Harrier LA, Watson CA (2004) The potential role of Arbuscular Mycorrhizal (AM) fungi in the bioprotection of plants against soil-borne pathogens in organic and/or other sustainable farming systems. Pest Manag Sci 60:149–157

Hawkins HJ, Cargill RI, Van Nuland ME, Hagen SC, Field KJ, Sheldrake M, Soudzilovskaia NA, Kiers ET (2023) Mycorrhizal mycelium as a global carbon pool. Curr Biol 33(11):R560–R573. https://doi.org/10.1016/j.cub.2023.02.027

Hu C, Xu T, Wang S, Bian H, Dai H (2024) Effect of aminating lignin loading with arbuscular mycorrhizal fungi on soil aggregate structure improvement. Polymers 16(12):1701. https://doi.org/10.3390/polym16121701

Huang D, Wang Q, Zhang Z, Jing G, Ma M, Ma F, Li C (2021) Silencing mdgh3-2/12 in apple reduces drought resistance by regulating am colonization. Hortic Res 8(1). https://doi.org/10.1038/s41438-021-00524-z

Ivanov S, Austin J, Berg RH, Harrison MJ (2019) Extensive membrane systems at the host–arbuscular mycorrhizal fungus interface. Nat Plants 5(2):194–203

Jabborova D, Annapurna K, Al-Sadi AM, Alharbi SA, Datta R, Zuan ATK (2021) Biochar and arbuscular mycorrhizal fungi mediated enhanced drought tolerance in okra (*Abelmoschus esculentus*) plant growth, root morphological traits and physiological properties. Saudi J Biol Sci 28(10):5490–5499. https://doi.org/10.1016/j.sjbs.2021.08.016

Jia Y, van der Heijden MG, Wagg C, Feng G, Walder F (2020) Symbiotic soil fungi enhance resistance and resilience of an experimental grassland to drought and nitrogen deposition. J Ecol 109(9):3171–3181. https://doi.org/10.1111/1365-2745.13521

Jiang D, Tan M, Wu S, Zheng L, Wang Q, Wang G, Yan S (2021) Defense responses of arbuscular mycorrhizal fungus-colonized poplar seedlings against gypsy moth larvae: a multiomics study. Hortic Res 8(1). https://doi.org/10.1038/s41438-021-00671-3

Jo I, Fei S, Oswalt CM, Domke GM, Phillips RP (2019) Shifts in dominant tree mycorrhizal associations in response to anthropogenic impacts. Sci Adv 5(4). https://doi.org/10.1126/sciadv.aav6358

Kakouridis A, Hagen JA, Kan MP, Mambelli S, Feldman LJ, Herman DJ, Weber PK, Pett-Ridge J, Firestone MK (2022) Routes to roots: direct evidence of water transport by arbuscular mycorrhizal fungi to host plants. New Phytol 236(1):210–221. https://doi.org/10.1111/nph.18281

Kebede TG, Birhane E, Kiros-Meles A, Egziabher YG (2023) Arbuscular mycorrhizal fungi improve biomass, photosynthesis, and water use efficiency of *Opuntia ficus-indica* (L.) Miller under different water levels. J Arid Land 15(8):975–988. https://doi.org/10.21203/rs.3.rs-3985792/v1

Khatoon H, Solanki P, Narayan M, Tewari L, Rai J, Hina Khatoon C (2017) Role of microbes in organic carbon decomposition and maintenance of soil ecosystem. Int J Chem Studies 5(6):1648–1656

Kivlin SN (2020) Global mycorrhizal fungal range sizes vary within and among mycorrhizal guilds but are not correlated with dispersal traits. J Biogeogr 47(9):1994–2001. https://doi.org/10.1111/jbi.13866

Klinsukon C, Lumyong S, Kuyper TW, Boonlue S (2021) Colonization by arbuscular mycorrhizal fungi improves salinity tolerance of eucalyptus (*Eucalyptus camaldulensis*) seedlings. Sci Rep 11(1):4362

Lambers H, Raven JA, Shaver GR, Smith SE (2008) Plant nutrient acquisition strategies change with soil age. Trends Ecol Evol 23:95–103. https://doi.org/10.1016/j.tree.2007.10.008

Ledford WC, Silvestri A, Fiorilli V, Roth R, Rubio-Somoza I, Lanfranco L (2023) A journey into the world of small RNAs in the arbuscular mycorrhizal symbiosis. New Phytol 242(4):1534–1544. https://doi.org/10.1111/nph.19394

Lehmann J, Kleber M (2015) The contentious nature of soil organic matter. Nature 528(7580):60–68. https://doi.org/10.1038/nature16069

Leopold DR, Peay KG, Vitousek PM, Fukami T (2020) Ericoid mycorrhizal diversity increases with soil age and progressive phosphorus limitation across a 4.1 million-year chronosequence. BioRxiv 2020-08. https://doi.org/10.1101/2020.08.27.270413

Li J, Meng B, Chai H, Yang X, Song W, Li S, Lu A, Zhang T, Sun W (2019a) Arbuscular mycorrhizal fungi alleviate drought stress in C3 (*Leymus chinensis*) and c4 (*Lemarthria altissima*) grasses via altering antioxidant enzyme activities and photosynthesis. Front Plant Sci 10:499. https://doi.org/10.3389/fpls.2019.00499

Li Z, Ngwene B, Hong T, George E (2019b) Effects of nitrogen feeding for extraradical mycelium of rhizophagus irregularis maize symbiosis incorporated with phosphorus availability. J Plant Nutr Soil Sci 182(4):647–655. https://doi.org/10.1002/jpln.201800674

Li M, Hou L, Liu J, Yang J, Zuo Y, Zhao L, He X (2021) Growth-promoting effects of dark septate endophytes on the non-mycorrhizal plant *Isatis indigotica* under different water conditions. Symbiosis 85(3):291–303. https://doi.org/10.1007/s13199-021-00813-0

Li Y, Xu J, Hu J, Zhang T, Wu X, Yang Y (2022) Arbuscular mycorrhizal fungi and glomalin play a crucial role in soil aggregate stability in Pb-contaminated soil. Int J Env Res Public Health 19(9):5029

Liang M, Liu X, Parker IM, Johnson D, Zheng Y, Luo S, Gilbert GS, Yu S (2019) Soil microbes drive phylogenetic diversity-productivity relationships in a subtropical forest. Sci Adv 5(10). https://doi.org/10.1126/sciadv.aax5088

Liang J, Zou R, Huang Y, Qin H, Tang J, Wei X, Liang Y, Chai S (2022) Structure and diversity of mycorrhizal fungi communities of different part of *Bulbophyllum tianguii* in three terrestrial environments. Front Plant Sci 13:992184. https://doi.org/10.3389/fpls.2022.992184

Linderman RG (1994) Role of VAM fungi in biocontrol. Mycorrhizae and plant health

Ma J, Wang W, Yang J, Qin S, Yang Y, Sun C, Pei G, Zeeshan M, Liao H, Liu L, Huang J (2022) Mycorrhizal symbiosis promotes the nutrient content accumulation and effects the root exudates in maize. BMC Plant Biol 22:64. https://doi.org/10.1186/s12870-021-03370-2

Mafa-Attoye T, Borden KA, Obregón D, Thevathasan NV, Isaac ME, Dunfield KE (2022) Roots alter soil microbial diversity and interkingdom interactions in diversified agricultural landscapes. Oikos 2023(1). https://doi.org/10.1111/oik.08717

Martin FM, van der Heijden MGA (2024) The mycorrhizal symbiosis: research frontiers in genomics, ecology, and agricultural application. New Phytol 242(4):1486–1506. https://doi.org/10.1111/nph.19541

Metwally A (2019) Arbuscular mycorrhizal symbiosis alleviates drought stress imposed on wheat plants (*Triticum aestivum* L.). Appl Ecol Environ Res 17(6):13713–13727. https://doi.org/10.15666/aeer/1706_137131372

Mohammadi K, Khalesro S, Sohrabi Y, Heidari G (2011) A review: beneficial effects of the mycorrhizal fungi for plant growth. J Appl Environ Biol Sci 1(9):310–319

Mummey DL, Antunes PM, Rillig MC (2009) Arbuscular mycorrhizal fungi pre-inoculant identity determines community composition in roots. Soil Biol Biochem 41(6):1173–1179. https://doi.org/10.1016/j.soilbio.2009.02.027

Nuske S, Anslan S, Tedersoo L, Congdon BC, Abell-Davis SE (2019) Ectomycorrhizal fungal communities are dominated by mammalian dispersed truffle-like taxa in north-east australian woodlands. Mycorrhiza 29(3):181–193. https://doi.org/10.1007/s00572-019-00886-2

Oliveira TC, Cabral JS, Santana LR, Tavares GG, Santos LD, Paim TP, Müller C, Silva FG, Costa AC, Souchie EL, Mendes GC (2022) The arbuscular mycorrhizal fungus *Rhizophagus clarus* improves physiological tolerance to drought stress in soybean plants. Sci Rep 12(1):9044. https://doi.org/10.1038/s41598-022-13059-7

Olson KR, Al-Kaisi M, Lal R, Lowery B (2014) Examining the paired comparison method approach for determining soil organic carbon sequestration rates. J Soil Water Conserv 69(6):193A–197A

Peterson RL, Massicotte HB, Melville LH (2004) Mycorrhizas: anatomy and cell biology. NCR Research Press, Ottawa

Phillips M (2017) Mycorrhizal planet: how symbiotic fungi work with roots to support plant health and build soil fertility. Chelsea Green Publishing

Prasad S, Malav LC, Choudhary J, Kannojiya S, Kundu M, Kumar S, Yadav AN (2021) Soil microbiomes for healthy nutrient recycling. In: Current trends in microbial biotechnology for sustainable agriculture, pp 1–21

Raklami A, Bechtaoui N, Tahiri A, Anli M, Meddich A, Oufdou K (2019) Use of rhizobacteria and mycorrhizae consortium in the open field as a strategy for improving crop nutrition, productivity and soil fertility. Front Microbiol 10. https://doi.org/10.3389/fmicb.2019.01106

Rasouli F, Amini T, Skrovankova S, Asadi M, Hassanpouraghdam MB, Ercisli S, Buckova M, Mrazkova M, Mlcek J (2023) Influence of drought stress and mycorrhizal (*Funneliformis mosseae*) symbiosis on growth parameters, chlorophyll fluorescence, antioxidant activity, and essential oil composition of summer savory (*Satureja hortensis* L.) plants. Front Plant Sci 14:1–15. https://doi.org/10.3389/fpls.2023.1151467

Reddy GSV, Bokka SKR, Chellem SR, Kalpana K, Indrani K, Bindu TPLH, Navya DS (2024) Mutualistic relationships between plants and mycorrhizal fungi impacts on ecosystem functioning: a review. Uttar Pradesh J Zool 45(13):242–261. https://doi.org/10.56557/upjoz/2024/v45i134151

Reverkar K, Singh AB, Ganguli TK (2005) Mycorrhiza: role and applications. Allied Publishers Pvt Ltd, New Delhi, pp 53–58

Rillig MC, Maestre FT, Lamit LJ (2003) Microsite differences in fungal hyphal length, glomalin, and soil aggregate stability in semiarid Mediterranean steppes. Soil Biol Biochem 35(9):1257–1260

Rini VM (2001) Effect of arbuscular mycorrhiza on soil palm seedling growth and development basal stem rot disease caused by *Ganoderma boninense*. Universiti Putra Malaysia

Rodriguez-Ramos JC, Cale JA, Cahill JF, Simard SW, Karst J, Erbilgin N (2020) Changes in soil fungal community composition depend on functional group and forest disturbance type. New Phytol 229(2):1105–1117. https://doi.org/10.1111/nph.16749

Rosliani R, Prathama M, Sulastrini I, Amisnaipa Sugiartini E, Dahlan Santosa A (2023) The effect of arbuscular mycorrhizae inoculation on growth and yield of red chili on the highland andisol soil. IOP Conf Ser Earth Environ Sci 1287(1):012005. https://doi.org/10.1088/1755-1315/1287/1/012005

Rostami T, Abbasdokht H, Makarian H, Gholipoor M, Karimzadehasl K (2024) Investigation of phytochemical characteristics of quinoa in an intercropping production system with maize in response to mycorrhiza and phosphorus fertilizer. 1–26. https://doi.org/10.21203/rs.3.rs-3930581/v1

Rui W, Mao Z, Li Z (2022) The roles of phosphorus and nitrogen nutrient transporters in the arbuscular mycorrhizal symbiosis. Int J Mol Sci 3(19):11027

Saboor A, Ali MA, Husain S, Tahir MS, Irfan M, Bilal M, Baig KS, Datta R, Ahmed N, Danish S, Glick BR (2021) Regulation of phosphorus and zinc uptake in relation to arbuscular mycorrhizal fungi for better maize growth. Agronomy 11(11):2322

Sachsenmaier L, Schnabel F, Dietrich P, Eisenhauer N, Ferlian O, Quosh J, Richter R, Wirth C (2024) Forest growth resistance and resilience to the 2018-2020 drought depend on tree diversity and mycorrhizal type. J Ecol 112(8):1787–1803. https://doi.org/10.1101/2024.01.23.576797

Salim MA (2023) Effect of *Glomus manihotis* inoculation and salt stress on antioxidant and biochemical properties of chia (*Salvia hispanica* L.). SAINS TANAH—J Soil Sci Agroclimat 20(2):172. https://doi.org/10.20961/stjssa.v20i2.66681

Santos CS, Habyarimana E, Vasconcelos MW (2023) Editorial: the impact of climate change on nutrient composition of staple foods and the role of diversification in increasing food system resilience. Front Plant Sci 14:1087712. https://doi.org/10.3389/fpls.2023.1087712

Sathiyadash K, Jaffer AJ, Karthigai SE, Eswaran PU (2022) Diversity of arbuscular mycorrhizal fungi in shola forests of Kodaikanal, Southern India. Int J Life Sci Pharm Res 10(4):L109–L116. https://doi.org/10.22376/ijpbs/lpr.2020.10.4.1109-116

Shao S, Wurzburger N, Sulman B, Hicks Pries C (2023) Ectomycorrhizal effects on decomposition are highly dependent on fungal traits, climate, and litter properties: a model-based assessment. Soil Biol Biochem 184:109073. https://doi.org/10.1016/j.soilbio.2023.109073

Shi J, Zhao B, Zheng S, Zhang X, Wang X, Dong W, Xie Q, Wang G, Xiao Y, Chen F, Yu N, Wang E (2021) A phosphate starvation response-centered network regulates mycorrhizal symbiosis. Cell 184:5527–5540. https://doi.org/10.1016/j.cell.2021.09.030

Shi J, Wang X, Wang E (2023) Mycorrhizal symbiosis in plant growth and stress adaptation: from genes to ecosystems. Ann Rev Plant Biol 74(1):569–607. https://doi.org/10.1146/annurev-arplant-061722-090342

Smith SE, Read D (2008) Mycorrhizal symbiosis, 3rd edn. Academic, pp 637–768

Solaiman ZM (2014) Contribution of arbuscular mycorrhizal fungi to soil carbon sequestration. In: Mycorrhizal fungi: use in sustainable agriculture and land restoration. Springer, Berlin/Heidelberg, pp 287–296

Spatafora JW, Chang Y, Benny GL, Lazarus K, Smith ME, Berbee ML (2016) A phylum-level phylogenetic classification of zygomycete fungi based on genome-scale data. Mycologia 108:1028–1046

Strack D, Fester B, Hause B, Schliemann W, Walter MH (2003) Arbuscular mycorrhiza: biological chemical and molecular aspects. J Chem Ecol 9:1955–1979

Szentpéteri L, Irmes K, Kristó I, Rácz A, Vályi-Nagy M, Tar M (2023) The effect of arbuscular mycorrhiza on the yield of different winter wheat varieties at four nutrient levels. Rev Agric Rural Dev 12(1–2):88–94. https://doi.org/10.14232/rard.2023.1-2.88-94

Tahat MM, Sijam K, Othman R (2010) Mycorrhizal fungi as a biocontrol agent. Plant Pathol J 9(4):198–207

Tao J, Dong F, Wang Y, Chen H, Ming T (2022) Arbuscular mycorrhizal fungi enhance photosynthesis and drought tolerance by regulating MAPK genes expressions of *Populus simonii* × *P. nigra*. Physiol Plant 174(6). https://doi.org/10.1111/ppl.13829

Tedersoo L, Bahram M, Zobel M (2020) How mycorrhizal associations drive plant population and community biology. Science 367(6480). https://doi.org/10.1126/science.aba1223

Thirkell TJ, Pastok D, Field KJ (2020) Carbon for nutrient exchange between arbuscular mycorrhizal fungi and wheat varies according to cultivar and changes in atmospheric carbon dioxide concentration. Glob Chang Biol 26(3):1725–1738

Tisdall JM, Smith SE, Rengasamy P (1997) Aggregation of soil by fungal hyphae. Soil Res 35(1):55–60

Torre MMPL, Méndez-Morán L, Casarrubias-Castillo K (2022) Application of arbuscular mycorrhizal fungi and organic fertilizers in tomato (*Solanum lycopersicum* L.) plants. J Natl Agric Sci 9:18–23. https://doi.org/10.35429/jnas.2022.25.9.18.23

Usman M, Ho-Plágaro T, Frank HE, Gaillard I, Garcia K, Zimmermann SD (2021) Mycorrhizal symbiosis for better adaptation of trees to abiotic stress caused by climate change in temperate and boreal forest. Front For Glob Change 4:742392. https://doi.org/10.3389/ffgc.2021.742392

Van Der Heijden MGA, Klironomos JN, Ursic M, Moutoglis P, Streitwolf-Engel R, Boller T, Wiemkem A, Sanders IR (1998) Mycorrhizal fungal diversity determines plant biodiversity, ecosystem variability and productivity. Nature 396:69–72

Vassilev N, Mendes GD (2024) Soil fungi in sustainable agriculture. Microorganisms 12(1):163. https://doi.org/10.3390/microorganisms12010163

Wahab A, Muhammad M, Munir A, Abdi G, Zaman W, Ayaz A, Khizar C, Reddy SPP (2023) Role of arbuscular mycorrhizal fungi in regulating growth, enhancing productivity, and potentially influencing ecosystems under abiotic and biotic stresses. Plants 12(17):3102. https://doi.org/10.3390/plants12173102

Wang XX, Hoffland E, Feng G, Kuyper TW (2020) Arbuscular mycorrhizal symbiosis increases phosphorus uptake and productivity of mixtures of maize varieties compared to monocultures. J Appl Ecol 57(11):2203–2211

Wang S, Ren Y, Han L, Nie Y, Zhang S, Xie X, Hu W, Chen H, Tang M (2023) Insights on the impact of arbuscular mycorrhizal symbiosis on *Eucalyptus grandis* tolerance to drought stress. Microbiol Spect 11(2). https://doi.org/10.1128/spectrum.04381-22

Wang S, Han L, Ren Y, Hu W, Xie X, Chen H, Tang M (2024) The receptor kinase rishol in *Rhizophagus irregularis* regulates arbuscule development and drought tolerance during arbuscular mycorrhizal symbiosis. New Phytol 242(5):2207–2222. https://doi.org/10.1111/nph.19677

Watts-Williams SJ (2022) Track and trace: how soil labelling techniques have revealed the secrets of resource transport in the arbuscular mycorrhizal symbiosis. Mycorrhiza 32(3):257–267

Wing IS, De Cian E, Mistry MN (2021) Global vulnerability of crop yields to climate change. J Environ Econ Manage 109:102462. https://doi.org/10.1016/j.jeem.2021.102462

Wu QS, Cao MQ, Zou YN, He XH (2014) Direct and indirect effects of glomalin, mycorrhizal hyphae and roots on aggregate stability in rhizosphere of trifoliate orange. Sci Rep 4(1):5823

Xie L, Lin G, Wang Q, Yin C (2024) Tree species-specific differences in carbon allocation to ectomycorrhizal symbionts tightly relate to ectomycorrhizal fungal community composition. For Ecol Manag 562:121935. https://doi.org/10.1016/j.foreco.2024.121935

Yan G, Bongers FJ, Trogisch S, Li Y, Chen G, Yan H, Deng X, Ma K, Liu X (2022) Climate and mycorrhizae mediate the relationship of tree species diversity and carbon stocks in subtropical forests. J Ecol 110(10):2462–2474. https://doi.org/10.1111/1365-2745.13962

Yang Y, He C, Huang L, Ban Y, Tang M (2017) The effects of arbuscular mycorrhizal fungi on glomalin-related soil protein distribution, aggregate stability and their relationships with soil properties at different soil depths in lead-zinc contaminated area. PLoS One 12(8):e0182264

Zhang L, Chu Q, Zhou J, Rengel Z, Feng G (2021) Soil phosphorus availability determines the preference for direct or mycorrhizal phosphorus uptake pathway in maize. Geoderma 403:115261

Zhang W, Yu L, Han B, Liu K, Shao X (2023) Mycorrhizal inoculation enhances nutrient absorption and induces insect-resistant defense of *Elymus nutans*. Front Plant Sci 13. https://doi.org/10.3389/fpls.2022.898969

Zhang X, Zhang J, He J, Li M, Matsushita N, Geng Q, Lian C, Zhang S (2024) Physiological and transcriptome responses of *Pinus massoniana* seedlings inoculated by various ecotypes of the ectomycorrhizal fungus *Cenococcum geophilum* during the early stage of drought stress. J Fungi 10(1):71. https://doi.org/10.3390/jof10010071

Zhou Y, Shi Z, Pang Q, Liang X, Li H, Sui X, Li C, Song F (2022) Responses of bacterial community structure, diversity, and chemical properties in the rhizosphere soil on fruiting-body formation of *Suillus luteus*. Microorganisms 10(10):2059. https://doi.org/10.3390/microorganisms10102059

Zong J, Zhang Z, Huang P, Yang Y (2023) Arbuscular mycorrhizal fungi alleviates salt stress in *Xanthoceras sorbifolium* through improved osmotic tolerance, antioxidant activity, and photosynthesis. Front Microbiol 14:1138771

Market Avenues and the Availability of Mycorrhizal Products for Use in Agriculture by Farmers

Vanitha Prasannath, Keerthika Shakthivel, Thibisha Rakulan, and Kandeeparoopan Prasannath

Abstract

Mycorrhizae, symbiotic associations between certain fungi and roots of most plants, play a vital role in enhancing plant growth by improving nutrient uptake, particularly phosphorus and nitrogen. This symbiotic association also helps plants withstand environmental stress, heavy metal toxicity, and harmful pathogens, making them essential to plant health and development. The rapidly growing organic food industry has led to a significant increase in demand for mycorrhizal products. As consumers become more aware of health and environmental risks linked to food residues, they are increasingly opting for chemical-free alternatives. Consequently, mycorrhizal products have emerged as an excellent substitute, leading to a notable increase in the production of these mycorrhizal fungal products by commercial enterprises catering to the burgeoning agricultural market in recent years. The chapter reviews the market overview and types of mycorrhizal products, their availability, supply chains, and distribution channels, as well as the pricing and affordability of these products. Further, the chapter discusses available regulatory frameworks and future prospects and opportunities for mycorrhizal products. The availability of various types of mycorrhizal products, ranging from endomycorrhizal to ectomycorrhizal formulations in liquid, powder, and granular forms, is discussed in detail. In addition, the global market for mycorrhizal biofertilizers is classified into three aspects,

V. Prasannath (✉)
Department of Biosystems Technology, Faculty of Technology, Eastern University, Chenkalady, Sri Lanka
e-mail: vanithap@esn.ac.lk

K. Shakthivel · T. Rakulan · K. Prasannath
Department of Agricultural Biology, Faculty of Agriculture, Eastern University, Chenkalady, Sri Lanka
e-mail: keerthikas@esn.ac.lk; prasannathk@esn.ac.lk

Y. Siddiqui et al. (eds.), *Sustainable Mycorrhizal Cultivation*,
https://doi.org/10.1007/978-981-95-1778-7_15

namely seed treatment, soil application, and fertilization treatment, according to the method of application. The mycorrhizal product sector faces several challenges, including limited market knowledge, insufficient distribution strategies to meet the diverse needs of various clients, rising demand for modified products, retaining quality, the impact of farmers' management practices in the field, traceability and longevity of inoculants, and production techniques and application. By adopting proper technological advancements and research approaches, the challenges can be overcome with time, thus leading to an enormous demand for mycorrhizal products in the future. This book chapter provides valuable insights for researchers, policymakers, businesses, and agricultural professionals interested in the market and availability of mycorrhizal products.

Keywords

Biofertilizers · Farmers · Inoculants · Market · Mycorrhizal products

Abbreviations

AMF Arbuscular Mycorrhizal Fungi
AM Arbuscular Mycorrhizal

1 Introduction

Mycorrhizae are among the most prevalent soil microbial species, functioning in association with most terrestrial plants to promote their growth, development, and longevity. Gaining insight into their working mechanisms may help boost agricultural yields, particularly under challenging circumstances. The state of the soil is optimized by Arbuscular mycorrhizal fungi (AMF) through the enhancement of its biological, chemical, and physical properties. Hyphae, around a hundred meters long and formed by the AMF, establish networks that join soil particles and create strong aggregates, thereby preventing soil erosion (Mardhiah et al. 2016). Mycorrhizae cling to plant roots to create a web of filaments that extract nutrients from the ground that root systems typically cannot. The literature also states that these fungi aid in the absorption of calcium, iron, manganese, sulfur, potassium, nitrogen, zinc, copper, and magnesium. As a result, plants with strong mycorrhizal colonization are expected to be less dependent on external inputs and more resilient to potential nutrient shortages compared to those with minimal symbiosis (Milton et al. 2021). Fungi and plants work together to improve root development and increase plant growth (Navarro-Fernandez et al. 2016). Furthermore, it lowers the plant's vulnerability to environmental stressors like salinity and drought as well as soil-borne diseases and pest attacks (Wahab et al. 2023). These microbes help plants absorb more water, thereby rendering them more resistant to drought. Mycorrhizae-associated plants have higher stomatal conductivity and moisture, which encourage transpiration, the primary process that cools the leaves. Particularly, the fungi prevent the passive intake of harmful substances that limit the host plant's sensitivity to heavy metals. Apart from providing nutritional advantages, AMF colonization can increase host plant resistance to pests and diseases (Whipps 2004). Figure 1 illustrates a concise overview of the role of mycorrhizal fungi colonization in agriculture.

Since AMF have been shown to form symbiotic connections with more than 80% of plant species, they are particularly utilized in the majority of bioinoculant production. The market for organic food is growing quickly, which is increasing the demand for biofertilizers based on mycorrhizae. Chemical-free food options are gaining more attention as people become more conscious of health dangers, residues of pesticides, food spoilage, and environmental issues. As a result, the trend toward organic farming practices is driving an increase in the usage of mycorrhizal biofertilizers (Antunes et al. 2012).

The global market for mycorrhizal fungal products is experiencing significant growth, with numerous companies producing mycorrhizal fungal inocula for a variety of applications, including crop production, forestry, gardening, and environmental rehabilitation (Basiru et al. 2021). Estimates suggest that the mycorrhiza market will reach USD 1.18 billion by 2024 and USD 1.87 billion by 2029, reflecting a compound annual growth rate of 9.65% over this period (Salomon et al. 2022a, b). This valuation highlights the growing market for mycorrhizae-based biofertilizers.

In the United States, funding for the research and development of mycorrhizal biofertilizers is increasing, driven by the growing demand for sustainable farming solutions. This demand encourages large agricultural companies to add these products to their offerings. In China, the adoption of mycorrhizal biofertilizers is rising swiftly due to government initiatives supporting natural farming practices. To optimize resource utilization, recent advancements have focused on expanding production capacities and integrating these biofertilizers with precision farming techniques.

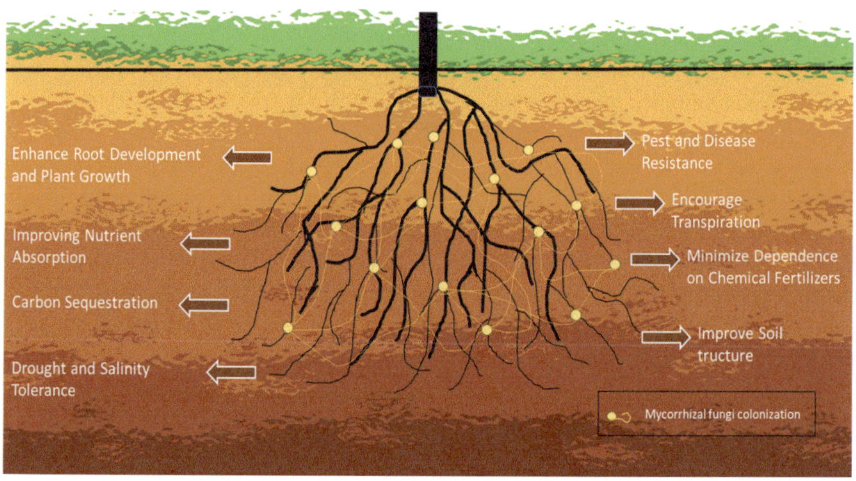

Fig. 1 Role of mycorrhizal fungal colonization in agriculture

In India, the need for sustainable agricultural practices to support its extensive farming sector is fueling a surge in the use of mycorrhizal biofertilizers. Government subsidies and partnerships with foreign companies are helping to make these products more affordable and accessible. Major players in the mycorrhizal biofertilizer market include Valent BioSciences and Helena Agri Enterprises LLC from the United States, Novozymes from Denmark, Lallemand and Premier Tech from Canada, and Groundwork BioAg from Israel (Noe and Kiers 2018). Meanwhile, in Japan, consumer interest in natural products is driving the popularity of mycorrhizal biofertilizers. Current initiatives aim to boost efficiency by developing customized formulations for economically important crops and combining them with advanced agricultural technologies (Ullah et al. 2024).

2 Market Overview

With prospects in the farming and nonagricultural sectors, the global market for mycorrhizal biofertilizers appears to have a bright future. At a compound annual growth rate of 14.5% from 2024 to 2030, the global market for mycorrhizae-based biofertilizers is projected to surpass approximately $1.6 billion (Market Research Report 2021). In 2022, Europe accounted for 55.2% of the global biofertilizer market, establishing itself as the leading region. With an estimated amount of 1.1×10^3 Mt. in 2022, mycorrhiza biofertilizer had a market value of $0.0398 billion in the European region. The key participants in the mycorrhizae-based biofertilizer market include Biolchim SPA, Symborg, Inc., Green Vision Biosciences, Atlantica Agrícola, Unicrop Biochem, Koppert Biological Systems Inc., Agrinos, and Gujarat State Fertilizers and Chemicals Ltd. (Smith and Read 2008).

North America holds a substantial market share and dominates the international market as a result of the early adoption of cutting-edge and novel goods in nations like the United States of America and Canada, as well as greater government investment in organic agricultural techniques. Nearly a hundred companies manufacture and distribute mycorrhizal products in the European region alone, establishing it as a thriving marketplace for mycorrhizal inoculants. Furthermore, the market share of mycorrhizae-based products, especially the biofertilizers, was the third highest in Asia Pacific due to the growing consumer preference for organic food and the numerous legislative changes and projects implemented by the governments of various developing Asian nations, especially India and China. Moreover, the region's market is expanding due to increasing consumer knowledge of the advantages of eating organic food and agricultural communities' general comprehension of the enduring benefits of these fertilizers (Chen et al. 2018; Basiru and Hijri 2022; Market Research Report 2022).

The need for novel and ecologically friendly farming techniques is growing along with the organic food sector. Amidst these innovative methods, biofertilizers based on mycorrhizae have become a potent tool that is improving the process of growing organic food while nourishing the environment. Mycorrhizal fungi play a crucial role in delivering nutrients, such as micronutrients and phosphorus, directly to plant roots. This process enhances the nutritional quality of organic produce, boosting the nutrient levels in crops (Auge 2001). Consequently, the rising demand for organic food has driven growth in the market for biofertilizers that utilize mycorrhizae.

With a higher AMF signifying better soil health and increased crop yields, the agriculture sector and related businesses have seen growth, leading to a booming market for mycorrhizae-based biofertilizers. Chemical fertilizers reduce the levels of organic matter and microbial life in the soil, incur high costs, harm the environment, pose health risks to humans, and result in eutrophication (Luty et al. 2021). In order to ensure that crops get nourishment in a way that promotes optimal growth without the need for excessive synthetic fertilization, mycorrhizae help plants acquire elements from the ground. Aligned with the principles of environmentally conscious consumers, mycorrhizae promote sustainable agricultural practices by enhancing soil health and reducing reliance on synthetic chemicals (Smith and Read 2008). This support for a thriving soil ecosystem contributes to more eco-friendly farming methods. The demand is expected to be driven during the forecast time frame by the numerous advantages of mycorrhizae-based biofertilizers.

For a variety of reasons, mycorrhizae-based biological fertilizers can be extremely expensive. These include the necessity of creating ideal methods for manufacturing, and discovering and choosing mycorrhizal genotypes that work effectively. Substantial mycorrhizal fungal production while retaining viability is a challenging undertaking requiring certain resources and knowledge. Strict inspection and testing procedures are also necessary to guarantee the viability and efficacy of these fungi in the finished product (Ghorui et al. 2024). Another difficulty in marketing mycorrhizal commodities is the shorter longevity of biofertilizers made from mycorrhizae. The potential benefits for soil and plants are undermined by the short shelf life of mycorrhizae-based biofertilizers, which also become less effective when applied (Berruti et al. 2016). The market for mycorrhizae-based biofertilizers has changed, and an array of significant developments is driving its expansion. These trends indicate a wider movement toward sustainable agriculture, driven by technological innovations and a growing awareness of environmental effects. Scientists are exploring a wide variety of mycorrhizal fungi to find variants that establish strong alliances with different kinds of plants. Additionally, advances in genetic and molecular research are enhancing the development of more effective biofertilizer formulations by revealing the intricate genetic aspects of mycorrhizal relationships (Cameron et al. 2013). In an effort to maximize results, researchers are also developing customized mycorrhizal fungal mixtures to satisfy the unique requirements of various plants, soils, and environmental circumstances.

3 Types of Mycorrhizal Products Available

Endo and ectomycorrhizae are the two major groups of fungi frequently used in the development of mycorrhizal products, which are crucial for the growth and development of crops (Pal et al. 2016). Endomycorrhizae produce hyphae inside the roots, while ectomycorrhizae produce hyphae outside the roots. Endomycorrhizae are primarily classified into arbuscular, arbutoid, orchidaceous, ericoid, and monotropoid types. Ectomycorrhizae consist of fungal species from the Ascomycete, Zygomycete, and Basidiomycete groups (Basiru et al. 2021). The majority of ectomycorrhizae are found in trees, including eucalyptus, fir, willow, hemlock, poplar, oak, pine, larch, birch, spruce, and beech (Churchland and Grayston 2014). Deja-Sikora et al. (2019) stated that endomycorrhizae are found in plants such as palms, beets, carnations, grapes, wheat, dates, potatoes, broccoli, cassava, cabbage, beans, alfalfa, corn, mustard, and sugarcane.

Mycorrhizal product formulations typically include a carrier, inoculant, and additive compounds (Benami et al. 2020). They can be customized to meet the specific requirements of the host plant and application method. Several types of formulations are available, including granules, liquids, powders, fine powders, wettable powders, fillers, and binders. Mycorrhizal products are applied through coating, dusting, mixing, and drip irrigation methods for use as fertilizers, seed treatments, and root treatments (Basiru et al. 2021). Mycorrhizal multiplication, on a vast scale, aimed at producing inoculants for agricultural use, is usually carried out via substrate-based, substrate-free, and in vitro methods (Mukhongo et al. 2016). Commercial-level inoculants were produced in many countries, especially in Europe and Asia. Nevertheless, farmers and nursery owners bear the costs associated with the inoculum production technique, which includes manufacturing the carrier substrate, transportation, processing, and establishing single cultures of fungi. Several countries produce mycorrhizal products in different ways and under various trade names, using inoculants (Goswami et al. 2019). Table 1 lists the manufacturers and their various mycorrhizal products worldwide. Among them, approximately 60% of the inoculants are available in powder form, with the remaining 25% and 10% coming in granular and liquid formulations, respectively (Rouphael et al. 2015). The majority of the commercial inoculants consist of species from the *Glomus*, *Septoglomus*, *Rhizophagus*, and *Funneliformis* genera (Sturmer et al. 2018). Each plant consists of unique mycorrhizal inoculants, which are listed in Table 2.

4 Availability of Mycorrhizal Products

Recently, there has been increasing demand for organic-based farming systems and organic products with minimal use of chemicals. Consequently, the manufacturing and distribution of mycorrhizal products have increased globally (Baar, 2008). According to Sudheer et al. (2023), the growing demand for mycorrhizal products has led to an increase in challenges to the production of high-quality mycorrhizal products. The mycorrhizal production sector faces several challenges, including the need for product modification, inadequate consumer awareness and market demand, high labor and equipment costs, capital intensiveness, selection of suitable cultural media, the production of large quantities of inoculants in nursery beds and pots, limited resources for the production and storage of mycorrhizal inoculum, a shortage of trained personnel and technical support, determining the proper formulation

Table 1 The details of specific mycorrhizal product manufacturers worldwide

Country	Company	Product	References
USA	Agbio Development Inc.	MycoApply	Basiru and Hijri (2022)
	Mycorrhizal Applications Inc.	MycoApply soluble, MycoApply endonet, and MycoApply Micronized endo	
	Becker Underwood Biologicals	Rhizanova™	
	Sustane Corporate Headquarters	Sustane®	
	Helena Agri-Enterprises, LLC	Myco-sol, myco-vam plus 6–3-3, myco-vam bare root, and myco-vam granular	
	Becker Underwood, Ames	Myke®	
	Bio-Organics	Bio-organics™, endomycorrhizal inoculant, bio-organics™ Mycorrhizal root dip inoculant	
	Reforestation Technologies International	MycoPaks™, AM120™, and Silva Dip™	
India	VAMTech, Inc.	Myconate® and Mycoform®	Akhtar et al. (2024)
	TARI Biotech	Vesicular-arbuscular mycorrhiza	
	KCP Sugar and Industries Corporation Ltd.	Mycorrhiza VAM	
	Ambika Biotech	Root care	
	Cadila Pharmaceuticals, Ltd.	Josh	
	Dr. Rajan Laboratories	Mycorrhizas	
	Krishidhan Seeds Pvt. Ltd.	Arbuscular mycorrhiza	
Spain	Odd Distributions	Root Webstar	Mukhongo et al. (2016)
	Arvensis Agro	Arbuscular mycorrhiza	
Canada	Mikro-Tek, Inc. Timmins	MIKRO-CONE® and MIKRO-VAM®	Antunes et al. (2009)
	Premier Tech	Myke®	
Mexico	BIOMIC	TM-73	Basiru and Hijri (2022)
	Plant Health Care	MycorTree-Injectable, Hotic Plus, and PHC VAM.PWI	
	Biokrone	Glumix	
	OBA	HIPER-GLOM	
	INIFAP	Micorriza INIFAP	
Japan	Idemitsu Kosan Co. Ltd	R-10	Pal et al. (2016)
	Central Glass Co. Chemicals Section	Cerakinkong Tokyo	

(continued)

Table 1 (continued)

Country	Company	Product	References
United Kingdom	Zander Corporation	Arbuscular mycorrhiza	Goswami et al. (2019)
	Biological Crop Protection Limited	Mycorrhizal inoculants	
	Crop Intellect Limited	Arbuscular mycorrhiza	
	PlantWorks Ltd.	Root grow	
China	Qingdao Jingling Ocean Technology Co., Limited	Biofertilizer	Meng et al. (2022)
	Bio Plus Fertilizer		
	Xiamen Vastland Chemical Co., Limited		
	Beijing Leili Marine Bioindustry Inc.	Bio compost fertilizer	
	Qingdao Future Group	High-quality granular biofertilizer	
Netherland	Global Horticare	BioRoot	Pal et al. (2016)
Kenya	Dudutech, Naivasha	Rhizatech	Kavoo-Mwangi et al. (2013)

Table 2 Details of mycorrhizal inoculants and their host plants

Mycorrhizal inoculants	Host plant	References
Glomus intraradices	*Zea mays* *Capsicum annuum* *Anthurium andraeanum* and *Vigna radiata*	Castillo et al. (2009), Synowiec et al. (2011), Habibzadeh et al. (2014) and Nunes et al. (2014)
Rhizophagus irregularis	*Oryza sativa* and *Brassica rapa*	Drasek (2014) and Campo et al. (2020)
Rhizophagus clarus	*Glycine max* and *Gossypium hirsutum*	Cely et al. (2016)
Claroideoglomus etunicatum	*Lycopersicon esculentum*	Jamiolkowska et al. (2020)
Glomus fasciculatum	*Cicer arietinum, Solanum melongena, Arachis hypogaea,* and *Abelmoschus esculentus*	Doley and Jite (2012), Darade (2015) and Phad et al. (2016)
Aculospora laevis	*Cicer arietinum*	Kumar et al. (1998)
Glomus etunicatum	*Cucumis sativus* and *Anthurium andraeanum*	Hao et al. (2005) and Nunes et al. (2014)
Diversispora versiformis, Funneliformis mosseae, and Rhizophagus intraradices	*Citrus reticulata*	Cao et al. (2021)
Glomus clarum	*Sorghum bicolor*	Simpson and Daft (1990)
Glomus mosseae	*Vigna radiata,* rosa hybrid, *Tagetes erecta, Allium cepa* and *Carica papaya*	Jaizme-Vega et al. (2006), Lopez and Martin (2012), Habibzadeh et al. (2014) and Shuab et al. (2014)
Glomus coronatum, Glomus caledonium, and Glomus geosporum	*Allium cepa*	Shuab et al. (2014)
Glomus epigaeum and Acaulospora morroweae	*Vigna mungo*	Devi and Sitaramaiah (2001)

and application techniques, difficulty in attaining different climate and soil requirements, facing difficulties during field-level application, some mycorrhizal products may interact with and affect the native species already present in that particular ecosystem, absence of proper legislation for manufacture and distribution, some soil-based inoculants may spread diseases, the longevity of the inoculum, its inconsistent performance at the field level, and the lack of procedures and techniques for producing high-quality inoculants on a large scale (Fester and Sawers 2011; Vosatka et al. 2012; Benami et al. 2020; Madawala 2021).

Taking into account the following, these bottlenecks can be overcome by using native AMF isolates. Adding soil containing mycorrhizal media makes the inoculum heavier and more difficult to transfer, which will prevent the inoculants from spreading disease. To avoid contamination, certain agricultural chemicals should be included with the inoculum (Wu et al. 2019). Implementing molecular-level approaches to identify and characterize mycorrhizal communities, especially next-generation sequencing, will yield more satisfactory results for fungal surveillance within the host (Werner and Kiers 2015). It is necessary to develop appropriate regulatory frameworks and inoculum quality control procedures (Owen et al. 2015). Adapting in vitro production techniques, particularly transformed root organ culture, is one of the recent technical advancements aimed at addressing production challenges such as expansion, contamination, and consistency in performance across various environments (Kumar et al. 2017). Extending collaboration with mycorrhizal scientists and manufacturers will result in ongoing research to overcome hindrances and provide compelling data for production-related decision-making (Vosatka et al. 2012). This includes choosing the right inoculant for formulations based on the region and available crops, adhering to scientific principles and standard protocols to maintain quality and consistency (Mukhongo et al. 2016), and adapting the use of native strains to fit specific climates and soils.

Additionally, this will reduce the cost of shifting inoculants from one place to another, modify application methods appropriate for a given farming system, and raise awareness to educate people about adopting mycorrhizal products (Veresoglou and Agathokleous 2023). It will also involve selecting suitable substances for formulation to enhance quality, storing inoculum in cold storage conditions for the short term, and using dry and liquid storage conditions at 5 °C for long-term periods. Routine monitoring of the production environment and educating people about mycorrhizas and their role in ensuring global food security will minimize producers' insecurities about investing in mycorrhizal technology (Fester and Sawers 2011).

Regulation is essential to ensure the effectiveness, quality, and safety of mycorrhizal inoculants (Herrmann and Lesueur 2013). To manage potential hazards, assure product efficacy, and promote consumer confidence in using these products, proper regulation is crucial. Regulations include standardization protocols in production, inspection for live propagules, and accurate labeling of product components (Salomon et al. 2022a, b). Regulation facilitates the evaluation of mycorrhizal inoculants' performance and effectiveness in diverse crops and environments (Hijri 2016).

The European Union Fertilizer Rule 2019/1009 was amended more recently, becoming effective in April 2019 (Salomon et al. 2022a, b). This law addresses the quality handling of mycorrhizal products, and presently, standard procedures for product certification of mycorrhizal inoculants are being established by the European Standardization Committee. Independent laboratories will conduct ring tests across Europe to validate these standards. These techniques focus on measuring the value of live microorganisms present in the products and using standardized testing to verify the claimed benefits. The mycorrhizal community varies with seasonal changes; thus, understanding these fluctuations will give an idea for designing the quantity of inoculum supply during production. Hewins et al. (2015) claimed that *Allium tricoccum* roots exhibited the greatest amount of the mycorrhizal community in the latter part of winter and early spring. According to a different study, mycorrhizal colonies in wetland regions peaked in the spring and decreased in late summer, with plant phenology being the primary factor driving seasonal fluctuations rather than abiotic factors (Bohrer et al. 2004). Varela-Cervero et al. (2016) reported that the composition of the mycorrhizal community in plant roots was considerably varied between the fall and spring. Future research should incorporate a seasonal component to more accurately evaluate the supply and production of mycorrhizal inoculum for various environments.

New developments in mycorrhizal symbiosis have expanded the possible applications of mycorrhizal biotechnology in phytoremediation, horticulture, forestry, landscaping, and agriculture. However, several unresolved issues in mycorrhizal production remain, such as inoculum purity, quality control, and unreliable traceability of fungal strains (Madawala 2021). Increased collaboration between enterprises and the academic sector is promoting the industry's ongoing development. Further research is needed to facilitate the economic exploitation of mycorrhizal fungi, with collaboration between science and industry being essential to overcoming these challenges (Vosatka et al. 2012).

5 Supply Chain and Distribution Channels

Green technologies with minimal impact on the environment and human health are becoming increasingly popular, diminishing the market share of agrochemicals. Plant biostimulants enhance plant nutrition independent of the product's nutrient content, improving nutrient use efficiency and tolerance to abiotic stress. Biostimulants may originate from microbial or nonmicrobial sources, including AMF (Chen et al. 2018). However, the adoption of AMF-based biofertilizers has been hampered due to the prevalence of poor-quality commercial products on the market. At the same time, awareness is growing within the farming community about the benefits of AMF. Numerous local and global ventures are developing AMF inoculants/products that are user-friendly, easily transportable, and applicable across various crops. Although large-scale adoption is still in progress, the number of companies marketing mycorrhizal products has surged significantly in recent years. The primary companies specializing in mycorrhizal products are located in

Europe, Asia, North America, and Latin America. Currently, the leading exporters of mycorrhizal products are based in the United States, Italy, Canada, and Spain (Khaliq et al. 2022). The European market leads in biostimulant sales, with companies selling AMF products growing from 10 to over 75 between 1990 and 2017, primarily based in Italy, Spain, Belgium, France, Austria, and Switzerland (Chen et al. 2018; Khaliq et al. 2022).

AMF products are widely used in agriculture, forestry, horticulture, and landscaping, with various formulations including biofertilizers, biostimulants, and bioprotectants that enhance plant growth and resilience. Registration of these products follows EU national regulations, though the process can be costly. Organizations like the International Mycorrhiza Society and the European Bio-stimulant Industry Council promote biostimulants. Key players in AMF production include Canada, Spain, Italy, the Czech Republic, the United States, the United Kingdom, Germany, China, Taiwan, India, South Africa, and Kenya, though there is a need to expand production and awareness in developing countries (Khaliq et al. 2022).

Basiru et al. (2021) summarized seven countries to be major shareholders in AMF commercial production, including Canada, Spain, Italy, the Czech Republic, the United States, the United Kingdom, and Germany; however, AMF commercial production data are not available for Australia. After North America and Europe, Asia Pacific, including China, Taiwan, and India, are leading players in the AMF market. Moreover, South Africa and Kenya are leading in this market in Africa. Nevertheless, further efforts are needed to expand AMF production companies in developing countries and to raise awareness about these products (Khaliq et al. 2022). Major obstacles to purchasing include high prices and limited awareness. Companies should strategize promotional activities, such as field demonstrations and farmer meetings conducted by sales representatives, to boost sales (Chandel et al. 2017).

Several suppliers sell mycorrhizal products directly to customers, either at gardening exhibitions and trade fairs or through their websites. Online retailers also allow consumers to purchase these products from the comfort of their homes. Many wholesalers stock these goods, making them accessible to local gardeners through gardening shops and farm supply centers (Vosatka et al. 2008). It is challenging to define and study a fungal "individual" as an independent agent with its own trading strategy. Additionally, since nutrient exchanges occur through shared structures, competition and transaction costs are temporarily reduced, making these exchanges more akin to transactions between divisions of a company than to typical market trade. Fungal nuclei might coordinate their trading strategies while maintaining a degree of independence, similar to human cooperatives (Noe and Kiers 2018).

6 Regulatory Framework

Regulation is essential for ensuring the quality, safety, and efficacy of arbuscular mycorrhizal (AM) inoculants. Effective regulatory frameworks address potential risks, confirm product functionality, and foster consumer confidence in AM inoculants. By standardizing production processes, verifying viable propagule counts, and enforcing accurate labeling of product constituents, regulations ensure consistent product quality, composition, viability, and efficacy across different batches and manufacturers (Herrmann and Lesueur 2013; Ghorui et al. 2024). Additionally, regulatory oversight is crucial for assessing the effectiveness and performance of AM inoculants across diverse environmental conditions and crop systems. Field trials and data collection on metrics like root colonization rates, nutrient uptake, and crop yield improvements play a central role in these evaluations (Hijri 2016).

Moreover, regulations protect against misleading claims regarding the benefits of AM inoculants, requiring manufacturers to substantiate their product claims with scientific evidence. This approach discourages exaggerated or unsupported marketing practices (Hijri 2016; Salomon et al. 2022a, b). Regulatory frameworks also support sustainable agriculture by promoting AM inoculant use, which can reduce reliance on chemical fertilizers, enhance soil health, and minimize environmental impact (Ceballos 2013). Ensuring product safety is another key regulatory function, involving hazard evaluations for introducing non-native AMF strains and assessing potential adverse effects on soil microorganisms and ecosystems (Mummey et al. 2009). Mandated safety assessments help mitigate risks and prevent unintended ecological consequences of AM inoculant use.

Further, global harmonization of regulations simplifies international trade by establishing common standards, facilitating distribution across countries, and expanding product accessibility (Ghorui et al. 2024). Regulatory oversight also spurs research and development within the AM inoculant sector, as companies strive to meet regulatory requirements and demonstrate product efficacy. Through well-structured regulatory frameworks that promote responsible AM inoculant use, governments and industry stakeholders can drive agricultural productivity and environmental sustainability. Currently, a lack of universal regulations for AM inoculants persists, though existing compliance standards address many essential aspects, such as legal terminology, regulatory authorities, relevant orders or acts, year of enactment, and product specifications like spore count, pH, inoculum potential, moisture content, contamination levels, heavy metal content, testing methodologies, and allowable product deviations (Ghorui et al. 2024).

The AM inoculants are promising tools for advancing sustainable agriculture and ecosystem management. By forming beneficial symbiotic relationships with plant roots, these inoculants offer significant advantages, including enhanced nutrient uptake and improved resilience to environmental stresses. However, commercial AM inoculants currently face challenges, such as inconsistent field performance, ecological risks from non-native strains, and an absence of standardized global regulations. Regulatory frameworks aim to establish standards for composition, safety, and efficacy, yet challenges remain, including scientific data gaps, standardization issues, field testing, and ecological considerations. To address these challenges and fully leverage the potential of AM inoculants, increased research funding, public-private partnerships, monitoring, education, and ecosystem impact studies are essential.

Future regulatory frameworks have the potential to greatly enhance product quality, soil health, and crop productivity while reducing dependence on chemical inputs and benefiting the environment. Achieving these goals will require addressing challenges related to compliance, standardization, awareness, certification, monitoring, and cost. Mandating the submission of scientific data will help ensure the safety and efficacy of AM inoculants, fostering transparency within the industry. Strengthened regulations will improve product quality, support sustainability, and provide substantial environmental benefits. The new regulatory objectives for AMF aim to maximize agricultural and environmental benefits while mitigating risks. Ghorui et al. (2024) suggested the following scenarios. Environmental stewardship initiatives will encourage AMF use to enhance soil health and reduce chemical dependence, supporting sustainable farming. Quality control measures will ensure AMF product efficacy and safety, building trust among farmers and consumers. To promote wider adoption, regulations will drive research and innovation through collaborative partnerships. Environmental and human safety will be prioritized through studies on AMF's ecological impact and toxicity, guiding responsible use. Transparency will be improved through clear labeling, providing farmers with essential product information to make informed decisions. Global standardization efforts will unify AMF regulations and definitions worldwide, facilitating international trade by establishing consistent standards and testing parameters, thus enhancing market access and promoting fair trade. Moreover, global harmonization and collaborative public-private partnerships will maximize the positive impact of AM inoculants on agriculture and ecosystem management, promoting healthier soils, higher crop yields, and a more sustainable agricultural sector.

7 Adoption by Farmers

The production and adoption of mycorrhizal systems are still insufficient as a result of a lack of understanding of concepts and awareness among producers and distributors (Mukhongo et al. 2016). Distributors and farmers need different kinds of pre- and post-sale assistance, including information on the right mycorrhizal formulation, how to apply it, and guidance on suitable hosts and care. Despite growing public awareness of the advantages of mycorrhizal technology, its widespread use is still hindered by ineffective production and inadequate marketing, distribution, and promotion (Benami et al. 2020). Law regulation, financial aid, infrastructure, market trends, community networks, and firsthand experience are the factors that can influence farmers' decisions to adapt mycorrhizal products (Klerkx et al. 2012). There are several ways to encourage farmers to adopt new technologies in a variety of ways, such as by educating them about technological advances, showcasing some successful field-level trials, motivating them to publish the results of their research, connecting local farmers with foreign and external networks from different regions, and demonstrating new practices and products to farmers who have still stuck to traditional farming systems (Maertens and Barrett 2012; Suri 2011). To remain competitive in a constantly evolving market, farmers must successfully implement new technology to increase their income and output.

Numerous international case studies demonstrate how mycorrhizal products can be successfully included in agricultural farming systems. Ceballos et al. (2013) claimed that inoculating cassava plants with *Rhizophagus irregularis* enhanced yields, particularly with lower phosphorus levels. Another field study was in Israel; it was discovered that combining mycorrhiza with pepper cultivation produced an extra $86,000 yield per farm. According to a different study, *Aster amellus* plants rapidly acclimated to their native soil when they were combined with mycorrhizal inoculum instead of alone (Pankova et al. 2014). Bauer et al. (2020) reported that when mycorrhizal inoculum was locally adapted, it improved the diversity and biomass of plant populations, especially in earlier successional crops.

8 Future Prospects and Opportunities of Mycorrhizal Products in Sustainable Agriculture

Over the past 10 years, mycorrhizal inoculations have emerged as a groundbreaking and effective biotechnological approach in plant production. This has led to significant advancements across the entire sector, highlighting its various intricacies and developments. The manufacturing process for AM inoculum has now been successfully scaled to an industrial level, driven by the increasing focus on the environmentally sustainable aspects of mycorrhizae. This growth is fueled by widespread interest and significant commercial applications. Developing mycorrhizal technology might be the exclusive emphasis of practical studies on mycorrhiza in sustainable farming (Vosatka et al. 2012). The goal of feeding the increasing population and ensuring the food available under the changing climate patterns is quite a challenge, but mycorrhizal products ease this goal and promote a greener environment. The development of new molecular techniques to identify inoculated fungi in the field, investigate mycorrhizal population interaction and adaptation mechanisms for different environments, and gain a comprehensive understanding of mycorrhizae's interactions with other microorganisms is made possible by biotechnological advancements. Recent developments in the bioengineering of mycorrhizal plant symbiosis have enhanced the prospective uses in several fields such as landscape design, phytoremediation, horticulture, agriculture, and other plant sectors (Kumar et al. 2017).

When compared with traditional inoculum production, the transformed root organ culture technique (Kumar et al. 2017) has created a positive impact on the production of mycorrhizal propagules under sterile circumstances. That is to enhance the market demand for mycorrhizal products. Among all the biofertilizers, mycorrhizal products have become widely accepted. This highlights future trends and opens a window for mycorrhizal-related enterprises and research initiatives (Vosatka et al. 2012). Furthermore, the use of mycorrhizal products resulted in a decrease in the use of phosphorus-based pesticides in agricultural activities, according to the global economy. This offers the possibility for future studies on phosphorus restoration, monitoring, and detection (Albrechtova et al. 2012). According to recent research, some mycorrhizal populations can improve food quality by increasing sugar content, vital nutrients, and antioxidants while reducing the transport of heavy metals to aboveground plant components. Furthermore, certain microorganisms support phytoextraction in contaminated soils, which aids in soil restoration. This will open up possibilities for future research (Gianinazzi et al. 2010; Cabral et al. 2015). The mycorrhizal enterprises have made some highly encouraging advancements, despite a number of unresolved barriers. However, further research is necessary to utilize the potential of mycorrhizal fungi.

9 Conclusion

The market for mycorrhizal products in agriculture is expanding as more farmers recognize the advantages of mycorrhizae in increasing crop yields, enhancing nutrient absorption, and improving soil health. Mycorrhizal fungal-based biofertilizers, in particular, are gaining popularity due to their sustainability and minimal environmental impact, offering an alternative to traditional chemical fertilizers. By forming symbiotic relationships with plant roots, mycorrhizae improve water retention, boost disease resistance, and help plants access nutrients in the soil that would otherwise remain unavailable. These benefits are crucial as global agriculture seeks more eco-friendly and resilient solutions to meet the demands of a growing population. The Asia-Pacific region, especially China and India, currently holds the largest market share for mycorrhizal products, driven by rising consumer demand for organic foods and government initiatives that promote sustainable farming practices. In China, for example, government policies encourage natural farming methods, supporting the adoption of mycorrhizal biofertilizers. Similarly, India's emphasis on sustainability in agriculture and food security has fueled a rapid increase in the use of these biofertilizers. Products are now more widely available through traditional agricultural supply channels as well as online platforms, which allow small-scale farmers greater access to mycorrhizal inoculants and other biofertilizers.

Despite this growth, challenges remain. High costs, limited awareness, and inconsistent application methods hinder wider adoption. Production inefficiencies, along with limited advertising and distribution networks, further restrict market penetration. Addressing these issues requires a multifaceted approach. Educating farmers on the benefits of mycorrhizal products through field trials, technical workshops, and case studies can build awareness and demonstrate effectiveness. Partnerships with international organizations for technical support can help overcome production limitations and improve product accessibility. Efforts to enhance product quality and application techniques, along with affordable pricing, could make mycorrhizal products a cornerstone in sustainable agriculture. As the market matures and awareness grows, these biofertilizers could play a central role in the global transition to resilient, environmentally friendly farming practices.

References

Akhtar O, Zoomi I, Pandey D (2024) New formulations, product development, and commercialization of arbuscular mycorrhizal fungi. In: Parihar M (ed) Arbuscular mycorrhizal fungi in sustainable agriculture: inoculum production and application. Springer, Singapore, pp 186–203

Albrechtova J, Latr A, Nedorost L, Pokluda R, Posta K, Vosatka M (2012) Dual inoculation with mycorrhizal and saprotrophic fungi applicable in sustainable cultivation improves the yield and nutritive value of onion. Sci World J 2012:1–8

Antunes PM, Koch AM, Dunfield KE, Hart MM, Downing A, Rillig MC, Klironomos JN (2009) Influence of commercial inoculation with *Glomus intraradices* on the structure and functioning of an AM fungal community from an agricultural site. Plant Soil 317:257–266

Antunes PM, Lehmann A, Hart MM, Baumecker M, Rillig MC (2012) Long-term effects of soil nutrient deficiency on arbuscular mycorrhizal communities. Funct Ecol 26(2):532–540

Auge R (2001) Water relations, drought and vesicular-arbuscular mycorrhizal symbiosis. Mycorrhiza 11:3–42

Baar J (2008) From production to application of arbuscular mycorrhizal fungi in agricultural systems: requirements and needs. In: Varma A (ed) Mycorrhiza. Springer, Berlin, pp 361–373

Basiru S, Hijri M (2022) The potential applications of commercial arbuscular mycorrhizal fungal inoculants and their ecological consequences. Microorganisms 10:1897

Basiru S, Mwanza HP, Hijri M (2021) Analysis of arbuscular mycorrhizal fungal inoculant benchmarks. Microorganisms 9(1):81

Bauer JT, Koziol L, Bever JD (2020) Local adaptation of mycorrhizae communities changes plant community composition and increases aboveground productivity. Oecologia 192:735–744

Benami M, Isack Y, Grotsky D, Levy D, Kofman Y (2020) The economic potential of arbuscular mycorrhizal fungi in agriculture. In: Nevalainen H (ed) Grand challenges in fungal biotechnology. Springer, Cham, pp 239–281

Berruti A, Lumini E, Balestrini R, Bianciotto V (2016) Arbuscular mycorrhizal fungi as natural biofertilizers: let's benefit from past successes. Front Microbiol 6:1559

Bohrer KE, Friese CF, Amon JP (2004) Seasonal dynamics of arbuscular mycorrhizal fungi in differing wetland habitats. Mycorrhiza 14:329–337

Cabral L, Soares CRFS, Giachini AJ, Siqueira JO (2015) Arbuscular mycorrhizal fungi in phytoremediation of contaminated areas by trace elements: mechanisms and major benefits of their applications. World J Microbiol Biotechnol 31(11):1655–1664

Cameron DD, Neal AL, van Wees SCM, Ton J (2013) Mycorrhiza-induced resistance: more than the sum of its parts. Trends Plant Sci 18(10):539–545

Campo S, Martin-Cardoso H, Olive M, Pla E, Catala-Forner M, Martinez-Eixarch M, San Segundo B (2020) Effect of root colonization by arbuscular mycorrhizal fungi on growth, productivity and blast resistance in rice. Rice 13(1):42

Cao MA, Wang P, Hashem A, Wirth S, Abd-Allah EF, Wu QS (2021) Field inoculation of arbuscular mycorrhizal fungi improves fruit quality and root physiological activity of citrus. Agriculture 11(12):1297

Castillo C, Sotomayor L, Ortiz C, Leonelli G, Borie F, Rubio R (2009) Effect of arbuscular mycorrhizal fungi on an ecological crop of chili peppers (*Capsicum annuum* L.). Chil. J Agric Res 69(1):79–87

Ceballos I, Ruiz M, Fernandez C, Pena R, Rodriguez A, Sanders IR (2013) The in vitro mass-produced model mycorrhizal fungus, *Rhizophagus irregularis,* significantly increases yields of the globally important food security crop cassava. PLoS One 8(8):e70633

Cely MV, de Oliveira AG, de Freitas VF, de Luca MB, Barazetti AR, dos Santos IM, Gionco B, Garcia GV, Prete CE, Andrade G (2016) Inoculant of arbuscular mycorrhizal fungi (*Rhizophagus clarus*) increase yield of soybean and cotton under field conditions. Front Microbiol 7:720

Chandel A, Prabhavathi Y, Rajeswari S, Reddy BR (2017) Buying behaviour and perception on mycorrhizal biofertilizer product – a case study. J Res ANGRAU 45:95–98

Chen M, Arato M, Borghi L, Nouri E, Reinhardt D (2018) Beneficial services of arbuscular mycorrhizal fungi—from ecology to application. Front Plant Sci 9:1270
Churchland C, Grayston SJ (2014) Specificity of plant-microbe interactions in the tree mycorrhizosphere biome and consequences for soil C cycling. Front Microbiol 5:261
Darade MS (2015) Effect of inoculation of vam fungi on enhancement of biomass and yield in okra. IJISET 2(8):859–865
Deja-Sikora E, Mercy L, Baum C, Hrynkiewicz K (2019) The contribution of endomycorrhiza to the performance of potato virus *y*-infected solanaceous plants: disease alleviation or exacerbation. Front Microbiol 10:516
Devi GU, Sitaramaiah K (2001) Response of endomycorrhizal fungi at different levels of inoculum in blackgram. Indian Phytopathol 54(3):377–379
Doley K, Jite PK (2012) Response. of groundnut ('JL-24') cultivar to mycorrhiza inoculation and phosphorous application. Not Sci Biol 4(3):118–125
Drasek JV (2014) The effects of mycorrhizae on rapid-grow radish, Brassica rapa. Honors Council of the Illinois Region Papers 1
Fester T, Sawers R (2011) Progress and challenges in agricultural applications of arbuscular mycorrhizal fungi. Crit Rev Plant Sci 30(5):459–470
Ghorui M, Chowdhury S, Balu P, Burla S (2024) Arbuscular mycorrhizal inoculants and its regulatory landscape. Heliyon 10(9):e30359
Gianinazzi S, Gollotte A, Binet MN, Van Tuinen D, Redecker D, Wipf D (2010) Agroecology: the key role of arbuscular mycorrhizas in ecosystem services. Mycorrhiza 20(8):519–530
Goswami A, Ghosh L, Banerjee R (2019) The lacuna between randd and technological commercialization of biofertilizers in south Asian countries. Int J Adv Res Micro Immunol 1(1):25–44
Habibzadeh Y, Evazi AR, Abedi M (2014) Alleviation drought stress of mungbean (*Vigna radiata* L.) plants by using arbuscular mycorrhizal fungi. Int J Agric Nat Res 1(1):1–6
Hao Z, Christie P, Qin L, Wang C, Li X (2005) Control of fusarium wilt of cucumber seedlings by inoculation with an arbuscular mycorrhical fungus. J Plant Nutr 28(11):1961–1974
Herrmann L, Lesueur D (2013) Challenges of formulation and quality of biofertilizers for successful inoculation. Appl Microbiol Biotechnol 97:8859–8873
Hewins CR, Carrino-Kyker SR, Burke DJ (2015) Seasonal variation in mycorrhizal fungi colonizing roots of *Allium tricoccum* (wild leek) in a mature mixed hardwood forest. Mycorrhiza 25:469–483
Hijri M (2016) Analysis of a large dataset of mycorrhiza inoculation field trials on potato shows highly significant increases in yield. Mycorrhiza 26:209–214
Jaizme-Vega M, Del C, Rodriguez-Romero AS, Barroso Nunez LA (2006) Effect of the combined inoculation of arbuscular mycorrhizal fungi and plant growth-promoting rhizobacteria on papaya (*Carica papaya* L.) infected with the root-knot nematode *Meloidogyne incognita*. Fruits 61(3):151–162
Jamiolkowska A, Thanoon AH, Skwarylo-Bednarz B, Patkowska E, Mielniczuk E (2020) Mycorrhizal inoculation as an alternative in the ecological production of tomato (*Lycopersicon esculentum* mill.). Int Agrophys 34(2):253–264
Kavoo-Mwangi AM, Kahangi EM, Ateka E, Onguso J, Mukhongo RW, Mwangi EK, Jefwa JM (2013) Growth effects of microorganisms based commercial products inoculated to tissue cultured banana cultivated in three different soils in Kenya. Appl Soil Ecol 64:152–162
Khaliq A, Perveen S, Alamer KH, Ul Haq M, Rafique Z, Alsudays IM, Althobaiti AT, Saleh MA, Hussain S, Attia H (2022) Arbuscular mycorrhizal fungi symbiosis to enhance plant–soil interaction. Sustainability 14(13):7840
Klerkx L, Van Mierlo B, Leeuwis C (2012) Evolution of systems approaches to agricultural innovation: concepts, analysis and interventions. In: Darnhofer I, Gibbon D, Dedieu B (eds) Farming systems research into the 21st century: the new dynamic. Springer, Dordrecht, pp 457–483
Kumar M, Yadav K, Thakur SK, Mandal K (1998) Effect of vescicular-arbuscular mycorrhizal fungi and rllizobium inoculation on nodulation, root colonization, nitrogen fixation and yield of chick pea. J Indian Soc Soil Sci 46(3):375–378

Kumar A, Singh R, Adholeya A (2017) Biotechnological advancements in industrial production of arbuscular mycorrhizal fungi: achievements, challenges, and future prospects. In: Developments in fungal biology and applied mycology. Springer, Singapore, pp 413–431

Lopez IG, Martin VJM (2012) Application of arbuscular mycorrhizal fungi on the production of cut flower roses under commercial-like conditions. SJAR 10(1):166–174

Luty L, Musial K, Ziolo M (2021) The role of selected ecosystem services in different farming systems in Poland regarding the differentiation of agricultural land structure. Sustainability 13:6673

Madawala HMSP (2021) Arbuscular mycorrhizal fungi as biofertilizers: current trends, challenges, and future prospects. In: Rakshit A, Meena VS, Parihar M, Singh HB, Singh AK (eds) Biofertilizers. Woodhead Publishing, Cambridge, pp 83–93

Maertens A, Barrett CB (2012) Measuring social networks' effects on agricultural technology adoption. Am J Agric Econ 95(2):353–359

Mardhiah U, Caruso T, Gurnell A, Rillig MC (2016) Arbuscular mycorrhizal fungal hyphae reduce soil erosion by surface water flow in a greenhouse experiment. Appl Soil Ecol 99:137–140

Market Research Report (2021) Accessed 28 October 2024. https://www.fortunebusinessinsights.com/industry-reports/agricultural-microbial-market-100412

Market Research Report (2022) Accessed 28 October 2024. https://www.fortunebusinessinsights.com/industry-reports/agricultural-microbial-market100412

Meng J, Zhang X, Han X, Fan B (2022) Application and development of biocontrol agents in China. Pathogens 11(10):1120

Milton M, Bisarya D, Kumar V, Kumar S, Singh AK (2021) Mycorrhizae and their importance in agriculture. J Emerg Technol Innov Res 8:ISSN-2349-5162

Mukhongo RW, Tumuhairwe JB, Ebanyat P, Abdelgadir AH, Thuita M, Masso C (2016) Review article production and use of arbuscular mycorrhizal fungi inoculum in sub-saharan Africa: challenges and ways of improving. Int J Soil Sci 11(3):108–122

Mummey DL, Antunes PM, Rillig MC (2009) Arbuscular mycorrhizal fungi pre-inoculant identity determines community composition in roots. Soil Biol Biochem 41(6):1173–1179

Navarro-Fernandez CM, Perez-Ramos IM, de la Riva EG, Vera JR, Roumet C, Villar R, Maranon T (2016) Functional responses of mediterranean plant communities to soil resource heterogeneity: a mycorrhizal trait-based approach. J Veg Sci 27(6):1243–1253

Noe R, Kiers ET (2018) Mycorrhizal markets, firms, and co-ops. Trends Ecol Evol 33(10):777–789

Nunes CEP, Stancato GC, Da Silveira APD (2014) Anthurium growth responses to phosphate fertilisation and inoculation with an arbuscular mycorrhizal fungus. J Hortic Sci Biotechnol 89(3):261–267

Owen D, Williams AP, Griffith GW, Withers PJA (2015) Use of commercial bio-inoculants to increase agricultural production through improved phosphorus acquisition. Appl Soil Eco 86:41–54

Pal S, Singh HB, Farooqui A, Rakshit A (2016) Commercialization of arbuscular mycorrhizal technology in agriculture and forestry. In: Singh H, Sarma B, Keswani C (eds) Agriculturally important microorganisms. Springer, Singapore, pp 97–105

Pankova H, Munzbergova Z, Rydlova J, Vosatka M (2014) Co-adaptation of plants and communities of arbuscular mycorrhizal fungi to their soil conditions. Folia Geobot 49:521–540

Phad NV, Kumbhar CT, Khadatare RM, Khot GG (2016) Dual inoculation of *Glomus fasciculatum* and *Azotobacter chroococcum* improves growth and yield of brinjal (*Solarium melongena* L.). Crop Res 51:134–139

Rouphael Y, Franken P, Schneider C, Schwarz D, Giovannetti M, Agnolucci M, Pascale SD, Bonini P, Colla G (2015) Arbuscular mycorrhizal fungi act as biostimulants in horticultural crops. Sci Hortic 196:91–108

Salomon M, Demarmels R, Watts-Williams SJ, McLaughlin MJ, Kafle A, Ketelsen C, Soupir A, Bücking H, Cavagnaro TR, Van Der Heijden MGA (2022a) Global evaluation of commercial arbuscular mycorrhizal inoculants under greenhouse and field conditions. Appl Soil Ecol 169:104225

Salomon MJ, Watts-Williams SJ, McLaughlin MJ, Bucking H, Singh BK, Hutter I, Schneider C, Martin FM, Vosatka M, Guo L, Ezawa T, Saito M, Declerck S, Zhu YG, Bowles T, Abbott

LK, Smith FA, Cavagnaro TR, van der Heijden MGA (2022b) Establishing a quality management framework for commercial inoculants containing arbuscular mycorrhizal fungi. iScience 25(7):104636

Shuab R, Lone R, Naidu J, Sharma V, Imtiyaz S, Koul KK (2014) Benefits of inoculation of arbuscular mycorrhizal fungi on growth and development of onion (*Allium cepa*) plant. Am Eurasian J Agric Environ Sci 14(6):527–535

Simpson D, Daft MJ (1990) Interactions between water-stress and different mycorrhizal inocula on plant growth and mycorrhizal development in maize and sorghum. Plant Soil 121:179–186

Smith SE, Read D (2008) Mycorrhizal symbiosis, 3rd edn. Academic Press, London

Sturmer SL, Oliveira LZ, Morton JB (2018) *Gigasporaceae* versus *Glomeraceae* (phylum Glomeromycota): a biogeographic tale of dominance in maritime sand dunes. Fungal Ecol 32:49–56

Sudheer S, Johny L, Srivastava S, Adholeya A (2023) The trade-in-trade: multifunctionalities, current market and challenges for arbuscular mycorrhizal fungal inoculants. Symbiosis 89:259–272

Suri T (2011) Selection and comparative advantage in technology adoption. Econometrica 79:159–209

Synowiec A, Nandanavanam R, Puczel U, Upadhyaya MK (2011) Influence of isoxaflutole on colonization of corn (*Zea mays* L.) roots with arbuscular mycorrhizal fungus *Glomus intraradices*. Can J Plant Sci 91:143–145

Ullah A, Gao D, Wu F (2024) Common mycorrhizal network: the predominant socialist and capitalist responses of possible plant–plant and plant–microbe interactions for sustainable agriculture. Front Microbiol 15:1183024

Varela-Cervero S, Lopez-Garcia A, Barea JM, Azcon-Aguilar C (2016) Spring to autumn changes in the arbuscular mycorrhizal fungal community composition in the different propagule types associated to a Mediterranean shrubland. Plant Soil 408(1):107–120

Veresoglou S, Agathokleous E (2023) Pertinent challenges in mycorrhizal ecopreneurship for sustainable food production. Sustain Horiz 8:100064

Vosatka M, Albrechtova J, Patten R (2008) The international market development for mycorrhizal technology. In: Varma A (ed) Mycorrhiza. Springer, Berlin, Heidelberg, 419–438

Vosatka M, Latr A, Gianinazzi S, Albrechtova J (2012) Development of arbuscular mycorrhizal biotechnology and industry: current achievements and bottlenecks. Symbiosis 58:29–37

Wahab A, Muhammad M, Munir A, Abdi G, Zaman W, Ayaz A, Khizar C, Reddy SPP (2023) Role of arbuscular mycorrhizal fungi in regulating growth, enhancing productivity, and potentially influencing ecosystems under abiotic and biotic stresses. Plants 12(17):3102

Werner GDA, Kiers ET (2015) Partner selection in the mycorrhizal mutualism. New Phytol 205:1437–1442

Whipps JM (2004) Prospects and limitations for mycorrhizas in biocontrol of root pathogens. Can J Bot 82(8):1198–1227

Wu QS, He JD, Srivastava AK, Zhang FEI, Zou YN (2019) Development of propagation technique of indigenous AMF and their inoculation response in citrus. Indian J Agric Sci 89(1):1190–1194